# HANDBOOK OF APPLICABLE MATHEMATICS

Volume VI: Statistics

PART B

# HANDBOOK OF APPLICABLE MATHEMATICS

# HANDBOOK OF

# APPLICABLE MATHEMATICS

Chief Editor: Walter Ledermann

## Volume VI: Statistics

### PART B

Edited by

Emlyn Lloyd

*University of Lancaster*

A Wiley–Interscience Publication

## JOHN WILEY & SONS

Chichester – New York – Brisbane – Toronto – Singapore

QA
36
.H36
vol. 6
pt. B

*Library of Congress Cataloging in Publication Data*:
(Revised for vol. 6)
Main entry under title:
Handbook of applicable mathematics.
 'A Wiley–Interscience publication.'
 Includes bibliographies and indexes.
 Contents:—v. 2. Probability—v. 3. Numerical
methods—v. 4. Analysis—[etc.]—v. 6 Statistics.
1. 1. Mathematics—1961–    . I. Ledermann, Walter,
1911
QA36.H36    510    79-42724
ISBN 0 471 27821 1 (v. 2)
ISBN 0 471 90274 8 (PART A)
ISBN 0 471 90272 1 (PART B)
ISBN 0 471 90024 9 (SET)


*British Library Cataloguing in Publication Data*:

Handbook of applicable mathematics.
  Vol. 6: Statistics
  1. Mathematics
  I. Ledermann, Walter   II. Lloyd, Emlyn
  510    QA36
ISBN 0 471 90274 8 (PART A)
ISBN 0 471 90272 1 (PART B)
ISBN 0 471 90024 9 (SET)
Typeset and printed by J. W. Arrowsmith Ltd, Bristol BS3 2NT

# Contributing Authors

*David Cooper*, Institute of Hydrology, Wallingford, U.K.

*Ian Dunsmore*, The University, Sheffield, U.K.

*John Gower*, Rothamsted Experimental Station, Harpenden, Hertfordshire, U.K.

*Peter Jones*, University of Keele, Staffordshire, U.K.

*Emlyn Lloyd*, University of Lancaster, Lancaster, U.K.

*Enda O'Connell*, Institute of Hydrology, Wallingford, U.K.

*Adrian Smith*, University of Nottingham, Nottingham, U.K.

*Granville Tunnicliffe-Wilson*, University of Lancaster, Lancaster, U.K.

*David Warren*, University of Lancaster, Lancaster, U.K.

*Joe Whittaker*, University of Lancaster, Lancaster, U.K.

# Contents

## PART B

vii

Chapters 1–10 are contained in Part A.

# Introduction
# to the
# Handbook of Applicable Mathematics

Today, more than ever before, mathematics enters the lives of every one of us. Whereas, thirty years ago, it was supposed that mathematics was only needed by somebody planning to work in one of the 'hard' sciences (physics, chemistry), or to become an engineer, a professional statistician, an actuary or an accountant, it is recognized today that there are very few professions in which an understanding of mathematics is irrelevant. In the biological sciences, in the social sciences (especially economics, town planning, psychology), in medicine, mathematical methods of some sophistication are increasingly being used and practitioners in these fields are handicapped if their mathematical background does not include the requisite ideas and skills.

Yet it is a fact that there are many working in these professions who do find themselves at a disadvantage in trying to understand technical articles employing mathematical formulations, and who cannot perhaps fulfil their own potential as professionals, and advance in their professions at the rate that their talent would merit, for want of this basic understanding. Such people are rarely in a position to resume their formal education, and the study of some of the available textbooks may, at best, serve to give them some acquaintance with mathematical techniques, of a more or less formal nature, appropriate to current technology. Among such people, academic workers in disciplines which are coming increasingly to depend on mathematics constitute a very significant and important group.

Some years ago, the Editors of the present Handbook, all of them actively concerned with the teaching of mathematics with a view to its usefulness for today's and tomorrow's citizens, got together to discuss the problems faced by mature people already embarked on careers in professions which were taking on an increasingly mathematical aspect. To be sure, the discussion ranged more widely than that—the problem of 'mathematics avoidance' or 'mathematics anxiety', as it is often called today, is one of the most serious problems of modern civilization and affects, in principle, the entire community—but it was decided to concentrate on the problem as it affected professional effectiveness. There emerged from those discussions a novel format for presenting mathematics to this very specific audience. The intervening years have been spent in putting this novel conception into practice, and the result is the Handbook of Applicable Mathematics.

## THE PLAN OF THE HANDBOOK

The 'Handbook' consists of two sets of books. On the one hand, there are (or will be!) a number of *guide books*, written by experts in various fields in which mathematics is used (e.g. medicine, sociology, management, economics). These guide books are by no means comprehensive treatises; each is intended to treat a small number of particular topics within the field, employing, where appropriate, mathematical formulations and mathematical reasoning. In fact, a typical guide book will consist of a discussion of a particular problem, or related set of problems, and will show how the use of mathematical models serves to solve the problem. Wherever any mathematics is used in a guide book, it is cross-referenced to an article (or articles) in the *core volumes*.

There are 6 core volumes devoted respectively to Algebra, Probability, Numerical Methods, Analysis, Geometry and Combinatorics, and Statistics. These volumes are texts of mathematics—but they are no ordinary mathematical texts. They have been designed specifically for the needs of the professional adult (though we believe they should be suitable for any intelligent adult!) and they stand or fall by their success in explaining the nature and importance of key mathematical ideas to those who need to grasp and to use those ideas. Either through their reading of a guide book or through their own work or outside reading, professional adults will find themselves needing to understand a particular mathematical idea (e.g. linear programming, statistical robustness, vector product, probability density, round-off error); and they will then be able to turn to the appropriate article in the core volume in question and *find out just what they want to know*—this, at any rate, is our hope and our intention.

How then do the content and style of the core volumes differ from a standard mathematical text? First, the articles are designed to be read by somebody who has been referred to a particular mathematical topic and would prefer not to have to do a great deal of preparatory reading; thus each article is, to the greatest extent possible, self-contained (though, of course, there is considerable cross-referencing within the set of core volumes). Second, the articles are designed to be read by somebody who wants to get hold of the mathematical ideas and who does not want to be submerged in difficult details of mathematical proof. Each article is followed by a bibliography indicating where the unusually assiduous reader can acquire that sort of 'study in depth'. Third, the topics in the core volumes have been chosen for their relevance to a number of different fields of application, so that the treatment of those topics is not biased in favour of a particular application. Our thought is that the reader—unlike the typical college student—will already be motivated, through some particular problem or the study of some particular new technique, to acquire the necessary mathematical knowledge. Fourth, this is a handbook, not an encyclopedia—if we do not think that a particular aspect of a mathematical topic is likely to be useful or interesting to the kind of reader we have in mind, we have omitted it. We have not set out to include everything known on a particular topic, and we are *not* catering for the professional mathematician! The Handbook has been written as a contribution to the practice of mathematics, not to the theory.

The reader will readily appreciate that such a novel departure from standard textbook writing—this is neither 'pure' mathematics nor 'applied' mathematics as traditionally interpreted—was not easily achieved. Even after the basic concept of the Handbook had been formulated by the Editors, and the complicated system of cross-referencing had been developed, there was a very serious problem of finding authors who would write the sort of material we wanted. This is by no means the way in which mathematicians and experts in mathematical applications are used to writing. Thus we do not apologize for the fact that the Handbook has lain so long in the womb; we were trying to do something new and we had to try, to the best of our ability, to get it right. We are sure we have not been uniformly successful; but we can at least comfort ourselves that the result would have been much worse, and far less suitable for those whose needs we are trying to meet, had we been more hasty and less conscientious.

It is, however, not only our task which has not been easy. Mathematics itself is not easy! The reader is not to suppose that, even with his or her strong motivation and the best endeavours of the editors and authors, the mathematical material contained in the core volumes can be grasped without considerable effort. Were mathematics an elementary affair, it would not provide the key to so many problems of science, technology and human affairs. It is universal, in the sense that significant mathematical ideas and mathematical results are relevant to very different 'concrete' applications—a single algorithm serves to enable the travelling salesman to design his itinerary, and the refrigerator manufacturing company to plan a sequence of modifications of a given model; and could conceivably enable an intelligence unit to improve its techniques for decoding the secret messages of a foreign power. Given this universality, mathematics cannot be trivial! And, if it is not trivial, then some parts of mathematics are bound to be substantially more difficult than others.

This difference in level of difficulty has been faced squarely in the Handbook. The reader should not be surprised that certain articles require a great deal of effort for their comprehension and may well involve much study of related material provided in other referenced articles in the core volumes—while other articles can be digested almost effortlessly. In any case, different readers will approach the Handbook from different levels of mathematical competence and we have been very much concerned to cater for all levels.

## THE REFERENCING AND CROSS-REFERENCING SYSTEM

To use the Handbook effectively, the reader will need a clear understanding of our numbering and referencing system, so we will explain it here. Important items in the core volumes or the guidebooks—such as definitions of mathematical terms or statements of key results—are assigned sets of numbers according to the following scheme. There are six categories of such mathematical items, namely:

(i) Definitions
(ii) Theorems, Propositions, Lemmas and Corollaries
(iii) Equations and other Displayed Formulae
(iv) Examples
(v) Figures
(vi) Tables

Items in any one of these six categories carry a triple designation a.b.c. of arabic numerals, where 'a' gives the *chapter* number, 'b' the *section* number, and 'c' the number of the individual *item*. Thus items belonging to a given category, for example, definitions are numbered in sequence within a section, but the numbering is independent as between categories. For example, in Section 5 of Chapter 3 (of a given volume), we may find a displayed formula labelled (5.3.7) and also Lemma 5.3.7. followed by Theorem 5.3.8. Even where sections are further divided into *subsections*, our numbering system is as described above, and takes no account of the particular subsection in which the item occurs.

As we have already indicated, a crucial feature of the Handbook is the comprehensive cross-referencing system which enables the reader of any part of any core volume or guide book to find his or her way quickly and easily to the place or places where a particular idea is introduced or discussed in detail. If, for example, reading the core volume on Statistics, the reader finds that the notion of a *matrix* is playing a vital role, and if the reader wishes to refresh his or her understanding of this concept, then it is important that an immediate reference be available to the place in the core volume on Algebra where the notion is first introduced and its basic properties and uses discussed.

Such ready access is achieved by the adoption of the following system. There are six core volumes, enumerated by the Roman numerals as follows:

I Algebra
II Probability
III Numerical Methods
IV Analysis
V Geometry and Combinatorics
VI Statistics: Parts A and B

A reference to an item will appear in square brackets and will *typically* consist of a pair of entries [see A, B] where A is the volume number and B is the triple designating the item in that volume to which reference is being made. Thus '[see II, (3.4.5)]' refers to equation (3.4.5) of Volume II (Probability). There are, however, two exceptions to this rule. The first is simply a matter of economy!—if the reference is to an item in the same volume, the volume number designation (A, above) is suppressed; thus '[see Theorem 2.4.6]', appearing in Volume III, refers to Theorem 2.4.6. of Volume III.

The second exception is more fundamental and, we contend, wholly natural. It may be that we feel the need to refer to a substantial discussion rather than to a single mathematical item (this could well have been the case in the reference to 'matrix', given as an example above). If we judge that such a comprehensive reference is appropriate, then the second entry B of the reference may carry only two numerals—or even, in an extreme case, only one. Thus the reference '[see I, 2.3]' refers to Section 3 of Chapter 2 of Volume I and recommends the reader to study that entire section to get a complete picture of the idea being presented.

Bibliographies are to be found at the end of each chapter of the core volumes and at the end of each guide book. References to these bibliographies appear in the text as '(Smith (1979))'.

It should perhaps be explained that, while the referencing *within* a chapter of a core volume or *within* a guide book is substantially the responsibility of the author of that part of the text, the cross-referencing has been the responsibility of the editors as a whole. Indeed, it is fair to say that it has been one of their heaviest and most exacting responsibilities. Any defects in putting the referencing principles into practice must be borne by the editors. The successes of the system must be attributed to the excellent and wholehearted work of our invaluable colleague, Carol van der Ploeg.

# CHAPTER 11

# *Linear Models I*

## 11.1. DESCRIBING THE MODEL

Science is concerned with relationships between variables, the simplest of which is linear according to which, if the level of one variable increases by one unit the level of the other inexorably changes by a corresponding constant. If measurements on variables were totally accurate and freely available there would be little need for statistical analysis but as they are generally subject to error and costly to obtain the relationships between variables have to be examined in an atmosphere of uncertainty, approximation and suspicion. The statistical theory of linear models is an area of applied mathematics whose growth has been stimulated by the needs of working scientists in biology, economics and many other areas. The theory of linear regression began in the nineteenth century with Galton's interest in heredity. The techniques of analysis of variance [see Chapter 8] were proposed in the 1920's to answer issues concerned with the improvement of agricultural crops. Log-linear models were developed in the 1960s to provide quantitative techniques to deal with qualitative data arising in the medical and social sciences. The theory of generalized linear models sets a frame-work in which these and related techniques are special cases.

The construction of a linear model is an attempt to elucidate the linear relationships between variables in the presence of uncertainty. There are several ingredients.

The variables are labelled $Y, x_1, x_2, \ldots, x_p$ and we wish to choose the optimal linear combination of $x_1, x_2, \ldots, x_p$ (the explanatory) variables to best approximate $Y$.

Specification of a linear model requires:

(1) a probability density function for $Y$,
(2) a parameter of this density function to depend linearly on $x_1, x_2, \ldots, x_p$ (the linear predictor),
(3) distinct observations on the variables (the data), and
(4) a sampling model for the observations.

499

The specification of the density function and the linear predictor give the probability model for $Y$. If the optimal linear combination of the explanatory variables is known, the statisticians job is done and the model is ready for use. More usually, the data (3), and the way in which it has been collected, provide (4), the means to estimate the linear predictor as well as possible and to check the assumptions made about the density function and its connection with the linear predictor.

When the probability model is completely known it can be applied in several ways: to forecast the most probable value, or range of values, for $Y$ given values of $x_1, x_2, \ldots, x_p$; to assess the relative influence of one of the variables $x_1$, say, on $Y$; to determine which combinations of values of the explanatory variables give rise to a fixed value for the mean value of $Y$; to compare the relationship between a set of variables in one system with the relationship between a set in another system.

When the model is determined from the observed data set by the usual mixture of assumption, estimation, testing and checking [see Chapter I], the fitted model also serves as a concise summary of the data and as a smoothed version of the data set where chance variations have been suppressed.

The probability model is described in section 11.1 and its data base in section 11.2. Further details of the models appropriate to regression, and the analysis of variance and covariance are examined in section 11.3 and contingency tables are discussed in section 11.4. The methods of statistical inference are based on the likelihood function [see § 6.2.1], and one of the unifying themes in the analysis of generalized linear models is the common technique of likelihood inference. The mathematical technique of least squares described in Chapter 8 is re-examined in section 12.1 and the sampling properties of the estimators are indicated in section 12.2. These two themes are brought together in section 12.3 in the analysis of the likelihood function of the models. The emphasis of Chapter 11.8 is towards application while that of Chapter 12 is towards statistical theory.

In writing this material, I have tried to fix my mind on the needs and abilities of the working scientist. Abstract mathematical argument is minimized wherever possible and numerical examples are used to provide coherence. However, the concepts of a vector and of a vector space provide an extremely powerful method for dealing with linear models. The advantage of this approach is to (a) unify the treatment of regression, analysis of variance, covariance and contingency table analysis, (b) allow a concise description of models in terms of vector subspaces, (c) eliminate the complicated formulae for coefficients based in a coordinate notation while preserving more information than the overstreamlined approach of matrix algebra and to (d) relate linear models to other areas such as multivariate analysis [see Chapters 16, 17], time series analysis [Chapter 18] and experimental design [Chapter 9].

The necessary linear algebra is contained in sections 11.1.1 and 11.3.1. The other prerequisite is a familiarity with probability and statistics to the level of the other chapters in this volume [see Volume II].

The numerical examples are meant to illustrate the argument. They are not intended to be general computing recipes. Since many of the techniques require iterative numerical procedures, it is expected that the reader has access to the relevant computer software. In fact much of the notation, and the choice of which material to include in those chapters on linear models has been motivated by the existence of the statistical package GLIM cf. Baker and Nelder (1978).

### 11.1.1.  Elements of Linear Algebra

The concepts of linear algebra [see I: Chapter 5] have a natural application to the theory of linear statistical models. They comprise the language in which to describe the linear model and the theoretical machinery to solve the numerical estimation problem. Important notions are those of vector addition and scalar multiplication, vector spaces and subspaces, transformations and projections, inner products and norms. The only additional concept required by statisticians that might not be familiar to linear algebraists is the notion of indicator vectors with their associated pointwise multiplication required to describe factorial models [see § 9.8].

*Vectors*

A point on a line can be represented by a number, e.g. $x$. A point in a plane can be represented by an ordered pair of numbers $(x_1, x_2)$. A point in three-dimensional space can be represented by $(x_1, x_2, x_3)$. Generalizing, $(x_1, x_2, \ldots, x_n)$ gives a point in $n$ dimensional space though, unfortunately, pictures are only available for $n \leq 3$.

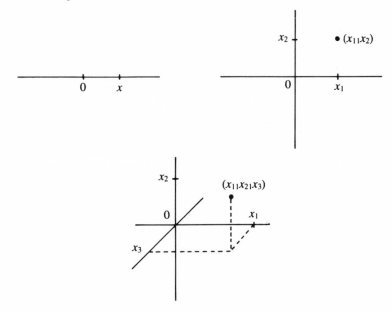

DEFINITION 11.1.1.    *Vector in n-space.*    A vector in *n*-space is an ordered array $(x_1, x_2, \ldots, x_n)$ ($=\mathbf{x}$, say) of real numbers $x_1, x_2, \ldots, x_n$, which are called the coordinates of the vector **x**. (All vectors are taken to be column vectors but for typographical convenience are sometimes written out as rows.)

For example, suppose the age, height, weight and intelligence of a child are, respectively, 7 years, 1·10 m, 35 kg and 122 IQ points. Then these attributes of this child can be represented by the point $(7, 1·10, 35, 122)$ in four-space. Constructions of this kind are, however uncommon in linear models. A more representative example of the kinds of construction that are used in linear models is the following: Suppose four children are $7, 5, 6$ and $5$ years old respectively. These four children can be represented by the point $(7, 5, 6, 5)$ in four-space.

Vector addition and scalar multiplication are straightforward.

*Addition.* If $\mathbf{x} = (x_1, x_2, \ldots, x_n)$ and $\mathbf{y} = (y_1, y_2, \ldots, y_n)$ then

$$\mathbf{x} + \mathbf{y} = (x_1 + y_1, x_2 + y_2, \ldots, x_n + y_n).$$

*Scalar multiplication.* If $\mathbf{x} = (x_1, x_2, \ldots, x_n)$ and if $\alpha$ is a real number then $\alpha \mathbf{x}$ is given by

$$\alpha \mathbf{x} = (\alpha x_1, \alpha x_2, \ldots, \alpha x_n).$$

These operations are exemplified in the diagrams below:

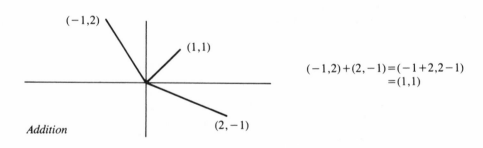

$(-1,2)+(2,-1)=(-1+2,2-1)$
$\qquad\qquad\quad =(1,1)$

*Addition*

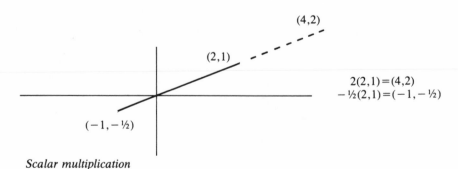

$2(2,1)=(4,2)$
$-\tfrac{1}{2}(2,1)=(-1,-\tfrac{1}{2})$

*Scalar multiplication*

*Vector space.* Vectors satisfy certain rules based on addition and scalar multiplication in $n$-space. In fact, any set of vectors which satisfy these rules constitute a vector space. [See I, §§ 5.1, 5.2.]

If **x**, **y** and **z** are any vectors in a vector space $V$ and $\alpha$ and $\beta$ are scalars then

$$\mathbf{x}+\mathbf{y}\in V \qquad\qquad \alpha\mathbf{x}\in V$$
$$\mathbf{x}+\mathbf{y}=\mathbf{y}+\mathbf{x} \qquad\qquad (\alpha\beta)\mathbf{x}=\alpha(\beta\mathbf{x})$$
$$(\mathbf{x}+\mathbf{y})+\mathbf{z}=\mathbf{x}+(\mathbf{y}+\mathbf{z}) \qquad 1\cdot\mathbf{x}=\mathbf{x}$$
$$\mathbf{x}+\mathbf{0}=\mathbf{x} \qquad\qquad (\alpha+\beta)\mathbf{x}=\alpha\mathbf{x}+\beta\mathbf{x}$$
$$\mathbf{x}-\mathbf{x}=\mathbf{0} \qquad\qquad \alpha(\mathbf{x}+\mathbf{y})=\alpha\mathbf{x}+\alpha\mathbf{y}.$$

*Inner product.* The notion of an inner product is the algebraic correspondent of the geometric notions of length and angle. The usual inner product of **x** and **y** in $n$-space is

$$[\mathbf{x},\mathbf{y}]=x_1y_1+x_2y_2+\ldots+x_ny_n$$

Thus, if $\mathbf{x}=(-2,3,1)$ and $\mathbf{y}=(1,0,4)$ then $[x,y]=-2\cdot1+3\cdot0+1\cdot4=2$. The following are important properties of the inner product.

Symmetry:     $[\mathbf{x},\mathbf{y}]=[\mathbf{y},\mathbf{x}]$;

Linearity:     $\begin{cases}[\mathbf{x},\mathbf{y}+\mathbf{z}]=[\mathbf{x},\mathbf{y}]+[\mathbf{x},\mathbf{z}]\\ [\mathbf{x},\alpha\mathbf{y}]=\alpha[\mathbf{x},\mathbf{y}];\end{cases}$

Positivity:     $[\mathbf{x},\mathbf{x}]\geq0$; $[\mathbf{x},\mathbf{x}]=0$ implies $\mathbf{x}=\mathbf{0}$,

for any vectors $x$, $y$, $z$ in $n$-space.

*Norm.* The norm of a vector corresponds to geometric concept of length. If $\mathbf{x}=(x_1,x_2,\ldots,x_n)$ is a vector in $n$-space then the norm of **x** is the non-negative quantity given by

$$\|\mathbf{x}\|=[\mathbf{x},\mathbf{x}]^{1/2}=(x_1^2+x_2^2+\ldots+x_n^2)^{1/2}.$$

Since $[\mathbf{x},\mathbf{x}]\geq0$ the norm is always real.

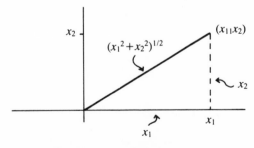

The figure explains why the
norm corresponds to length.

For example, if $x = (2, 3)$ then $\|x\| = (2^2 + 3^2)^{1/2} = \sqrt{13}$; similarly, if $x = (-2, 3, 1)$ then $\|x\| = ((-2)^2 + 3^2 + 1^2)^{1/2} = \sqrt{14}$.

*Unit vector.* If $\|x\| = 1$ then $x$ is said to be a unit vector. If $x$ is any arbitrary vector other than the zero vector $0 = (0, 0, \ldots, 0)$ then $\|x\|^{-1}x$ is a unit vector.

*Remark.* It turns out that all correlations and regression coefficients have very simple structures when expressed in terms of inner products. If $x$ and $y$ are two vectors of observations (measured from their mean) their sample correlation is

$$\frac{[x, y]}{\|x\|\|y\|}$$

and the sample regression coefficient of $y$ on $x$ is

$$\frac{[x, y]}{\|x\|^2}.$$

*Indicator vectors.* An indicator vector is one whose coordinates can only take the values 0 and 1. In six-space examples are $(1, 1, 0, 1, 0, 0)$ or $(0, 0, 0, 1, 1, 0)$. The zero vector $0 = (0, 0, \ldots, 0)$ and the one vector $1 = (1, 1, \ldots, 1)$ are both indicator vectors. In six-space there are $2^6 = 64$ such indicator vectors. An interesting subset consists of the unit indicator vectors; in six-space these are

$$e_1 = (1, 0, 0, 0, 0, 0) \qquad e_2 = (0, 1, 0, 0, 0, 0) \ldots e_6 = (0, 0, 0, 0, 0, 1).$$

Here $[e_1, e_1] = 1$ and $[e_1, e_2] = 0$. Most important is that any vector $x$ can be written as a linear combination of the $e$'s. Thus $x = (x_1, x_2, \ldots, x_6) = x_1 e_1 + x_2 e_2 + \ldots + x_6 e_6$.

*Pointwise multiplication.* If $x$ and $y$ are two vectors in $n$-space then define

$$x y = (x_1 y_1, x_2 y_2, \ldots, x_n y_n)$$

so that $x y$ is a vector in $n$-space constructed by multiplying the coordinates of $x$ and $y$ together. It is clear that

$$x y = y x, \qquad x 1 = x, \qquad x (y + z) = x y + x z$$

and that $[x, y 1] = [x, y]$. If $a$ and $b$ are indicator vectors then $a b$ is also an indicator vector. This last condition is the main motive for introducing pointwise multiplication.

(Note: the pointwise product of two vectors must not be confused with the 'dot product' $d.b$ as used in the theory of cartesian vectors in geometry, kinematics, etc. The latter is the same as our inner product $[a, b]$.)

## 11.1.2.   A Binomial-Logistic Model

The lethal dose of a drug is estimated by observing deaths in groups of mice injected with different dosages of the drug. [see § 6.6] Let $x$ denote the quantity injected and $p(x)$ the probability that a mouse dies given an injection of $x$

units. Each member of each group of five mice is injected with $x$ units. Let $Y$ denote the number of mice to die. If the survival of any one mouse is independent of the survival of any other, the binomial probability [see II, § 5.2.2] gives

$$P(Y = y) = \binom{5}{y} p(x)^{y}(1 - p(x))^{5-y} \quad \text{for } y = 0, 1, \ldots, 5.$$

A plausible candidate for the death rate, $p(x)$ is the logistic function [see II, § 11.10; for tables see Owen (1962), Bibliography G];

$$p(x) = \exp(-1 + 3x)/(1 + \exp(-1 + 3x)).$$

This is illustrated in Figure 11.1.1.

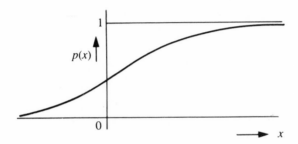

Figure 11.1.1: The logistic function.

By taking logs this can be written as

$$\log\{p(x)/(1 - p(x))\} = -1 + 3x,$$

which is linear in $x$ and so belongs to our category of *linear models*.

The use of such a model is illustrated by its answers to the following questions. What is the probability that a mouse will die if $x = 0$?

$$p(0) = e^{-1}/(1 + e^{-1}) = 0 \cdot 269.$$

What is the probability that all five mice will die when $x = 1$?

$$P(Y = 5] = \binom{5}{5} p(1)^{5}(1 - p(1))^{5-5}$$

$$= p(1)^{5} = [e^{-1+3}/(1 + e^{-1+3})]^{5} = (0 \cdot 881)^{5} = 0 \cdot 530.$$

What value of $x$ gives $p(x) = \frac{1}{2}$?

$$\log\{\tfrac{1}{2}/(1 - \tfrac{1}{2})\} = 0 = -1 + 3x \quad \text{whence } x = \tfrac{1}{3}.$$

Suppose 50 mice were injected at $x = \frac{1}{3}$ and 100 at $x = 0$, how many mice would be expected to die?

$$50p(\tfrac{1}{3}) + 100p(0) = 25 \cdot 0 + 26 \cdot 9 = 51 \cdot 9.$$

### 11.1.3.  The General Linear Model

The simple Binomial Logistic example above gives insight into the probability structure of a linear model and the use to which it can be put.

Interest lies in the relationship of one variable $Y$, the *dependent* variable, to another variable or set of variables $x_1, x_2, \ldots, x_p$ known as the *explanatory* variables.

The dependent variable $Y$ is a random variable with a probability density function, $f$, that is assumed to be a member of the exponential family of density functions [see § 1.4.2], indexed by at most two parameters, namely $\eta$, the *linear predictor*, and $\phi$, a 'nuisance' parameter. The density function may be written as

$$f(y|\eta, \phi),$$

where the linear predictor is a linear combination of the explanatory variables $x_1, x_2, \ldots, x_p$, that is

$$\eta = \beta_1 x_1 + \beta_2 x_2 + \ldots + \beta_p x_p; \tag{11.1.1}$$

and $\phi$ is a constant that does not depend on the explanatory variables. It is assumed that there is a functional relationship between $EY$, the expected value of the dependent variable, and the linear predictor:

$$\eta = g(EY). \tag{11.1.2}$$

The function $g$ is known as the *link function*.

In the Binomial-Logistic example, $Y$ is the number of dead mice in a group of five; there is just one explanatory variable $x$, the dose level of a drug; the density function of $Y$ is binomial, a member of the exponential family; the linear predictor is $\eta = -1 + 3x$; there is no nuisance or scale parameter; and the link function is the function $\log\{EY/(5-EY)\} = \eta$.

In this example all the component parts (density, link, predictor) are regarded as known and the model is ready for use. In reality, when the only available information on the relationship between the dependent and explanatory variables is contained in $n$ observations on the variables alone, the probability structure is unknown and may not even exist. In the theory of linear models we compromise and assume $f(y|\eta, \phi)$ to be completely known apart from the values of the parameters $\beta_1, \beta_2, \ldots, \beta_p$ and $\phi$.

Though this might err towards requiring omniscience, it appears to work fairly well in practice. The context in which the data arise is usually such that the choice of density function and of the link function is relatively uncontroversial, while the freedom to select suitable parameter values allows flexibility in applying linear models to different bodies of data.

In the Binomial-Logistic example, this means that the parameters $\alpha$ and $\beta$ in the linear predictor $\eta = \alpha + \beta x$ are to be estimated from the data. Otherwise the structure remains as before.

An example of a dosage-mortality analysis employing a different link function from this has been given in section 6.6. In that case, the dependent variable corresponding to $Y$ is the random variable $R_j$ induced by $r_j$, the number of insects killed, out of a group of $n_j$ insects, by the application of the $j$th level of insecticide [see Table 6.6.1]; there is again just one explanatory variable, namely the logarithm $(x_j)$ of the dose of insecticide administered; there is no nuisance parameter; the density function of the dependent variable $R_j$ is Binomial $(n_j, \pi_j)$ where $\pi_j = \Phi(\alpha + \beta x_j)$. [Here $\Phi$ denotes the standard Normal c.d.f., so that

$$\Phi(u) = (2\pi)^{-1/2} \int_{-\infty}^{u} \exp\left(-\tfrac{1}{2}y^2\right) dy$$

see II, § 11.4.1]. The linear predictor is $\eta = \alpha + \beta x_j$, and the link function

$$g\{E(R_j)\} = \eta$$

is given by

$$E(R_j) = n_j \pi_j = n_j \Phi(\alpha + \beta x_j)$$

whence

$$\alpha + \beta x_j = \Phi^{-1}\left\{\frac{1}{n_j} E(R_j)\right\}.$$

[Here $\Phi^{-1}$ denotes the *inverse* of the function $\Phi$. See IV, § 2.7.]

### 11.1.4.  A Normal-Identity Model

The volume of timber that can be extracted from a tree varies with the height and radius. If the tree were a perfect cylinder we should have: vol $= \pi(\text{radius})^2(\text{height})$, and

$$\log \text{vol} = \log(\pi) + 2 \log \text{radius} + \log \text{height}.$$

The right hand side gives the linear predictor. Put $x_1 = 1$, $x_2 = \log$ radius and $x_3 = \log$ height, and put

$$\eta = \log(\pi)x_1 + 2x_2 + x_3.$$

There are two obvious choices for the density function of volume; either to assume that measured volume, $Y$, is Normally distributed with mean $EY$ given by

$$\log EY = \eta,$$

or to assume that $Y = \log(\text{measured volume})$ is Normally distributed with mean

$$EY = \eta.$$

Without further information I prefer the latter assumption. Volume only takes positive values, while a Normal random variable takes both positive and negative values and there is a possibility that the variance of measured volume increases with the size of a tree while the variance of measured log volume might stay constant.

Thus, the density function of $Y$ = measured log volume is taken to be

$$f(y|\eta, \phi) = (2\pi\phi)^{-1/2} \exp\{-\tfrac{1}{2}(y-\mu)^2/\phi\}.$$

$$-\infty < y < \infty, \qquad -\infty < \mu < \infty, \qquad \phi > 0$$

where $\mu = EY$, the overall average log volume and $\phi = \text{var}(Y)$ is the scale parameter. The linear predictor is

$$\eta = \beta_1 x_1 + \beta_2 x_2 + \beta_3 x_3 \qquad (\beta_1 = \log \pi, \beta_2 = 2, \beta_3 = 1),$$

and the link function is the identity function, so that

$$E(Y) = \mu = \eta.$$

Given values of $\beta_1, \beta_2, \beta_3$ and $\phi$ the model is ready for use. For example, what is the range of the 95% most likely values for log volume when $x_1 = 1$, $x_2 = \log 10$ and $x_3 = \log 2$? Taking the above values for $\beta_1, \beta_2, \beta_3$ and taking $\phi = 0\cdot 5$ gives

$$\mu = \log(\pi) + 2 \log 10 + \log 2 = 6\cdot 443.$$

Requiring $P\{-y < \phi^{-1/2}(Y-\mu) < y\} = 0\cdot 95$ gives $y = 1\cdot 96$ and hence the 95% interval of most likely values is the central 95% probability interval

$$\mu \pm 1\cdot 96\phi^{1/2} = 6\cdot 443 \pm 1\cdot 96\sqrt{0\cdot 5} = (5\cdot 06, 8\cdot 93),$$

[see § 4.1.3].

### 11.1.5.   The Linear Predictor and the Link Function

The linear predictor plays a key role in the theory and application of linear models. The relationship between the dependent variables and the explanatory variables is mediated through the action of the linear predictor alone. This is one of the unifying themes in linear models. Values of $x_1, x_2, \ldots, x_p$ determine the linear predictor $\eta$ which in turn determines the density of $Y$ and consequently the relative probabilities of different values of $Y$. Schematically

$$x_1, x_2, \ldots, x_p \to \eta \to f \to Y.$$

The predictor is given by

$$\eta = \beta_1 x_1 + \beta_2 x_2 + \ldots + \beta_p x_p$$

where $\beta_1, \beta_2, \ldots, \beta_p$ are fixed parameters. The important features of this are that:

(a) all explanatory variables are numerically valued, so that information on qualitative variables (e.g. colour) has to be specially coded;

(b) the explanatory variables are numerically exchangeable. For example, if $\eta = 3 + 4x_2 + x_3$, then a rise of one unit in $x_2$ is equivalent to a rise of four units in $x_3$. In this sense the variables can be traded between themselves.

(c) The coefficients $\beta_1, \beta_2, \ldots, \beta_p$ in the linear predictor can be interpreted as partial derivatives,

$$\frac{\partial \eta}{\partial x_1} = \beta_1, \qquad \frac{\partial \eta}{\partial x_2} = \beta_2, \ldots, \frac{\partial \eta}{\partial x_p} = \beta_p.$$

This says that if the value of $x_j$, rises by exactly one unit, while the other variables are held constant then the linear predictor will change by $\beta_j$ units. (This may, or may not, be a realistic interpretation. In the tree volume example if the radius of a tree increases then it is likely that the height will also increase.)

The expected value of $Y$ is related to the linear predictor $\eta$ through the link function $g(EY) = \eta$. In certain cases there will be a natural candidate for $g$, in others the choice will be more or less arbitrary. In regression and analysis of variance the natural choice is the identity function, in contingency tables the natural choice is the logarithmic function.

An interesting example is the Exponential distribution [see II, § 11.2]. This is sometimes used when the dependent variable represents the survival time of an observation. If the density of $Y$ is

$$f(y) = \frac{1}{\mu} e^{-y/\mu}, \qquad y > 0, \qquad \mu > 0$$

then $EY = \mu$. Two candidates for the link function are the identity with $EY = \eta$, and the reciprocal with $1/EY = \eta$.

The first implies that the average survival time depends linearly on the explanatory variables. The second implies that the average death rate is linear in the explanatory variables. Both are meaningful though both suffer the disadvantage that the left hand side must be positive while the right hand side may be positive or negative. (For this reason the link function $\log(EY) = \eta$ is often fitted in practice.)

The link function is not to be confused with a transformation of the dependent variable. That is, the model in which $Y$ has density $f$ with $g(EY) = \eta$ is not the same as the model in which $g(Y)$ has density $f$ and $Eg(Y) = \eta$. These two models may not even be 'near' each other.

### 11.1.6.  The Density Function

The probability density function of $Y$ is assumed to belong to the family of density functions that can be written as

$$f(y | \theta, \phi) = \exp\left\{ \frac{1}{a(\phi)} (\theta y - b(\theta)) + c(y, \phi) \right\}. \tag{11.1.3}$$

This family is part of the 'Exponential family' of densities [see § 3.4.2]. There are two parameters; the 'natural' parameter $\theta$ and the 'nuisance' scale parameter $\phi$. The functions $a$, $b$ and $c$ are specific functions of their arguments. The following are familiar examples.

Normal

$$(2\pi)^{-1/2}\sigma^{-1}\exp-(y-\mu)^2/2\sigma^2 = \exp\left\{\frac{1}{\sigma^2}\left(\mu y - \frac{1}{2}\mu^2\right) - \frac{y^2}{2\sigma^2} - \frac{1}{2}\log 2\pi\sigma^2\right\}$$

[see II, § 11.4.3]

Poisson

$$\mu^y e^{-\mu}/y! = \exp\{y\log\mu - \mu - \log y!\} \qquad \text{[see II, § 5.4]}$$

Exponential

$$\mu^{-1}e^{-y/\mu} = \exp\{-y/\mu - \log\mu\} \qquad \text{[see II, § 11.2]}$$

Binomial

$$\binom{k}{y}p^y(1-p)^{k-y} = \exp\left\{y\log\frac{p}{1-p} + k\log(1-p) + \log\binom{k}{y}\right\}.$$

[see II, § 5.2.2].

The status of this Exponential family rests on two features; it includes the important densities required in practical application such as those listed above and it is sufficiently general to provide a unifying theoretical framework.

The classical theories of regression, the analysis of variances and the analysis of contingency tables fall neatly into this framework. For each the linear predictor is $\eta = \beta_1 x_1 + \beta_2 x_2 + \ldots + \beta_p x_p$. In regression analysis, the $x$'s are continuous while in the analysis of variance and contingency table analysis they are indicator (dummy) variables. For regression and the analysis of variance, the density is Normal and the link between $\mu = EY$ and $\eta$ is the identity. For contingency tables the density is Poisson and the link function is the log.

The relationship between the parameters in this exponential family of density functions can be structured according to the diagram in Figure 11.1.2.

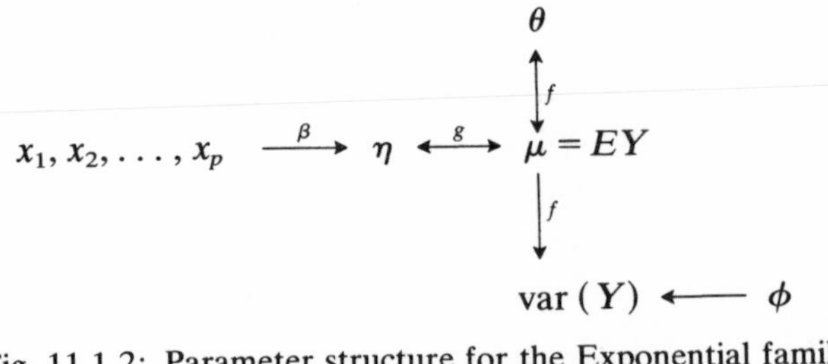

Fig. 11.1.2: Parameter structure for the Exponential family.

The relationships between $\theta$ and $\mu$ and between $\eta$ and $\mu$ are one to one, that between $\theta$ and $\mu$ is determined by the density function and that between $\mu$ and $\eta$ by the link function. The variance of the dependent variable, $Y$, depends on the expected value $\mu$ and the scale parameter $\phi$, if it exists.

### 11.1.7.  The Likelihood Function for the Exponential Family

Likelihood methods provide the statistician with a general and a coherent theory of inference. In particular they provide a methodology for parameter estimation (model fitting) [see Chapter 6] and hypothesis testing (model selection) [see Chapter 5]. The development of these topics requires a preliminary analysis of the likelihood [or log-likelihood: see § 6.2.1] based on a single observation $y$, alone. When we are concerned with the estimation of the parameter $\theta$, and not that of $\phi$, in the exponential family (11.1.3) we shall consider the log-likelihood as being a function of $\theta$ only and write:

$$\lambda(\theta) = \log f(y|\theta, \phi) = \text{const} + \frac{1}{a(\phi)}\{\theta y - b(\theta)\}, \qquad (11.1.4)$$

though its value will usually depend also on that of $\phi$. The estimation problems posed by $\phi$ will be examined separately.

The primary interest of the statistician is in the action of the explanatory variables, in the form of $\eta$, on the dependent variable $Y$. The behaviour of the likelihood function as the linear predictor varies is given by the following important results

$$\frac{d\lambda}{d\eta} = \frac{(y-\mu)}{\text{var}(Y)}\frac{d\mu}{d\eta} \qquad (11.1.5)$$

and

$$E\frac{d^2\lambda}{d\eta^2} = -\frac{1}{\text{var}(Y)}\left(\frac{d\mu}{d\eta}\right)^2. \qquad (11.1.6)$$

The derivative and the curvature of the log-likelihood function have simple expressions in terms of $y - \mu$, var $Y$ and the derivative $d\mu/d\eta$ of the link. When incorporated into the likelihood for the full sample these expressions usually indicate that the solution to the likelihood equations [see Example 6.2.6] is unique and, as the curvature is always negative, give a maximum [cf. § 6.2.2]. Since these results hold for all densities included in the exponential family it allows a single numerical algorithm to solve the likelihood equations.

The remainder of this section outlines a proof of these results based on assuming two classic properties of the log-likelihood function

$$E\frac{d\lambda}{d\theta} = 0 \quad \text{and} \quad E\frac{d^2\lambda}{d\theta^2} = -E\left(\frac{d\lambda}{d\theta}\right)^2$$

[see (6.2.10) and (6.2.11)].

Differentiating the log-likelihood function (11.1.4) gives

$$\frac{d\lambda}{d\theta} = \frac{y - b'(\theta)}{a(\phi)} \quad \text{and} \quad \frac{d^2\lambda}{d\theta^2} = -\frac{b''(\theta)}{a(\phi)}.$$

Applying the first property gives $\mu = EY = b'(\theta)$. The second property gives var $(Y) = b''(\theta) \cdot a(\phi)$ and consequently

$$\frac{d\mu}{d\theta} = \frac{\text{var }(Y)}{a(\phi)}.$$

The chain rule of the differential calculus [see IV, Theorem 3.2.1] gives

$$\frac{d\lambda}{d\eta} = \frac{d\lambda}{d\theta}\frac{d\theta}{d\eta} \quad \text{and} \quad \frac{d\theta}{d\eta} = \frac{d\theta}{d\mu}\frac{d\mu}{d\eta}.$$

Substituting for $d\lambda/d\theta$ and $d\theta/d\mu$ gives the result (11.1.5), the expression for $d\lambda/d\eta$. The result (11.1.6) can be obtained from $E\, d^2\lambda/d\eta^2 = -E(d\lambda/d\eta)^2$ or with a little more information from the repeated chain rule

$$\frac{d^2\lambda}{d\eta^2} = \frac{d^2\lambda}{d\theta^2}\left(\frac{d\theta}{d\eta}\right)^2 + \frac{d\lambda}{d\theta}\frac{d^2\theta}{d\eta^2}.$$

### 11.1.8.  The Binomial-Logistic Example

Returning to the poisoned mice example of section 11.1.2, a natural candidate for the linear model has the following components:

p.d.f.:          $f(y|p) = \binom{5}{y} p^y (1-p)^{5-y}, \qquad (p = p(x)),$

Linear predictor:    $\eta = \alpha + \beta x,$

Link:           $\eta = \log \dfrac{p}{1-p}.$

The consequences of this are

(i)              $\mu = EY = 5p,\ \text{var }(Y) = 5p(1-p),$

(ii)             $p = \dfrac{e^\eta}{1+e^\eta} \quad \text{and} \quad \dfrac{d\mu}{d\eta} = \text{var }(Y),$

Since

$$5\, d\eta/d\mu = d\{\log p - \log(1-p)\}/dp = 1/p + 1/(1-p) = 1/p(1-p)$$
$$= 5/\text{var }(Y).$$

Hence

$$\frac{d\lambda}{d\eta} = \frac{y - \mu}{\text{var }(Y)}\left(\frac{d\mu}{d\eta}\right) = y - \mu = y - 5p$$

and

$$E\frac{d^2\lambda}{d\eta^2} = -\frac{1}{\text{var}(Y)}\left(\frac{d\mu}{d\eta}\right)^2 = -\text{var}(Y) = -5p(1-p).$$

To find the turning points for $\alpha$ and $\beta$ in the linear predictor use

$$\frac{\partial\eta}{\partial\alpha} = 1 \quad \text{and} \quad \frac{\partial\eta}{\partial\beta} = x$$

so that

$$\frac{\partial\lambda}{\partial\alpha} = (y-\mu) \quad \text{and} \quad \frac{\partial\lambda}{\partial\beta} = (y-\mu)x.$$

Note that the curvature is greatest when var $(Y)$ has its largest value. This occurs at $p = \frac{1}{2}$, so most information is gained by experimenting at those values of $x$ for which $p = \frac{1}{2}$.

## 11.2. FITTING A MODEL TO THE DATA

### 11.2.1. The Data Structure

The structure of a linear model has strong implications for the structure of the data. The data array must conform to the following pattern. On each of $t = 1, 2, \ldots, n$ distinct, homogeneous *units*, the observed values of a dependent variable and $p$ explanatory variables are recorded. Each row of the rectangular data array corresponds to a unit and each column to a variable.

Data array:

| Variable | $t$ | $y$ | $x_1$ | $x_2$ | $\cdots$ | $x_p$ |
|----------|-----|-----|-------|-------|----------|-------|
| Unit | 1 | $y_1$ | $x_{11}$ | $x_{12}$ | | $x_{1p}$ |
| | 2 | $y_2$ | $x_{21}$ | $x_{22}$ | | $x_{2p}$ |
| | $\vdots$ | | | | | |
| | $n$ | $y_n$ | $x_{n1}$ | $x_{n2}$ | | $x_{np}$ |

This array can be described as set of column vectors $\mathbf{y}, \mathbf{x}_1, \mathbf{x}_2, \ldots, \mathbf{x}_p$ with $\mathbf{y} = (y_t)$ and $\mathbf{x}_j = (x_{tj})$, or as a data matrix $(\mathbf{y} : \mathbf{X})$ with $\mathbf{X} = (x_{tj})$. The array has $n(p+1)$ entries.

Many data sets are excluded by this specification; for example, data arrays with missing entries, arrays with more than one dependent variable (in the province of multivariate analysis) and arrays with no dependent variables. The assumption of homogeneity is also restrictive. It implies that the subsequent data analysis would reach the same conclusion if the rows of the array were permuted among themselves. This requirement excludes data on time series where successive units are correlated; it excludes data where different units have a different weight, as for example in stratified sampling; it excludes data where groups of units are correlated as in cluster sampling or household-survey data structures.

Similarly the data analysis would remain unchanged if the explanatory variables were permuted among themselves. The ordering of the columns within the data array should not contain any pertinent information.

Even though these restrictions are severe the scope and flexibility of the applications of linear model theory are surprising.

The parameter structure of the model is affected in the following way. On each unit, the expected value $EY_t$ is given by

$$EY_t = \mu_t \overset{g}{\leftrightarrow} \eta_t = \beta_1 x_{t1} + \beta_2 x_{t2} + \ldots + \beta_p x_{tp}$$

or in vectors

$$E\mathbf{Y} = \mathbf{\mu} \overset{g}{\leftrightarrow} \mathbf{\eta} = \beta_1 \mathbf{x}_1 + \beta_2 \mathbf{x}_2 + \ldots + \beta_p \mathbf{x}_p.$$

Both the expected value $\mu_t$ and the linear predictor $\eta_t$ depend on $t$. However the coefficients of the explanatory variables $\beta_1, \beta_2 \ldots \beta_p$ are constant across all units. Fitting the model is equivalent to estimating these parameters, together with $\phi$.

The scale parameter $\phi$ does not vary with $t$, but the theory extends easily to weighting $\phi$, by replacing it with $\phi/w_t$ where the $w_t$'s are known weights. This includes the case of a Normally distributed dependent variable which is in fact the average of $w_t$ independent measurements. This extension somewhat alleviates the restriction imposed by requiring homogeneous units in the data array.

### 11.2.2.   The Likelihood Function and its Maximum

The simplest set of assumptions concerning the method of data collection is to assume that the units constitute a simple random sample from the population of all units that could have been potentially observed and that given the explanatory variables, the $n$ observations on the dependent variable, are mutually independent. The sample so drawn is representative of that population. By symmetry, the inference procedures give equal weight to each unit in the sample.

These two assumptions of random sampling and independent observations lead to the additivity of the log-likelihood function

$$\lambda(\mathbf{\beta}, \phi) = \lambda(\mathbf{\eta}, \phi) = \sum_{t=1}^{n} \lambda(\eta_t, \phi)$$

where $\lambda(\eta_t, \phi)$, the log-likelihood of a single observation from the exponential family is given in (11.1.4).

Differentiating with respect to $\beta_j$ we have

$$\frac{\partial \lambda}{\partial \beta_j} = \sum_t \frac{\partial \lambda(\eta_t, \phi)}{\partial \eta_t} \frac{\partial \eta_t}{\partial \beta_j}$$

Define the vector $\nabla \lambda$ and the diagonal matrix $\mathbf{H}$ as follows:

$$\nabla \lambda = \left(\frac{d\lambda(\eta_t, \phi)}{d\eta_t}\right), \qquad \mathbf{H} = \text{diag}\left(\frac{d^2\lambda(\eta_t, \phi)}{d\eta_t^2}\right);$$

then, as

$$\frac{\partial \eta_t}{\partial \beta_j} = x_{tj},$$

the first and second derivatives of $\lambda(\boldsymbol{\eta}, \phi)$ can be written

$$\frac{\partial \lambda}{\partial \beta_j} = [\nabla \lambda, \mathbf{x}_j]$$

and

$$\frac{\partial^2 \lambda}{\partial \beta_j \partial \beta_k} = [\mathbf{x}_j, \mathbf{H}\mathbf{x}_k]$$

where $[\,\cdot\,,\,\cdot\,]$ is the standard inner product. These equations are remarkable for their simplicity and generality.

The value $\hat{\beta}$ of $\boldsymbol{\beta}$ which leads to a maximum is a solution of

$$[\nabla \lambda, \mathbf{x}_j] = 0 \quad \text{for } j = 1, 2, \ldots, p$$

for which the $p \times p$ matrix $[\mathbf{x}_j, \mathbf{H}\mathbf{x}_k]$ is negative definite. An iterative weighted least squares algorithm solves this and is discussed later in section 12.3.5. As $\mathbf{H}$ is diagonal, the solution will be a maximum if all elements of $\mathbf{H}$ are negative, and this is usually the case with linear models from the Exponential family.

Having solved the likelihood equations for $\hat{\boldsymbol{\beta}}$, the vector of fitted values, $\hat{\boldsymbol{\mu}}$ is computed from

$$\hat{\boldsymbol{\mu}} = g^{-1}(\hat{\boldsymbol{\eta}})$$

where $\hat{\boldsymbol{\eta}} = \hat{\beta}_1 \mathbf{x}_1 + \hat{\beta}_2 \mathbf{x}_2 + \ldots + \hat{\beta}_p \mathbf{x}_p$.

EXAMPLE. *The Binomial-Logistic.* The experiment recorded the deaths ($y$) in six groups of five mice, each group having been injected at equally spaced dosage levels with a drug $(x)$. The data array is:

| Variables | | $y$ | 1 | $x$ |
|---|---|---|---|---|
| Units | $t=1$ | 1 | 1 | 0 |
| | 2 | 0 | 1 | 2 |
| | 3 | 2 | 1 | 4 |
| | 4 | 4 | 1 | 6 |
| | 5 | 3 | 1 | 8 |
| $n=6$ | 6 | 4 | 1 | 10 |

Model:

$$\text{Linear predictor } \eta = \alpha + \beta x,$$

$$\text{Link} \qquad \mu = 5p \quad \text{and} \quad p = e^{\eta}/(1+e^{\eta})$$

so that $\dfrac{d\mu}{d\eta} = 5p(1-p) = \text{var}\,(Y)$,

$$\text{p.d.f. of } Y \colon \text{Bin}\,(5, p);$$

Log-likelihood:

$$\lambda(\eta) = y \log p + (5-y) \log(1-p) + \text{const.}$$

$$\frac{d\lambda}{d\eta} = \frac{y-\mu}{\text{var}\,(Y)} \cdot \frac{d\mu}{d\eta} = y - 5p;$$

Likelihood equations:

$$[\nabla\lambda, \mathbf{1}] = 0 = \sum_t (y_t - 5p_t) \cdot 1 = \sum y_t - 5 \sum p_t$$

$$= 14 - 5 \left( \frac{e^{\alpha}}{1+e^{\alpha}} + \frac{e^{\alpha+2\beta}}{1+e^{\alpha+2\beta}} + \ldots + \frac{e^{\alpha+10\beta}}{1+e^{\alpha+10\beta}} \right),$$

$$[\nabla\lambda, \mathbf{x}] = 0 = \sum_t (y_t - 5p_t)x_t = \sum y_t x_t - 5 \cdot \sum p_t x_t$$

$$= 96 - 5 \left( \frac{e^{\alpha+2\beta}}{1+e^{\alpha+2\beta^2}} + \ldots + \frac{e^{\alpha+10\beta}}{1+e^{\alpha+10\beta}} \cdot 10 \right);$$

Solution:             $\hat{\alpha} = -1{\cdot}974 \qquad \hat{\beta} = 0{\cdot}359$ (check by substitution);

Fitted values:      $\hat{\mu}_t = 5\hat{p}_t = 5e^{\hat{\alpha}+\hat{\beta}x_t}/(1+e^{\hat{\alpha}+\hat{\beta}x_t})$,

gives   $\hat{\mu} = (0{\cdot}61, 1{\cdot}11, 1{\cdot}84, 2{\cdot}72, 3{\cdot}55, 4{\cdot}17)$.

There is no need to compute the hessian $([\mathbf{x}_j, \mathbf{Hx}_k])$ to see if this is a maximum in this example [see IV, § 5.7]. Clearly

$$\frac{d^2\lambda}{d\eta^2} = \frac{d}{d\eta}(y-\mu) = -\frac{d\mu}{d\eta} = -\text{var}(Y) < 0$$

and thus the elements of $H$ are always negative.

### 11.2.3.  The Deviance

How well a model fits a set of data is judged by the discrepancy between the observation vector **y** and the vector $\hat{\boldsymbol{\mu}}$ of fitted values. There are various ways to measure this. In line with our adherence to likelihood inference the appropriate measure is the (maximized) log-likelihood ratio test statistic (LLRTS) constructed in the following manner [cf. § 5.5].

Models with a larger set of explanatory variables should give fitted values which more closely approximate the data; in the extreme case of including as many variables as the number of units in the linear predictor the fitted model should reproduce the data exactly. This *saturated* model, $S$, has the property that

$$\hat{\boldsymbol{\mu}}(S) = \mathbf{y}.$$

Denote a particular model with its set of explanatory variables by $M$ and its fitted values by $\hat{\boldsymbol{\mu}}(M)$. The *deviance* of $M$ is defined as follows:

$$\tfrac{1}{2}\,\text{dev}\,(M) = \lambda(\hat{\boldsymbol{\mu}}(S), \phi) - \lambda(\hat{\boldsymbol{\mu}}(M), \phi).$$

As the maximized log-likelihood $\lambda$ can only grow over a larger set of variables and as $M$ is a subset of $S$, the deviance of $M$ must always be positive. The closer the deviance is to zero, the closer is $\hat{\boldsymbol{\mu}}(M)$ to **y**.

Examples:                       Deviance

Normal $N(\mu, \phi)$
    with identity link          $\dfrac{1}{\phi}\sum_t (y_t - \hat{\mu}_t)^2$

Poisson $P(\mu)$
    with log link               $2\sum_t y_t \log y_t/\hat{\mu}_t$

Binomial $B(k, p)$               $2\sum (y_t \log y_t/\hat{\mu}_t + (k - y_t) \log (k - y_t)/(k - \hat{\mu}_t))$
    with logistic link

With data from the Binomial-Logistic mouse deaths example we have

$$\text{dev}\,(M) = 2\{1 + \log (1/0\cdot61) + 4 \log (4/4\cdot39) + \ldots + 4 \log (4/4\cdot17)$$

$$+ \log (1/0\cdot83)\}$$

$$= 4\cdot519.$$

In the Normal example the deviance is a sum of squares and the analysis of deviance is a generalisation of the techniques of the analysis of variance. This example also illustrates that the role of $\phi$, when it occurs, is not the same as $\mu$; in fact, the deviance depends on the value of $\phi$ and can be used to provide an estimate of $\phi$.

The deviance plays a large part in the model fitting procedure.

(a) It provides a summary statistic to examine the adequacy of fit of a particular model and with its sampling distribution it can provide an omnibus goodness of fit test.

(b) Alternatively, by equating the deviance to its expected value it can be used to estimate the nuisance-scale parameter $\phi$. It cannot, of course, perform functions (a) and (b) simultaneously.

(c) The comparison of deviance is the basis of the likelihood ratio test for testing whether a particular explanatory variable or set of variables can be included or excluded from the model.

(d) It can be used to select the 'best subset' of explanatory variables from all possible ones.

Under the assumption that $M$ is the true model, the sampling distribution of the deviance dev $(M)$ is chi-squared [see 2.5.4(a)] with degrees of freedom df $(M) = n - $ number of explanatory variables in $M$.

For the mouse-deaths example

$$\text{dev } (M) \sim \chi^2(6-2) = \chi^2(4)$$

and the observed value is 4·52, which is not significant at the 5% level, indicating a reasonable fit.

### 11.2.4.  Regression Through the Origin

A regression model that passes through the origin is unlikely to be useful in practice but it does provide a simple analytic illustration to the above discussion. Here is an example:

Data:

| $y$ | -2 | -3 | 2 | 5 | 8 | $n = 5$ |
|---|---|---|---|---|---|---|
| $x$ | -2 | -1 | 0 | 1 | 2 | |

Model:

p.d.f. of $Y$: Normal, mean $\mu$, variance $\phi$, $\phi$ unknown;

linear predictor $\mu_t = Ey_t = \eta_t$ and link $\eta_t = \beta x_t$, $\beta$ unknown

Log-likelihood:

$$\lambda(\beta) = \sum_t \log f(y_t)$$

$$= \sum_t \left(-\tfrac{1}{2}\log 2\pi\phi - \tfrac{1}{2}\phi^{-1}(y_t - \beta x_t)^2\right)$$

$$= -\tfrac{1}{2}n \log 2\pi\phi - \frac{1}{2\phi}\sum_t (y_t - \beta x_t)^2$$

Likelihood equation:

$$\sum_t (y_t - \beta x_t)x_t = 0 \quad \text{so that}$$

$$\hat{\beta} = \sum x_t y_t / \sum x_t^2 = (-2 . -2 + \ldots + 8 . 2)/(-2^2 + \ldots + 2^2)$$

$$= 28/10 = 2.8$$

Fitted values:

$$\hat{y} = 2.8x$$

Residuals:

| $\hat{y}$ | $-5{\cdot}6$ | $-2{\cdot}8$ | $0{\cdot}0$ | $2{\cdot}8$ | $5{\cdot}6$ |
|---|---|---|---|---|---|
| $y - \hat{y}$ | $3{\cdot}6$ | $-0{\cdot}2$ | $2{\cdot}0$ | $2{\cdot}2$ | $2{\cdot}4$ |

Graph:

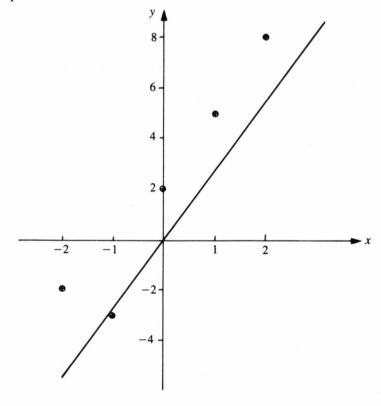

Deviance:

$$\lambda(S) = -\frac{1}{2}n \log 2\pi\phi$$

$$\lambda(\beta) = -\frac{1}{2}n \log 2\pi\phi - \frac{1}{2\phi}\sum_t (y_t - \hat{y}_t)^2$$

$$= -\frac{1}{2}n \log 2\pi\phi - \frac{27\cdot6}{2\phi}$$

$$\text{dev}(\beta) = \frac{27\cdot6}{\phi}.$$

There are various noteworthy features of this analysis. The problem of maximizing the likelihood function became the least squares problem of minimizing $\sum (y_t - \beta x_t)^2 = \sum y_t^2 - 2\beta \sum x_t y_t + \beta^2 \sum x_t$. The deviance is scaled by $\phi$, which is the direct consequence of assuming a Normal density. Equating the deviance to its expected value (number of degrees of freedom) gives the estimate $\hat{\phi} = 27\cdot6/(5-1) = 6\cdot90$. Examining the residuals on the graph suggests a better fit would be attained by fitting the model $\eta_t = \alpha + \beta x_t$.

### 11.2.5.  Partitioning the Deviance

To see if a particular explanatory variable can be dropped from a model without substantially reducing its explanatory power one fits the model twice, once with the variable $(M)$ and once without $(N)$. The difference in deviance is a statistic for assessing the contribution of the variable to the model. Consider the Binomial-Logistic example again. The experimenter wishes to test the effect of drug dosage $(x)$ on deaths $(y)$.
Model:

$$\text{p.d.f. of } Y : \text{Bin}(5, p)$$

$$\mu = 5p, \qquad p = e^\eta/(1+e^\eta)$$

$$M : \eta = \alpha + \beta x \qquad N : \eta = \alpha$$

Results:

| $x$ | 0 | 2 | 4 | 6 | 8 | 10 |
|---|---|---|---|---|---|---|
| $y$ | 1 | 0 | 2 | 4 | 3 | 4 |
| $\hat{y}(M)$ | 0·6 | 1·1 | 1·8 | 2·7 | 3·6 | 4·2 |
| $\hat{y}(N)$ | 2·3 | 2·3 | 2·3 | 2·3 | 2·3 | 2·3 |

$$\text{dev}(M) = 4\cdot52 \qquad\qquad \text{dev}(N) = 12\cdot98$$

The deviance under the larger model, dev $(M)$ is smaller than dev $(N)$, as it must be. The assessment of the size of this reduction requires its sampling distribution. The large-sample approximation for the distribution of likelihood ratio test statistics is that if $N$ is the 'true' model (null model) and if $N$ is contained in $M$ (nested) then dev $(N)-$ dev $(M)$ is chi-squared with df $(N)-$ df $(M)$ degrees of freedom (where df $(N)=$ degrees of freedom of $N$, etc). Furthermore it is distributed independently of dev $(M)$. In our case dev $(N)-$ dev $(M)=8\cdot46$. On the hypothesis that $\beta=0$ this is a realization of the chi-squared distribution on $2-1=1$ d.f., and its significance level with respect to the hypothesis is about $0\cdot004$. The hypothesis that $\beta=0$ is thus emphatically discredited [cf. Table 5.2.1].

In the case of data sampled from a Normal density the large sample result is exact but some modification is required due to the presence of the scale parameter $\phi$. This is illustrated by reanalysing the data used in regression through the origin.

Data:

$$
\begin{array}{c|ccccc}
y & -2 & -3 & 2 & 5 & 8 \\
x & -2 & -1 & 0 & 1 & 2
\end{array}
$$

Cross product matrix:

$$n = 5 \quad \Sigma\, x_t = 0 \qquad \Sigma\, y_t = 10$$
$$\Sigma\, x_t^2 = 10 \quad \Sigma\, x_t y_t = 28$$
$$\Sigma\, y_t^2 = 106$$

Model:

p.d.f. of $Y$ : Normal $(\mu,\ \phi)$

Identity link $\mu = \eta$

Linear predictor $M : \eta = \alpha + \beta x \qquad N : \eta = \alpha;$

Log-likelihood:

$$\lambda = \text{const} - \frac{1}{2\phi} \sum_t (y_t - \eta_t)^2;$$

Likelihood equations for $M$:

$$\Sigma\, (y_t - \alpha - \beta x_t) = 0$$
$$\Sigma\, (y_t - \alpha - \beta x_t) x_t = 0:$$

Solution:

$$\hat{\alpha} = \frac{\Sigma\, y_t}{n} = \frac{10}{5} = 2;$$

$$\hat{\beta} = \frac{\Sigma\, y_t (x_t - \bar{x})}{\Sigma\, (x_t - \bar{x})^2} = \frac{28}{10} = 2\cdot8;$$

$$\text{dev}\,(M) = \frac{1}{\phi} \sum (y_t - \hat{\mu}_t(M))^2 = \frac{1}{\phi}\, 5\cdot6;$$

$$\text{df}\,(M) = 3;$$

Likelihood equations for $N$:

$$\sum (y_t - \alpha);$$

Solution:

$$\hat{\alpha} = \frac{\sum y_t}{n} = 2;$$

$$\text{dev}(N) = \frac{1}{\phi}\sum(y_t - \hat{\mu}_t(N))^2 = \frac{1}{\phi}86{\cdot}0;$$

$$\text{df}(N) = 4.$$

The scale parameter $\phi$ is estimated by equating the deviance of the larger model to its expected value. As the expected value of chi-squared is simply the number of degrees of freedom, this gives

$$\hat{\phi} = 5{\cdot}6/3 = 1{\cdot}87.$$

A test for $\beta = 0$ is then

$$F = \hat{\phi}^{-1}[\text{dev}(N) - \text{dev}(M)]/[\text{df}(N) - \text{df}(M)].$$

This statistic is structured as a ratio of independent chi-squared variates and so has an $F$ distribution. This analysis can be laid out in the following *analysis of deviance* table:

| Source | df | Deviance | Mean deviance | F |
|---|---|---|---|---|
| M$-$N  regression | 1 | 81·4 | 81·4 | 43·61 |
| M    residual | 3 | 5·6 | 1·87 | |
| N    total | 4 | 86·0 | | |

As the observed value of $F$ lies between the 1% value 34·1 and the 0·1% value 167 of the $F_{1,3}$ distribution its significance level with respect to the hypothesis of zero slope is between 0·01 and 0·001. The test result is therefore highly significant. The hypothesis is firmly discredited.

## 11.3.  MODEL SPECIFICATION AND SELECTION

### 11.3.1.  Subspaces

*Linear combinations.*   If $x_1, x_2, \ldots, x_p$ are $p$ vectors in $n$-space [see Definition 11.1.1] and $\alpha_1, \alpha_2, \ldots, \alpha_p$ are $p$ scalars (real numbers) then

$$\alpha_1 x_1 + \alpha_2 x_2 + \ldots + \alpha_p x_p$$

is a linear combination of $x_1, x_2, \ldots, x_p$, as in the following examples:

(i) $2\mathbf{x}_1 + \mathbf{x}_2 - 7\mathbf{x}_3, \mathbf{x}_1 + \mathbf{x}_3, \alpha_1\mathbf{x}_1 + \alpha_3\mathbf{x}_3$ are all linear combinations of $\mathbf{x}_1, \mathbf{x}_2, \mathbf{x}_3$,
(ii) In six-space any vector $\mathbf{a} = (a_1, a_2, \ldots, a_6)$ can be written as a linear combination of the six unit indicator vectors $\mathbf{e}_1, \mathbf{e}_2 \ldots \mathbf{e}_6$.
(iii) The linear predictor is a linear combination of the explanatory vectors.

*Span.*  The set of all linear combinations of $\mathbf{x}_1, \mathbf{x}_2, \ldots, \mathbf{x}_p$ is known as the span $(\mathbf{x}_1, \mathbf{x}_2, \ldots, \mathbf{x}_p)$. Any vector in this set can be written as $\alpha_1\mathbf{x}_1 + \ldots + \alpha_p\mathbf{x}_p$ for some $\alpha_1, \alpha_2, \ldots, \alpha_p$. Examples are:

(i) In six-space, the first two unit indicator vectors are $\mathbf{e}_1 = (1, 0, \ldots, 0)$ and $\mathbf{e}_2 = (0, 1, \ldots, 0)$. Any vector in span $(\mathbf{e}_1, \mathbf{e}_2)$ can be written as $(\alpha_1, \alpha_2, 0, \ldots, 0)$, but, for instance, span $(\mathbf{e}_1, \mathbf{e}_2)$ does not contain $(0, 0, \ldots, 0, 1)$. The zero vector is of course in span $(\mathbf{e}_1, \mathbf{e}_2)$.
(ii) To see if an explanatory vector $\mathbf{x}_p$ contributes to the model fit, test if the linear predictor falls in span $(\mathbf{x}_1, \mathbf{x}_2, \ldots, \mathbf{x}_{p-1})$ or in span $(\mathbf{x}_1, \mathbf{x}_2, \ldots, \mathbf{x}_{p-1}, \mathbf{x}_p)$.

*Subspaces.*  Let $S$ denote a subset of vectors in $n$-space and suppose that $\mathbf{s}_1$ and $\mathbf{s}_2$ are in $S$. Then $S$ is a subspace if (a) $\mathbf{s}_1 + \mathbf{s}_2$ is in $S$ and (b) $\alpha\mathbf{s}_1$ is in $S$.
*Example.*  In three-space, $S = \text{span} (\mathbf{e}_1, \mathbf{e}_2)$ is a subspace because if $\mathbf{s}_1 = \alpha_1\mathbf{e}_1 + \alpha_2\mathbf{e}_2$ and $\mathbf{s}_2 = \beta_1\mathbf{e}_1 + \beta_2\mathbf{e}_2$ then $\mathbf{s}_1 + \mathbf{s}_2 = (\alpha_1 + \beta_1)\mathbf{e}_1 + (\alpha_2 + \beta_2)\mathbf{e}_2$, which is a linear combination of $\mathbf{e}_1, \mathbf{e}_2$ and hence is in $S$. Similarly $\alpha\mathbf{s}_1 = (\alpha\alpha_1)\mathbf{e}_1 + (\alpha\alpha_2)\mathbf{e}_2$ is in $S$. Thus $S$ is a subspace. Geometrically it is a plane that passes through the origin at right angles to the third axis. As another example, it is easy to see that span $(\mathbf{x}_1, \mathbf{x}_2, \ldots, \mathbf{x}_p)$ is always a subspace.
*Dimension.*  The dimension of a sub-space $S$, denoted by dim $(S)$, is the minimum number of vectors required to span it. In three-space, dim span $(\mathbf{e}_1, \mathbf{e}_2) = 2$, dim span $(\mathbf{e}_1, \mathbf{e}_2, \mathbf{e}_3) = 3$ dim span $(\mathbf{1}, \mathbf{e}_1, \mathbf{e}_2, \mathbf{e}_3) = 3$.
*The sum of two subspaces.*  If $S_1$ and $S_2$ are two sub-spaces then the sum $S = S_1 + S_2$ is the set $S$ of vectors that can be written as $\mathbf{s} = \mathbf{s}_1 + \mathbf{s}_2$ where $\mathbf{s}_1 \in S_1$ and $\mathbf{s}_2 \in S_2$. It is fairly easy to see that this sum is also a subspace, as in the following examples.

(i) If $S_1 = \text{span} (\mathbf{e}_1, \mathbf{e}_2, \mathbf{e}_3)$ and $S_2 = \text{span} (\mathbf{e}_2, \mathbf{e}_3, \mathbf{e}_4)$ are sub-spaces in six-space composed of the unit indicators then $S_1 + S_2 = \text{span} (\mathbf{e}_1, \mathbf{e}_2, \mathbf{e}_3, \mathbf{e}_4)$. Note that dim $(S_1 + S_2) \le \text{dim } S_1 + \text{dim } S_2$.
(ii) In general if $S_1 = \text{span} (\mathbf{x}_1, \mathbf{x}_2, \ldots, \mathbf{x}_p)$ and $S_2 = \text{span} (\mathbf{z}_1, \mathbf{z}_2, \ldots, \mathbf{z}_q)$ then $S_1 + S_2 = \text{span} (\mathbf{x}_1, \ldots, \mathbf{x}_p, \mathbf{z}_1, \ldots, \mathbf{z}_p)$.

*The product of two subspaces.*  If $S_1$ and $S_2$ are two subspaces then the product $S = S_1 * S_2$ is the span of the set of vectors, $\mathbf{s}$, that can be written as the pointwise product $\mathbf{s} = \mathbf{s}_1\mathbf{s}_2$ where $\mathbf{s}_1 \in S_1$ and $\mathbf{s}_2 \in S_2$ [see § 11.1.1]. In particular, if $S_1 = \text{span} (\mathbf{x}_1, \mathbf{x}_2, \ldots, \mathbf{x}_p)$ and $S_2 = \text{span} (\mathbf{z}_1, \mathbf{z}_2, \ldots, \mathbf{z}_q)$, the product $S_1 * S_2$ is

$$S_1 * S_2 = \text{span} (\mathbf{x}_1\mathbf{z}_1, \mathbf{x}_1\mathbf{z}_2, \ldots, \mathbf{x}_1\mathbf{z}_q, \mathbf{x}_2\mathbf{z}_1, \ldots, \mathbf{x}_2\mathbf{z}_q, \ldots, \mathbf{x}_p\mathbf{z}_q).$$

For example, if in six-space $S_1 = \text{span}(\mathbf{e}_1, \mathbf{e}_2)$ and $S_2 = \text{span}(\mathbf{e}_2, \mathbf{e}_3)$ then $S_1 * S_2 = \text{span}(\mathbf{e}_2)$, since $\mathbf{e}_1\mathbf{e}_2 = \mathbf{e}_1\mathbf{e}_3 = \mathbf{e}_1\mathbf{e}_3 = \mathbf{0}$ and $\mathbf{e}_2 = \mathbf{e}_2\mathbf{e}_2$. Similarly, in four-space if $\mathbf{a}_1 = (1, 1, 0, 0)$, $\mathbf{a}_2 = (0, 0, 1, 1)$, $\mathbf{b}_1 = (1, 0, 1, 0)$ and $\mathbf{b}_2 = (0, 1, 0, 1)$ and if $A = \text{span}(\mathbf{a}_1, \mathbf{a}_2)$, $B = \text{span}(\mathbf{b}_1, \mathbf{b}_2)$ then $A * B = \text{span}(\mathbf{e}_1, \mathbf{e}_2, \mathbf{e}_3, \mathbf{e}_4)$. To see this note $\mathbf{e}_1 = \mathbf{a}_1\mathbf{b}_1$, $\mathbf{e}_2 = \mathbf{a}_1\mathbf{b}_2$, $\mathbf{e}_3 = \mathbf{a}_2\mathbf{b}_1$, $\mathbf{e}_4 = \mathbf{a}_1\mathbf{b}_2$.

Note that the set of vectors $S$, for which $\mathbf{s} = \mathbf{s}_1\mathbf{s}_2$ with $\mathbf{s}_1 \in S_1$ and $\mathbf{s}_2 \in S_2$ does not in general constitute a subspace. For example, return to four-space with $A = \text{span}(\mathbf{a}_1, \mathbf{a}_2)$ and $B = \text{span}(\mathbf{b}_1, \mathbf{b}_2)$ in the example above. Then $\mathbf{e}_1 = \mathbf{a}_1\mathbf{b}_1$ and $\mathbf{e}_4 = \mathbf{a}_2\mathbf{b}_2$ so that $\mathbf{e}_1$ and $\mathbf{e}_4$ can both be written as products. However $\mathbf{e}_1 + \mathbf{e}_4 = (1, 0, 0, 1)$. But $(\alpha_1\mathbf{a}_1 + \alpha_2\mathbf{a}_2)(\beta_1\mathbf{b}_1 + \beta_2\mathbf{b}_2) = \alpha_1\beta_1\mathbf{e}_1 + \alpha_1\beta_2\mathbf{e}_2 + \alpha_2\beta_1\mathbf{e}_3 + \alpha_2\beta_2\mathbf{e}_4$ and there is no solution to $\alpha_1\beta_1 = 1$, $\alpha_1\beta_2 = 0$, $\alpha_2\beta_1 = 0$, $\alpha_2\beta_2 = 1$.

### 11.3.2.  Model Formulae for the Linear Predictor

The linear predictor is a linear combination of the explanatory variables given by $\eta = \beta_1\mathbf{x}_1 + \beta_2\mathbf{x}_2 + \ldots + \beta_p\mathbf{x}_p$. It is extremely useful to develop a more concise notation for linear predictor, in particular one that suppresses reference to the coefficients of the combination. An alternative way of describing $\eta$ is

$$\eta \in \text{span}(\mathbf{x}_1, \mathbf{x}_2, \ldots, \mathbf{x}_p)$$

where span $(\cdot)$ generates the subspace of all linear combinations of the given vectors.

The specification, $\eta = \alpha + \beta x$ or in vectors $\eta = \alpha\mathbf{1} + \beta\mathbf{x}$, adopted in the Binomial-Logistic formulation for the mouse-death data can be written as

$$\eta \in \text{span}(\mathbf{1}, \mathbf{x}).$$

Now, it may be that the relationship between expected deaths $EY$ and drug dosage $x$ is more complicated than this. A reasonable procedure might be to see if the fit of the data is improved by enlarging the model to include a quadratic term in $x$, i.e. to fit

$$\eta \in \text{span}(\mathbf{1}, \mathbf{x}, \mathbf{x}^2)$$

where $\mathbf{x}^2 = \{x_t^2\}$. A comparison of the deviance of the fits of these two models provides a test for non-linearity.

The notation for these subspaces can be streamlined by writing $X = \text{span}(\mathbf{1}, \mathbf{x})$ and $X^2 = \text{span}(\mathbf{1}, \mathbf{x}^2)$. (This is a departure from the standard statistical usage in which a capital Latin letter such as $X$ denotes a random variable.) The quadratic model above can then be written as

$$\eta \in X + X^2.$$

In general if $X_j = \text{span}(\mathbf{1}, \mathbf{x}_j)$ then the model $\eta = \beta_0\mathbf{1} + \beta_1\mathbf{x}_1 + \ldots + \beta_p\mathbf{x}_p$ is equivalent to $\eta \in X_1 + X_2 + \ldots + X_p$. The reason for requiring $\mathbf{1} \in X_j$ is concerned with indicator variables and will emerge later. This notation highlights the view of linear models as the specification of a subspace to which the linear predictor belongs.

*Standard models*

| | |
|---|---|
| Simple linear regression | $\eta \in X$ |
| Quadratic regression | $\eta \in X + X^2$ |
| Polynomial regression | $\eta \in X + X^2 + \ldots + X^k$ |
| Regression through the origin | $\eta \in \text{span}\,(\mathbf{x})$ |
| Bivariate regression | $\eta \in X_1 + X_2$ |
| Multiple regression | $\eta \in X_1 + X_2 + \ldots + X_p.$ |

EXAMPLE. *Bivariate regression.* Consider the example of trees and timber volume again. If $\mathbf{y}$ = vector of log volume, $\mathbf{x}_1$ = vector of log radius, $\mathbf{x}_2$ = vector of log height then the linear predictor for a group of 'perfect' trees is

$$\eta = \log \pi \mathbf{1} + 2\mathbf{x}_1 + \mathbf{x}_2.$$

To test if these coefficients adequately describe our data, fit $\eta \in X_1 + X_2$ and compare the fitted coefficients $\hat\beta_0$, $\hat\beta_1$ and $\hat\beta_2$ with $\log \pi$, 2 and 1. One might ask a different question: If radius and height were perfectly correlated then knowledge of radius would determine height; in this case the model $\eta \in X_1$ has as much information as the model $\eta \in X_2$ or as the model $\eta \in X_1 + X_2$; which model gives the best predictor when they are not perfectly correlated?

This example leads to considering the lattice of four models

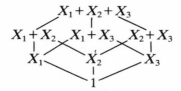

where the connecting lines indicate which subspaces are contained in which. In practice, these diagrams provide a convenient format to summarise which models have been fitted.

With an increasing number of variables these lattices rapidly become complex. The diagram

$$
\begin{array}{c}
X_1 + X_2 + X_3 \\
X_1 + X_2 \quad X_1 + X_3 \quad X_2 + X_3 \\
X_1 \quad X_2 \quad X_3 \\
1
\end{array}
$$

gives all submodels for a linear predictor based on three variables.

The degrees of freedom associated with any model $M$ is just $n\text{-dim}\,(M)$ where dim $(M)$ is the minimum number of vectors required to span $M$.

### 11.3.3.  Models for Qualitative Data

Linear models explain the values of the dependent variable by means of a linear combination of the explanatory variables, the linear predictor. Suppose interest lies in relating weight to height among a population of schoolchildren. A priori the relationship will depend on the sex of the child and so the model has to take account of this. At first sight it appears difficult to make any sense of a linear combination $0 \cdot 25$ height $+ 1 \cdot 6$ sex, as sex unlike height is not measured numerically.

*Factor.*   A qualitative variable that takes a finite number of non-numerical values will be called a *factor*. The values are called *levels*. Thus, sex is a factor with two levels, male and female; similarly, if trees are classified by species then species is a factor with as many levels as there are species.

The distinction between factors and numerical variables is usually straightforward but some variables can fall in the middle. For instance, if height took the values, short, medium and tall then it could be regarded either as a factor with three levels or as a numerical variable with values say $-1$, $0$, $1$.

Suppose $A$ is a factor with four levels, labelled $A_1$, $A_2$, $A_3$, $A_4$, and that $B$ and $C$ are factors with two levels, $B_1$, $B_2$, and $C_1$, $C_2$ respectively. Each unit takes one and only one level of each factor. Suppose $n = 6$ and that part of the data array contains the following information:

| Unit | $A$ | $B$ | $C$ |
|------|-----|-----|-----|
| 1 | $A_2$ | $B_2$ | $C_1$ |
| 2 | $A_1$ | $B_2$ | $C_2$ |
| 3 | $A_1$ | $B_2$ | $C_1$ |
| 4 | $A_2$ | $B_1$ | $C_2$ |
| 5 | $A_4$ | $B_1$ | $C_1$ |
| 6 | $A_3$ | $B_1$ | $C_2$ |

This information can be converted into indicator vectors;

| Unit | $a_1$ | $a_2$ | $a_3$ | $a_4$ | $b_1$ | $b_2$ | $c_1$ | $c_2$ |
|------|-------|-------|-------|-------|-------|-------|-------|-------|
| 1 | 0 | 1 | 0 | 0 | 0 | 1 | 1 | 0 |
| 2 | 1 | 0 | 0 | 0 | 0 | 1 | 0 | 1 |
| 3 | 1 | 0 | 0 | 0 | 0 | 1 | 1 | 0 |
| 4 | 0 | 1 | 0 | 0 | 1 | 0 | 0 | 1 |
| 5 | 0 | 0 | 0 | 1 | 1 | 0 | 1 | 0 |
| 6 | 0 | 0 | 1 | 0 | 1 | 0 | 0 | 1 |

The indicator vector $\mathbf{a}_1$, is an indicator of level $A_1$ of factor $A$, which occurs on the second and third units. The indicator vector $\mathbf{c}_2$ tells us that level $C_2$ of factor $C$ occurred on observations 2, 4 and 6. This procedure generalizes to an arbitrary number of factors with arbitrary numbers of levels in the obvious way.

In this context indicator vectors follow some straightforward rules. In the example

$$\mathbf{a}_1 + \mathbf{a}_2 + \mathbf{a}_3 + \mathbf{a}_4 = 1, \qquad \mathbf{b}_1 + \mathbf{b}_2 = 1, \qquad \mathbf{c}_1 + \mathbf{c}_2 = 1.$$

Also, the pointwise product of two vectors on the same factor is the zero vector. For example

$$\mathbf{a}_1 \mathbf{a}_2 = 0, \qquad \mathbf{a}_3 \mathbf{a}_4 = 0, \qquad \mathbf{b}_1 \mathbf{b}_2 = 0, \qquad \mathbf{c}_1 \mathbf{c}_2 = 0.$$

The pointwise product of two vectors on different factors indicates which observations are at specified levels of *both* factors and is itself an indicator vector. For example

$$\mathbf{a}_2 \mathbf{b}_2 = (1, 0, 0, 0, 0, 0)$$

indicates that only the first observation is both at level $A_2$ of $A$ and $B_2$ of $B$. Similarly $\mathbf{a}_2 \mathbf{b}_2 \mathbf{c}_1$ indicates which observations are at levels $A_2$, $B_2$ and $C_1$. The inner product $[\mathbf{1}, \cdot\,]$ gives the number of units on each level. So $[\mathbf{1}, \mathbf{a}_2] = 2$ and $[\mathbf{1}, \mathbf{a}_2 \mathbf{b}_2] = 1$.

More formally, let $A$ be a factor with $I$ levels and $\mathbf{a}_1, \mathbf{a}_2, \ldots, \mathbf{a}_I$ their indicator vectors. Similarly define $B$, $J$ and $\mathbf{b}_1, \mathbf{b}_2, \ldots, \mathbf{b}_J$. Then it is easy to verify

$$\sum_i \mathbf{a}_i = 1, \qquad \sum_j \mathbf{b}_i = 1;$$

also

$$\mathbf{a}_i \mathbf{a}_{i'} = \begin{cases} \mathbf{a}_i & \text{if } i = i' \\ 0 & \text{if } i \neq i', \end{cases}$$

also

$$\mathbf{1} \mathbf{a}_i = \mathbf{a}_i;$$

$$\sum_i \mathbf{a}_i \mathbf{b}_j = \mathbf{b}_j, \qquad \sum_j \mathbf{a}_i \mathbf{b}_j = \mathbf{a}_i.$$

*The linear predictor.*   We have now set up sufficient machinery to turn qualitative factors into numbers. Suppose that six children are observed, that the first three are girls each with weight of 40 kg and the last three are boys each with a weight of 45 kg. If $A$ is the factor denoting sex, it has two levels with indicator vectors $\mathbf{a}_1$ for girls, with $\mathbf{a}_1 = (1, 1, 1, 0, 0, 0)$ and $\mathbf{a}_2$ for boys, with $\mathbf{a}_2 = (0, 0, 0, 1, 1, 1)$. Taking

$$\boldsymbol{\eta} = 40\, \mathbf{a}_1 + 45\, \mathbf{a}_2$$

gives the vector $(40, 40, 40, 45, 45, 45)$ which reproduces the weights for girls and for boys. The linear predictor is thus a linear combination of the indicator vectors $\mathbf{a}_1$ and $\mathbf{a}_2$.

In essence this is how arbitrary factors are treated though there is one slight modification that simplifies the subsequent development. The linear predictor in this example could also have been written as a linear combination of $\mathbf{1}$ and $\mathbf{a}_2$, namely

$$\boldsymbol{\eta} = 40\,\mathbf{1} + 5\,\mathbf{a}_2.$$

The coefficient of the vector $\mathbf{1}$ gives the value of the predictor on the first level of $A$ (girls). The coefficient of the indicator vector $\mathbf{a}_2$ now gives the *difference* between the values on the second level and the first. Here $\boldsymbol{\eta} \in$ span $(\mathbf{1}, \mathbf{a}_2)$.

In general, if $A$ is a factor with $I$ levels then interest lies in linear models for which the predictor $\boldsymbol{\eta}$ lies in span $(\mathbf{1}, \mathbf{a}_2, \mathbf{a}_3, \ldots, \mathbf{a}_I)$. For example, in comparing four different species of potato suppose the linear predictor for yield is

$$\boldsymbol{\eta} = 20\,\mathbf{1} + 2\,\mathbf{a}_2 - 4\,\mathbf{a}_3 + \mathbf{a}_4;$$

then the yield for the first species would be 20, for the second, 22, for the third 16 and for the fourth 21. If there were no differences between species and all had yield 20 then

$$\boldsymbol{\eta} = 20\,\mathbf{1} + 0\,\mathbf{a}_2 + 0\,\mathbf{a}_3 + 0\,\mathbf{a}_4 = 20\,\mathbf{1}$$

so that $\boldsymbol{\eta} \in$ span $(\mathbf{1})$.

Testing for differences between levels of factor $A$ is equivalent to comparing the fit of the model $\boldsymbol{\eta} \in$ span $(\mathbf{1}, \mathbf{a}_2, \ldots, \mathbf{a}_I)$ with $\boldsymbol{\eta} \in$ span $(\mathbf{1})$.

The symbol $A$ represents a factor (i.e. a qualitative variable). Without confusion, it can also be used to denote the subspace generated by the indicator vectors of $A$. So

$$A = \text{span } (\mathbf{1}, \mathbf{a}_2, \mathbf{a}_3, \ldots, \mathbf{a}_I).$$

This allows reference to the appropriate subspace of $n$-space without the awful tedium of listing the vectors that generate the subspace. Also without confusion let $1 = \text{span }\{\mathbf{1}\}$.

### 11.3.4.   Two Factors: Main Effects and Interaction

We start with a simple example. Suppose $A$ represents two species of potato and $B$ two varieties of fertilizer. Suppose the true potato yield $\mu = \eta$ is measured under these four conditions and gives

|        | Fertilizer | $B_1$ | $B_2$ |
|--------|------------|-------|-------|
| Potato | $A_1$      | 20    | 22    |
|        | $A_2$      | 25    | 27    |

This converts to the following array

| $\mathbf{n}$ | $\mathbf{1}$ | $\mathbf{a}_2$ | $\mathbf{b}_2$ | $\alpha_2\mathbf{b}_2$ |
|---|---|---|---|---|
| 20 | 1 | 0 | 0 | 0 |
| 22 | 1 | 0 | 1 | 0 |
| 25 | 1 | 1 | 0 | 0 |
| 27 | 1 | 1 | 1 | 1 |

and with these values the linear predictor is given by

$$\eta = 20\,\mathbf{1} + 5\,\mathbf{a}_2 + 2\,\mathbf{b}_2.$$

The essential point about this set of values is that the increase in yield due to species $A_2$ compared to species $A_1$ is the same at fertilizer $B_1$ $(25-20=5)$ as at fertilizer $B_2$ $(27-22=5)$. Similarly the increase due to fertilizer $B_2$ over fertilizer $B_1$ is the same for both species $(22-20=2=27-25)$. This makes it possible to talk about a species effect without having to specify which fertilizer is used. Similarly it is possible to talk about a fertilizer *effect* without specifying which species.

Alternatively suppose that yield is given by

| | Fertilizer | $B_1$ | $B_2$ |
|---|---|---|---|
| Potato | $A_1$ | 20 | 22 |
| | $A_2$ | 25 | 26 |

then $\eta$ is given by

$$\eta = 20\,\mathbf{1} + 5\,\mathbf{a}_2 + 2\,\mathbf{b}_2 - 1\,\mathbf{a}_2\mathbf{b}_2.$$

The difference between species is 5 at $B_1$ but 4 at $B_2$; the difference between fertilizers is 2 at $A_1$ but 1 at $A_2$. The difference between species depends on which fertilizer and vice versa. This second example exhibits *interaction* between $A$ and $B$. The former does not. In terms of the linear predictor, interaction between $A$ and $B$ is present if it is necessary to include pointwise product vectors of the form $\mathbf{a}\,\mathbf{b}$ in the linear predictor. With just two levels, $A = \mathrm{span}\,(\mathbf{1}, \mathbf{a}_2)$ and $B = \mathrm{span}\,(\mathbf{1}, \mathbf{b}_2)$ so that $A + B = \mathrm{span}\,(\mathbf{1}, \mathbf{a}_2, \mathbf{b}_2)$ and $A*B = \mathrm{span}\,(\mathbf{1}, \mathbf{a}_2, \mathbf{b}_2, \mathbf{a}_2\mathbf{b}_2)$. The linear predictor for the first example, $\eta = 20\,\mathbf{1} + 5\,\mathbf{a}_2 + 2\,\mathbf{b}_2$ is a linear combination of $\mathbf{1}, \mathbf{a}_2$ and $\mathbf{b}_2$ and hence $\eta \in A + B$. The linear predictor for the second example, $\eta = 20\,\mathbf{1} + 5\,\mathbf{a}_2 + 2\,\mathbf{b}_2 - \mathbf{a}_2\mathbf{b}_2$, is a linear combination of $\mathbf{1}, \mathbf{a}_2, \mathbf{b}_2$ and $\mathbf{a}_2\mathbf{b}_2$ so $\eta \in A*B$.

In general, for two factors, we say there is no interaction between the factors $A$ and $B$ if $\boldsymbol{\eta} \in A + B$. We call the model $\boldsymbol{\eta} \in A + B$ a *main effects* model and we call the model $\boldsymbol{\eta} \in A * B$ a *main effects with interaction* model [cf. § 9.8.1]. [see Figure 11.3.1.]

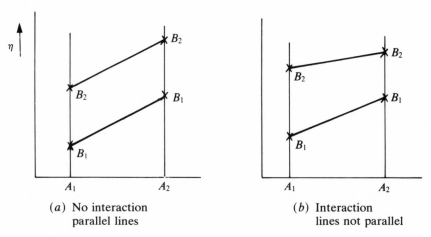

(a) No interaction
parallel lines

(b) Interaction
lines not parallel

Figure 11.3.1: Illustrating interaction.

The importance of having no interaction is that it permits a parsimonious (economical) description of the data set. It means that indicator vectors of the form $\mathbf{a}_i \mathbf{b}_j$, which indicate the observations at level $A_i$ *and* $B_j$, are not necessary to describe the behaviour of the linear predictor. The presence of interaction awkwardly complicates inferences from data.

Testing for no interaction is equivalent to comparing the fit of $\boldsymbol{\eta} \in A * B$ with the fit of $\boldsymbol{\eta} \in A + B$. Note this is a real reduction as $A + B \subset A * B$ and there are vectors in $A * B$ which are not contained in $A + B$. Recall that

$$A + B = \mathrm{span}\,(\mathbf{1}, \mathbf{a}_2, \ldots, \mathbf{a}_I, \mathbf{b}_2, \ldots, \mathbf{b}_J)$$

and

$$A * B = \mathrm{span}\,(\mathbf{1}, \mathbf{a}_2, \ldots, \mathbf{a}_I, \mathbf{b}_2, \ldots, \mathbf{b}_J, \mathbf{a}_2 \mathbf{b}_2, \ldots, \mathbf{a}_I \mathbf{b}_J).$$

### 11.3.5.  A Two-factor Model for a Two-way Analysis of Variance Model

This example is included here to illustrate the application of linear models to the analysis of designed experiments. In point of fact much of the theory of linear models was stimulated by the growth of experimentation in the agricultural and medical sciences.

The following experiment was designed to investigate the effect of sulphate on wheat yield. The randomized block design adopted consisted of six blocks each with four plots. [see Example 9.7.1.] The four treatment levels of manuring ($A$ = unmanured, $B$ = 20 lb, $C$ = 40 lb, $D$ = 60 lb) were applied to one plot within each block. The yields are given below:

Wheat yield in a randomized block experiment

| Block | 1 | 2 | 3 | 4 | 5 | 6 |
|-------|---|---|---|---|---|---|
| Plot | 17·1 (D) | 17·5 (B) | 15·3 (A) | 13·1 (A) | 14·0 (D) | 20·5 (C) |
|  | 18·6 (C) | 19·3 (D) | 19·5 (D) | 18·0 (D) | 12·6 (C) | 16·5 (B) |
|  | 17·1 (B) | 19·7 (C) | 21·6 (C) | 15·1 (C) | 13·1 (A) | 15·4 (A) |
|  | 13·3 (A) | 14·4 (A) | 19·9 (B) | 15·9 (B) | 16·0 (B) | 18·2 (D) |

The basic idea of the design is to obtain six measurement on each treatment which are not confounded [see § 9.9] by the differences between the blocks (which could represent different soil types or wheat species).

For the analysis of this data the natural candidate for a model assumes that yield, $Y$, is Normal $(\mu, \phi)$ with the identity function linking the mean and the linear predictor. The explanatory variables are the indicator vectors for the treatments $t_1, t_2, t_3, t_4$ and the indicator vectors for the blocks $b_1, b_2, \ldots, b_6$. The data array contains 24 units. The obvious choice for the linear predictor is $\mu \in T + B$. To see this; let $\mu_{ij}$ denote the expected yield for the $i$th treatment and $j$th block. Suppose there were no differences between treatments or between blocks then $\mu_{ij}$ would be constant, say $\lambda$, on all 24 units. Equivalent specifications are

$$\mu_{ij} = \lambda \quad \text{or} \quad \mu = \lambda \mathbf{1} \quad \text{or} \quad \mu \in \text{span } (\mathbf{1}).$$

If there are differences between blocks (as is likely, for this is the reason for *blocking* [see § 9.3]) then $\mu_{ij}$ would depend on $j$ as well. If $\lambda._j$ is the additional effect on yield of being in the $j$th block then $\mu_{ij} = \lambda + \lambda._j$ or equivalently

$$\mu = \lambda \mathbf{1} + \sum_j \lambda._j \mathbf{b}_j \quad \text{or} \quad \mu \in \text{span } (\mathbf{1}, \mathbf{b}_1, \ldots, \mathbf{b}_6) \quad \text{or} \quad \mu \in B.$$

If there are treatment differences, $\lambda_i.$ as well then

$$\mu_{ij} = \lambda + \lambda_i. + \lambda._j \quad \text{or} \quad \mu = \lambda \mathbf{1} + \sum_i \lambda_i. \mathbf{t}_i + \sum_j \lambda._j \mathbf{b}_j \quad \text{or} \quad \mu \in T + B.$$

It is supposed that there is no interaction between treatments and blocks, i.e. treatment and block effects are additive.

Details of the estimation procedure are omitted, apart from noting that the fitted values for the models are straightforwardly given by

$$\hat{\mu}_{ij}(B) = \bar{y}._j$$

and

$$\hat{\mu}_{ij}(T + B) = \bar{y}_i. + \bar{y}._j - \bar{y}.. .$$

The test for treatment differences computes

$$\text{dev}(B) = \frac{1}{\phi} \sum_{ij} \{y_{ij} - \hat{\mu}_{ij}(B)\}^2 = 88{\cdot}23/\phi,$$

and

$$\text{dev}(T+B) = \frac{1}{\phi} \sum_{ij} \{y_{ij} - \hat{\mu}_{ij}(T+B)\}^2 = 30{\cdot}29/\phi.$$

The assumption of additive treatment and block effects is crucial to further analysis as $\phi$ can only be eliminated by equating dev $(T+B)$ to its degrees of freedom (df $(T+B) = 24 - (1+3+5) = 15$ and so $\hat{\phi} = 30{\cdot}29/15 = 2{\cdot}02$). The model that contains all interactions of treatments and blocks as well as an unknown scale parameter is unidentifiable. To see this note that the model with interaction $T*B$ contains $6{\cdot}4 = 24 = 1+4-1+6-1+(4-1)(6-1)$ indicator vectors; but there are only 24 units. So the number of parameters equal the number of observations, $\hat{\mu}(T*B) = \mathbf{y}$ and so $T*B$ is the saturated model. Consequently df $(T*B) = 0$ and there are no degrees of freedom remaining to estimate $\phi$. If treatment $\times$ block interaction is suspected it must be allowed for at the design stage by requiring replication on the treatment-block combinations.

Under additivity, $\hat{\phi} = \text{dev}(T+B)/15 = 20{\cdot}29/15 = 2{\cdot}02$. The statistic to test for difference between the treatments is

$$F = \frac{[\text{dev}(B) - \text{dev}(T+B)]/3}{\hat{\phi}} = \frac{57{\cdot}94/3}{2{\cdot}02} = 9{\cdot}56.$$

This ratio has the $F$ distribution with 3 and 15 degrees of freedom and thus is significant at the 5% point ($9{\cdot}56 > 3{\cdot}29$). Clearly there are treatment differences.

The following deviance graph together with the accompanying analysis of variance (deviance) table gives a more systematic presentation of these results.

*Deviance tree*

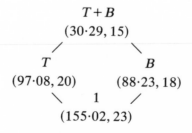

*Anova table*

| Source of variation | df | s.s. | m.s. | F |
|---|---|---|---|---|
| Treatments $T$ | 3 | 57·94 | 19·31 | 9·56 |
| Blocks $B$ | 5 | 66·79 | 13·36 | |
| Residual | 15 | 30·29 | 2·02 | |
| Total | 23 | 155·02 | | |

The investigation of this data and model would continue with an examination of the separate treatment levels and inspection of the residuals. [From a plot it appears that the treatment effect could be approximated by a quadratic; the yield rises steeply from $A$ to reach a maximum at $C$ and then declines slightly. The yield on the plot treated with $C$ in block 5 appears too small is indicated by calculating the residuals $\mathbf{y} - \hat{\boldsymbol{\mu}}(T + B)$].

*Balance.* The deviances of the models 1, $T$, $B$, $T + B$ are given in the deviance tree and in this experiment they satisfy

$$\text{dev}\,(1) - \text{dev}\,(T) = 57·94 = \text{dev}\,(B) - \text{dev}\,(T + B)$$

so that the deviance ascribed to treatments is the same whether blocks have been adjusted for or not. This feature is a consequence of balance in both the design of the experiment and in the model used for its analysis. At the design stage each treatment has been evaluated exactly once on each block. The additive model $T + B$ and the assumption that each unit has the same variance together imply balance in the model.

### 11.3.6. Factorial Models for Cross-classified Data

An informative presentation of a data array that has factors as explanatory variables is as a table cross-classified by the levels of the factors. Here are two examples.

*Smoking and sex*

| | $B_1$ (smokers) | $B_2$ (non-smokers) |
|---|---|---|
| $A_1$ (girls) | 10 | 40 |
| $A_2$ (boys) | 45 | 25 |

Lizard lengths by species and region

|                | $B_1$ (species 1) | $B_2$ (species 2) |
|----------------|-------------------|-------------------|
| $A_1$ (north)  | 15·2              | 17·7              |
| $A_2$ (south)  | 16·3              | 18·6              |

The tabular structure of these two examples is identical and this leads to fitting models which have the same linear predictor, but to begin with the differences between the two examples require discussion.

First, note there is an ambiguity about the size of the data array for the smoking and sex example. There could be four units in the array corresponding to the four cells of the table or there could be 120 units corresponding to the 120 children. In the other example, the lizard length data is balanced; there is exactly one lizard for each cell of the table and there are four units.

The second difference is the choice of a probability model for the dependent variable. The number of children falling in a cell is the dependent variable for the first example. A plausible model is Multinomial [see II, § 6.4] with parameters $k = 120$ and the probabilities attached to the four cells. An alternative, though in some senses equivalent, is the Poisson density function [see II, § 5.4]. Both of these are discrete and neither has an unknown scale parameter. In the other table, a lizard's length is a continuous variable and a reasonable choice is the Normal density [see II, § 11.4], which does involve a scale parameter.

The third difference is a difference in the appropriate scale of measurement. The natural question to ask about the smoking and sex table is whether the ratio of smokers to non-smokers is the same for both sexes i.e. is 10/40 different from 45/25? The question for lizards is whether the regional difference in lengths is the same for both species, i.e. is 16·3–15·2 equal to 18·6–17·7? In the former example the natural scale of computation is multiplicative while in the latter it is additive.

This third difference can be made to disappear by adopting a log scale in the smoking and sex example. The comparison is $(\log 10 - \log 40) - (\log 45 - \log 25)$ which has the same additive structure as in the lizards example. Note that $(\log 10 - \log 40) - (\log 45 - \log 25) = (\log 10 - \log 45) - (\log 40 - \log 25)$ so that the question whether the ratio of smokers to non-smokers is the same for both sexes has the same answer as does the question whether the ratio of girls to boys is the same for both smokers and non-smokers.

The linear predictor for both examples is

$$\boldsymbol{\eta} = \lambda \, \mathbf{1} + \alpha \, \mathbf{a}_2 + \beta \, \mathbf{b}_2 + \gamma \, \mathbf{a}_2 \mathbf{b}_2$$

and both analyses query whether $\gamma = 0$. The smoking and sex example has a log link function $\log(\mu) = \eta$ while the lizards length example has the identity link $\mu = \eta$, but in both examples the interesting question is the presence or absence of interaction and so lead to comparing the fit of $\eta \in A*B$ with $\eta \in A + B$.

The model lattice for the linear predictor for factorial models with two factors is

where $A$ and $B$ are the two (subspaces corresponding to the) factors cross classifying the data. Comparing $\eta \in A*B$ with $\eta \in A + B$ tests for interaction: is there a relationship between smoking and sex? is the difference between species affected by region? Comparing $\eta \in A + B$ with $\eta \in A$ tests for differences between the levels of $B$ at each level of $A$; is the number of smokers equal to the number of non-smokers for girls and for boys? are the two species the same length in the north and in the south? Comparing $\eta \in B$ with $\eta \in 1$ tests for differences between the levels of $B$ overall; are the number of smokers equal to the number of non-smokers irrespective of sex? Are the two species the same length irrespective of region?

Care and caution has to be exercised in selecting models for comparison. Thus in the smoking and sex example the only comparison of real interest is between $A*B$ and $A + B$, while the interesting comparisons in the lizard length example are between $A + B$, $A$ and $B$.

A major benefit of this notation is the ease with which reference can be made to different models. In particular there is no reference made to the number of levels of a factor and so the same hierarchy of models would be appropriate had there been five species of lizard and four regions. The notation easily extends to three dimensions. Had observations been taken on a third factor then the two-way tables above would become three way tables.

*Bartlett's data on propagating plum root stocks*

| Time of planting (A) | Length of cutting (B) | Condition (C) Alive | Dead |
|---|---|---|---|
| At once | Long | 156 | 84 |
|  | Short | 107 | 133 |
| In spring | Long | 84 | 156 |
|  | Short | 31 | 209 |

An exhaustive hierarchy of factorial models for the linear predictor pertaining to a three-way table cross classified by the factors $A$, $B$ and $C$ is given.

Model lattice for three factors

Three-way interaction $\qquad\qquad\qquad\qquad$ $A*B*C$

$\qquad\qquad\qquad\qquad\qquad\qquad\qquad\qquad\quad$ |

Two-way interaction $\qquad\qquad\qquad\qquad\quad$ $A*B+B*C+C*A$

$\qquad\qquad\qquad\qquad\qquad\qquad\qquad\qquad\quad$ |

$\qquad\qquad\qquad\qquad\qquad\quad A*B+B*C$

$\qquad\qquad\qquad\qquad\qquad\qquad\quad$ |

$\qquad\qquad\qquad\qquad\qquad\quad A*B+C$

$\qquad\qquad\qquad\qquad\qquad\qquad\quad$ |

$\qquad\qquad\qquad\qquad\qquad\quad A*B$

$\qquad\qquad\qquad\qquad\qquad\qquad\qquad\qquad A+B+C$

Main effects $\qquad\qquad\qquad\qquad\quad A + B$

$\qquad\qquad\qquad\qquad\qquad\qquad\quad$ |

$\qquad\qquad\qquad\qquad\qquad\qquad A$

$\qquad\qquad\qquad\qquad\qquad\qquad\qquad 1$

The other models are obtained by permuting the labels $A$, $B$ and $C$.

In all there is one model containing all three-way interactions i.e. indicator vectors for the simultaneous specification of the levels of $A$, $B$ and $C$ ($\mathbf{a}_i\mathbf{b}_j\mathbf{c}_k$), there are ten models which have two-way interactions but no three-way interactions, there are seven models with just main effects, and there is one model with just a constant.

Which models are fitted depends on the way in which the data has been collected, prior knowledge about the relationships involved, which questions are important, and the ease with which a model can be interpreted in the context. It is a subjective procedure. To illustrate the complexities of interpretation that arise in just three dimensions, suppose it is known that a relationship between $A$ and $C$ existed so that $\boldsymbol{\eta} \in A*C$ cannot be reduced to $\boldsymbol{\eta} \in A+C$. An explanation of this relationship might imply that if the level of $B$ were held constant then this interaction would disappear. In this case either of the models $\boldsymbol{\eta} \in A*B*C$ or $\boldsymbol{\eta} \in A*B+B*C+C*A$ should reduce to $\boldsymbol{\eta} \in A*B+B*C$.

As the number of factors increase the number of factorial models rapidly increases. The reader is left to sketch out the lattice of models for four factors.

### 11.3.7.  Mixed Models

An interesting set of models are those for which the linear predictor contains both quantitative and qualitative variables. This set includes the analysis of covariance models and the models used to test the homogeneity of regression lines between groups.

If $X$ is the model formula generated by the quantitative vector $\mathbf{x}$ and $A$ is the model formula for the indicator vectors of the levels of the factor $A$ then

$$A + X = \text{span}\,(\mathbf{1}, \mathbf{a}_2, \ldots, \mathbf{a}_I, \mathbf{x})$$

and

$$A*X = \text{span } (\mathbf{1}, \mathbf{a}_2, \ldots, \mathbf{a}_I, \mathbf{x}, \mathbf{a}_2\mathbf{x}, \ldots, \mathbf{a}_I\mathbf{x}).$$

These are the two basic terms for mixed models. In the equivalent coordinate notation, the linear predictor for a unit at the $i$th level of $A$ would be written $\alpha_i + \beta x$ if $\boldsymbol{\eta} \in A + X$, and $\alpha_i + \beta_i x$ if $\boldsymbol{\eta} \in A*X$.

The *analysis of covariance* is quickly explained in this notation by way of example. Recall the randomized block experiment that tested manuring effects, $T$, on wheat yield [see § 11.3.5]. In essentials, the test for the hypothesis of no difference between the treatments was conducted by fitting $\boldsymbol{\eta} \in B$ (blocks) and $\boldsymbol{\eta} \in T + B$ and comparing the reduction in deviance to tabulated values of significance (though slight complications arose because the scale parameter $\phi$ also had to be estimated). Now suppose that together with wheat yield, $y$, the number of plants to germinate, $x$, in the 24 plots was observed. As an increase in the number that germinate should result in a larger yield this variable might well have an effect on the experimental conclusions. Specifically, if the number to germinate is correlated with manure level then the comparison of $B$ with $T + B$ will be biased. The principal aims of the analysis of covariance is to adjust for this bias. (Even if the number is not correlated with manure level the analysis of covariance should lead to a more powerful test by estimating $\phi$ more efficiently).

The idea of the analysis is to adjust the predictor by taking $\boldsymbol{\eta} - \beta \mathbf{x}$, and then compare the fits of $\boldsymbol{\eta} - \beta \mathbf{x} \in B$ with $\boldsymbol{\eta} - \beta \mathbf{x} \in T + B$. As $\beta$ is unknown, this is equivalent to comparing the fits of $\boldsymbol{\eta} \in B + X$ with $\boldsymbol{\eta} \in T + B + X$.

The idealized figures in Figure 11.3.2 below illustrate the motivation for analysis of covariance models.

In both (a) and (b) in the figure (Figure 11.3.2) the inference made concerning the significance of the treatment effects is different according to whether the number of plants germinating has been adjusted for or not. Comparing the fit of $B$ with $T + B$ leads to significant $T$ for (a) and non-significant $T$ for

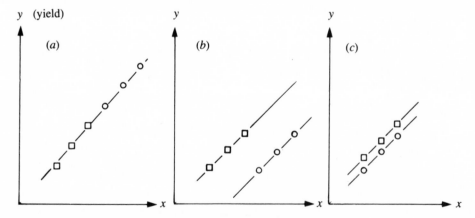

Figure 11.3.2: Illustrations for the analysis of covariance: two treatments ($\square$, $\bigcirc$) on six units.

(b); while comparing the fit of $B+X$ with $T+B+X$ leads to the opposite conclusions. Only in figure (c), where the distribution of $x$ is the same for both treatments do the two comparisons lead to the same conclusion that treatment effect is (just) significant.

Similarly, testing for the homogeneity of regression lines, or more generally of linear predictors, has a simple description in this framework. In the previously considered example of deaths $(Y)$ in groups of five mice dosed with amount $x$ of drug, together with its Binomial-Logistic model, $Y$ was Bino $(5, p)$ and $p = e^{\eta}/(1+e^{\eta})$ and $\eta = \alpha + \beta x$. Suppose now that each group of mice is single sexed; then it is possible to test whether the drug affects females in a different way to males, i.e. to test if $\eta = \alpha_1 + \beta_1 x$ for females and $\eta = \alpha_2 + \beta_2 x$ for males. If $S$ is the sex factor this is accomplished by fitting

$$\eta \in S*X, \qquad \eta \in S+X, \qquad \eta \in X.$$

Comparison of the deviance of $S*X$ and $S+X$ enables us to test whether the linear predictors are parallel; if they are, then the further comparison checks whether they have a common intercept. If so, they are identical.

### 11.3.8.  Factors with Ordered Levels

Factors with ordered levels are susceptible to more detailed analysis. The randomized block experiment in section 11.3.5 concerning the effect of manuring levels on wheat yield provides a good example. The two-factor main effects model can be written

$$\mu_{ij} = \lambda + \lambda_j + \tau_i \quad \text{or} \quad \mu \in B+T$$

where $\tau_i$ denotes the effect of the $i$th-level of manuring on expected wheat yield $\mu_{ij}$ and the $\lambda_j$ denote the block effects. The levels of manure are $x_1 = 0$, $x_2 = 20$, $x_3 = 40$ and $x_4 = 60$ and if the relationship between wheat yield and manuring level is quadratic

$$\mu_{ij} = \lambda + \lambda_j + \beta x_i + \gamma x_i^2.$$

The $3 = 4-1$ parameters $\tau_2, \tau_3, \tau_4$ required to span $T$ are reduced to two corresponding to the linear and quadratic component of treatment effect. Modelling the yield by a quadratic provides an approximation to the location of the maximum yield; equating $\partial \mu / \partial x = \beta + 2\gamma x$ to zero gives $x = -\beta/2\gamma$. Put $\mathbf{x} = \sum x_i \mathbf{a}_i$, $\mathbf{x}^2 = \sum x_i^2 \mathbf{a}_i$ and put $X = \text{span } (\mathbf{1}, x)$, $X^2 = \text{span } (\mathbf{1}, \mathbf{x}^2)$; then this model can be written $\mu \in B+X+X^2$.

An alternative application is to use the linear component of the treatment factor, $\mathbf{x}$, to test if there is any interaction between treatments and blocks. The original wheat yield experiment was designed with each treatment applied only once to each block. As there is no replication it is impossible to obtain a separate estimate of the error; equivalently the $T*B$ interaction is confounded with estimating the scale parameter $\phi$. However, it is possible to test whether the linear component of $T$ varies with the block, i.e. whether there is a (linear $T*B$ interaction). This can be accomplished by comparing the deviance of

$$\mu_{ij} = \lambda + \lambda_j + \tau_i + \beta_j x_i \quad \text{or} \quad \mu \in B + T + B*X$$

with the deviance of $\mu \in B + T$.

Whether there is any interaction and whether the treatment main effects can be modelled by a quadratic can be simultaneously examined by fitting the models in the lattice

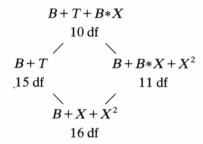

$$B + T + B*X$$
$$10 \text{ df}$$

$$B + T \qquad\qquad B + B*X + X^2$$
$$15 \text{ df} \qquad\qquad 11 \text{ df}$$

$$B + X + X^2$$
$$16 \text{ df}$$

Note though that this lattice could be made more complex by including (quadratic $T$) $\times B$ interaction, that is the term $B*X^2$.

More generally, if $A$ is a factor with $I$ levels, and a score $x_i$ corresponds to the $i$th level of $A$, put $\mathbf{x}^k = \sum x_i^k \mathbf{a}_i$. Then the $k$th component of $A$ is $X^k = \text{span} (\mathbf{1}, \mathbf{x}^k)$. If the $x_i$ are distinct then $A$ can always be expressed as

$$A = X + X^2 + \ldots X^{I-1}$$

and this is the decomposition of $A$ into its linear, quadratic, ... etc., components. When the scores are equally spaced ($x_i = i$) orthogonal polynomials are easily constructed by choosing the linear combinations [c.f. § 11.3.1] in $\mathbf{z}_0 = \mathbf{1}$, $\mathbf{z}_1 = \text{lin comb} (\mathbf{1}, \mathbf{x}_1)$, $\mathbf{z}_2 = \text{lin comb} (\mathbf{1}, \mathbf{x}_1, \mathbf{x}_2)$ etc. so that the $\mathbf{z}$'s are mutually orthogonal.

EXAMPLE. Suppose $\mathbf{n} = 6$ and $A$ has $I = 3$ levels with indicator vectors $\mathbf{a}_1, \mathbf{a}_2, \mathbf{a}_3$ and scores $x_1 = 1$, $x_2 = 2$, $x_3 = 3$ then

| $\mathbf{a}_1$ | $\mathbf{a}_2$ | $\mathbf{a}_3$ | $\mathbf{1}$ | $\mathbf{x}_1$ | $\mathbf{x}_2$ | $\mathbf{z}_0$ | $\mathbf{z}_1$ | $\mathbf{z}_2$ |
|---|---|---|---|---|---|---|---|---|
| 1 | 0 | 0 | 1 | 1 | 1 | 1 | -1 | 1 |
| 1 | 0 | 0 | 1 | 1 | 1 | 1 | -1 | 1 |
| 0 | 1 | 0 | 1 | 2 | 4 | 1 | 0 | -2 |
| 0 | 1 | 0 | 1 | 2 | 4 | 1 | 0 | -2 |
| 0 | 0 | 1 | 1 | 3 | 9 | 1 | 1 | 1 |
| 0 | 0 | 1 | 1 | 3 | 9 | 1 | 1 | 1 |

and it is easy to check that $X^1 + X^2 = \text{span} (\mathbf{1}, \mathbf{x}_1, \mathbf{x}_2) = A = \text{span} (\mathbf{1}, \mathbf{a}_2, \mathbf{a}_3)$.

Though a factor may have ordered levels it may not be possible to assign scores to them unequivocally. For example, with eyesight classified into poor, average, good the scores $-1, 0, 1$ may be just as appropriate as $-1, 0, 100$. There are methods to fit models where the scores are only assumed to be ranked $(x_1 < x_2 < x_3)$ but they are too complicated to discuss here.

## 11.4.  CONTINGENCY TABLES

### 11.4.1.  Sampling Models for Contingency Tables

A contingency table is a form of summarizing a data array appropriate when the dependent and explanatory variables are indicator vectors. If $A_i$ is the $i$th level of $A$ and $y_{it} = 1$ if $t \in A_i$ and $y_{it} = 0$ otherwise, so that $\mathbf{y}_i$ is the indicator vector of $A_i$, then the total $y_{i+} = \sum_t y_{it}$ gives the number of units at level $A_i$ in the whole array. A one-way contingency table is just the table of totals;

| $A_1$ | $A_2$ | $\ldots$ | $A_I$ | Total |
|-------|-------|----------|-------|-------|
| $y_{1+}$ | $y_{2+}$ | $\ldots$ | $y_{I+}$ | $n$ |

A two-way table is constructed from the indicator vectors, $\mathbf{y}_i$ of $A_i$ and $\mathbf{y}_j$ of $B_j$ by pointwise multiplication giving $\mathbf{y}_{ij} = \mathbf{y}_i \mathbf{y}_j$ so that $y_{ijt} = 1$ if $t \in A_i$ *and* $t \in B_j$. The typical element in the table is $Y_{ij+}$. To illustrate

| $I = 2 = J$ | | $B_1$ | $B_2$ | | | Smokers | Non-smokers |
|-------------|------|-------|-------|---|------|---------|-------------|
| | $A_1$ | $y_{11+}$ | $y_{12+}$ | | Girls | 10 | 40 |
| | $A_2$ | $y_{21+}$ | $y_{22+}$ | | Boys | 45 | 25 |

$$n = 120$$

An easy extension leads to three and higher-way contingency tables.

As there is nearly always more than one dependent variable, contingency tables really fall in the province of multivariate analysis [see Chapter 6]; but as this extension only involves indicator variables it is possible to accommodate them within our conceptual framework by rearranging the data array. In the example of smoking and sex the original array contained 120 rows with each row corresponding to a child. In the new array (shown in the table) each unit corresponds to a cell in the smoking by sex table.

This transformation substantially reduces the volume of data but under the assumptions of the linear model it does not reduce its information content.

| Old units | G | B | S | −S | New units | y | G | B | S | −S |
|---|---|---|---|---|---|---|---|---|---|---|
| 1 | 0 | 1 | 1 | 0 | 1 | 10 | 1 | 0 | 1 | 0 |
| 2 | 1 | 0 | 0 | 1 | 2 | 40 | 1 | 0 | 0 | 1 |
| ⋮ | ⋮ | ⋮ | ⋮ | ⋮ | 3 | 45 | 0 | 1 | 1 | 0 |
| $n = 120$ | 1 | 0 | 0 | 1 | $k = 4$ | 25 | 0 | 1 | 0 | 1 |

Table 11.4.1: Table showing transformed data array.

The old array can be reconstructed from the new one apart from the order of the rows, and order is immaterial by the hypothesis that data arrays have permutable rows. In statistical terminology the indicator totals for the $k$ cells are sufficient statistics [see § 3.4].

This vector of cell totals $\mathbf{y} = (y_1, y_2, \ldots, y_k)$ now forms the dependent variable but unfortunately it does not have independently distributed elements as the new $y_t$'s must sum to $n$, the size of the original array. However it turns out that $\mathbf{y}$ has a conditional Poisson distribution and that fortunately the appropriate model fitting procedure is identical to that for independent Poisson sampling as long as conditionality is respected.

We outline the derivation of this result for simple Multinomial sampling where the original units $t = 1, 2, \ldots, n$ are independently distributed and each and every one has the same set of probabilities $p_1, p_2, \ldots, p_k$ of falling in one of the $k$ cells. A well known result of standard distribution theory is that the probability density of the sum of the units, that is of the cell totals $y_1, y_2, \ldots, y_k$, is Multinomial with parameters $n = y_+$ and $p_1, p_2, \ldots, p_k$ [see § 2.9]. Another standard result is that if $y_1, y_2, \ldots, y_k$ are independently Poisson distributed with parameters $\mu_1, \mu_2, \ldots, \mu_k$ and if $y_+ = y_1 + y_2 + \ldots + y_k$ and $\mu_+ = \mu_1 + \mu_2 + \ldots + \mu_k$ then $y_+$ is Poisson $(\mu_+)$ [see Table 2.4.1]; and the conditional joint distribution of $y_1, y_2, \ldots, y_k$ given $y_+ = n$ is Multinomial $(n; \mu_1/\mu_+, \ldots, \mu_k/\mu_+)$ [see § 2.9]. Hence the Multinomial density function of the cell totals $y_1, y_2, \ldots, y_k$ can be represented as conditionally Poisson. A further consequence is that their joint density can be written as the ratio of Poisson densities so that the log-likelihood of $p_1, p_2, \ldots, p_k$ is given by

$$\lambda(p_1, p_2, \ldots, p_k) = \lambda_{ip}(np_1, np_2, \ldots, np_k) - \lambda_p(n)$$

where $\lambda_{ip}$ is the log-likelihood function based on an independent Poisson sample and $\lambda_p$ is the log-likelihood function for a single Poisson variable. The term $\lambda_p$ does not vary with $p_1, p_2, \ldots, p_k$ so that inference for $p_1, p_2, \ldots, p_k$ is only based on $\lambda_{ip}$. This is equivalent to assuming that $y_1, y_2, \ldots, y_k$ are *independently* Poisson with means $np_1, np_2, \ldots, np_k$.

There are other sampling schemes for contingency tables, that is, different ways of sampling the units all of which give rise to inference procedures based on the Poisson likelihood function. One of the most important is the so called product Multinomial. Among many other applications it is extensively used in clinical trials for the analysis of retrospective or prospective experiments.

A simple example of a prospective design occurs where individuals (units) are allocated either to the treatment group or to the control group. Both groups are then exposed to some clinical condition and subsequently tested to see if they have become ill or not. The resulting two-way table is

|                      | $B_1$ = ill | $B_2$ = not ill | Total    |
|----------------------|-------------|-----------------|----------|
| $A_1$ = treated      | $y_{11}$    | $y_{12}$        | $y_{1+}$ |
| $A_2$ = control      | $y_{21}$    | $y_{22}$        | $y_{2+}$ |

with $n = y_{++}$ and $k = 4$. Note that the number of individuals treated and in the control group are fixed in advance and an analysis must condition on the values taken by $y_{1+}$ and $y_{2+}$; these are not outcomes of the experiment. For this reason, $A$ is known as an explanatory factor (or treatment factor or exogenous factor) and $B$ is known as a response factor.

It is easy to see that this is an extension of the simple Multinomial above, to two independent Multinomials (in fact two Binomials as there are only two levels of the response factor $B$). The log likelihood function is just

$$\lambda(p_{11}, p_{12}, p_{21}, p_{22}) = \lambda_{ip}(y_{1+}p_{1|1}, y_{1+}p_{2|1}) - \lambda_p(y_{1+})$$
$$+ \lambda_{ip}(y_{2+}p_{1|2}, y_{2+}p_{2|2}) - \lambda_p(y_{2+})$$

where $p_{1|1} = P(B_1|A_1)$, $p_{2|1} = P(B_2|A_1)$, etc., are the conditional probabilities and so $p_{1|1} + p_{2|1} = 1$, $p_{1|2} + p_{2|2} = 1$. This reduces to the same form as before,

$$\lambda_{ip}(y_{++}p_{11}, y_{++}p_{12}, y_{++}p_{21}, y_{++}p_{22}) - \lambda_p(y_{1+}) - \lambda_p(y_{2+}),$$

subject to the constraint $y_{++}(p_{11} + p_{12}) = y_{1+}$, $y_{++}(p_{21} + p_{22}) = y_{2+}$, and hence can be fitted as though the sampling distribution of the counts in the cell were genuinely Poisson but with a constraint on the parameters.

It is worth pointing out that genuine Poisson sampling does occur in practice. An example is the number of connections made between telephone subscribers; this is a random variable which may well be Poisson due to the characterization of the Poisson distribution as the frequency of 'rare' events. A two-way square table classified by the originator and recipient of the call giving the number of connections made is a contingency table.

### 11.4.2.   Independence Models for the Linear Predictor

Factors $A$ and $B$ are said to be independent if and only if the probability that a unit is classified into both the $i$th level of $A$ and the $j$th level of $B$ factorizes into the product of the probabilities. Thus

$$P(T \in A_i \text{ and } T \in B_j) = P(T \in A_i)P(T \in B_j)   \text{ for all } i \text{ and } j,$$

or, equivalently, $p_{ij} = p_{i+}p_{+j}$, and in an obvious shorthand $P(A \cap B) = P(A)P(B)$. [see II; Definition 4.4.1.]

The original data array with $n$ units, collapsed to a two-way contingency table with $IJ$ cells, has cell entries $y_{ij}$ with expected values $EY_{ij} = n\,p_{ij}$. If $A$ and $B$ are independent

$$EY_{ij} = n\,p_{i+}p_{+j}.$$

With a logarithmic link function, the linear predictor is

$$\eta_{ij} = \log EY_{ij} = \log n + \log p_{i+} + \log p_{+j}$$

and expressed in vectors

$$\boldsymbol{\eta} = \log n\,\mathbf{1} + \sum_i \log p_{i+}\mathbf{a}_i + \sum_j \log p_{+j}\mathbf{b}_j.$$

Thus $\boldsymbol{\eta}$ is a linear combination of the indicator vectors for the levels of $A$ and $B$, $\mathbf{a}_i$ and $\mathbf{b}_j$ and thus is in the subspace $A+B$. The converse follows by reworking the argument backwards, and these together imply the important result that the factors $A$ and $B$ are independent if and only if $\boldsymbol{\eta} \in A+B$. That $\boldsymbol{\eta} = \log EY$ falls in the linear subspace $A+B$ under independence accounts for the importance of *log-linear* models.

*Example.* This table of expected counts exhibits independence and the linear predictor is in $A+B$.

|       | $B_1$ | $B_2$ |
|-------|-------|-------|
| $A_1$ | 32    | 8     |
| $A_2$ | 16    | 4     |

| Unit | $\boldsymbol{\eta}$ | $= 5{\cdot}1$ | $+ -1\mathbf{a}_2$ | $+ -2\mathbf{b}_2$ |
|------|------|------|------|------|
| 1 | 5 | 1 | 0 | 0 |
| 2 | 3 | 1 | 0 | 1 |
| 3 | 4 | 1 | 1 | 0 |
| 4 | 2 | 1 | 1 | 1 |

Logs have been taken to base 2 for convenience in this example.

Both factors have been taken to be response factors. Now consider the case when $B$ is fixed so that the number $n_j$ of observations at level $B_j$ is fixed in advance. The entries in the two-way table must satisfy

$$y_{+j} = \sum_i y_{ij} = n_j$$

and so the expected counts are given by

$$EY_{ij} = n_j p_{i|j}$$

where $p_{i|j}$ is the conditional probability [see II, Definition 3.9.1] that a unit is classified at level $i$ of $A$ given that it is at level $j$ of $B$.

An equivalent definition of independence between $A$ and $B$ is that the conditional probability of $A$ given $B$ is equal to the marginal probability of

*A* [see II, § 3.9.3]. That is $P(A|B) = P(A)$, or $p_{i|j} = p_{i+}$; the appropriate independence model for the expected counts with a fixed factor is then

$$EY_{ij} = n_j p_{i+}.$$

This again is equivalent to $\boldsymbol{\eta} \in A + B$, as can be seen by taking logs.

In the example of smoking and sex considered in section 11.3.6, the two-way contingency table classifying 120 children by sex and smoking exhibits an association; smoking is much higher among boys, and the cross product ratio cpr $= 10 \times 25 / 40 \times 45 = 0 \cdot 139$ is very different from unity. A formal model fitting procedure to test for independence consists of the following:

Model: the p.d.f. of *Y* is Poisson, conditional on total of 120

$$\text{log link, } \eta = \log (EY)$$

linear predictor for independence, $\boldsymbol{\eta} \in A + B$.

Fitted values: details are omitted apart from noting that

$$\hat{\mu}_{ij}(A + B) = y_{i+} y_{+j} / y_{++}$$

Data and (fitted values):

|         | Smokers      | Non-smokers  |           |
|---------|--------------|--------------|-----------|
| Girls   | 10 (22·9)    | 40 (27·1)    |           |
| Boys    | 45 (32·1)    | 25 (37·9)    | $n = 120$ |

Deviance:

$$\text{dev} (A + B) = 2 \left( 10 \log \frac{10}{22 \cdot 9} + \ldots + 25 \log \frac{25}{37 \cdot 9} \right) = 24 \cdot 23$$

$$\text{df} (A + B) = 1.$$

The significance level of this value of the deviance is well beyond the $0 \cdot 1\%$ value of $x^2_{(1)}$ and so is extremely unlikely to have occurred by chance if smoking and sex were independent. The independence hypothesis is very emphatically discredited.

*The log-cross product ratio in* $2 \times 2$ *tables.*

The fact that $\boldsymbol{\eta} \in A + B = \text{span} (\mathbf{1}, \mathbf{a}_2, \mathbf{b}_2)$ is equivalent to independence can be used to construct a measure of association. If the factors are dependent then $\boldsymbol{\eta} \in A * B = \text{span} (\mathbf{1}, \mathbf{a}_2, \mathbf{b}_2, \mathbf{a}_2 \mathbf{b}_2)$, and the coefficient of the interaction terms gives a measure of dependence. This is known as the *log-cross product* *ratio.* Evaluation gives

$$\text{coeff of } \mathbf{a}_2\mathbf{b}_2 = \eta_{22} - \eta_{21} - \eta_{12} + \eta_{11}$$

$$= \log \frac{EY_{11}EY_{22}}{EY_{12}EY_{12}} = \log \frac{p_{11}p_{22}}{p_{12}p_{21}}.$$

### 11.4.3.  Three-way Independence Models.

Independence models are substantially more interesting in three dimensions than in two, as the interrelationships between three factors are more complicated than those between two.

We define various categories of independence between three response factors $A$, $B$ and $C$, as follows:

DEFINITION 11.4.1.  $A$ and $B$ are pairwise independent if $P(A \cap B) = P(A)P(B)$; $A, B$ and $C$ are mutually independent if $P(A \cap B \cap C) = P(A)P(B)P(C)$; $A$ and $B$ are conditionally independent, given $C$, if $P(A \cap B|C) = P(A|C)P(B|C)$.

From these definitions it is straightforward to show that when $A$, $B$ and $C$ are mutually independent any pair must be pairwise independent, but that if all pairs are pairwise independent it does not follow that $A$, $B$ and $C$ are mutually independent. Similarly that $A$ and $B$ are conditionally independent on $C$ does not imply that $A$ and $B$ are pairwise independent, nor is the converse true.

In applied work one of the more interesting features is to note that an association between $A$ and $B$ might be explained by $C$ if the pairwise dependency in $A$ and $B$ becomes a conditional independence given $C$.

Here is a numerical example of conditional independence:

| $A \times B$ | | $A \times B|C_1$ | | $A \times B|C_2$ | |
|:---:|:---:|:---:|:---:|:---:|:---:|
| 35 | 25 | 15 | 5 | 20 | 20 |
| 25 | 15 | 15 | 5 | 10 | 10 |
| $n = 100$ | | $n_1 = 40$ | | $n_2 = 60$ | |

This illustrates that the association between $A$ and $B$ exhibited in the marginal table $A \times B$ can be derived by adding the three-way table over the levels of $C$, even though there is no association between $A$ and $B$ at any level of $C$.

The correspondence between independence models and (log) linear models is close and it is fairly straightforward to show the correspondence

|       | Probability model for $P(A \cap B \cap C)$ | Linear model |
|-------|:-----:|:-----:|
| (1) | $1/IJK$ | $1$ |
| (2) | $P(A)/JK$ | $A$ |
| (3) | $P(A)P(B)P(C)$ | $A + B + C$ |
| (4) | $P(A \cap B)P(C)$ | $A*B + C$ |
| (5) | $P(A|C)P(B|C)P(C)$ | $A*C + B*C$ |
| (6) | ? | $A*B + B*C + C*A$ |
| (7) | $P(C|A \cap B)P(A)P(B)$ | ? |
| (8) | $P(A \cap B \cap C)$ | $A*B*C$ |
| (9) | (7) | (6) |

The first two models specify equiprobabilities and have little practical significance. Mutual independence corresponds to model (3) and conditional independence to model (5). Model (6) is a model with nö three-way but all two-way interactions included, for which it is not possible to give a closed form expression for the probability. Model (7) specifies that $A$ and $B$ are pairwise independent. In the two-way marginal table obtained by collapsing over $C$ the log-linear model for (7) would be $A + B$. The saturated model is at (8) and model (9) is obtained from $P(A \cap B \cap C) = P(C|A \cap B)P(A \cap B)$ by supposing that $P(A \cap B) = P(A)P(B)$ as in (7) while the log-linear model for $P(C|A \cap B)$ is given by (6).

## 11.4.4.  Response and Treatment Factors

The context of the data determines the models of interest. As in regression analysis where there is no automatic set of rules to determine which explanatory variables should be included in the linear predictor so too it is true that there is no rule book for contingency tables either. In fact the situation is more complex due to the multivariate aspect of contingency tables. Certain simplifications follow from a division of the factors into response factors and treatment factors as described above. We shall only consider tables with one response factor.

In the following three-way table, *M*ortality of cuttings from plum root stocks is the response factor and the *T*ime of planting and the *L*ength of the cutting are treatment factors.

Contingency table of plum root cuttings.

|        |          | M | | |
| :---: | :---: | :---: | :---: | :---: |
| L | T | Alive | Dead | Total |
| Long | at once | 156 | 84 | 240 |
|      | in spring | 84 | 156 | 240 |
| Short | at once | 107 | 133 | 240 |
|       | in spring | 31 | 209 | 240 |

This table can be modelled as Bino $(240, p_{ij})$ with a logit [see § 2.7.3(d)] link for $EY_{ij} = 240p_{ij}$. Equivalently, it is a conditional Poisson with the four marginal totals of 240 fixed. The p.d.f. is Poisson, with a log link. The linear predictor $\eta$ contains $L*T$ to satisfy the parameter constraints imposed by conditioning on this margin. It contains $M$ to reflect the overall probability of a cutting staying alive and then it contains interaction terms between $M$ and $L, T$ to investigate the dependence of mortality on length and time. The plausible models to fit are

$$\begin{array}{cc} & \text{Deviance, df} \\ L*T+M & (151{\cdot}0, 3) \\ L*T+M*L \quad L*T+M*T & (105{\cdot}2, 2) \quad (53{\cdot}4, 2) \\ L*T+M*(L+T) & (2{\cdot}29, 1) \\ L*T+M*L*T & (0, 0) \end{array}$$

Analysis of this hierarchy suggests that the no three-way interaction model gives a good fit (deviance $= 2{\cdot}29$, df $= 1$) with the conclusion that mortality depends both on the time of planting and the length of the cutting but those two factors do not have a catalytic effect. That is, there is no interaction between the decrease in mortality caused by planting at once with the decrease due to taking a longer cutting.

J.C.W.

## 11.5.   FURTHER READING AND REFERENCES

The approach to linear models outlined here originates in the paper Nelder, J. A. and Wedderborn, R. W. (1972), Generalised Linear Models, J. Roy. Statist. Soc. A. **135**, 370–384.

For a guide to further reading see section 12.4.

Details of the GLIM manual referred to in the chapter are as follows:

Baker, R. J. and Nelder, J. (1978) *The GLIM System*, Release 3, N.A.G.

# CHAPTER 12

# *Linear Models II*

The Introduction to Chapter 11 gives a brief survey of the contents of the present chapter and their relation to Chapter 11.

## 12.1.  LEAST SQUARES

### 12.1.1.  Introductory Ideas on Least Squares

Least squares [see § 3.5.2, § 3.5.5, Chapter 8, Chapter 10] solves the exercise of choosing a straight line that 'best' fits a data array $(x_1, y_1) \ldots (x_n, y_n)$. For simplicity take the line $y = \beta x$ which is forced to pass through the origin and is determined by the choice of $\beta$.

Measure the discrepancy between a point $(x_t, y_t)$ of the data array and the line by the vertical distance $\varepsilon_t = y_t - \beta x_t$, as in Figure 12.1.1. The smaller these are the better the fit. The *sum of squares* $\varepsilon_1^2 + \varepsilon_2^2 + \ldots + \varepsilon_n^2$ gives an overall measure of discrepancy between the points and the line, and the principle of least squares is to choose the line to minimize this quantity.

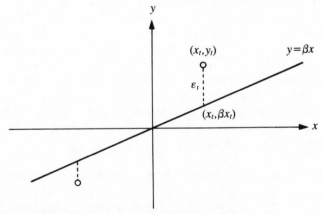

Figure 12.1.1: Diagram showing a typical observed point $(x_t, y_t)$, the corresponding 'fitted' point $(x_t, \beta x_t)$, and the discrepancy $\varepsilon_t$.

In $n$-space [see § 11.1.1], with $\mathbf{y} = (y_1, y_2, \ldots, y_n)$, $\mathbf{x} = (x_1, x_2, \ldots, x_n)$, and $\boldsymbol{\varepsilon} = (\varepsilon_1, \varepsilon_2, \ldots, \varepsilon_n) = \mathbf{y} - \beta\mathbf{x}$, the sum of squares is $\|\boldsymbol{\varepsilon}\|^2 = \|\mathbf{y} - \beta\mathbf{x}\|^2$.

If $b$ is the value of $\beta$ that minimizes $\|\mathbf{y} - \beta\mathbf{x}\|^2$ then $\hat{\mathbf{y}} = b\mathbf{x}$ is the vector of fitted values. Its derivation and consequences can be structured in the following way, which requires the concept of orthogonal vectors; a concept we now define:

DEFINITION 12.1.1.   *Orthogonality of vectors.*   The vectors $\mathbf{x}$ and $\mathbf{y}$, both in $n$-space, are *orthogonal* $(\mathbf{x} \perp \mathbf{y})$ if

$$[\mathbf{x}, \mathbf{y}] = 0.$$

We now give the structure:

the value of $\beta = b$ minimizes $\|\mathbf{y} - \beta\mathbf{x}\|^2$

$$\Updownarrow$$

$$\left.\begin{array}{l} \mathbf{y} - b\mathbf{x} \perp \mathbf{x} \Rightarrow b = [\mathbf{y}, \mathbf{x}][\mathbf{x}, \mathbf{x}]^{-1} \\ \hat{\mathbf{y}} = [\mathbf{y}, \mathbf{x}][\mathbf{x}, \mathbf{x}]^{-1}\mathbf{x} \end{array}\right\} \qquad \text{(a)}$$

$$\left.\begin{array}{l} \mathbf{y} - \hat{\mathbf{y}} \perp \hat{\mathbf{y}} \\ [\hat{\mathbf{y}}, \mathbf{x}] = [\mathbf{y}, \mathbf{x}] \end{array}\right\} \qquad \text{(b)}$$

$$\left.\begin{array}{l} \|\hat{\mathbf{y}}\|^2 = [\mathbf{y}, \mathbf{x}]^2\|\mathbf{x}\|^{-2} = b[\mathbf{y}, \mathbf{x}] \\ \|\mathbf{y}\|^2 = \|\hat{\mathbf{y}}\|^2 + \|\mathbf{y} - \hat{\mathbf{y}}\|^2 \end{array}\right\} \qquad \text{(c)}$$

$$R = [\mathbf{y}, \hat{\mathbf{y}}]/\|\mathbf{y}\| \cdot \|\hat{\mathbf{y}}\| = \|\hat{\mathbf{y}}\|/\|\mathbf{y}\|.] \qquad \text{(d)}$$

The equivalence states that the value of $\beta$ which minimizes $\|\mathbf{y} - \beta\mathbf{x}\|^2$ is the value that makes the residual vector $\mathbf{y} - b\mathbf{x}$ orthogonal to $\mathbf{x}$. Rephrased as $[\mathbf{y} - b\mathbf{x}, \mathbf{x}] = 0$ this condition is known as the normal equation. The converse is that if $b$ satisfies the normal equation it minimizes the sum of squares. A proof of this equivalence is deferred to the next section.

The results in (a) and (b) concerning the least squares coefficient $b$ and the fitted values $\hat{\mathbf{y}}$ flow immediately from the normal equation $[\mathbf{y} - b\mathbf{x}, \mathbf{x}] = 0$. The sum of squares attributable to the fitted line is $\|\hat{\mathbf{y}}\|^2$ and its expression in (c) comes from (a). The sum of squares partition for $\|\mathbf{y}\|^2$ is derived by noting that $\mathbf{y}$ can be written as $\mathbf{y} = \hat{\mathbf{y}} + (\mathbf{y} - \hat{\mathbf{y}})$ and that from (b) these two components are orthogonal. The *multiple correlation coefficient $R$* is defined as the correlation between $\mathbf{y}$ and the fitted values of $\hat{\mathbf{y}}$. It turns out to equal a ratio of sums of squares given in (d).

A numerical example follows.

EXAMPLE 12.1.1.   *Least squares fit of a line through the origin.*
   Data:

| $\mathbf{y}$ | $-3$ | $2$ | $1$ | $4$ | $8$ |
|---|---|---|---|---|---|
| $\mathbf{x}$ | $-2$ | $-1$ | $0$ | $1$ | $2$ |

Cross products:

$$\begin{pmatrix} [\mathbf{x}, \mathbf{x}] & [\mathbf{x}, \mathbf{y}] \\ & [\mathbf{y}, \mathbf{y}] \end{pmatrix} = \begin{pmatrix} 10 & 24 \\ & 94 \end{pmatrix}$$

Least squares estimate:

$$b = [\mathbf{y}, \mathbf{x}][\mathbf{x}, \mathbf{x}]^{-1} = 24/10 = 2 \cdot 4$$

$$(\hat{\mathbf{y}} = 2 \cdot 4x \text{ is the fitted line})$$

Fitted values:

$$\hat{\mathbf{y}} \big| \quad -4 \cdot 8 \quad -2 \cdot 4 \quad 0 \cdot 0 \quad 2 \cdot 4 \quad 4 \cdot 8$$

Sum of squares partition:

$$\|\mathbf{y}\|^2 = 94$$

$$\|\hat{\mathbf{y}}\|^2 = (2 \cdot 4)24 = 57 \cdot 6$$

$$\|\mathbf{y} - \hat{\mathbf{y}}\|^2 = \|\mathbf{y}\|^2 - \|\hat{\mathbf{y}}\|^2 = 94 - 57 \cdot 6 = 36 \cdot 4$$

Multiple correlation:

$$R^2 = 57 \cdot 6/94 = 61 \cdot 3\%.$$

### 12.1.2.   Least Squares in 2-space

Representing the vectors as points in $n$-space diagramatically provides a certain insight into the least squares solution. For $n = 2$ the vectors $\mathbf{y} = (y_1, y_2)$ and $\mathbf{x} = (x_1, x_2)$ are shown. The vector $\beta\mathbf{x}$ is a scalar multiple of $\mathbf{x}$ and it is clear from the diagram that the value $b$ of $\beta$ which minimizes the distance from $\mathbf{y}$ to $\beta\mathbf{x}$ is that value $b$ of $\beta$ for which $\mathbf{y} - b\mathbf{x}$ orthogonal (i.e. at right angles) to $\mathbf{x}$. (See Figure 12.1.2.)

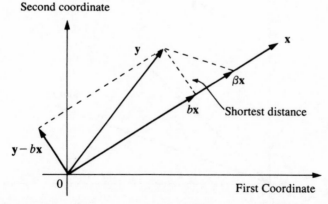

Figure 12.1.2: Geometrical representation of vectors in $n$-space. The vector $b\mathbf{x}$ is the orthogonal projection of $\mathbf{y}$ on $\mathbf{x}$. The residual $\mathbf{y} - b\mathbf{x}$ is perpendicular to $\mathbf{x}$.

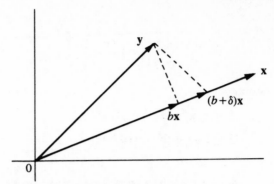

Figure 12.1.3: Showing the orthogonal projection $b\mathbf{x}$ and a non-orthogonal projection $(b+\delta)\mathbf{x}$.

To prove that the condition $\mathbf{y} - b\mathbf{x} \perp \mathbf{x}$ is equivalent to the condition that $\|\mathbf{y} - b\mathbf{x}\|^2$ minimizes $\|y - \beta x\|^2$, consider a pertubation of $b$ to $b + \delta$. (See Figure 12.1.3.) Then

$$\|\mathbf{y} - (b+\delta)\mathbf{x}\|^2 - \|\mathbf{y} - b\mathbf{x}\|^2 = -2\delta[\mathbf{y} - b\mathbf{x}, \mathbf{x}] + \delta^2\|\mathbf{x}\|^2.$$

Let us assume that $b$ is the minimizing value, in which case the right hand side must be non-negative for all $\delta$. As this is a quadratic with a positive coefficient of $\delta^2$ and one root at zero, both roots must be zero. Hence the coefficient of $\delta$ is zero and so $[\mathbf{y} - b\mathbf{x}, \mathbf{x}] = 0$. Conversely, if the normal equation is satisfied the right-hand side is never negative and thus $b$ is the minimizing value.

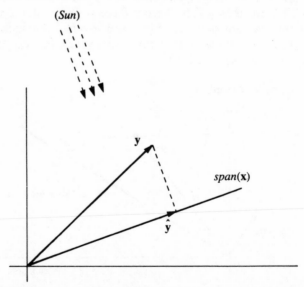

Figure 12.1.4: If the sun's rays are at right angles to span $(\mathbf{x})$, $\hat{\mathbf{y}}$ is the shadow of $\mathbf{y}$ on span $(\mathbf{x})$.

Further insight accrues by noting that $\hat{\mathbf{y}}$ is the projection of $\mathbf{y}$ onto the subspace spanned [see § 11.3.1] by $\mathbf{x}$. If the sun shone from the direction at right angles to span $(\mathbf{x})$, $\hat{\mathbf{y}}$ would be the shadow of $\mathbf{y}$. (See Figure 12.1.4.) This notion is expressed algebraically by $\hat{\hat{\mathbf{y}}} = \hat{\mathbf{y}}$.

*Proof*:

$$\hat{\mathbf{y}} = [\mathbf{y}, \mathbf{x}][\mathbf{x}, \mathbf{x}]^{-1}\mathbf{x} = b\mathbf{x}$$

so that

$$\hat{\hat{\mathbf{y}}} = [\hat{\mathbf{y}}, \mathbf{x}][\mathbf{x}, \mathbf{x}]^{-1}\mathbf{x} = [b\mathbf{x}, \mathbf{x}][\mathbf{x}, \mathbf{x}]^{-1}x$$

$$= b[\mathbf{x}, \mathbf{x}][\mathbf{x}, \mathbf{x}]^{-1}\mathbf{x} = b\mathbf{x} = \hat{\mathbf{y}}.$$

Furthermore it is an orthogonal projection as $\mathbf{y} - \hat{\mathbf{y}} \perp \hat{\mathbf{y}}$. The visual representation of the sum of squares partition is just Pythagoras's theorem for right angled triangles. (See Figure 12.1.5.)

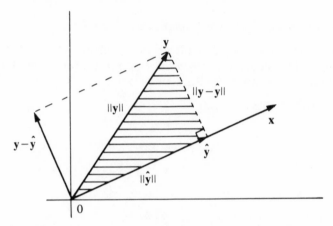

Figure 12.1.5: Pythagoras' Theorem applied to the right-angled triangle gives the sum of squares partition:

$$\|\mathbf{y}\|^2 = \|\hat{\mathbf{y}}\|^2 + \|\mathbf{y} - \hat{\mathbf{y}}\|^2.$$

### 12.1.3. Maximum Likelihood Estimation for the Normal Density

These least squares results are immediately applicable when sampling from a Normal density [see II, § 11.4]. The log-likelihood function [see Examples 6.2.4 and 6.5.1] for an independent random sample $y_1, y_2, \ldots, y_n$ where $y_i$ comes from a Normal p.d.f. with mean $\mu_i = \beta x_i$ and variance $\phi$ is

$$\lambda(\boldsymbol{\mu}) = \text{const} - \sum_i (y_i - \mu_i)^2 / 2\phi$$

$$= \text{const} - \|\mathbf{y} - \boldsymbol{\mu}\|^2 / 2\phi = \text{const} - \|\mathbf{y} - \beta\mathbf{x}\|^2 / 2\phi.$$

Thus maximum likelihood is equivalent to least squares in this situation [see § 3.5.5]. Differentiating gives

$$\frac{\partial \lambda}{\partial \beta} = \sum_i (y_i - \beta x_i) x_i / \phi = [\mathbf{y} - b\mathbf{x}, \mathbf{x}] / \phi,$$

and $\partial \lambda / \partial \beta = 0$ gives the normal equation $[\mathbf{y} - b\mathbf{x}, \mathbf{x}] = 0$. The fitted values $\hat{\mu}$ are given by $\hat{y}$.

The deviance of a fitted model $\mu \in \mathrm{span}\,(\mathbf{x})$ is

$$\mathrm{dev} = 2\{\lambda\,(\mathbf{y}) - \lambda\,(\hat{\mu})\} = \|\mathbf{y} - \hat{\mu}\|^2 / \phi,$$

and the deviance associated with the null model $\mu = \mathbf{0}$ is

$$\mathrm{dev} = 2\{\lambda\,(\mathbf{y}) - \lambda\,(\mathbf{0})\} = \|\mathbf{y}\|^2 / \phi.$$

Thus the sum of squares partition corresponds to a partition of the deviance.

### 12.1.4.  Least Squares on Two Explanatory Variables

Least squares theory easily generalizes to $p$ explanatory variables but most of the additional concepts in the general case are well illustrated with $p = 2$. The argument has a similar structure when $p = 1$.

$b_1, b_2$ minimizes $\|\mathbf{y} - \beta_1 \mathbf{x}_1 - \beta_2 \mathbf{x}_2\|^2$

$\Updownarrow$

$\mathbf{y} - b_1\mathbf{x}_1 - b_2\mathbf{x}_2 \perp \mathbf{x}_1, \mathbf{x}_2 \Rightarrow b_1, b_2$ solution of simultaneous equations:

$$b_1[\mathbf{x}_1, \mathbf{x}_j] + b_2[\mathbf{x}_2, \mathbf{x}_j] = [\mathbf{y}, \mathbf{x}_j], \qquad j = 1, 2.$$

$$\hat{\mathbf{y}} = b_1\mathbf{x}_1 + b_2\mathbf{x}_2$$

$$\mathbf{y} - \hat{\mathbf{y}} \perp \hat{\mathbf{y}}$$

$$[\hat{\mathbf{y}}, \mathbf{x}_1] = [\mathbf{y}, \mathbf{x}_1], [\hat{\mathbf{y}}, \mathbf{x}_2] = [\mathbf{y}, \mathbf{x}_2]$$

$$\|\hat{\mathbf{y}}\|^2 = \|b_1\mathbf{x}_1 + b_2\mathbf{x}_2\|^2 = b_1[\mathbf{x}_1, \mathbf{y}] + b_2[\mathbf{x}_2, \mathbf{y}]$$

$$\|\mathbf{y}\|^2 = \|\hat{\mathbf{y}}\|^2 + \|\mathbf{y} - \hat{\mathbf{y}}\|^2$$

$$R = \|\hat{\mathbf{y}}\| / \|\mathbf{y}\|.$$

The orthogonality condition gives the normal equations

$$0 = [\mathbf{y} - b_1\mathbf{x}_1 - b_2\mathbf{x}_2, \mathbf{x}_1] = [\mathbf{y}, \mathbf{x}_1] - b_1[\mathbf{x}_1, \mathbf{x}_1] - b_2[\mathbf{x}_2, \mathbf{x}_1]$$

$$0 = [\mathbf{y} - b_1\mathbf{x}_1 - b_2\mathbf{x}_2, \mathbf{x}_2] = [\mathbf{y}, \mathbf{x}_2] - b_1[\mathbf{x}_1, \mathbf{x}_2] - b_2[\mathbf{x}_2, \mathbf{x}_2]$$

and $b_1, b_2$ is the solution to this pair of simultaneous equations, whence

$$\binom{b_1}{b_2} = \begin{pmatrix} [\mathbf{x}_1, \mathbf{x}_1] & [\mathbf{x}_2, \mathbf{x}_1] \\ [\mathbf{x}_1, \mathbf{x}_2] & [\mathbf{x}_2, \mathbf{x}_2] \end{pmatrix}^{-1} \begin{pmatrix} [\mathbf{x}_1, \mathbf{y}] \\ [\mathbf{x}_2, \mathbf{y}] \end{pmatrix}$$

EXAMPLE 12.1.2.  *Least squares fit of a 2-parameter line*
  Data:

$$
\begin{array}{c|ccccc}
y & -3 & 2 & -1 & 4 & 8 \\
x_1 & -2 & -1 & 0 & 1 & 2 \\
x_2 & 1 & 1 & -4 & 1 & 1
\end{array}
$$

Least squares estimates:

$$
\begin{bmatrix} b_1 \\ b_2 \end{bmatrix} = \begin{bmatrix} 10 & 0 \\ 0 & 20 \end{bmatrix}^{-1} \begin{bmatrix} 24 \\ 15 \end{bmatrix} = \begin{bmatrix} 2 \cdot 4 \\ 0 \cdot 75 \end{bmatrix}
$$

Fitted line:

$$
y = 2 \cdot 4 x_1 + 0 \cdot 75 x_2.
$$

EXAMPLE 12.1.3.  *Ordinary linear regression.*  The least squares estimates of the coefficients $\alpha, \beta$ of the simple linear regression model

$$
y_t = \alpha + \beta x_t + \varepsilon_t \qquad t = 1, 2, \dots, n
$$

are given by noting $x_1 = 1, x_2 = x$. As

$$
[1, 1] = n, \qquad [1, x] = \sum x_t, \qquad [x, x] = \sum x_t^2
$$

the estimates are

$$
\begin{bmatrix} b_1 \\ b_2 \end{bmatrix} = \begin{bmatrix} n & \sum x_t \\ \sum x_t & \sum x_t^2 \end{bmatrix}^{-1} \begin{bmatrix} \sum y_t \\ \sum x_t y_t \end{bmatrix}.
$$

Put

$$
b = (\sum x_t y_t - n^{-1} \sum x_t \sum y_t) / (\sum x_t^2 - n^{-1}(\sum x_t)^2)
$$

then this expression for $b_1$ and $b_2$ simplifies to

$$
\begin{pmatrix} b_1 \\ b_2 \end{pmatrix} = (n \sum x_t^2 - (\sum x_t)^2)^{-1} \begin{pmatrix} \sum x_t^2 \sum y_t - \sum x_t \sum x_t y_t \\ n \sum x_t y_t - \sum x_t \sum y_t \end{pmatrix}
$$

$$
= \begin{pmatrix} \bar{y} - b\bar{x} \\ b \end{pmatrix}.
$$

The fitted line is $y = \bar{y} + b(x - \bar{x})$ which passes through the centre of gravity $(\bar{x}, \bar{y})$ with gradient $b = \sum (x_t - \bar{x})(y_t - \bar{y}) / \sum (x_t - \bar{x})^2$.
  The proof of the equivalence of the normal equations and the minimization problem is almost identical to the proof when $p = 1$. The least squares estimates $b_1, b_2$ are obtained by solving the normal equations, which usually involves a matrix inversion. The residual vector $\mathbf{y} - \hat{\mathbf{y}}$ must be orthogonal to the fitted values $\hat{\mathbf{y}}$ and this leads to the sum of squares partition as before.

The additional feature in the case $p=2$ is in attributing part of the sum of squares, $\|\hat{\mathbf{y}}\|^2$, to $\mathbf{x}_1$ and part to $\mathbf{x}_2$. Consider the diagram in Figure 12.1.6. This is drawn in the plane spanned by $\mathbf{x}_1$ and $\mathbf{x}_2$, which therefore also contains $\hat{\mathbf{y}}$. How much of $\|\hat{\mathbf{y}}\|^2$ is attributable to $\mathbf{x}_1$ and how much to $\mathbf{x}_2$? To answer this the notation has to be extended slightly.

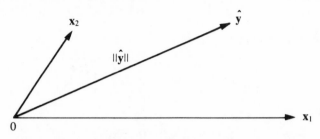

Figure 12.1.6: $\hat{\mathbf{y}}$ lies in the plane spanned by $\mathbf{x}_1$ and $\mathbf{x}_2$.

Write $\hat{\mathbf{y}}(\mathbf{x}_1)$ for a fitted value obtained by minimizing $\|\mathbf{y}-\beta_1\mathbf{x}_1\|^2$, so that $\hat{\mathbf{y}}(\mathbf{x}_1) \in \text{span}(\mathbf{x}_1)$. Similarly $\hat{\mathbf{y}}(\mathbf{x}_2) \in \text{span}(\mathbf{x}_2)$, and the fitted values $\hat{\mathbf{y}}(\mathbf{x}_1, \mathbf{x}_2)$ are obtained by minimizing $\|\mathbf{y}-\beta_1\mathbf{x}_1-\beta_2\mathbf{x}_2\|^2$, so that $\hat{\mathbf{y}}(\mathbf{x}_1\mathbf{x}_2) \in \text{span}(\mathbf{x}_1, \mathbf{x}_2)$. The manipulations simplify when $[\mathbf{x}_1, \mathbf{x}_2]=0$:

*The orthogonal case* $[\mathbf{x}_1, \mathbf{x}_2]=0$
  In this case

$$\hat{\mathbf{y}}(\mathbf{x}_1, \mathbf{x}_2) = \hat{\mathbf{y}}(\mathbf{x}_1) + \hat{\mathbf{y}}(\mathbf{x}_2),$$

$$\|\hat{\mathbf{y}}(\mathbf{x}_1, \mathbf{x}_2)\|^2 = \|\hat{\mathbf{y}}(\mathbf{x}_1)\|^2 + \|\hat{\mathbf{y}}(\mathbf{x}_2)\|^2.$$

The easiest way to see this result is to note that the normal equation for $b_1$ is $[\mathbf{y}-b_1\mathbf{x}_1, \mathbf{x}_1]=0$ when $\hat{\mathbf{y}} \in \text{span}(\mathbf{x}_1)$ and it is $[\mathbf{y}-b_1\mathbf{x}_1-b_2\mathbf{x}_2, \mathbf{x}_1]=0$ when $\hat{\mathbf{y}} \in \text{span}(\mathbf{x}_1, \mathbf{x}_2)$. But if $[\mathbf{x}_1, \mathbf{x}_2]=0$ these both give the same value for $b_1$ and so $\hat{\mathbf{y}}(\mathbf{x}_1, \mathbf{x}_2) = b_1\mathbf{x}_1 + b_2\mathbf{x}_2 = \hat{\mathbf{y}}(\mathbf{x}_1) + b_2\mathbf{x}_2$. An identical argument holds for $b_2$ to give $\hat{\mathbf{y}}(\mathbf{x}_1, \mathbf{x}_2) = \hat{\mathbf{y}}(\mathbf{x}_1) + \hat{\mathbf{y}}(\mathbf{x}_2)$. Now note $\hat{\mathbf{y}}(\mathbf{x}_1) \perp \hat{\mathbf{y}}(\mathbf{x}_2)$, since $\mathbf{x}_1 \perp \mathbf{x}_2$, which establishes the sum of squares partition

$$\|\mathbf{y}\|^2 = \|\mathbf{y}-\hat{\mathbf{y}}(\mathbf{x}_1, \mathbf{x}_2)\|^2 + \|\hat{\mathbf{y}}(\mathbf{x}_1)\|^2 + \|\hat{\mathbf{y}}(\mathbf{x}_2)\|^2.$$

EXAMPLE 12.1.4.   *Fitting an orthogonal design.*
  Data: from Example 12.1.1.
  Cross-product matrix:

$$\begin{bmatrix} [\mathbf{x}_1, \mathbf{x}_1] & [\mathbf{x}_1, \mathbf{x}_2] & [\mathbf{x}_1, \mathbf{y}] \\ & [\mathbf{x}_2, \mathbf{x}_2] & [\mathbf{x}_2, \mathbf{y}] \\ & & [\mathbf{y}, \mathbf{y}] \end{bmatrix} = \begin{bmatrix} 10 & 0 & 24 \\ & 20 & 15 \\ & & 94 \end{bmatrix}$$

Sum of squares:

$$\|\mathbf{y}\|^2 = 94$$

$$\|\hat{\mathbf{y}}(\mathbf{x}_1, \mathbf{x}_2)\|^2 = \|\hat{\mathbf{y}}\|^2 = [\hat{\mathbf{y}}, b_1\mathbf{x}_1 + b_2\mathbf{x}_2]$$

$$= b_1[\mathbf{y}, \mathbf{x}_1] + b_2[\mathbf{y}, \mathbf{x}_2]$$

$$= (2\cdot4)24 + (0\cdot75)15 = 57\cdot6 + 11\cdot25 = 68\cdot85$$

$$\|\hat{\mathbf{y}}(\mathbf{x}_1)\|^2 = b_1[\mathbf{y}, \mathbf{x}_2] = 24^2/10 = 57\cdot6$$

$$\|\hat{\mathbf{y}}(\mathbf{x}_2)\|^2 = b_2[\mathbf{y}, \mathbf{x}_2] = 15^2/20 = 11\cdot25$$

$$\|\mathbf{y} - \hat{\mathbf{y}}\|^2 = 94 - 68\cdot85 = 25\cdot15.$$

Partition:

| Source | df | Sum of squares |
|---|---|---|
| Due to $x_1$ | 1 | 57·60 |
| Due to $x_2$ | 1 | 11·25 |
| Due to $x_1$ and $x_1$ | 2 | 68·85 |
| Residual | 3 | 25·15 |
| Total | 5 | 94·0 |

### 12.1.5. The Non-orthogonal Case

When $\mathbf{x}_1$ and $\mathbf{x}_2$ are not orthogonal, that is, when $[\mathbf{x}_1, \mathbf{x}_2] \neq 0$, difficulties of interpretation arise as there is no longer a unique way in which to partition the sum of squares due to fitting $\mathbf{x}_1$ and $\mathbf{x}_2$, $\|\hat{\mathbf{y}}(\mathbf{x}_1, \mathbf{x}_2)\|^2$, into separate components. The diagrams in Figure 12.1.7 illustrate this.

*Result*:

$$\hat{\mathbf{y}}(\mathbf{x}_1, \mathbf{x}_2) = \hat{\mathbf{y}}(\mathbf{x}_1) + \hat{\mathbf{y}}(\mathbf{e}_2) = \hat{\mathbf{y}}(\mathbf{e}_1) + \hat{\mathbf{y}}(\mathbf{x}_2)$$

$$\|\hat{\mathbf{y}}(\mathbf{x}_1, \mathbf{x}_2)\|^2 = \|\hat{\mathbf{y}}(\mathbf{x}_1)\|^2 + \|\hat{\mathbf{y}}(\mathbf{e}_2)\|^2 = \|\hat{\mathbf{y}}(\mathbf{e}_1)\|^2 + \|\hat{\mathbf{y}}(\mathbf{x}_2)\|^2.$$

The vectors $\mathbf{x}_1$ and $\mathbf{x}_2$ in (a) of Figure 12.1.7 can be transformed into the orthogonal pair $\mathbf{x}_1$ and $\mathbf{e}_2 = \mathbf{x}_2 - [\mathbf{x}_2, \mathbf{x}_1][\mathbf{x}_1, \mathbf{x}_1]^{-1}\mathbf{x}_1$ in Figure 12.1.7 (b(i)) or into the orthogonal pair $\mathbf{x}_2$ and $\mathbf{e}_1 = \mathbf{x}_1 - [\mathbf{x}_1, \mathbf{x}_2][\mathbf{x}_2, \mathbf{x}_2]^{-1}\mathbf{x}_2$ in Figure 12.1.7(b(ii)). The results for the orthogonal case above now apply giving two partitions for the sum of squares. The total sum of squares $\|\hat{\mathbf{y}}(\mathbf{x}_1, \mathbf{x}_2)\|^2$ attributable to fitting $\mathbf{y}$ on $\mathbf{x}_1$ and $\mathbf{x}_2$ can be partitioned into that due to fitting $\mathbf{y}$ on $\mathbf{x}_1$ alone and that due to fitting $\mathbf{y}$ on $\mathbf{x}_2$ adjusted for $\mathbf{x}_1$; or it can be partitioned into that due to fitting $\mathbf{y}$ on $\mathbf{x}_2$ alone and that due to fitting $\mathbf{y}$ on $\mathbf{x}_1$ adjusted for $\mathbf{x}_2$.

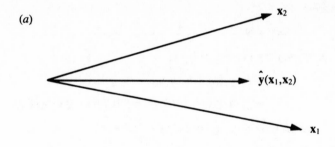

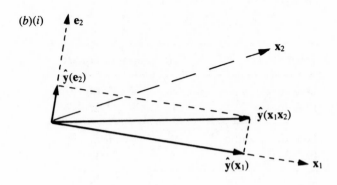

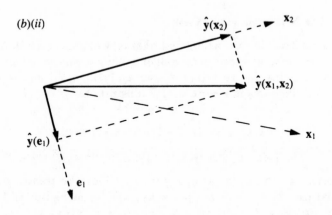

Figure 12.1.7: (b) shows two ways of partitioning a fitted vector $\hat{\mathbf{y}}(\mathbf{x}_1, \mathbf{x}_2)$.

In (i), $\hat{\mathbf{y}}(\mathbf{x}_1, \mathbf{x}_2) = \hat{\mathbf{y}}(\mathbf{x}_1) + \hat{\mathbf{y}}(\mathbf{e}_2)$;

In (ii), $\hat{\mathbf{y}}(\mathbf{x}_1, \mathbf{x}_2) = \hat{\mathbf{y}}(\mathbf{e}_1) + \hat{\mathbf{y}}(\mathbf{x}_2)$.

This is the content of the result above. A formal proof requires us to show that $[x_1, e_2] = 0$ (easy, as $e_2$ is just the residual of fitting $x_2$ on $x_1$) and to demonstrate that $\hat{y}(x_1, x_2)$ equals $\hat{y}(x_1, e_2)$.

EXAMPLE 12.1.5.  *Least squares in the non-orthogonal case.*
Data: a modification of the previous example
  Cross product matrix:

$$\begin{bmatrix} [x_1, x_2] & [x_1, x_2] & [x_1, y] \\ & [x_2, x_2] & [x_2, y] \\ & & [y, y] \end{bmatrix} = \begin{bmatrix} 10 & 8 & 24 \\ & 20 & 15 \\ & & 94 \end{bmatrix}$$

Sums of squares:

$$\|y\|^2 = 94$$

$$\|\hat{y}(x_1)\|^2 = (2 \cdot 4)24 = 57 \cdot 6$$

$$\|\hat{y}(x_2)\|^2 = (0 \cdot 75)15 = 11 \cdot 25$$

$$\begin{bmatrix} b_1 \\ b_2 \end{bmatrix} = \begin{bmatrix} 10 & 8 \\ 8 & 20 \end{bmatrix}^{-1} \begin{bmatrix} 24 \\ 15 \end{bmatrix} = \begin{bmatrix} 2 \cdot 65 \\ -0 \cdot 34 \end{bmatrix}$$

$$\|\hat{y}(x_1, x_2)\|^2 = b_1[y, x_1] + b_2[y, x_2] = (24, 15) \begin{pmatrix} 10 & 8 \\ 8 & 20 \end{pmatrix}^{-1} \begin{pmatrix} 24 \\ 15 \end{pmatrix}$$

$$= 2 \cdot 65(24) + -31(15)$$

$$= 63 \cdot 60 - 5 \cdot 10$$

$$= 58 \cdot 90.$$

Partition:

| Source | df | ss | Source | df | ss |
|---|---|---|---|---|---|
| $x_1$ alone | 1 | 57·60 | $x_1$ given $x_2$ | 1 | 47·65 |
| $x_2$ given $x_1$ | 1 | 1·30 | $x_2$ along | 1 | 11·25 |
| $x_1$ and $x_2$ | 2 | 58·90 | $x_1$ and $x_2$ | 2 | 58·90 |
| Residual | | 35·10 | | | 35·10 |
| Total | | 94 | | | 94 |

These sums of squares partitioned can be rearranged into a lattice of scaled deviances.

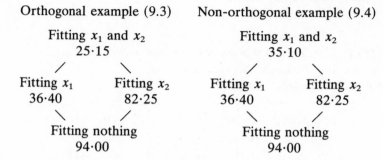

Orthogonal example (9.3)        Non-orthogonal example (9.4)

As $[\mathbf{x}_1, \mathbf{x}_2]$ changes from 0, the values of the coefficients, $b_1$ and $b_2$ change. In the above example $b_2$ alters from $0\cdot75$ to $-0\cdot34$, actually changing sign. Interpreting the coefficients must be done carefully as their values depend on the values of the other explanatory variables. The sum of squares attributable to $\mathbf{x}_1$ and $\mathbf{x}_2$ is no longer the sum of that attributable to $\mathbf{x}_1$ and that attributable to $\mathbf{x}_2$. In fact it is usually smaller than this quantity; in the example $58\cdot90 < 68\cdot85$.

If $[\mathbf{x}_1, \mathbf{x}_2]$ increases so that the matrix

$$\begin{pmatrix} [\mathbf{x}_1, \mathbf{x}_1] & [\mathbf{x}_1, \mathbf{x}_2] \\ [\mathbf{x}_2, \mathbf{x}_1] & [\mathbf{x}_2, \mathbf{x}_2] \end{pmatrix}$$

is not invertible, it turns out that $\mathbf{x}_1$ must be linearly related to $\mathbf{x}_2$. The variables $\mathbf{x}_1, \mathbf{x}_2$ are then said to be colinear. When this occurs, the correlation between $\mathbf{x}_1$ and $\mathbf{x}_2$ is unity and so one variable can be discarded without a loss of information. The case where the correlation between $\mathbf{x}_1$ and $\mathbf{x}_2$ is nearly but not quite equal to 1 poses the added difficulty of deciding what 'nearly 1' means and of deciding which variable to discard.

### 12.1.6.   Extensions to Further Explanatory Variables

The algebraic solution to the least squares problem extends immediately to $p$ explanatory variables. As in the cases already discussed with $p = 1$ or 2, all truth flows from the normal equations which arise from the $p$ orthogonalities:

$$\mathbf{y} - b_1\mathbf{x}_1 - b_2\mathbf{x}_2 \ldots - b_p\mathbf{x}_p \perp \mathbf{x}_1, \mathbf{x}_2, \ldots, \mathbf{x}_p.$$

When the $\mathbf{x}$'s are orthogonal to one another the solution is just $b_j = [\mathbf{y}, \mathbf{x}_j] \|\mathbf{x}_j\|^{-2}$, as when $p = 1$. More generally it is necessary to solve $p$ simultaneous equations in $p$ unknowns requiring a matrix inversion.

The sum of squares attributable to the explanatory variables $\|\mathbf{y}(\mathbf{x}_1, \mathbf{x}_2, \ldots, \mathbf{x}_p)\|^2$ can be partitioned further. Given an ordering of the variables, say $x_1, x_2, \ldots, x_p$ then define

$$e_1 = x_1$$

$$e_2 = x_2 - \hat{x}_2(e_1)$$

$$e_3 = x_3 - \hat{x}_3(e_1) - \hat{x}_3(e_2)$$

$$\vdots$$

$$e_p = x_p - \hat{x}_p(e_1) - \hat{x}_p(e_2) \ldots - \hat{x}_p(e_{p-1}).$$

The new variables are mutually orthogonal, so that

$$\|\hat{y}(x_1, x_2, \ldots, x_p)\|^2 = \|\hat{y}(e_1)\|^2 + \|\hat{y}(e_2)\|^2 + \ldots + \|\hat{y}(e_p)\|^2$$

The interpretation for $\|\hat{y}(e_j)\|^2$ is the sum of squares attributable to $x_j$ having adjusted for $x_1, x_2, \ldots, x_{j-1}$. In applications, the difficult part of this procedure is to decide which of the $p!$ orderings of the variables is the most appropriate.

Now the fitted values are a linear combination of $x_1, x_2, \ldots, x_p$, namely

$$\hat{y}(x_1, x_2, \ldots, x_p) = b_1 x_1 + b_2 x_2 + \ldots + b_p x_p.$$

Similarly they can be expressed as a linear combination of the orthogonal variables $e_1, e_2, \ldots, e_p$, namely $\hat{y}(e_1, e_2, \ldots, e_p)$. The coefficients of $x_1, x_2, \ldots,$ and in particular that of $x_p$, must be the same in both linear combinations. But $e_j$ is a function only of $x_1, x_2, \ldots$ up to $x_j$, so that only $e_p$ is a function of $x_p$. Hence the coefficients of $e_p$ and $x_p$ are the same and so

$$b_p = [y, e_p][e_p, e_p]^{-1}$$

where

$$e_p = x_p - \hat{x}_p(x_1, x_2, \ldots, x_p).$$

This is an analytic expression for $b_p$ in terms of $y$ and the residual of $x_p$ having adjusted for all the other variables in the predictor, useful for investigating the theoretical properties of the least squares coefficients.

## 12.2. REPEATED SAMPLING

### 12.2.1. Normal Distribution Theory

The results summarized here are necessary to derive the sampling distributions [see § 2.2] of least squares coefficients, fitted values, residuals and components of the sum of squares partition when sampling from an underlying Normal p.d.f.

Notation:

~means distributed according to;
$N(\mu, \phi)$ denotes a Normal p.d.f. with mean (expectation) $\mu$ and variance $\phi$ [see II, § 11.4.3 where, as elsewhere in this book, the variance is

denoted by $\sigma^2$. The notation $N(\mu, \phi)$ used here is elsewhere replaced by $N(\mu, \sqrt{\phi})$ or $N(\mu, \sigma)$];

$\chi^2(k, \lambda)$ denotes a non-central chi-squared p.d.f. with $k$ degrees of freedom and non-centrality parameter $\lambda$ [see § 2.8.1];

$F(k_1, k_2)$ denotes an $F$ distribution with degrees of freedom $k_1$ and $k_2$;

$t(k)$ denotes Student's distribution with $k$ df.

Facts:

1. If $z \sim N(\mu, \phi)$ then
   (a) $az + c \sim N(a\mu + c, a^2\phi)$ [see § 2.5.1];
   (b) $z^2/\phi \sim \chi^2(1, \mu^2/2\phi)$ [see § 2.8.1]
2. If $z_1, z_2$ are bivariate Normal [see II, § 13.4.6] then
   (a) $z_1 + z_2 \sim N$;
   (b) $z_1, z_2$ independent if and only if cov $(z_1, z_2) = 0$ [see II, Theorem 13.4.1].
3. If $z_1 \sim N(\mu_1, \phi_1)$, $z_2 \sim N(\mu_2, \phi_2)$ and if $z_1, z_2$ independent then
   (a) $z_1 + z_2 \sim N(\mu_1 + \mu_2, \phi_1 + \phi_2)$ [see § 2.5.3(a)];
   (b) $z_1^2 + z_2^2 \sim \chi^2 (2, \frac{1}{2}(\mu_1^2 + \mu_2^2))$ when $\phi_1 = 1 = \phi_2$ [see § 2.8.1].
4. (a) If $z_1^2 \sim \chi^2(k_1, \lambda_1)$, $z_2^2 \sim \chi^2(k_2, \lambda_2)$ and $z_1^2, z_2^2$ are independent then $z_1^2 + z_2^2 \sim \chi^2(k_1 + k_2, \lambda_1 + \lambda_2)$ [see § 2.8.1];
   (b) If $z_1^2 \sim \chi^2(k_1, \lambda_1)$, $z^2 \sim \chi^2(k, \lambda)$ and $z^2 = z_1^2 + z_2^2$ and $z_1^2, z_2^2$ independent, then $z_2^2 \sim \chi^2 (k - k_1, \lambda - \lambda_1)$.
5. (a) If $z_1 \sim N(\mu, \phi)$, $z_2^2/\phi \sim \chi^2(k, 0)$ and $z_1, z_2^2$ independent, then $k^{1/2}(z_1 - \mu)/z_2 \sim t(k)$.
   (b) If $z_1^2 \sim \chi^2(k_1, 0)$, $z_2^2 \sim \chi^2(k_2, 0)$ and $z_1^2, z_2^2$ independent, then $z_1^2 k_2 / z_2^2 k_1 \sim F(k_1, k_2)$.

Application: The sampling model for Normal linear models assumes that the $n$ observations on the dependent variables $y_1, y_2, \ldots, y_n$ are realizations of independent Normal random variables with means $\mu_1, \mu_2, \ldots, \mu_n$ and a common variance $\phi$. Put $\mathbf{y} = (y_1, y_2, \ldots, y_n)$. Then for any fixed vector $\mathbf{r}$,

$$[\mathbf{r}, \mathbf{y}] \sim N([\mathbf{r}, \boldsymbol{\mu}], \phi \|\mathbf{r}\|^2) \text{ [see § 2.5.3(a)]}$$

and

$$\phi^{-1}[\mathbf{y}, \mathbf{y}] \sim \chi^2(n, \|\boldsymbol{\mu}\|^2/2\phi) \text{ [see § 2.5.4(a)]}.$$

gives the distribution of linear and quadratic forms in Normal variables. These results follow as $[\mathbf{r}, \mathbf{y}] = r_1 y_1 + r_2 y_2 + \ldots r_n y_n$ is a linear combination of Normal random variables (fact 1(a) and 3(a)) while $\phi^{-1}[\mathbf{y}, \mathbf{y}] = \phi^{-1}(y_1^2 + y_2^2 + \ldots + y_n^2)$ is $\chi^2$ by 3(b).

## 12.2.2.   First and Second Moments

Our concern is to derive the mean and variance of the least squares estimates from assumptions about the first two moments of the sample $y_1, y_2, \ldots, y_n$.

In particular we assume that

$$Ey_t = \mu_t \quad \text{and} \quad \text{cov}(y_t, y_{t'}) = \begin{cases} 0 \text{ if } t' \neq t, \\ \phi \text{ if } t' = t \end{cases} \quad \text{for} \quad t = 1, 2, \ldots, n$$

Equivalently in terms of vectors this is

$$Ey = \mu \quad \text{and} \quad \text{cov}([r, y], [s, y]) = \phi[r, s]$$

for all fixed vectors $r$ and $s$ in $n$-space.

This is a standard set of assumptions appropriate to sampling from a Normal p.d.f. with constant variance $\phi$. Interest lies in obtaining expressions for the mean, variance and covariance of the least squares coefficient $b$, fitted values $\hat{y}$, residuals $y - \hat{y}$ and the expected values of the components of the partition $\|y\|^2 = \|\hat{y}\|^2 + \|y - \hat{y}\|^2$. The cases $p = 1$ and $p = 2$ are dealt with separately and main emphasis is given to the result; proofs, where given, are compressed.

The above assumptions imply the first two moments of the linear form $[r, y]$ are

$$E[r, y] = [r, \mu] \quad \text{and} \quad \text{var}[r, y] = \phi[r, r].$$

### 12.2.3. One Explanatory Variable

*Model*

For $p = 1$, $Ey = \mu = \beta x$

$$\text{cov}([r, y], [s, y]) = \phi[r, s]$$

$y = (y_1, y_2, \ldots, y_n)$ are independent and each $y_t$ is Normal.

*Mean and variance of b*

$$Eb = \beta, \text{var}(b) = \phi\|x\|^{-2}$$

*Proof*:

$$b = [x, y]\|x\|^{-2} \qquad \text{(from section 12.1.1)}$$
$$E[x, y] = [x, \mu] = \beta[x, x] = \beta\|x\|^2 \quad \text{(from above)}$$
$$\text{var}[x, y] = \phi[x, x] \qquad \text{(from above).}$$

*Gauss–Markov theorem*

In the class of all linear unbiased estimates of $\beta$, $b$ has minimum variance [cf. § 8.2].

*Proof.* Let $[a, y]$ be an estimate. It is linear in $y$ and

$$E[a, y] = [a, \mu] = \beta[a, x].$$

Unbiasedness requires $[\mathbf{a}, \mathbf{x}] = 1$. Now

$$\text{var}[\mathbf{a}, \mathbf{y}] = \phi[\mathbf{a}, \mathbf{a}] \geq \phi[\mathbf{a}, \mathbf{x}]^2/[\mathbf{x}, \mathbf{x}]^2 = \phi\|\mathbf{x}\|^{-2}$$

by the Cauchy–Schwarz inequality [see IV, 21.2.4]. Equality is attained when $\mathbf{a} = \|\mathbf{x}\|^{-2}\mathbf{x}$, giving the least squares estimate.

### Normality of b

As $b$ is a linear form in $y$, it is Normally distributed with mean and variance as given above. (This is used to set confidence intervals for the parameter.)

EXAMPLE.   If $[\mathbf{y}, \mathbf{x}] = 200$, $[\mathbf{x}, \mathbf{x}] = 400$ and $\phi = 4$ then

$$b = 200/400 = 0\cdot5, \qquad \text{var}(b) = 4/400 = 10^{-2}.$$

Now $(b - \beta)\,\text{var}(b)^{-1/2} \sim N(0, 1)$ so that a 95% confidence interval for $\beta$ is $0\cdot5 \pm 1\cdot96 \cdot 10^{-1} = (0\cdot304, 0\cdot696)$. [See Example 4.2.1.]

### Fitted values ŷ

$$E[\mathbf{r}, \hat{\mathbf{y}}] = [\mathbf{r}, \boldsymbol{\mu}], \qquad \text{cov}([\mathbf{r}, \hat{\mathbf{y}}], [\mathbf{s}, \hat{\mathbf{y}}]) = \phi[\mathbf{r}, \mathbf{x}][\mathbf{x}, \mathbf{s}]\|\mathbf{x}\|^{-2}.$$

*Proof.*   By definition $\hat{\mathbf{y}} = b\mathbf{x}$, so that $[\mathbf{r}, \hat{\mathbf{y}}] = b[\mathbf{r}, \mathbf{x}]$. Then use the above results for $b$.

EXAMPLE.   To find the variance of the predicted value at $t = 1$, choose $\mathbf{r} = \mathbf{a}_1 = (1, 0, \ldots, 0)$ giving $y_1 = [\mathbf{a}_1, \mathbf{y}]$ and $\text{var}(\hat{y}_1) = \phi x_1^2 / \sum_{t=1}^{n} x_t^2$.

### Residuals

$$E[\mathbf{r}, \mathbf{y} - \hat{\mathbf{y}}] = 0 \quad \text{and} \quad \text{cov}([\mathbf{r}, \mathbf{y} - \hat{\mathbf{y}}], [\mathbf{s}, \mathbf{y} - \hat{\mathbf{y}}]) = \phi([\mathbf{r}, \mathbf{s}] - [\mathbf{r}, \mathbf{x}][\mathbf{x}, \mathbf{s}]\|\mathbf{x}\|^{-2})$$

*Proof.*   Now

$$\begin{aligned}
\text{cov}([\mathbf{r}, \mathbf{y}], [\mathbf{s}, \hat{\mathbf{y}}]) &= \text{cov}([\mathbf{r}, \mathbf{y}], [\mathbf{x}, \mathbf{y}])[\mathbf{s}, \mathbf{x}]\|\mathbf{x}\|^{-2} \\
&= \phi[\mathbf{r}, \mathbf{x}][\mathbf{s}, \mathbf{x}]\|\mathbf{x}\|^{-2} \\
&= \text{cov}([\mathbf{r}, \hat{\mathbf{y}}], [\mathbf{s}, \hat{\mathbf{y}}]) \quad \text{(from above)}
\end{aligned}$$

so that

$$\text{cov}([\mathbf{r}, \mathbf{y} - \hat{\mathbf{y}}], [\mathbf{s}, \hat{\mathbf{y}}]) = 0$$

on subtraction. These give

$$\text{cov}([\mathbf{r}, \mathbf{y} - \hat{\mathbf{y}}], [\mathbf{s}, \mathbf{y} - \hat{\mathbf{y}}]) = \text{cov}([\mathbf{r}, \mathbf{y}], [\mathbf{s}, \mathbf{y}]) - \text{cov}([\mathbf{r}, \hat{\mathbf{y}}], [\mathbf{s}, \hat{\mathbf{y}}])$$

and hence the result on substitution from above.

EXAMPLE. The variance of the residual at $t = 1$ is

$$\text{var}\,(y_1 - \hat{y}_1) = \phi(1 - x_1^2/\textstyle\sum x_t^2).$$

*Normality of fitted values and residuals*

As $[\mathbf{r}, \hat{\mathbf{y}}]$ and $[\mathbf{s}, \mathbf{y} - \hat{\mathbf{y}}]$ are linear functions of $\mathbf{y}$ and the coordinates of $\mathbf{y}$ are Normally distributed, these combinations of the fitted values and the residuals are also Normally distributed. Furthermore since they are also uncorrelated they are independently distributed.

EXAMPLE. To test if $y_1 - \hat{y}_1$ is too large a discrepancy to be consistent with the model, compute $(y_1 - \hat{y}_1)/\phi^{1/2}\,(1 - x_1^2/\sum x_t^2)^{1/2}$ and check if it falls within the interval $(-1\cdot96, 1\cdot96)$.

*Analysis of variance*

Each component of the sum of squares partition has a non-central chi-squared distribution when scaled by $\phi$:

| Source | SS | df | Non-centrality |
|---|---|---|---|
| Due to $x$ | $\phi^{-1}\|\hat{\mathbf{y}}\|^2$ | 1 | $\|\boldsymbol{\mu}\|^2/2\phi = \beta^2\|\mathbf{x}\|^2/2\phi$ |
| Residual | $\phi^{-1}\|\mathbf{y} - \hat{\mathbf{y}}\|^2$ | $n - 1$ | 0 |
| Total | $\phi^{-1}\|\mathbf{y}\|^2$ | $n$ | $\|\boldsymbol{\mu}\|^2/2\phi$ |

As $[\mathbf{r}, \hat{\mathbf{y}}]$ and $[\mathbf{s}, \mathbf{y} - \hat{\mathbf{y}}]$ are independent for all $\mathbf{r}$ and $\mathbf{s}$ it follows that $\|\hat{\mathbf{y}}\|^2$ and $\|\mathbf{y} - \hat{\mathbf{y}}\|^2$ are independent as well. The proofs are omitted.

The expected value of a non-central chi-squared variable is $\text{df} + 2$ non-centrality [see § 2.8.1] so that $E\|\hat{\mathbf{y}}\|^2 = \phi + \beta^2\|\mathbf{x}\|^2$. This also gives the unbiased estimate for $\phi$,

$$\hat{\phi} = \|\mathbf{y} - \hat{\mathbf{y}}\|^2/(n-1).$$

EXAMPLE. If

$$\begin{pmatrix} [\mathbf{x}, \mathbf{x}] & [\mathbf{x}, \mathbf{y}] \\ & [\mathbf{y}, \mathbf{y}] \end{pmatrix} = \begin{pmatrix} 400 & 200 \\ & 120 \end{pmatrix} \quad \text{and} \quad n = 5$$

then

$$\|\hat{\mathbf{y}}\|^2 = b^2\|\mathbf{x}\|^2 = [\mathbf{x}, \mathbf{y}]^2\|\mathbf{x}\|^{-2} = (200)^2 \cdot 400^{-1} = 100$$

with the analysis of variance table

| Source | df | SS | ms | *F* ratio |
|--------|----|----|----|----|
| Due to *x* | 1 | 100 | 100 | 20 |
| Residual | 4 | 20 | $\hat{\phi} = 5$ | |
| Total | 5 | 120 | | |

As $7 \cdot 71$ is the 5% significance point of $F(1, 4)$ the contribution from *x* is significantly different from zero.

### 12.2.4.   Two Orthogonal Explanatory Variables

It is easier to separate the orthogonal from the non-orthogonal case and to deal with the former first. The proofs are omitted.

*Model*

For $p = 2$, $E\mathbf{y} = \boldsymbol{\mu} = \beta_1 \mathbf{x}_1 + \beta_2 \mathbf{x}_2$, $[\mathbf{x}_1, \mathbf{x}_2] = 0$

$$\text{cov}\,([\mathbf{r}, \mathbf{y}], [\mathbf{s}, \mathbf{y}]) = \phi[\mathbf{r}, \mathbf{s}]$$

$\mathbf{y} = (y_1, y_2, \ldots, y_n)$ are independent and each $y_t$ is Normal.

*Least squares estimates*

Given

$$b_1 = [\mathbf{x}_1, \mathbf{y}] \|\mathbf{x}_1\|^{-2}, \qquad b_2 = [\mathbf{x}_2, \mathbf{y}] \|\mathbf{x}_2\|^{-2}$$

then

$$E\begin{pmatrix} b_1 \\ b_2 \end{pmatrix} = \begin{pmatrix} \beta_1 \\ \beta_2 \end{pmatrix} \quad \text{and} \quad \begin{pmatrix} \text{var}\,(b_1) & \text{cov}\,(b_1, b_2) \\ & \text{var}\,(b_2) \end{pmatrix} = \phi \begin{pmatrix} \|\mathbf{x}\|^{-2} & 0 \\ & \|\mathbf{x}_2\|^{-2} \end{pmatrix}$$

and $b_1, b_2$ are independently and Normally distributed. The Gauss–Markov theorem asserts they have minimum variance in the class of all linear unbiased estimates.

*Fitted values and residuals*

Given fitted values $\hat{\mathbf{y}} = \hat{\mathbf{y}}(\mathbf{x}_1, \mathbf{x}_2) = \hat{\mathbf{y}}(\mathbf{x}_1) + \hat{\mathbf{y}}(\mathbf{x}_2)$ then

$$E[\mathbf{r}, \hat{\mathbf{y}}] = [\mathbf{r}, \boldsymbol{\mu}] \quad \text{and} \quad E[\mathbf{r}, \mathbf{y} - \hat{\mathbf{y}}] = 0 \quad \text{for all } \mathbf{r}$$

$$\text{cov}\,([\mathbf{r}, \hat{\mathbf{y}}], [\mathbf{s}, \hat{\mathbf{y}}]) = \phi([\mathbf{r}, \mathbf{x}_1] \|\mathbf{x}_1\|^{-2} [\mathbf{x}_1, \mathbf{s}] + [\mathbf{r}, \mathbf{x}_2] \|\mathbf{x}_2\|^{-2} [\mathbf{x}_2, \mathbf{s}])$$

$$\text{cov}\,([\mathbf{r}, \hat{\mathbf{y}}], [\mathbf{s}, \mathbf{y} - \hat{\mathbf{y}}]) = 0$$

$$\text{cov}\,([\mathbf{r}, \mathbf{y} - \hat{\mathbf{y}}], [\mathbf{s}, \mathbf{y} - \hat{\mathbf{y}}]) = \text{cov}\,([\mathbf{r}, \mathbf{y}], [\mathbf{s}, \mathbf{y}]) - \text{cov}\,([\mathbf{r}, \hat{\mathbf{y}}], [\mathbf{s}, \hat{\mathbf{y}}])$$

All these linear forms are Normally distributed.

*Analysis of variance*

Each component of the sum of squares partition has a non-central chi-squared distribution when scaled by $\phi$:

| Source | SS | df | Non-centrality |
|---|---|---|---|
| Due to $x_1$ alone | $\|\hat{\mathbf{y}}(\mathbf{x}_1)\|^2$ | 1 | $\beta_1^2\|\mathbf{x}_1\|^2/2\phi$ |
| Due to $x_2$ alone | $\|\hat{\mathbf{y}}(\mathbf{x}_2)\|^2$ | 1 | $\beta_2^2\|\mathbf{x}_2\|^2/2\phi$ |
| Due to $x_1, x_2$ | $\|\hat{\mathbf{y}}(\mathbf{x}_1, \mathbf{x}_2)\|^2$ | 2 | $\|\boldsymbol{\mu}\|^2/2\phi$ |
| Residual | $\|\mathbf{y}-\hat{\mathbf{y}}\|^2$ | $n-2$ | 0 |
| Total | $\|\mathbf{y}\|^2$ | $n$ | $\|\boldsymbol{\mu}\|^2/2\phi$ |

The first two components in this table are independent of each other and of the sum of squares attributable to the residual. An unbiased estimate of $\phi$ is $\hat{\phi} = \|\mathbf{y}-\hat{\mathbf{y}}\|^2/(n-2)$.

EXAMPLE 12.2.1.   *Analysis with two orthogonal explanatory variables.*   Return to the numerical example in section 12.1.4.

Data: $n = 10$; cross-products
$$\begin{pmatrix} [x_1, x_1] & [x_1, x_2] & [x_1, y] \\ & [x_2, x_2] & [x_2, y] \\ & & [y, y] \end{pmatrix} = \begin{pmatrix} 10 & 0 & 24 \\ & 20 & 15 \\ & & 94 \end{pmatrix}$$

first unit $(y_1, x_{11}, x_{12}) = (2{\cdot}1, 1, 2)$.

Model: $E\mathbf{y} = \beta_1\mathbf{x}_1 + \beta_2\mathbf{x}_2$, var $(y_t) = \phi$.

*Analysis:*

We obtain a confidence interval for $\beta_1 - \beta_2$, an estimate of $\phi$, the analysis of variance table to test if $\beta_2 = 0$, and an interval for the residual on the first unit.

*Confidence interval for $\beta_1 - \beta_2$*

Use $b_1 - b_2$. Require var $(b_1 - b_2) = $ var $(b_1) - 2$ cov $(b_1, b_2) + $ var $(b_2)$

$$b_1 = 24/10 = 2{\cdot}4, \qquad b_2 = 15/20 = 0{\cdot}75$$

$$b_1 - b_2 = 2{\cdot}4 - 0{\cdot}75 = 1{\cdot}65$$

$$\text{var } (b_1 - b_2) = \phi(10^{-1} + 2(0) + 20^{-1}) = 0{\cdot}15\phi$$

interval:  $1{\cdot}65 \pm 1{\cdot}96$  $(0{\cdot}15\phi)^{1/2}$. (If for example $\phi = 3$ this becomes $(0{\cdot}34, 2{\cdot}96)$.)

*Estimate of $\phi$*

Use $\hat{\phi} = \|\mathbf{y} - \hat{\mathbf{y}}\|^2 / (n - 2)$

$$= 2 \cdot 515 / 8 = 3 \cdot 14 \text{ (from section 12.1.4).}$$

To modify the above interval: $1 \cdot 65 \pm 2 \cdot 31 \ (0 \cdot 15 \ 3 \cdot 14)^{1/2} = (0 \cdot 07, 3 \cdot 21)$ on using a table of Student's distribution on 8 d.f. [see Appendix T5]

*Analysis of variance to test $\beta_2 = 0$*

| Source | df | SS | ms | F ratio |
|--------|----|----|----|---------|
| Due to $x_1$ | 1 | 57·6 | | |
| Due to $x_2$ | 1 | 11·25 | 11·25 | 3·58 |
| Residual | 8 | 25·15 | $\hat{\phi} = 3·14$ | |
| Total | 10 | 94 | | |

As the 5% significance point of $F(1, 8)$ is $5 \cdot 32 \ (> 3 \cdot 58)$ we conclude this is not significant and the hypothesis $\beta_2 = 0$ is accepted. In effect $\mathbf{x}_2$ can be dropped from the model.

*Residuals*

The fitted value at $x_1 = 1$, $x_2 = 2$ is $\hat{y} = 2 \cdot 4 + 0 \cdot 75.2 = 3 \cdot 9$ and so the residual is $2 \cdot 1 - 3 \cdot 9 = -1 \cdot 8$. Now

$$\text{var} \ (y_t - \hat{y}_t) = \text{var} \ (y_t) - \text{var} \ (\hat{y}_t)$$

$$= \phi - \phi(1^2/10 + 0 + 2^2/20)$$

$$= 0 \cdot 7\phi.$$

If $\phi = 3$, then $-1 \cdot 8 \pm 1 \cdot 96 \ (0 \cdot 7.3)^{1/2} = (-3 \cdot 64, 1 \cdot 04)$ as this includes zero, there is no reason to suspect this residual is discrepant.

## 12.2.5.  Non-orthogonal Explanatory Variables

Several important results change when going from the orthogonal to the non-orthogonal case. In particular the least squares coefficients are now correlated and allowance has to be made for this when setting intervals for contrasts etc. There is no unique way to attribute the sum of squares $\|\hat{\mathbf{y}}\|^2$ to $\mathbf{x}_1$ and $\mathbf{x}_2$, and hence tests for $\beta_2 = 0$ depend if $\mathbf{x}_1$ is to be included in the model or not. These changes are illustrated with the same data.

$$\text{Data: } n = 10 \begin{pmatrix} [\mathbf{x}_1, \mathbf{x}_1] & [\mathbf{x}_1, \mathbf{x}_2] & [\mathbf{x}_1, \mathbf{y}] \\ & [\mathbf{x}_2, \mathbf{x}_2] & [\mathbf{x}_2, \mathbf{y}] \\ & & [\mathbf{y}, \mathbf{y}] \end{pmatrix} = \begin{pmatrix} 10 & 8 & 24 \\ & 20 & 15 \\ & & 94 \end{pmatrix}$$

first observation: $y_1 = 2 \cdot 1$, $x_1 = 1$, $x_2 = 2$.

Model: $p = 2$   $E\mathbf{y} = \mu = \beta_1 \mathbf{x}_1 + \beta_2 \mathbf{x}_2$

$$\text{cov} ([\mathbf{r}, \mathbf{y}], [\mathbf{s}, \mathbf{y}]) = \phi [\mathbf{r}, \mathbf{s}]$$

$\mathbf{y} = (y_1, y_2, \ldots, y_n)$ are independent and each $y_t$ is normal.

Require: an interval for $\beta_1 - \beta_2$, an estimate of $\phi$, a test to see if $x_2$ can be excluded, residual at $t = 1$.

*Least squares estimates*

These are the solution to

$$\begin{pmatrix} b_1 \\ b_2 \end{pmatrix} = \begin{pmatrix} [\mathbf{x}_1, \mathbf{x}_1] & [\mathbf{x}_1, \mathbf{x}_2] \\ [\mathbf{x}_2, \mathbf{x}_1] & [\mathbf{x}_2, \mathbf{x}_2] \end{pmatrix}^{-1} \begin{pmatrix} [\mathbf{x}_1, \mathbf{y}] \\ [\mathbf{x}_2, \mathbf{y}] \end{pmatrix}$$

and can be simplified to

$$\begin{pmatrix} b_1 \\ b_2 \end{pmatrix} = \begin{pmatrix} [\mathbf{y}, \mathbf{e}_1] & \|\mathbf{e}_1\|^{-2} \\ [\mathbf{y}, \mathbf{e}_2] & \|\mathbf{e}_2\|^{-2} \end{pmatrix} \quad \text{where} \quad \begin{matrix} \mathbf{e}_1 = \mathbf{x}_1 - [\mathbf{x}_1, \mathbf{x}_2] \|\mathbf{x}_2\|^{-1} \mathbf{x}_2 \\ \mathbf{e}_2 = \mathbf{x}_2 - [\mathbf{x}_2, \mathbf{x}_1] \|\mathbf{x}_1\|^{-1} \mathbf{x}_1 \end{matrix}$$

are the residuals of fitting $\mathbf{x}_1$ on $\mathbf{x}_2$ and $\mathbf{x}_2$ on $\mathbf{x}_1$. Now

$$[\mathbf{y}, \mathbf{e}_1] = [\mathbf{x}_1, \mathbf{y}] - [\mathbf{x}_1, \mathbf{x}_2] \|\mathbf{x}_2\|^{-1} [\mathbf{x}_2, \mathbf{y}] = 24 - 8.15/20 = 18 \cdot 0$$

$$[\mathbf{y}, \mathbf{e}_2] = [\mathbf{x}_2, \mathbf{y}] - [\mathbf{x}_2, \mathbf{x}_1] \|\mathbf{x}_1\|^{-1} [\mathbf{x}_1, \mathbf{y}] = 15 - 8.24/10 = -4 \cdot 2$$

$$\|\mathbf{e}_1\|^2 = \|\mathbf{x}_1\|^2 - [\mathbf{x}_1, \mathbf{x}_2]^2 \|\mathbf{x}_2\|^{-2} = 10 - 8^2/20 = 6 \cdot 8$$

$$\|\mathbf{e}_2\|^2 = \|\mathbf{x}_2\|^2 - [\mathbf{x}_2, \mathbf{x}_1]^2 \|\mathbf{x}_1\|^{-2} = 20 - 8^2/10 = 13 \cdot 6$$

Thus

$$\begin{pmatrix} b_1 \\ b_2 \end{pmatrix} = \begin{pmatrix} 18/6 \cdot 8 \\ -4 \cdot 2/13 \cdot 6 \end{pmatrix} = \begin{pmatrix} 2 \cdot 647 \\ -0 \cdot 309 \end{pmatrix}.$$

The first two moments of $b_1$, $b_2$ are

$$E \begin{pmatrix} b_1 \\ b_2 \end{pmatrix} = \begin{pmatrix} \beta_1 \\ \beta_2 \end{pmatrix} \quad \text{and} \quad \begin{pmatrix} \text{var} (b_1) & \text{cov} (b_1, b_2) \\ & \text{var} (b_2) \end{pmatrix} = \phi \begin{pmatrix} [\mathbf{x}_1, \mathbf{x}_1] & [\mathbf{x}_1, \mathbf{x}_2] \\ [\mathbf{x}_2, \mathbf{x}_1] & [\mathbf{x}_2, \mathbf{x}_2] \end{pmatrix}^{-1}$$

$$= \phi \begin{pmatrix} \|\mathbf{e}_1\|^{-2} & -r\|\mathbf{e}_1\|^{-1}\|\mathbf{e}_2\|^{-1} \\ & \|\mathbf{e}_2\|^{-2} \end{pmatrix} = \phi \begin{pmatrix} 0 \cdot 15 & -0 \cdot 06 \\ & 0 \cdot 07 \end{pmatrix}$$

where $r = [\mathbf{x}_1, \mathbf{x}_2] \|\mathbf{x}_1\|^{-1} \|\mathbf{x}_2\|^{-1} = 8/(10.20)^{1/2} = 0 \cdot 566$. Hence

$$\text{var} (b_1 - b_2) = \phi (\text{var } b_1 - 2 \text{ cov } (b_1, b_2) + \text{var } b_2)$$

$$= \phi (0 \cdot 15 + 0 \cdot 12 + 0 \cdot 07) = 0 \cdot 34 \phi$$

and an interval for $\beta_1 - \beta_2$ is $2 \cdot 34 \pm 1 \cdot 96 \,(0 \cdot 34 . \phi)^{1/2} = (0 \cdot 36, 4 \cdot 32)$ if $\phi = 3$.

The effect of correlation between $x_1$ and $x_2$ is to reduce the efficiency of the estimates. For instance as $[\mathbf{x}_1, \mathbf{x}_2]$ goes from 0 to 8, var $(b_1)$ goes from $0\cdot10\phi$ to $0\cdot15\phi$.

*Fitted values and residuals*

Now

$$\hat{\mathbf{y}} = \hat{\mathbf{y}}(\mathbf{x}_1, \mathbf{x}_2) = b_1\mathbf{x}_1 + b_2\mathbf{x}_2 (\neq \hat{\mathbf{y}}(\mathbf{x}_1) + \hat{\mathbf{y}}(\mathbf{x}_2))$$
$$= \hat{\mathbf{y}}(\mathbf{x}_1) + \hat{\mathbf{y}}(\mathbf{e}_2)$$
$$= \hat{\mathbf{y}}(\mathbf{e}_1) + \hat{\mathbf{y}}(\mathbf{x}_2).$$

The moments of the fitted values and residuals are

$$E[\mathbf{r}, \hat{\mathbf{y}}] = [\mathbf{r}, \boldsymbol{\mu}] \quad \text{and} \quad E[\mathbf{r}, \mathbf{y} - \hat{\mathbf{y}}] = 0,$$

$$\text{cov}([\mathbf{r}, \hat{\mathbf{y}}], [\mathbf{s}, \hat{\mathbf{y}}]) = \phi([\mathbf{r}, \mathbf{x}_1], [\mathbf{r}, \mathbf{x}_2]) \begin{pmatrix} [\mathbf{x}_1, \mathbf{x}_1] & [\mathbf{x}_1, \mathbf{x}_2] \\ [\mathbf{x}_2, \mathbf{x}_2] \end{pmatrix}^{-1} \begin{pmatrix} [\mathbf{x}_1, \mathbf{s}] \\ [\mathbf{x}_2, \mathbf{s}] \end{pmatrix},$$

$$\text{cov}([\mathbf{r}, \hat{\mathbf{y}}], [\mathbf{s}, \mathbf{y} - \hat{\mathbf{y}}]) = 0,$$

$$\text{cov}([\mathbf{r}, \mathbf{y} - \hat{\mathbf{y}}], [\mathbf{s}, \mathbf{y} - \hat{\mathbf{y}}]) = \text{cov}([\mathbf{r}, \mathbf{y}], [\mathbf{s}, \mathbf{y}]) - \text{cov}([\mathbf{r}, \hat{\mathbf{y}}], [\mathbf{s}, \hat{\mathbf{y}}]),$$

and so the only change comes with the second moments of the fitted values. At $x_1 = 1$, $x_2 = 2$, $\hat{y}_1 = 2\cdot647 - 2\cdot309 = 2\cdot029$,

and

$$\text{var}(\hat{y}_1) = \phi(1, 2) \begin{pmatrix} 10 & 8 \\ 8 & 20 \end{pmatrix}^{-1} \begin{pmatrix} 1 \\ 2 \end{pmatrix} = 0\cdot67\phi,$$

and var $(y_1 - \hat{y}_1) = \phi - 0\cdot67\phi = 0\cdot33\phi$. If $\phi = 3$ this value of $\hat{y}_1$ is very close to $y_1 = 2\cdot1$.

*Analysis of variance*

All components of the sum of squares partition have non-central chi-squared distributions when scaled by $\phi$:

| Source | SS | df | Non-centrality |
|---|---|---|---|
| Due to $x_1$ alone | $\|\hat{\mathbf{y}}(\mathbf{x}_1)\|^2$ | 1 | $(\|\boldsymbol{\mu}\|^2 - \|\beta_2\mathbf{e}_2\|^2)/2\phi$ |
| Due to $x_2$ adjusted for $x_1$ | $\|\hat{\mathbf{y}}(\mathbf{e}_2)\|^2$ | 1 | $\beta_2^2\|\mathbf{e}_2\|^2/2\phi$ |
| Due to $x_1$ and $x_2$ | $\|\hat{\mathbf{y}}(\mathbf{x}_1, \mathbf{x}_1)\|^2$ | 2 | $\|\boldsymbol{\mu}\|^2/2\phi$ |
| Residual | $\|\mathbf{y} - \hat{\mathbf{y}}(\mathbf{x}_1, \mathbf{x}_2)\|^2$ | $n-2$ | 0 |
| Total | $\|\mathbf{y}\|^2$ | $n$ | $\|\boldsymbol{\mu}\|^2/2\phi$ |

The first two components in this table are independent of each other and of the sum of squares attributable to the residual. An unbiased estimate of $\phi$ is $\hat{\phi} = \|\mathbf{y} - \hat{\mathbf{y}}\|^2 / (n-2)$.

To test if $\mathbf{x}_2$ can be excluded from the model that contains $\mathbf{x}_1$, we wish to test if $\beta_2 = 0$. Hence

| Source | df | SS | ms | F ratio |
|--------|-----|-------|--------------|---------|
| Due to $x_1$ alone | 1 | 57·60 | | |
| $x_2$ adjusted for $x_1$ | 1 | 1·3 | 1·3 | 0·296 |
| Residual | 8 | 35·10 | $\hat{\phi} = 4·38$ | |
| Total | 10 | 94 | | |

from 12.1.5. Clearly, with the $F$ ratio not even reaching 1, we accept $\beta_2 = 0$.

## 12.3.   LEAST SQUARES AND THE LIKELIHOOD FUNCTION

### 12.3.1.   Weighted Least Squares

When the underlying p.d.f. of $y_t$ is Normal with variance $\phi / w_t$ rather than $\phi$, choosing estimators to minimize sums of *weighted* squared deviations is the appropriate procedure, as this criterion reflects the varying amounts of information in the observations. A weighted least squares approach is also necessary to approximate the likelihood function of a p.d.f. from the exponential family.

The extension is almost trivial. Define a new inner product by

$$[\mathbf{x}, \mathbf{y}]_w = w_1 x_1 y_1 + w_2 x_2 y_2 + \ldots + w_n x_n y_n$$

where $w_t \geq 0$ for all $t$ and now consider minimizing

$$\|\mathbf{y} - \boldsymbol{\mu}\|_w^2 = [\mathbf{y} - \boldsymbol{\mu}, \mathbf{y} - \boldsymbol{\mu}]_w, \quad \text{with} \quad \boldsymbol{\mu} = \beta \mathbf{x} \quad \text{or} \quad \boldsymbol{\mu} = \beta_1 \mathbf{x}_1 + \beta_2 \mathbf{x}_2.$$

All the algebraic results of 12.1 hold with $[ , ]_w$ replacing $[ , ]$.

Thus for $p = 1$, the normal equation is

$$[\mathbf{y} - \hat{\mathbf{y}}, \mathbf{x}]_w = 0,$$

which leads to $b = [\mathbf{x}, \mathbf{y}]_w \|\mathbf{x}\|_w^{-2}$ for the least squares estimate and to the sum of squares partition

$$\|\mathbf{y}\|_w^2 = \|\hat{\mathbf{y}}\|_w^2 + \|\mathbf{y} - \hat{\mathbf{y}}\|_w^2.$$

EXAMPLE

Data:

| | | | | | |
|---|---|---|---|---|---|
| $\mathbf{y}$ | −3 | 2 | 1 | 4 | 8 |
| $\mathbf{x}$ | −2 | −1 | 0 | 1 | 2 |
| $\mathbf{w}$ | 1 | 1 | 2 | 2 | 0 |

Units 3 and 4 are given twice as much weight as units 1 and 2 while unit 5 is suppressed altogether.

Crossproducts:

$$\begin{pmatrix} [\mathbf{x}, \mathbf{x}]_w & [\mathbf{x}, \mathbf{y}]_w \\ & [\mathbf{y}, \mathbf{y}]_w \end{pmatrix} = \begin{pmatrix} 7 & 12 \\ & 47 \end{pmatrix}$$

so that $b = 12/7 = 1 \cdot 7$ and the other results of Section 12.1 follow directly.

Similarly when there are two or more explanatory variables weighted least squares follow the same algebra as ordinary least squares but with the weighted inner product.

The repeated sampling results of 12.2 go through without alteration if the second moment assumption is modified to

$$\text{cov}([\mathbf{r}, \mathbf{y}]_w, [\mathbf{s}, \mathbf{y}]_w) = \phi[\mathbf{r}, \mathbf{s}]_w.$$

To see that this is equivalent to assuming var $y_t = \phi / w_t$ put $\mathbf{r} = \mathbf{s} = (1, 0, \ldots, 0)$ so that

$$\text{cov}(w_1 y_1, w_1 y_1) = w_1^2 \text{ var}(y_1) = \phi w_1.$$

This then ensures the weighted least squares estimates are unbiased and have minimum variance in the class of all linear unbiased estimates. All the sampling distributions remain the same when $y_1, y_2, \ldots, y_n$ are assumed to have independent normal distributions as long as $w_t > 0$ for all $t$. If any $w_t = 0$, there is a loss of one degree of freedom for the residual sum of squares, but otherwise results are identical.

However, for most of the p.d.f's from the exponential family, the variance of the density is a function of the mean so that the appropriate weight is $\mathbf{w} = \mathbf{w}(\mathbf{\mu})$ rather than $\mathbf{w}$ fixed. This can lead to difficulties.

EXAMPLE.   If $E\mathbf{y} = \mu = \beta x = \text{var}(y)$, put $w_t = 1/\mu_t$ so that $\mathbf{w}(\mathbf{\mu}) = \beta^{-1}\mathbf{x}^{-1}$, with a slight abuse of notation. Now

$$\|\mathbf{y} - \mathbf{\mu}\|^2_{\mathbf{w}(\mathbf{\mu})} = \beta^{-1}\|\mathbf{y} - \mathbf{\mu}\|^2_{\mathbf{x}^{-1}}$$

$$= \beta^{-1}\|\mathbf{y}\|^2_{\mathbf{x}^{-1}} + \beta\|\mathbf{x}\|^2_{\mathbf{x}^{-1}} - 2[\mathbf{y}, \mathbf{x}]_{\mathbf{x}^{-1}}.$$

This is minimized when

$$b = (\|\mathbf{y}\|^2_{\mathbf{x}^{-1}}/\|\mathbf{x}\|^2_{\mathbf{x}^{-1}})^{1/2} = (\sum y_t^2 x_t^{-1}/\sum x_t)^{1/2}.$$

which is highly nonlinear in $\mathbf{y}$. For $p > 1$ there is no analytic solution to this problem; but fortunately, the connection between least squares and the likelihood function leads to more tractable results than this.

### 12.3.2.  Approximating the Likelihood Function

This section provides the groundwork to employ an iteratively reweighted least squares procedure to maximize the likelihood for any generalized linear model.

We first recall the derivation of the normal equations; if $M =$ span $(\mathbf{x}_1, \mathbf{x}_2, \ldots \mathbf{x}_p)$ and $\mathbf{d}$ is a vector such that $\boldsymbol{\mu} \in M$ and $\mathbf{d} \in M$, then $\boldsymbol{\mu} + \delta\mathbf{d} \in M$ for any scalar $\delta$, and

$$\|\mathbf{y} - (\boldsymbol{\mu} + \delta\mathbf{d})\|_w^2 - \|\mathbf{y} - \boldsymbol{\mu}\|_w^2 = \delta[\mathbf{y} - \boldsymbol{\mu}, \mathbf{d}]_w + \delta^2\|\mathbf{d}\|_w^2$$

is an identity. Now $\hat{\boldsymbol{\mu}}$ is the weighted least squares fitted value if the left hand side is always non-negative when evaluated at $\boldsymbol{\mu} = \hat{\boldsymbol{\mu}}$. The right hand side is non-negative for any $\delta$ if and only if the coefficient of $\delta$ is zero. That is $[\mathbf{y} - \boldsymbol{\mu}, \mathbf{d}]_w = 0$ for all $\mathbf{d} \in M$.

The trick is now to show that the likelihood function can be approximated by an expression with the same structure as the right hand side.

Recall from section 11.2.1, that for a unit of the data array, $y$ has p.d.f. $f$ and expected value $\mu$. The linear predictor $\eta$ is related to $\mu$ via a specified link, $g(\mu) = \eta$. The log likelihood function is $\lambda(\eta) = \log f(y|\eta, \phi)$ and has derivatives

$$\frac{d\lambda}{d\eta} = \frac{y - \mu}{\text{var}(y)} \frac{d\mu}{d\eta}, \qquad E\frac{d^2\lambda}{d\eta^2} = \frac{1}{\text{var}(y)}\left(\frac{d\mu}{d\eta}\right)^2.$$

These equations are linearized by solving for $\eta$ rather than $\mu$. This linearization is accomplished by defining

$$z = \eta + (y - \mu)\frac{d\eta}{d\mu}$$

so that

$$Ez = \eta \quad \text{and} \quad \text{var}(z) = \text{var}(y)\left(\frac{d\eta}{d\mu}\right)^2.$$

Substituting for $y - \mu$ gives

$$\frac{d\lambda}{d\eta} = \frac{z - \eta}{\text{var}(z)}, \qquad E\frac{d^2\lambda}{d\eta^2} = -\frac{1}{\text{var}(z)}.$$

The first two terms of a Taylor series expansion [see IV, § 3.6] for the log-likelihood give

$$\lambda(\eta + \delta d) \simeq \lambda(\eta) + \delta d\frac{(z - \eta)}{\text{var}(z)} - \frac{1}{2}\frac{(\delta d)^2}{\text{var}(z)}.$$

(Actually the coefficient of $\delta^2$ should have employed $d^2\lambda/d\eta^2$ rather than $E\,d^2\lambda/d\eta^2$ but the difference is not substantial.)

As the units in the sample are independent, the log-likelihood for the whole sample is additive and

$$\lambda(\boldsymbol{\eta}) = \sum_t \lambda(\eta_t)$$

so that

$$\lambda(\boldsymbol{\eta} + \delta\mathbf{d}) - \lambda(\boldsymbol{\eta}) \approx \delta[\mathbf{z} - \boldsymbol{\eta}, \mathbf{d}]_w - \tfrac{1}{2}\delta^2\|\mathbf{d}\|_w^2$$

using the Taylor series approximation and with weights for the inner product

$$w_t = 1/\mathrm{var}\,(z_t).$$

### 12.3.3.   Example: Fitting an Independence Model

A two-way contingency table, classified by factors $A$ and $B$, has counts $y_{ij}$ in cell $(i, j)$ which have a Poisson distribution and expected value

$$\mu_{ij} = Ey_{ij} = np_{ij} = np_{i+}p_{+j}$$

(cf. § 11.4.1). The linear predictor, $\eta_{ij} = \log Ey_{ij}$, can be written as

$$\boldsymbol{\eta} = \log n\, \mathbf{1} + \sum_i \log p_{i+}\mathbf{a}_i + \sum_j \log p_{+j}\mathbf{b}_j$$

so that $\boldsymbol{\eta} \in A + B = \mathrm{span}\,(\mathbf{a}_1, \ldots, \mathbf{a}_I, \mathbf{b}_1, \ldots, \mathbf{b}_J)$.

The log-likelihood for a single unit is

$$\lambda(\eta) = -\mu + y \log \mu - \log y!$$

and has derivatives

$$\frac{d\lambda}{d\eta} = -e^\eta + y, \qquad \frac{d^2\lambda}{d\eta^2} = -e^\eta.$$

Put $z = \eta + (y - \mu)\, d\eta/d\mu = \eta + ye^{-\eta} - 1$ and substitute for $y$ to get

$$\lambda(\eta + \delta d) - \lambda(\eta) = \delta d(z - \eta)e^\eta - \tfrac{1}{2}(\delta d)^2\, e^\eta.$$

Aggregating over the sample gives the approximation

$$\lambda(\boldsymbol{\eta} + \delta\mathbf{d}) - \lambda(\boldsymbol{\eta}) = \delta[\mathbf{d}, \mathbf{z} - \boldsymbol{\eta}]_w - \tfrac{1}{2}\delta^2\|\mathbf{d}\|_w^2$$

with $w_t = e^{\eta_t}$.

In this example, there is no term involving $\mathbf{y}$ in the second derivative. Also it is possible to solve the likelihood equations explicitly;

$$[\mathbf{d}, \mathbf{z} - \boldsymbol{\eta}]_w = [\mathbf{d}, \mathbf{y} - \boldsymbol{\mu}] \quad \text{for } \mathbf{d} \in A + B$$

$$= \sum_{i,j} d_{ij}(y_{ij} - \mu_{ij}).$$

Choosing $\mathbf{d} = \mathbf{a}_i \in A + B$ and setting this to zero gives

$$0 = \sum_j (y_{ij} - \mu_{ij}) \quad \text{so that } \mu_{i+} = y_{i+},$$

and similarly $\mu_{+j} = y_{+j}$. As $\mu_{ij} = \mu_{i+}\mu_{+j}/\mu_{++}$ the maximum likelihood fitted values are $\hat{\mu}_{ij} = y_{i+}y_{+j}/y_{++}$.

### 12.3.4.   An Exponential Example

Consider observations on an exponential p.d.f.

$$f(y|\mu) = \exp(-y/\mu - \log \mu), \qquad y > 0,$$

for which $Ey = \mu$, var $y = \mu^2$. Suppose the link function is $\log \mu = \eta$, so that $\mu = e^\eta$, $d\mu/d\eta = e^\eta$, and the linear predictor is $\eta = \beta x$.

The log likelihood of $\eta$ is

$$\lambda(\eta) = -y e^{-\eta} - \eta,$$

with derivatives

$$\frac{d\lambda}{d\eta} = y e^{-\eta} - 1, \qquad \frac{d^2\lambda}{d\eta^2} = -y e^{-\eta} \quad \text{and} \quad E\left(\frac{d^2\lambda}{d\eta^2}\right) = -1.$$

Put

$$z = \eta + (y - \mu)\frac{d\eta}{d\mu}$$

$$= \eta + (y e^{-\eta} - 1);$$

then

$$\frac{d\lambda}{d\eta} = (z - \eta), \qquad \frac{d^2\lambda}{d\eta^2} = -1 - (z - \eta).$$

The Taylor series approximation is, with $d = x$,

$$\lambda(\eta + \delta x) - \lambda(\eta) = \delta x(z - \eta) - \tfrac{1}{2}(\delta x)^2(1 + z - \eta).$$

Aggregating over the sample $t = 1, 2, \ldots, n$ gives

$$\lambda(\boldsymbol{\eta} + \delta\mathbf{x}) - \lambda(\boldsymbol{\eta}) = \delta[\mathbf{x}, \mathbf{z} - \boldsymbol{\eta}] - \tfrac{1}{2}\delta^2[\mathbf{x}, \mathbf{x}] - \tfrac{1}{2}\delta^2[\mathbf{x}^2, \mathbf{z} - \boldsymbol{\eta}]$$

where the inner product has a constant weight.

The normal equation is $[\mathbf{x}, \mathbf{z} - \boldsymbol{\eta}] = 0$, or, in coordinate notation

$$\sum_t x_t(z_t - \eta_t) = \sum_t x_t(y_t e^{-\beta x_t} - 1) = 0,$$

to which there is no analytic solution for $\beta$. An iteration is described in the next section.

If as the sample size increases $[\mathbf{x}, \mathbf{x}]$ grows indefinitely then

$$[\mathbf{x}^2, \mathbf{z} - \boldsymbol{\eta}]/[\mathbf{x}, \mathbf{x}] = \sum x_t^2(z_t - \eta_t)/\sum x_t^2 \to 0$$

(its expected value) by the laws of large numbers. Hence this term can be neglected in the Taylor series.

## 12.3.5.  An Iterative Procedure

The approximation to the log-likelihood for a sample from the exponential family is

$$\lambda(\boldsymbol{\eta} + \delta\mathbf{d}) - \lambda(\boldsymbol{\eta}) \simeq \delta[\mathbf{z} - \boldsymbol{\eta}, \mathbf{d}]_w - \tfrac{1}{2}\delta^2 \|\mathbf{d}\|_w^2.$$

For a turning point at $\hat{\boldsymbol{\eta}}$, we must have

$$[\mathbf{z} - \boldsymbol{\eta}, \mathbf{d}]_w = 0 \quad \text{at} \quad \boldsymbol{\eta} = \hat{\boldsymbol{\eta}} \quad \text{for all } \mathbf{d} \in M.$$

Since the coefficient of $\delta^2$ is negative, this turning point will be a maximum. As the coefficient is always negative all turning points are maxima. But if the function is sufficiently smooth maxima must be separated by a minimum. Hence the maximum is unique.

This argument is actually rather coarse and may be fallacious; it has ignored part of the second derivative, $d^2\lambda / d\eta^2$; it assumes the remainder term in the Taylor series is negligible; it assumes that a feasible solution to the normal equations always exists; but it does indicate the type of results potentially available.

The normal equations suggest the following iterative scheme. Write $\mathbf{z}(\boldsymbol{\eta})$ for $\mathbf{z}$, and:

(1) Choose $\boldsymbol{\eta}_0$
(2) Compute $\mathbf{z}(\boldsymbol{\eta}_0)$ and $\mathbf{w}(\boldsymbol{\eta}_0)$
(3) Solve $[\mathbf{z}(\boldsymbol{\eta}_0) - \boldsymbol{\eta}_1, \mathbf{d}]_{w(\boldsymbol{\eta}_0)} = 0$ when $\mathbf{d} = \mathbf{x}_1, \mathbf{x}_2, \ldots, \mathbf{x}_p$ giving $\boldsymbol{\eta}_1$
(4) Put $\boldsymbol{\eta}_0 = \boldsymbol{\eta}_1$ and return to (2) unless convergence has occurred.

Note that at (3) these are just the normal equations for the weighted least squares problem. For $p = 2$ they require solving

$$\begin{pmatrix} [\mathbf{x}_1, \mathbf{x}_1]_w & [\mathbf{x}_1, \mathbf{x}_2]_w \\ & [\mathbf{x}_2, \mathbf{x}_2]_w \end{pmatrix} \begin{pmatrix} b_1 \\ b_2 \end{pmatrix} = \begin{pmatrix} [\mathbf{x}_1, z]_w \\ [\mathbf{x}_2, z]_w \end{pmatrix}$$

with $\mathbf{w} = \mathbf{w}(\boldsymbol{\eta}_0)$ and $\mathbf{z} = \mathbf{z}(\boldsymbol{\eta}_0)$. Then $\boldsymbol{\eta}_1 = b_1\mathbf{x}_1 + b_2\mathbf{x}_2$.

Each iteration requires inverting a matrix of order $p \times p$ and recomputing $\mathbf{z}$, $\mathbf{w}$ and $\tfrac{1}{2}p(p+1)$ cross products of the form $[\mathbf{x}_i, \mathbf{x}_j]_w$. The procedure is known as *iteratively reweighted least squares*.

## 12.3.6.  Asymptotic Sampling Theory

It would be satisfying to round off the theory of linear models with a section describing the sampling theory of the general model to parallel section 12.2 on repeated sampling applicable to the Normal identity model. Unfortunately such exact results are not available. At best, there are approximations valid in large samples based on the asymptotic sampling theory of maximum likelihood estimates [see § 6.2.5(a)].

The paradigm for these results is the weighted Normal identity model section 12.3.1. The log-likelihood function for an arbitrary linear model was approximated, for $\mathbf{d} \in \mathrm{span}\,(M)$, by the following, in which both

$$z = z(\eta) = \eta + (y - \mu)d\mu/d\eta$$

and

$$w = w(\eta) = 1/\mathrm{var}\,(z)$$

depend on unknown parameters $\eta$:

$$\lambda(\boldsymbol{\eta} + \delta\mathbf{d}) - \lambda(\boldsymbol{\eta}) = \delta[\mathbf{z} - \boldsymbol{\eta}, \mathbf{d}]_w - \tfrac{1}{2}\delta^2[\mathbf{d}, \mathbf{d}]_w.$$

The right-hand side is exact for the likelihood function of the weighted Normal identity model. Further, for this model $\mathbf{z} - \boldsymbol{\eta} = \mathbf{y} - \boldsymbol{\mu}$ so that $[\mathbf{z} - \boldsymbol{\eta}, \mathbf{d}]_w = [\mathbf{y} - \boldsymbol{\mu}, \mathbf{d}]_w$ which must be Normally distributed as $\mathbf{y}$ is Normal by assumption. The exact results for this model flow from this observation.

In the general case, $[\mathbf{z} - \boldsymbol{\eta}, \mathbf{d}]_w$ will be asymptotically Normal by virtue of the celebrated central limit theorems [see II, § 11.4.2, § 17.3]; for $[\mathbf{z} - \boldsymbol{\eta}, \mathbf{d}]_w = \sum_t (z_t - \eta_t)d_t w_t$ and the $z_t$'s are independently distributed. This observation leads to the asymptotic sampling distribution for the estimates $\hat{\beta}$ of $\beta$ and of the deviance; they are the same as for the weighted Normal identity model.

To illustrate, reconsider the Exponential density discussed in 12.3.4. Here $p = 1$, $\eta = \beta x$ and $\hat{\beta}$ satisfied

$$\sum_t x_t y_t \exp\,(-\hat{\beta}x_t) = \sum x_t,$$

This estimate $\hat{\beta}$ can also be written as

$$\hat{\beta} = [\mathbf{x}, \mathbf{x}]^{-1}[\mathbf{x}, \mathbf{z}]$$

where $z = z(\hat{\eta}) = \hat{\eta} + y\,e^{-\hat{\eta}} - 1$ as $\hat{\beta}$ is the solution to the iteratively reweighted least squares procedure. The asymptotic sampling distribution of $\hat{\beta}$ is the same as that of $[\mathbf{x}, \mathbf{x}]^{-1}[\mathbf{x}, \mathbf{z}_0]$ where the coordinates of $\mathbf{z}_0$ are taken to be independent Normal random variables.

These sampling results concerning the estimates and the deviances give enough guidance to be useful practical tools in many instances. However because they are not exact some important problems still remain intractable; for example there is not an adequate theory covering the analysis of residuals from an arbitrary linear model.

J.C.W.

## 12.4. FURTHER READING AND REFERENCES

For a detailed study of our particular case of the models studied in Chapters 11 and 12, namely that of the dosage-mortality analysis referred to in section

11.1.3, see Finney (1971). For excellent general account of the linear model see, e.g. Fraser (1979), Graybill (1976—Bibliography C). For a study of the special problems associated with contingency tables, see Bishop, Y., Fienberg, S. and Holland, P. (1975), *Discrete Multivariate Analysis*, MIT Press.

Finney, D. J. (1971). *Probit Analysis*, Third edn., Cambridge University Press.
Fraser, D. A. S. (1979). *Inference and Linear Models*, McGraw-Hill.

# CHAPTER 13

# *Sequential Analysis*

## INTRODUCTION

Consider the sampling inspection scheme [see Example 1.2.1 and § 5.12.1] where items (e.g. light bulbs) are packed in batches of 100 and the whole batch is rejected if there are at least five defectives in the batch and the batch is accepted if there are four or fewer defectives.

The *fixed sample size* scheme tests all 100 items and rejects the batch if, at the end of the test, five or more defectives are found. The *sequential sampling* scheme, however, tests the items one by one and sampling stops as soon as five defectives have been found or when the number of effectives is 96. Hence sequential sampling is more efficient than fixed sample size sampling since it is possible that not all 100 items need to be tested.

If there are actually 12 defectives in the batch of 100 then the sample size of the sequential scheme will be an integer between five (the first five sampled are all defective) and 93 (the remaining seven items are all defective). The sample size depends on the order in which the items are tested, and it is therefore a random variable. In this case there is a definite saving in the number of items which need to be tested to reach a final decision about the batch.

To illustrate another aspect of sequential sampling, consider an experiment to test the strengths of glass pipes. Ball bearings of the same dimensions and weight are dropped from various heights onto glass pipes until the pipe fractures, and the height at which this occurs is noted. The experimental procedure should be sequential, since rather than plan a whole series of experiments at arbitrarily chosen heights (a fixed sample size scheme), it seems sensible to carry out the experiments one by one so that if there is no break at a particular height the ball bearing is dropped from a greater height and so on, until a break occurs. The experimental design is such that the result of a previous experiment determines the level at which the next experiment takes place.

Sequential sampling procedures, as in the first example, where only a *stopping rule* is specified, are usually known as *sequential analysis*. The procedure in

579

the second example where both a *stopping rule* and a *sampling rule* are specified is known as *sequential design*. This chapter will concentrate on sequential analysis.

The obvious efficiency of sequential sampling schemes as compared to the equivalent fixed sample size schemes in the above two examples does not mean that in all cases sequential sampling should be used. Sequential sampling is obviously most efficient where observations occur naturally in a sequential manner, for example items produced on a factory production line. It will be less efficient than fixed sample size schemes where there is a long time between obtaining results of each individual experiment, for instance in agricultural trials.

## 13.1.  SEQUENTIAL TESTS OF HYPOTHESES

In previous chapters, inferences concerning the unknown parameters of a population, or probability distributions describing those populations, have been based on random samples of fixed size. Suppose a sequential sample is taken from the population where observations are obtained one at a time and, after the result of each is known, a decision is either made to stop sampling and use the current sample or sample statistics to make inferences about the parameters, or to continue sampling and take one further observation. In this way sampling stops once sufficient information, in terms of past observations, is obtained about the unknown parameters.

As outlined in section 5.12, the Neyman–Pearson fixed-sample-size test of two hypotheses $H_0$, $H_1$ is carried out by finding a set of values of a chosen statistic in which $H_0$ is rejected and $H_1$ accepted, known as a rejection region. To find this region, the maximum value of the level of significance of the test, that is the probability of a Type I error, is fixed at $\alpha$, and the test that is chosen is the one with the smallest probability of a Type II error, $\beta$, that is, the largest power $(1 - \beta)$. In fixed sample size tests therefore, $\alpha$ and the sample size $n$ are fixed, and $\beta$ is minimized.

In sequential tests of the hypotheses, sampling stops when the sample contains sufficient information to accept or reject $H_0$. If there is not sufficient information then one further observation is taken. Hence in a sequential test of a hypothesis, the current sample values may fall into one of three regions, a rejection region and an acceptance region, as in the fixed sample size case, and, in addition, a *continuation* region where a further observation must be taken. In defining these three regions, $\alpha$ is fixed, and $\beta$, the maximum desired probability of a Type II error, is fixed and the sample size $N$ is a random variable.

### 13.1.1.  The Operating Characteristic (O.C.) Function

Suppose that the observations $x_1, x_2, \ldots$ are from a probability distribution with an unknown parameter, $\theta$, and that the three regions above have been specified, giving a sequential test. The overall performance of a fixed sample size test is assessed by the calculation of the power function, which gives the probability of rejecting $H_0$, as a function of $\theta$. For a sequential test, one works with the *operating characteristic* (O.C.) function, $P(\theta)$. This is the probability of accepting $H_0$ as a function of $\theta$. (The 'power function' [see § 5.12.2] is $1 - P(\theta)$.) If the null hypothesis $H_0$ and the alternative $H_1$ are both simple, $H_0: \theta = \theta_0$ and $H_1: \theta = \theta_1$ $(\theta_1 > \theta_0)$ [cf. § 5.2.1(c)], then

$$P(\theta_0) = (1 - \alpha) \text{ and } P(\theta_1) = \beta.$$

### 13.1.2. Expected Sample Size

Since the sample size in a sequential test is a random variable, it is of interest to find its expectation. If any comparison is to be made with fixed sample size tests then it should be on the basis of this expected sample size rather than the observed sample size for a particular set of observations.

The expected sample size is a function of the true value of the parameter. For example, if the two hypotheses are simple as above and the true value of $\theta$ is near $(\theta_0 + \theta_1)/2$, then the expected sample size will be larger than that in the case where $|\theta - \theta_1|$ is small compared with $|\theta - \theta_0|$. The expected sample size function is also known as the Average Sample Number (ASN) function.

### 13.1.3. Examples of Sampling Inspection Schemes

The two following sampling inspection schemes are such that the unknown proportion, $\theta$, of defectives in a batch is unknown and each item in the batch is either defective or effective. It is required to accept or reject the batch based on a sequential sample of items. It is assumed that the batch is made up of a large number of items, $n$, and that the observations are independent.

EXAMPLE 13.1.1. *Simple sequential sampling (Wald).* Fix an integer $n_0$ such that if the first $n_0$ items are effective, sampling ceases and the batch is accepted. If for some sample size $m \le n_0$, the $m$th item is defective, the batch is rejected. Suppose the hypotheses are $H_0:$ accept the batch, $H_1:$ reject the batch.

The O.C. function $P(\theta)$ is $P(\theta) = P(\text{accepting } H_0 | \theta) = P(\text{first } n_0 \text{ observations are effective}) = (1 - \theta)^{n_0}$.

Note that $P(0) = 1$ (all items in the batch are effective) and $P(1) = 0$ (all items in the batch are defective). The function $P(\theta)$ may now be plotted against $\theta$, $0 < \theta < 1$. This is also the O.C. function of the fixed sample size scheme with sample size $n_0$. [See Figure 5.12.1].

In this sequential scheme, when $1 \le n \le n_0$, the sample size, $N$ will take a value $n$ in the range $1 \le n \le n_0 - 1$ if the last item is defective and will take the value $n_0$ only if the first $(n_0 - 1)$ items are effective.

Therefore the expected sample size $E(N|\theta)$ [see § 1.4.2(vi)] is given by

$$E(N|\theta) = \sum_{m=1}^{n_0-1} mP(N=m) + n_0 P(N=n_0)$$

where

$$P(N=m) = (1-\theta)^{m-1}\theta, \qquad 1 \le m \le n_0-1,$$
$$P(N=n_0) = (1-\theta)^{n_0-1}.$$

[Expectation is defined as II, Definition 8.1.1.] This function may also be plotted against $\theta$, $0 < \theta < 1$.

For further details of the procedure outlined in this example, see Wald (1947).

EXAMPLE 13.1.2.   *Sequential sampling with bounded sample size.*   Consider a modification of the above scheme where the maximum number of observations is $n_1$ and sampling ceases and the batch is rejected if $c(\ge 1)$ or more defectives are found within the first $n_1$ inspected, $n_1$ and $c$ being fixed. The batch is accepted if there are $\ge n_1 - c + 1$ effectives within the first $n_1$.

The O.C. function of this scheme is given by

$$P(\theta) = P(c-1 \text{ or fewer defectives in } n_1 \text{ inspected})$$

$$= \sum_{r=0}^{c-1} \binom{n_1}{r} \theta^r (1-\theta)^{n_1-r}.$$

Again $P(0) = 1$ and $P(1) = 0$, and the O.C. function may be plotted as a function of $\theta$. It should be noted that this O.C. function coincides with the O.C. function of the fixed sample size test of sample size $n_1$.

To obtain the expected sample size the batches must be split into those rejected and those accepted. For rejected batches

$$P_r(N=m) = P(c\text{th defective occurs at the } m\text{th trial})$$

$$= \binom{m-1}{c-1} \theta^c (1-\theta)^{m-c}, \qquad m = c, c+1, \ldots, n_1.$$

For accepted batches, let $s = (n_1 - c + 1)$. Then

$$P_a(N=m) = P(s\text{th effective occurs at } m\text{th trial})$$

$$= \binom{m-1}{c-1} (1-\theta)^s \theta^{m-s}, \qquad m = s, s+1, \ldots, n_1.$$

Then

$$E(N|\theta) = \sum_{m=c}^{n_1} mP_r(N=m) + \sum_{m=s}^{n_1} mP_a(N=m).$$

For further details, see Wetherill (1966).

## 13.2. THE SEQUENTIAL PROBABILITY RATIO TEST (SPRT)

We consider a sequential random sample $x_1, x_2, \ldots$ from a probability distribution which depends on one unknown parameter $\theta$. It will be assumed throughout that the observations are independent. Let the probability density function at $x$ be $f(x; \theta)$. It is required to test the simple hypothesis $H_0: \theta = \theta_0$ against the simple alternative $H_1: \theta = \theta_1$ [see §§ 5.2.1(c), (d)]. The best (most powerful) fixed sample size test, with significance level $\alpha$ and sample size $n$, as given by the Neyman-Pearson Lemma [see § 5.12.2], is based on the likelihood ratio

$$\lambda_n = \prod_{i=1}^n \frac{f(x_i; \theta_1)}{f(x_i; \theta_0)}.$$

This rejects $H_0$ if $\lambda_n > k$, where $k$ is a constant such that $P(\text{Type I error}) = P(\lambda_n > k | \theta_0) = \alpha$.

The Sequential Probability Ratio Test (SPRT) of the two hypotheses, due to Wald (1947), also uses the likelihood ratio. The values of $P(\text{Type I error}) = \alpha$ and $P(\text{Type II error}) = \beta$ are fixed. From these, two constants $A$, $B$ are obtained such that, after $m$ observations the next move is given by the following:

(i) if $\lambda_m \leq B$, stop sampling and accept $H_0$;
(ii) if $\lambda_m \geq A$, stop sampling and accept $H_1$;
(iii) if $b < \lambda_m < A$, continue sampling.

### 13.2.1. Approximate Values of the Stopping Boundaries

Suppose that the $m$th observation leads to the rejection of $H_0$, that is that

$$B < \lambda_i < A, \qquad i = 1, 2, \ldots, m-1$$

and

$$\lambda_m \geq A,$$

the last inequality being equivalent to

$$\prod_{i=1}^m f(x_i; \theta_1) \geq A \prod_{i=1}^m f(x_i; \theta_0).$$

Denote by $R_m$ all possible values of the sample $(x_1, x_2, \ldots, x_m)$ resulting in $\lambda_m \geq A$, then, for continuous data,

$$\alpha = P(\text{Type I error}) = P(\lambda_m \geq A | \theta_0)$$

$$= \int\int_{R_m} \cdots \int f(x_1, x_2, \ldots, x_m; \theta_0)\, dx_1, dx_2, \ldots, dx_m.$$

$$= \int\int_{R_m} \cdots \int \prod_{i=1}^m f(x_i; \theta_0)\, dx_1, dx_2, \ldots, dx_m$$

$$\leq \int\int_{R_m} \cdots \int \frac{1}{A} \prod_{i=1}^{m} f(x_i; \theta_1) \, dx_1, dx_2, \ldots, dx_m.$$

$$= \frac{1}{A} P(\lambda_m \geq A | \theta_1)$$

$$= \frac{1}{A} [1 - P(\text{Type II error})]$$

$$= \frac{1}{A} (1 - \beta),$$

whence

$$A \leq (1 - \beta)/\alpha.$$

Similarly

$$B \geq \beta/(1 - \alpha).$$

(For discrete observations the multiple integral may be replaced by a summation over all values of the sample that result in the rejection of $H_0$.)

To obtain these results it is assumed that the test will eventually terminate. A proof that the SPRT terminates with probability 1 is given in Wald (1947, pp. 157–58).

We now have a lower limit for $B$ and an upper limit for $A$ in terms of the specified values of $\alpha, \beta$. In practice, the SPRT with fixed $(\alpha, \beta)$ is defined as follows:

   (i) if $\lambda_m \leq \beta/(1 - \alpha)$, stop sampling and accept $H_0$ (reject $H_1$);
   (ii) if $\lambda_m \geq (1 - \beta)/\alpha$, stop sampling and accept $H_1$ (reject $H_0$);
   (iii) if $\beta/(1 - \alpha) < \lambda_m < (1 - \beta)/\alpha$, continue sampling.

These alterations will change the probabilities of Type I and Type II errors to new values $\alpha', \beta'$ respectively. Using the above approach with $A = (1 - \beta)/\alpha$ and $B = \beta/(1 - \alpha)$ we have

$$A = (1 - \beta)/\alpha \leq (1 - \beta')/\alpha'$$

and

$$B = \beta/(1 - \alpha) \geq \beta'/(1 - \alpha').$$

This gives

$$\alpha' \leq \alpha(1 - \beta')/(1 - \beta) \leq \alpha/(1 - \beta),$$

and similarly

$$\beta' \leq \beta(1 - \alpha)/(1 - \alpha) \leq \beta/(1 - \alpha).$$

Since $\alpha, \beta$ are usually small, the increases caused by the modification in the probabilities $\alpha'$, $\beta'$ over the fixed values $\alpha$, $\beta$ are negligible; thus

$$\alpha' \cong \frac{\alpha}{1-\beta}, \qquad \beta' \cong \frac{\beta}{1-\alpha}. \tag{13.2.1}$$

It may be seen that

$$(\alpha' + \beta') \le (\alpha + \beta),$$

whence at least one of the inequalities $\alpha' \le \alpha$, $\beta' \le \beta$ must be valid.

### 13.2.2.   Examples

EXAMPLE 13.2.1.   *Binomial SPRT.*   Let the observations $x_1, x_2, \ldots$ come from a point Bernoulli distribution [see II, § 5.2.1] with unknown parameter $\theta$ with

$$f(1; \theta) = \theta; \qquad f(0; \theta) = 1 - \theta.$$

We shall refer to '$x = 1$' as a 'success', '$x = 0$' as a 'failure'. Suppose it is required to test $H_0: \theta = \theta_0$ against $H_1: \theta = \theta_1$, $(\theta_1 > \theta_0)$, with fixed probabilities $\alpha$, $\beta$ of Type I and Type II error respectively.

After $m$ observations, the likelihood ratio is

$$\lambda_m = \frac{\theta_1^r (1-\theta_1)^{m-r}}{\theta_0^r (1-\theta_0)^{m-r}},$$

where $r = \sum_{i=1}^{m} x_i = $ number of successes in $m$ trials. This may be simplified by taking logarithms, to obtain

$$\log \lambda_m = r \log [\theta_1(1-\theta_0)/\theta_0(1-\theta_1)] + m \log [(1-\theta_1)/(1-\theta_0)].$$

The SPRT procedure becomes the following:

   (i) if $\log \lambda_m \le \log [\beta/(1-\alpha)]$,   accept $H_0$;
   (ii) if $\log \lambda_m \ge \log [(1-\beta)/\alpha]$,   reject $H_0$;
   (iii) if $\log [\beta/(1-\alpha)] < \log \lambda_m < \log [(1-\beta)/\alpha]$,   continue sampling.

Each time an observation is obtained, a new value of $\log \lambda_m$ is computed. A 'success' results in $\log \lambda_m$ increasing in value by $\log (\theta_1/\theta_0)$ (positive since $\theta_1 > \theta_0$) and a failure gives an increase of amount $\log [(1-\theta_1)/(1-\theta_0)]$ (negative since $\theta_1 > \theta_0$). Note also that the limits $\log [\beta/(1-\alpha)] < 0$ and $\log [(1-\beta)/\alpha] > 0$ for $\alpha, \beta < 0.5$.

Instead of basing the test on $\log \lambda_m$, it may be based on $r = \sum_{i=1}^{m} x_i$, the number of successes in $m$ trials and then represented graphically. To illustrate this, let $H_0: \theta = \frac{1}{4}$ and $H_1: \theta = \frac{1}{2}$ and $\alpha = \beta = 0.05$; then

$$\log \lambda_m = r \log 3 + m \log (\tfrac{2}{3}),$$

$$\log \{\beta/(1-\alpha)\} = -\log 19 \quad \text{and} \quad \log \{(1-\beta)/\alpha\} = \log 19.$$

In this version the procedure is to continue sampling if

$$-\log 19/\log 3 + m\,\{\log\,(\tfrac{3}{2})/\log 3\} < r$$

$$= \sum_{i=1}^{m} x_i < \log 19/\log 3 + m\{\log\,(\tfrac{3}{2})/\log 3\},$$

and to stop and make the appropriate decision otherwise.

If the number of successes $r$ in $m$ trials is plotted against the number of observations for $m = 0, 1, 2, \ldots$, then the *stopping boundary* is made up of values of $r$ on a pair of parallel straight lines of the form $r = a + bm$, $r = -a + bm$ with slope $b = \{\log \tfrac{3}{2}/\log 3\}$ and intercepts $\pm a$ where $a = \log 19/\log 3$ (note the intercepts in this case are symmetric about zero since $\alpha$, $\beta$ take the same value). Hence the sequential sampling scheme may be represented graphically before any observations have been taken (see Figure 13.2.1). Starting at the

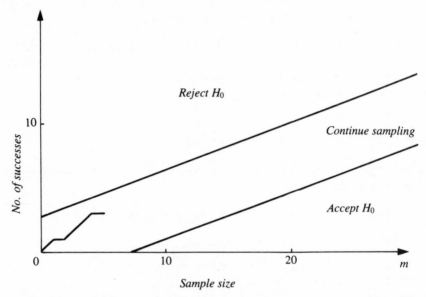

Figure 13.2.1: SPRT for Binomial parameter with $\theta_0 = \tfrac{1}{4}$, $\theta_1 = \tfrac{1}{2}$ and $\alpha = \beta = 0.05$.

origin ($m = 0, r = 0$), any sequential sample may now be plotted on the graph giving the *sample path*. As soon as one of the boundaries is hit or crossed for the first time, sampling stops and the appropriate decision is made. For example the sequence 1, 0, 1, 1, 0 would give rise to the sample path $(0, 0)$, $(1, 1)$, $(2, 1)$, $(3, 2)$, $(4, 3)$, $(5, 3)$ as shown in Figure 13.2.1.

The sequential sampling scheme in this case may also be presented in tabular form (see Table 13.2.1) as the minimum or maximum number of successes in $m$ trials where $H_0$ is accepted or rejected. The sequential scheme, in fact, consists of taking a fixed number of observations, 5, initially and then using the stopping points in the table. Since the number of successes is an integer, the sample path will in most cases overshoot the stopping boundaries at the termination of the test. For example the true limits for stopping for $m = 16$ are $r = 3.7758$ and $r = 9.1682$.

| Number of observations $m$ | Accept $H_0$ if number of successes $\gamma \leq r$ | Reject $H_0$ if number of successes $\gamma \geq r$ |
|:---:|:---:|:---:|
| 5 | — | 5 |
| 6 | — | 6 |
| 7 | 0 | 6 |
| 8 | 0 | 6 |
| 9 | 0 | 7 |
| 10 | 1 | 7 |
| 11 | 1 | 8 |
| 12 | 2 | 8 |
| 13 | 2 | 8 |
| 14 | 2 | 9 |
| 15 | 3 | 9 |
| 16 | 3 | 10 |
| 17 | 4 | 10 |
| 18 | 4 | 10 |
| 19 | 4 | 11 |
| 20 | 5 | 11 |

Table 13.2.1: Stopping points for SPRT for testing $H_0: \theta = \frac{1}{4}$ against $H_1: \theta = \frac{1}{2}$ for $\alpha = \beta = 0.05$, for sample size $m = 5(1)20$.

Unlike the sampling inspection schemes considered earlier, here there is no fixed maximum sample size. The stopping boundaries produced by the SPRT are 'open', hence it would seem to be possible for certain samples or sample paths that a decision to stop sampling might never be made. However, as mentioned earlier, Wald (1947) pp. 157–58 gives a proof that the test eventually terminates. It would be expected that the number of observations to termination for the true value of $\theta$ near $\frac{3}{8}$ would be larger than that for $\theta < \frac{1}{4}$ or $\theta > \frac{1}{2}$.

An alternative graphical representation of the SPRT is simply to plot log $\lambda_m$ against $m$ giving a 'sample path' where log $\lambda_m$ increases by log $(\theta_1/\theta_0) = $ log 2 for each success and by log $[(1 - \theta_1)/(1 - \theta_0)] = \log \frac{2}{3}$ for each failure. The stopping boundaries in this case will be two lines parallel to the axis, with log $\lambda_m = \pm \log 19$.

EXAMPLE 13.2.2.   *Normal SPRT.*   Let the random sample $x_1, x_2, \ldots$ be taken sequentially from a Normal distribution with unknown mean $\theta$ and variance 1. Suppose we wish to test $H_0: \theta = \theta_0$ against $H_1: \theta = \theta_1$, assuming $\theta_1 > \theta$.

After $m$ observations

$$\log \lambda_m = (\theta_1 - \theta_0) \sum_{i=1}^{m} x_i + \frac{m}{2}(\theta_0^2 - \theta_1^2)$$

[see Example 6.2.4] and the SPRT procedure is to continue sampling if

$$\log[(\beta/(1-\alpha)] < \log \lambda_m < \log[(1-\beta)/\alpha],$$

and to stop sampling and make the appropriate 'accept' or 'reject' decision otherwise. After each observation $x_i$, $\log \lambda_m$ increases by $(\theta_1 - \theta_0)x_i + \frac{1}{2}(\theta_0^2 - \theta_1^2)$.

This procedure may be simplified to the following:

continue sampling if

$$\log[\beta/(1-\alpha)]/(\theta_1 - \theta_0) + m(\theta_1 + \theta_0)/2 < \sum_{i=1}^{m} x_i$$

$$< \log[(1-\beta)/\alpha]/(\theta_1 - \theta_0) + m(\theta_1 + \theta_0)/2,$$

and stop sampling and make the appropriate decision otherwise. The current sum of the observations $\sum_{i=1}^{m} x_i$ may now be plotted against $m$, the number of observations, with the SPRT represented by stopping boundaries consisting of a pair of parallel straight lines with slope $(\theta_1 + \theta_0)/2$ and intercepts $\log[\beta/(1-\alpha)]/(\theta_1 - \theta_0)$ and $\log[(1-\beta)/\alpha]/(\theta_1 - \theta_0)$.

In the case where $\theta_0 = 0$, $\theta_1 = 1$ and $\alpha = \beta = 0 \cdot 05$, the slope of the stopping boundaries is $\frac{1}{2}$ in each case and the intercepts are $\pm \log 19$. The stopping boundaries for this case together with a sample path consisting of a sample of eight observations from a standard Normal distribution which terminates with the acceptance of $H_0$ are given in Figure 13.2.2.

Since the sum of the observations is a continuous random variable, the sample path will always overshoot the stopping boundary. The probability that the sum will hit the boundary exactly is zero.

## 13.3.   THE O.C. FUNCTION FOR THE SPRT

Given the sequential random sample $x_1, x_2, \ldots$ on the random variable $X$ with p.d.f. $f(x; \theta)$ at $x$, where $\theta$ is unknown, suppose that the simple hypotheses $H_0: \theta = \theta_0$, $H_1: \theta = \theta_1$ have to be tested under given values of $\alpha$ and $\beta$. As pointed out earlier, this simply fixes two points on the O.C. function, $P(\theta)$, namely $P(\theta_0) = 1 - \alpha$ and $P(\theta_1) = \beta$.

A derivation of the approximate form of the O.C. function of the SPRT is due to Wald (1947) pp. 48–50. This is based on the following ideas. The sampling expectation of the function

$$\left[ \frac{f(x; \theta_1)}{f(x; \theta_0)} \right]^{h(\theta)}$$

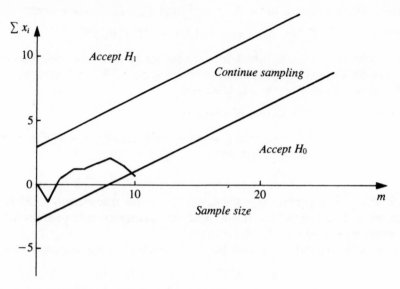

Figure 13.2.2: SPRT for Normal mean $\theta$ with $\theta_0 = 0$, $\theta_1 = 1$ and $\alpha = \beta = 0.05$.

is

$$E\left\{\left[\frac{f(X; \theta_1)}{f(X; \theta_0)}\right]^{h(\theta)}\right\} = \int_{-\infty}^{\infty} \left[\frac{f(x; \theta_1)}{f(x; \theta_0)}\right]^{h(\theta)} f(x; \theta) \, dx = 1$$

for $X$ continuous. (For $X$ discrete the integral may be replaced by a summation over all values of $x$.) This implies that $h(\theta_0) = 1$ and $h(\theta_1) = -1$.

It follows from (13.3.1) that the function

$$f^*(x; \theta) = \left[\frac{f(x; \theta_1)}{f(x; \theta_0)}\right]^{h(\theta)} f(x; \theta)$$

must be a probability density function. If $h(\theta) > 0$, the SPRT procedure for testing the hypothesis $H$, that $f(x; \theta)$ is the true density of $x$, against $H^*$, that $f^*(x; \theta)$ is the true density of $x$, may be expressed as follows:

(i) continue sampling if after $m$ observations

$$B_\theta < \lambda_m = \prod_{i=1}^{m} \left[\frac{f^*(x_i; \theta)}{f(x_i; \theta)}\right] < A_\theta;$$

(ii) stop sampling and accept $H$ if

$$\lambda_m \leq B_\theta;$$

(iii) stop sampling and accept $H^*$ if

$$\lambda_m \geq A_\theta.$$

If the limits are chosen to be $A_\theta = A^{h(\theta)}$ and $B_\theta = B^{h(\theta)}$, then, since

$$[f^*(x; \theta)/f(x; \theta)] = [f(x; \theta_1)/f(x; \theta_0)]^{h(\theta)}$$

the sample $x_1, x_2, \ldots, x_m$ which leads to the acceptance of $H$ will also accept $H_0$. The SPRT in (i), (ii), (iii) is simply the original SPRT with all terms raised to the power $h(\theta)$. For the modified test

$P(\text{Type I error}) = P(\text{rejecting } H \text{ when true})$

$\qquad = 1 - P(\text{accepting that the true density of } x \text{ is } f(x; \theta) \text{ when this is in fact true})$

$\qquad = 1 - P(\theta),$

since $P(\theta) = P(\text{accepting } H_0: \theta = \theta_0 | f(x; \theta)$ is the true density). [Note that here the vertical bar before '$f(x, \theta)$' does not signify a conditional probability. It simply means 'when'. See § 1.4.2(vi).]

After some algebra, this gives the O.C. function of the original test as

$$P(\theta) = (A^{h(\theta)} - 1)/(A^{h(\theta)} - B^{h(\theta)}).$$

Further, using the approximation to $A$ and $B$ given in section 13.2.1,

$$P(\theta) = \frac{\left(\dfrac{1-\beta}{\alpha}\right)^h - 1}{\left(\dfrac{1-\beta}{\alpha}\right)^h - \left(\dfrac{\beta}{1-\alpha}\right)^h},$$

approximately. The proof for $h(\theta) < 0$ is similar. As a check, note that, since $h(\theta_0) = 1$ and $h(\theta_1) = -1$, $P(\theta_0) = 1 - \alpha$ and $P(\theta_1) = \beta$. By finding values of $h(\theta)$ for some values of $\theta$, or otherwise, $P(\theta)$ may now be plotted against $\theta$.

## 13.4.   EXPECTED SAMPLE SIZE FOR THE SPRT

The likelihood ratio, $\lambda_m$, is the product of the terms $f(x_i; \theta_1)/f(x_i; \theta_0)$, $i = 1, 2, \ldots$ and therefore $\log \lambda_m$ is made up of the sum of the terms $\log[f(x_i; \theta_1)/f(x_i; \theta_0)] = z_i$. The SPRT could now be expressed in terms of $\sum_{i=1}^m z_i$, as follows:

continue sampling after the $m$th observation if $\log B < \sum_{i=1}^m z_i < \log A$, and stop

$\qquad\qquad\qquad\qquad$ otherwise.

It can be shown (see e.g., Wetherill (1975) pp. 19–21), that, if the random variable $N$ is the number of observations required to reach a decision to accept or to reject,

$$E\left(\sum_{i=1}^N Z_i | \theta\right) = E(N|\theta)E(Z|\theta)$$

where $E(N|\theta)$ is the expected sample size [see § 1.4.2(vi)] and

$$Z = \log [f(X; \theta_1)/f(X; \theta_0)].$$

The probability that the test leads to the acceptance of $H_0$ $(= P\sum_{i=1}^{n} Z_i \leq \log B)$ is $P(\theta)$, and the probability of rejecting $H_0$ $(= P\sum_{i=1}^{n} Z_i \geq \log A)$ is $1 - P(\theta)$. If the overshoot of the boundaries either by the sample path or $\sum_{i=1}^{n} z_i$ is ignored then

$$E\left(\sum_{i=1}^{n} Z_i|\theta\right) = P(\theta) \log B + [1 - P(\theta)] \log A$$

and

$$E(N|\theta) = \frac{P(\theta) \log B + [1 - P(\theta)] \log A}{E(Z|\theta)}.$$

In the case where $E(Z|\theta) = 0$, Wald suggests the approximation

$$E(N|\theta) = \frac{P(\theta) (\log B)^2 + [1 - P(\theta)] (\log B)^2}{E(Z^2|\theta)}.$$

$A$, $B$ may again be approximated as before.

## 13.5.   EXAMPLES

EXAMPLE 13.5.1.   *The O.C. function and expected sample size of the Binomial SPRT.*   The O.C. function for the Binomial SPRT will take the following values $P(0) = 1$, $P(1) = 0$ (assuming $\theta_0 < \theta_1$) and $P(\theta_0) = (1 - \alpha)$, $P(\theta_1) = \beta$. For other values of the O.C. function the procedure in section 13.3 has to be carried out.

It is required to find $h = h(\theta)$ satisfying

$$E\left\{\left[\frac{f(X; \theta_1)}{f(X; \theta_0)}\right]^h\right\} = \theta(\theta_1/\theta_0)^h + (1 - \theta)[(1 - \theta_1)/(1 - \theta_0)]^h$$

$$= 1.$$

The equation may be solved for $\theta$, giving

$$\theta = \frac{1 - [(1 - \theta_1)/(1 - \theta_0)]^h}{(\theta_1/\theta_0)^h - [(1 - \theta_1)/(1 - \theta_0)]^h},$$

and, on varying the value of $h$, pairs of values $(\theta, P(\theta))$ may be obtained. The expected sample size may then be calculated for all values of $\theta$ by first calculating $P(\theta)$ from above and then $E(Z|\theta)$ given by

$$E(Z|\theta) = \theta \log (\theta_1/\theta_0) + (1 - \theta) \log [(1 - \theta_1)/(1 - \theta_0)]$$

and finally $E(N|\theta)$.

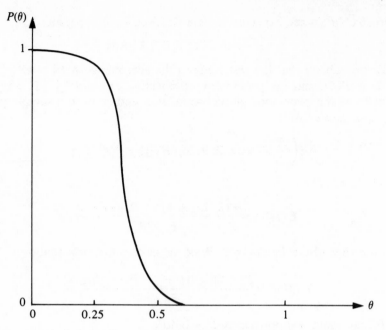

Figure 13.5.1: O.C. function for Binomial SPRT with $\theta_0 = \frac{1}{4}$, $\theta_1 = \frac{1}{2}$, $\alpha = \beta = 0 \cdot 5$.

The O.C. function for $\theta_0 = \frac{1}{4}$, $\theta_1 = \frac{1}{2}$, $\alpha = \beta = 0 \cdot 05$ is given in Figure 13.5.1, and $E(n|\theta)$ in Table 13.5.1.

| $\theta$ | $P(\theta)$ | $E(N|\theta)$ |
|---|---|---|
| 0 | 1 | 7·26 |
| $\frac{1}{4}$ | 0·95 | 20·24 |
| $\frac{3}{8}$ | 0·467 | 29·90 |
| $\frac{1}{2}$ | 0·05 | 18·43 |
| 1 | 0 | 4·25 |

Table 13.5.1: Expected sample size $E(N|\theta)$, $\theta_0 = \frac{1}{4}$, $\theta_1 = \frac{1}{2}$, $\alpha = \beta = 0 \cdot 05$.

EXAMPLE 13.5.2.    *The O.C. function and expected sample size of the Normal SPRT.*  For the Normal SPRT considered in Example 13.2.2 it may easily be shown that the function $h(\theta)$, which is required for the calculation of the O.C. function, is

$$h(\theta) = \frac{\theta_1 + \theta_0 - 2\theta}{\theta_1 - \theta},$$

whence $P(\theta)$ may be calculated for a range of values of $\theta$.

Further, to calculate $E(N|\theta)$, the value of $E(Z|\theta)$ is also required, and is given by

$$E(Z|\theta) = \tfrac{1}{2}[2(\theta_1 - \theta_0) + \theta_0^2 - \theta_1^2].$$

It would be interesting at this point to compare the expected sample size of a sequential test with the sample size of equivalent fixed sample size scheme. The equivalent fixed sample size scheme is one where both $\alpha$, $\beta$ are specified and it is required to find the value of the sample size and the range of values of the likelihood ratio or sample statistics which determine the final decision.

Consider the numerical example presented in Example 13.2.2 with $\theta_0 = 0$, $\theta_1 = 1$. For any values of $\alpha$ and $\beta$, it is known that the best fixed sample size test of $H_0: \theta = 0$ against $H_1: \theta = 1$ in the Neyman–Pearson sense [see § 5.12] reduces to the following:

$$\text{reject } H_0 \quad \text{if } \bar{x} > k$$

where $\bar{x}$ is the sample mean and $k$ is a constant which depends on $\alpha$ and $\beta$. To find $k$ and $n$ in this fixed sample size test, it is required to solve the following equations:

$$\alpha = P(\text{Type I error}) = P(\bar{X} > k|\theta = 0) = 1 - \Phi(\sqrt{n}k),$$

$$\beta = P(\text{Type II error}) = P(\bar{X} < k|\theta = 1) = \Phi(\sqrt{n}(k-1)),$$

where $\Phi(\cdot)$ is the tabulated distribution function of the standard Normal distribution.

Table 13.5.1 gives the expected sample sizes of the corresponding sequential test (assuming $\theta = 0$ is the true value of $\theta$) and the sample sizes of the equivalent fixed sample size tests, for a range of values of $\alpha$, $\beta$. The comparison assuming $\theta = 1$ is the true value follows easily. Note that $E(Z|\theta = 0) = -\tfrac{1}{2}$ and $E(Z|\theta = 1) = \tfrac{1}{2}$. In each case there is a considerable saving in observations. Wald (1947 p. 57) has shown that the SPRT produces a saving of about 47% over the equivalent fixed sample size test for any values of $\theta_0$, $\theta_1$. The fixed samples sizes have been rounded to integers.

| $\alpha$ | 0·01 | | 0·05 | | 0·1 | |
|---|---|---|---|---|---|---|
| $\beta$ | $E(N|\theta)$ | f.s.s. | $E(N|\theta)$ | f.s.s. | $E(N|\theta)$ | f.s.s. |
| 0·01 | 9·01 | 22 | 8·36 | 16 | 7·64 | 13 |
| 0·05 | 5·83 | 16 | 5·3 | 11 | 4·75 | 9 |
| 0·1 | 4·45 | 13 | 4·06 | 9 | 3·52 | 7 |

Table 13.5.2: Comparison of expected sample sizes ($E(N|\theta)$) and fixed sample sizes (f.s.s.) for the test of $H_0: \theta = 0$ against $H_1: \theta = 1$, assuming $\theta = 0$ is the true value of $\theta$.

## 13.6.  THE SPRT FOR COMPOSITE HYPOTHESES

Up to this point both hypotheses have been simple. The problem now is the construction of sequential tests where one or both hypotheses are composite.

Consider the case where $H_0$: $\theta = \theta_0$ and $H_1$: $\theta > \theta_0$, and $P$(Type I error) $= \alpha$. A solution which makes it possible to construct an SPRT is to replace the composite hypothesis $H_1$ by a simple hypothesis specifying a single value $\theta_1$ of $\theta$, where $\theta_1 > \theta_0$.

When $\theta$ is the unknown expectation of a Normal distribution with known variance, it is known from the Neyman-Pearson lemma [see § 5.12.2] that the best fixed sample size test of $H_0$: $\theta = \theta_0$ against $H_1$: $\theta > \theta_0$.

If we follow the same path for the sequential test, it is required that the O.C. function of the SPRT for fixed $\alpha$, $\beta$ satisfies $P(\theta) > 1 - \alpha$ for $\theta < \theta_0$ and $P(\theta) < \beta$ for $\theta > \theta_1$ with $P(\theta_1) = \beta$. However this still leaves the problem of the choice of the value $\theta_1$. The experimenter, choosing a specific value of $\theta_1$, has essentially stated that he is indifferent to which decision (accept or reject $H_0$) he makes for values of $\theta$ between $\theta_0$ and $\theta_1$, and for values of $\theta \leq \theta_0$ as $\geq \theta_1$ he would hope to make the decision to accept $H_0$ or reject $H_0$ respectively. The acceptance of $\theta = \theta_1$ at termination leads to the acceptance of $\theta > \theta_0$.

Suppose that the hypotheses are now $H_0$: $\theta < \theta'$ against $H_1$: $\theta > \theta'$ then again it is possible to convert both hypotheses to simple hypotheses by fixing two values of $\theta$, $\theta_0$, $\theta_1$ such that the experimenter is indifferent to which decision he makes when $\theta_0 < \theta < \theta_1$. He would however wish to make the further decision to accept $H_0$ when $\theta \leq \theta_0$, and to reject $H_0$ when $\theta > \theta_1$. By fixing the values $\theta_0$, $\theta_1$, $\alpha$, $\beta$, the experimenter is fixing two points in the O.C. curve $P(\theta_0) = 1 - \alpha$ and $P(\theta_1) = \beta$. The acceptance at termination of $\theta = \theta_0$ leads to the acceptance of $\theta < \theta'$.

Let $\theta$ be the unknown proportion of defective items produced by a manufacturing process. Batches containing a large number of items will be accepted if $\theta < \theta'$ and rejected if $\theta > \theta'$. The decision on each batch could be made by testing each item, in which case the correct decision would be made each time. A more efficient way is to subject the batch to sequential sampling and make one of the two decisions on termination. It should be possible for the manufacturer to find two values of $\theta$, $\theta_0$, $\theta_1$ such that he is indifferent to which decision is made for $\theta$ in the range $\theta_0$ to $\theta_1$ and, given these two values, to assess $P$(Type I error) $= P$(rejecting batch$|\theta = \theta_0) = \alpha$ and $P$(Type II error) $= P$(accepting batch$|\theta = \theta_1) = \beta$.

The value of $\alpha$ is a measure of manufacturer's desire to minimize the number of batches incorrectly rejected and $\beta$ is a measure of the manufacturer's desire not to dispatch batches to the customer which should be rejected.

It is unfortunate that the range of values of $\theta$ in the indifference region are precisely those which give rise to the largest expected sample sizes in the SPRT of $H_0$: $\theta = \theta_0$ against $H_1$: $\theta = \theta_1$.

If the hypothesis $H_0$ is tested against $H_1$: $\theta \neq \theta_0$, the construction of an SPRT is very difficult. The indifference region in this case could be obtained by specifying $\delta$ such that if $0 < |\theta - \theta_0| \leq \delta$ then the experimenter is indifferent to which final decision is made. If $|\theta - \theta_0| > \delta$ then the experimenter should reject $H_0$.

Suppose that the $P(\text{Type I error})$ is fixed at $\alpha$ and let the $P(\text{Type II error}) = \beta(\theta)$ for the values of $\theta$ in the range $|\theta - \theta_0| > \delta$, then it is required that $\beta(\theta) \leq \beta$ for some specified $\beta$.

Wald (1947) chapter 4, suggests an approach to the construction of an SPRT, based on what are called 'weight functions'. Let $\pi(\theta)$ be the weight function defined to be

$$\int_t \pi(\theta) \, d\theta = 1 \quad \text{and} \quad \pi(\theta) \geq 0 \quad \text{for all } \theta \in T$$

where $T$ is the region $\theta > \theta_0 + \delta$, $\theta < \theta - \delta$. For a given weight function let

$$\int_t \pi(\theta) \beta(\theta) \, d\theta = \beta.$$

The left-hand side is the weighted average of all possible values of $P(\text{Type II error})$, which means that the condition $\beta(\theta) \leq \beta$ no longer holds for all $\theta$.

Considering now only tests which satisfy this condition in $\beta(\theta)$, the equation above leads after $m$ observations to the condition

$$\int_T \pi(\theta) \left[ \iint_{R_m} \cdots \int f(x_1; \theta)(x_2; \theta) \ldots f(x_m; \theta) \, dx_1 \ldots dx_m \right] d\theta = \beta,$$

where $R_m$ represents the values of $x_1, x_2, \ldots, x_m$ leading to the rejection of $H_0$. On rearrangement this becomes

$$\iint \cdots \int \left[ \int \prod_{i=1}^{m} f(x_1; \theta) \pi(\theta) \, d\theta \right] dx_1 \ldots dx_m = \beta.$$

The term in the square brackets is simply a weighted average (using the same weights as above) of all the density functions of the sample for all values of $\theta$. Now since this is a weighted average of density functions, it is itself a density function. Under the modified conditions, $H_0: \theta = \theta_0$ so that, on $H_0$, $f(x_1 \ldots x_m; \theta_0)$ is the true density function. Now if $H_1$ is the hypothesis that $\int_T \prod(\theta) f(x_1 \ldots x_m: \theta) \, d\theta$ is the true density, then both hypotheses are simple, because the density is completely specified in each case. Hence the SPRT is based on

$$\lambda_m = \frac{f(x_1, x_2, \ldots, x_m)}{f(x_1, x_2, \ldots, x_m; \theta_0)},$$

where

$$f(x_1, x_2, \ldots, x_m) = \int \pi(\theta) f(x_1, x_2, \ldots, x_m; \theta) \, d\theta$$

and

$$f(x_1, x_2, \ldots, x_m; \theta) = \prod_{i=1}^{m} f(x_i; \theta),$$

with the usual limits calculated from $\alpha, \beta$.

The weight function $\pi(\theta)$, for example, could be a prior distribution of $\theta$ [see Chapter 15]. For the construction of a test without the modifications, the reader is referred to Chapter 4 of Wald (1947) and Chapter 4 of Wetherill (1975).

Obviously this approach may also be used for the composite hypotheses considered earlier and for situations where there are several unknown parameters and it is required to test a subset of those parameters (in the presence of nuisance parameters).

## 13.7.  TESTS INVOLVING TWO BINOMIALS

Suppose there are two binomial populations to be compared, with $p_1, p_2$ as the unknown probabilities of obtaining a success and $(1-p_1)$, $(1-p_2)$ as the probabilities of obtaining a failure at any one trial for populations 1, 2 respectively. It is required to test $H_0: p_1 \geq p_2$ against $H_1: p_1 < p_2$. It is assumed that observations from the two populations are independent of each other. The main difficulties in the construction of a sequential test in this situation are that the hypotheses compare two unknowns and no single value or range is mentioned for $p_1, p_2$, and that the observations come from two sources so that at each trial the population which provides the next observation must be chosen.

To overcome the second difficulty, suppose that the observations are taken in pairs, one from each population. The outcome of each single trial must be one of $(1, 0)$, $(0, 1)$, $(1, 1)$, $(0, 0)$, the first entry in each pair representing the observation from population 1 and the second from population 2. The four outcomes have probability $p_1(1-p_2)$, $(1-p_1)p_2$, $p_1p_2$, $(1-p_2)(1-p_2)$ respectively.

The two outcomes which give information about the differences between the two populations are $(1, 0)$ and $(0, 1)$ and it therefore seems sensible to base the test only on these outcomes, $(1, 0)$ supporting $H_0$ and $(0, 1)$ supporting $H_1$.

Considering only the pairs $(1, 0)$, $(0, 1)$ the conditional probability that $(0, 1)$ is observed is

$$\theta = \frac{(1-p_1)p_2}{p_1(1-p_2) + (1-p_1)p_2};$$

$(1, 0)$ is therefore observed with conditional probability $(1-\theta)$.

The series of trials where only $(1, 0)$ or $(0, 1)$ are observed form a new set of observations where a success is the occurrence of $(0, 1)$ with probability $\theta$

and a failure is the occurrence of $(1, 0)$ with probability $(1 - \theta)$. Hence in $m$ pairs observed where $m_1$ of them are $(0, 1)$, the likelihood is $\theta^{m_1}(1 - \theta)^{m - m_1}$.

If $p_1 = p_2$, $\theta = \frac{1}{2}$, if $p_1 > p_2((1, 0)$ is more likely than $(0, 1))$ $\theta < \frac{1}{2}$ and $p_1 < p_2$ gives $\theta > \frac{1}{2}$. We may now proceed by one of two methods, Wald's (§ 13.7.1) or Armitage's (§ 13.7.2)

### 13.7.1.  Wald's Method

The hypotheses $H_0$, $H_1$ above may now be reduced to $H_0' : \theta \leq \frac{1}{2}$ against $H_1' : \theta > \frac{1}{2}$ where the acceptance or rejection of $H_0'$ leads to the acceptance or rejection of $H_0$.

Applying the idea in the previous section, it is possible to simplify further and introduce two values of $\theta$, $\theta_0$, $\theta_1$, $\theta_0 < \frac{1}{2} < \theta_1$, such that for $\theta$ in the range $\theta_0 < \theta < \theta_1$, the experimenter is indifferent to which decision is made on termination. By fixing $\alpha$, $\beta$, $\theta_0$, $\theta_1$ the binomial SPRT considered earlier may now be used to test $H_0'' : \theta = \theta_0$ against $H_1'' : \theta = \theta_1$: $\theta = \theta_1(> \theta_0)$ and hence by extension, to test $H_0$: $p_1 \geq p_2$ against $H_1$: $p_1 < p_2$. See Wald (1947).

### 13.7.2.  Armitage's Method

Armitage (1975) discusses modifications of the Wald test with applications to the planning of sequential medical trials. The probabilities $p_1$, $p_2$ in this case may be interpreted as 'success rates' using two treatments 1, 2.

The problem now becomes one of testing three hypotheses (for simplicity, $H_1$, $H_{-1}$ are chosen to be symmetric).

$$H_0:\ \theta = \tfrac{1}{2},$$

$$H_1:\ \theta = \theta_1 > \tfrac{1}{2} \quad \text{(treatment 2 preferred to treatment 1)}$$

$$H_{-1}:\ \theta = 1 - \theta_1 \quad \text{(treatment 1 preferred to treatment 2).}$$

The resulting test is now a two-sided test, the alternative hypotheses being symmetric about $H_0$. The value of $P(\text{Type I error})$ is therefore

$$P(\text{accept } H_1 \text{ or } H_{-1} | \theta = \tfrac{1}{2}) = P(\text{accept } H_1 | \theta = \tfrac{1}{2}) + P(\text{accept } H_{-1} | \theta = \tfrac{1}{2}).$$

For convenience suppose that the total $P(\text{Type I error})$ is $2\alpha$, giving $P(\text{accept } H_1 | \theta = \tfrac{1}{2}) = P(\text{accept } H_{-1} | \theta = \tfrac{1}{2}) = \alpha$, and $P(\text{Type II error}) = (1 - \beta)$. There are two possible tests which have to be carried out, $H_0$ against $H_1$ and $H_0$ against $H_{-1}$.

The Armitage test is based on the statistic $d_m$ which is the difference between the number of $(0, 1)$ observations and $(1, 0)$ observations in $m$ pairs. Let $m_1$ be the number of $(0, 1)$'s observed and $m_2$ the number of $(1, 0)$'s observed. Then $m = m_1 + m_2$ and $d_m = m_1 - m_2)$, so that $d_m = 2m_1 - m$.

The likelihood ratio for testing $H_0$ against $H_1$ is

$$\lambda_m = \frac{\theta_1^{m_1}(1 - \theta_1)^{m - m_1}}{(\tfrac{1}{2})^m}.$$

This reduces to

$$\tfrac{1}{2}d_m \log\left[\theta_1/(1-\theta_1)\right] - m \log\left[\tfrac{1}{2}\theta_1^{-1/2}(1-\theta_1)^{-1/2}\right].$$

This particular test of $H_0$ against $H_1$ has fixed $\alpha$, $\beta$ whence the usual limits of $\log\left[\beta/(1-\alpha)\right]$ and $\log\left[(1-\beta)/\alpha)\right]$ apply to give the SPRT. The resulting stopping boundaries are two parallel lines $d_m = a_1 + bm$ and $d_m = a_2 + bm$ in a plot of $d_m$ against $m$ with

$$a_1 = \frac{2\log\left[(1-\beta)/\alpha\right]}{\log\left[\theta_1/(1-\theta_1)\right]}, \qquad a_2 = \frac{2\log\left[\beta/(1-\alpha)\right]}{\log\left[\theta_1/(1-\theta_1)\right]}$$

and

$$b = \frac{2\log\left[\tfrac{1}{2}\theta_1^{-1/2}(1-\theta_1)^{-1/2}\right]}{\log\left[\theta_1/(1-\theta_1)\right]}.$$

By a similar argument, using symmetry, the test of $H_0$ against $H_{-1}$ reduces to the stopping boundaries

$$d_m = -a_1 - bm \quad \text{and} \quad d_m = -a_2 - bm.$$

The boundaries for $\theta_1 = 5 \cdot 85$, $\alpha = 0 \cdot 025$, $\beta = 0 \cdot 05$ are given in Figure 13.7.1.

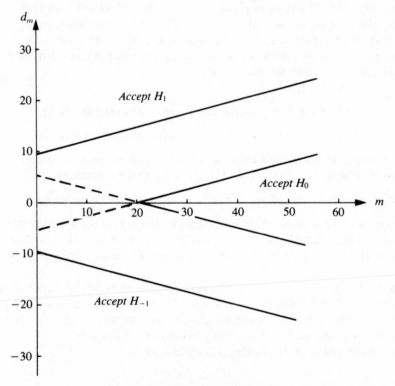

Figure 13.7.1: Open sequential scheme for testing $H_0$: $\theta = \tfrac{1}{2}$ against $H_1$: $\theta = 0 \cdot 75$, $H_{-1}$: $\theta = 0 \cdot 25$ and $2\alpha = 0 \cdot 01$, $\beta = 0 \cdot 05$, $N = 80$.

The region where the boundaries intersect (the dotted lines in the figure) presents obvious difficulties in interpretation hence it becomes a 'continue sampling' region. This produces a test with 'open' boundaries for the hypotheses.

To avoid having to continue sampling for a long period before termination, so using a large number of patients in the test, Armitage suggests a termination of the procedure where a predetermined number of $(0, 1)$ or $(1, 0)$ pairs has been reached. To accomplish this without violating the fixed $\alpha$, $\beta$ values, the middle boundaries of the open test described earlier are omitted and the maximum sample size, $n$, is chosen so that the new test has the given probabilities of error. By omitting the boundaries and fixing $n$, some points in the open test which were stopping points now become continue sampling points. Armitage (1957) describes the method of obtaining $n$ in practice.

Figure 13.7.2 gives the resulting closed boundary test when $\theta_1 = 0.75$, $\alpha = 0.005$, $\beta = 0.05$, which from Armitage (1975), gives the value of $n = 80$.

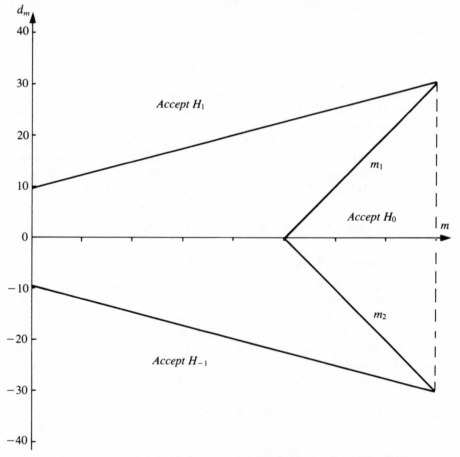

Figure 13.7.2: Closed sequential scheme for testing $H_0$: $\theta = \frac{1}{2}$ against $H_1$: $\theta = 0.75$, $H_{-1}$: $\theta = 0.25$ and $2\alpha = 0.01$, $\beta = 0.05$, $N = 80$.

The boundary at $n$ which is parallel to the $d_m$ axis may be replaced by the two boundaries shown $M_1$, $M_2$ enabling earlier stopping of the test to be achieved when $H_0$ is accepted. Once a point on these modified boundaries is reached, neither outer boundary will ever be reached even if all the remaining trials resulting in $(0, 1)$ or $(1, 0)$.

### 13.7.3.  Discussion

It is possible to truncate any SPRT. A maximum number of observations, $n$, could be fixed at the outset of the experiment. The usual parallel stopping boundaries are constructed but sampling is stopped once $n$ observations have been taken. The final decision at $n$ observations, provided neither stopping boundary has been crossed, could then be:

$$\text{accept} \quad H_0 \quad \text{if } \log B < \sum_{i=1}^{n} z_i \leq 0,$$

$$\text{reject} \quad H_0 \quad \text{if } 0 < \sum_{i=1}^{n} z_i < \log A,$$

where

$$z_i = \log \left[ \frac{f(x_i; \theta_1)}{f(x_i; \theta_0)} \right].$$

Unfortunately such a truncation changes the true probabilities of Type I and Type II errors from the original $\alpha$, $\beta$. The effect of truncation obviously becomes smaller as $n$ becomes larger.

By extension of the above test, the test of the hypotheses $H_0$: $\theta = \theta_0$ against $H_1$: $\theta \neq \theta_0$ could now be considered as a test of $H_0$: $\theta = \theta_0$ against a pair of alternative hypotheses $H_1$: $\theta > \theta_0$ and $H_{-1}$: $\theta < \theta_0$. A full discussion of sequential tests of three hypotheses may be found in Wetherill (1975) Chapter 3.

## 13.8.  OTHER SEQUENTIAL PROCEDURES

### 13.8.1.  Sequential Designs for the 'Two-Binomials' Problem

It was required in the last section that the observations should be taken in pairs. However there are several relatively simple sequential designs, with stopping rules, which could be used in place of the SPRT.

For example, observations could be taken one at a time and the population from which they are drawn could be chosen by a sampling rule known as the play the winner rule. This operates by choosing to sample a population at the next trial if the last trial on the same population resulted in a success and choosing the next trial to be on the other population if the last trial was a failure. This gives a sample which looks like

Population 1:   1   1   0       1   1   1   0                    1 . . .,

Population 2:                0                    1   1   0

where, as usual, 0 represents a failure, and 1 a success. The choice of the population for the first trial is made randomly. A possible stopping rule would be one where sampling stops once the difference between the number of trials from each population exceeds a prespecified value.

It has been claimed that using such a rule minimises the number of trials on the population with the smallest probability of success.

### 13.8.2. Bayesian Methods

If one uses Bayesian methods of analysis [see Chapter 5], the problems encountered earlier in dealing with composite hypotheses may be resolved. Suppose that the testing of hypotheses is reduced to a decision problem [see Chapter 19]. Consider the case where the observations $x_1, x_2, \ldots$ come from the p.d.f. $f(x; \theta)$ and where

$$H_0: \theta \leq \theta', \qquad H_1: \theta > \theta'.$$

Let the decision to accept $H_0$ be $d_1$ and the decision to accept $H_1$ be $d_2$. It should be possible to set up a series of loss functions which represent to the experimenter (or decision maker) the loss (in monetary terms) of making the incorrect decision. Let the loss in making final decision $d_i$ for all values of $\theta$ be $L(d_i; \theta)$ then a possible loss structure is:

$$L(d_1; \theta) = 0 \quad \text{if } \theta \leq \theta'$$

$$L(d_1; \theta) = k_1 \quad \text{if } \theta > \theta'$$

$$L(d_2; \theta) = k_2 \quad \text{if } \theta \leq \theta'$$

$$L(d_2; \theta) = 0 \quad \text{if } \theta > \theta'.$$

where $k_1$ and $k_2$ are constants.

If the unknown parameter $\theta$ is assigned a prior distribution then as the sequential sample is observed, the posterior distribution of $\theta$ may be obtained by Bayes' theorem. Furthermore the expected loss over the posterior distribution of making either final decision may be obtained. If the cost of each observation is $c$, then at each stage the reduction of the expected loss which could be obtained by buying an observation at cost $c$ is compared with $c$.

Details of the methods employed in this case may be found in Wetherill (1975) Chapter 7.

P.J.

## 13.9. FURTHER READING AND REFERENCES

Armitage, P. (1975). Restricted Sequential Procedures, *Biometrika* **44**, 9–26.
Armitage, P. (1975). *Sequential Medical Trials*, 2nd Edition. Oxford: Blackwell.
Wetherill, G. B. (1975). *Sequential Methods in Statistics*, 2nd Edition. London: Chapman and Hall.
Wald, A. (1974). *Sequential Analysis*. New York: J. Wiley.

# CHAPTER 14

# *Distribution-Free Methods*

## 14.1. INTRODUCTION

One important feature of much of the statistical inference so far encountered in this handbook is the assumption that the underlying distribution of the observations of interest belongs to some parametric family of distributions; for example the underlying random variables are Normal [see II, § 11.4] or Gamma [see II, § 11.3] or Poisson [see II, § 15.4], etc. Thus we assume that we know the form or family of distribution, although we may not know the exact member of the family; for example we may assume a Normal $(\mu, \sigma)$ model but with the parameters $\mu$ and $\sigma$ unknown. Estimation and significance testing techniques provide tools with which to draw conclusions about the unknown parameters.

The validity of any conclusions we draw must depend to some extent on the validity of our underlying assumption of parametric family. For example standard tests such as the $t$-test [see § 5.8.2] are strictly only valid if the assumption of Normality is true, although fortunately many of them are fairly robust to departures from Normality.

It would obviously be useful to be able to construct statistical models and tests which are less restrictive in that they do not depend on a particular underlying parametric family of distributions. Such models—known as *non-parametric* or *distribution-free* models—are the subject of this chapter. The only assumption that we make in most of the illustrative procedures introduced here is that the underlying variables are continuous.

The historical development of these methods has been somewhat ad hoc in that methods were suggested for particular problems. Only relatively recently has an overview of the subject been developed and the optimal properties of the procedures investigated. There are many excellent books on non-parametric and distribution-free methods, for example, Siegel (1956), Walsh (1962), Noether (1967), Hollander and Wolfe (1973) and Lehmann (1975).

Many of the test procedures given here have somewhat idiosyncratic and extensive associated tables. In most instances we have referred the reader to two sources, namely Siegel (1956) and Owen (1962 see Bibliography G).

## 14.2.   TESTS BASED ON THE EMPIRICAL DISTRIBUTION FUNCTION

As an introduction to distribution-free tests we consider the basic problem of questioning whether or not we can assume that a random sample of observations $x_1, x_2, \ldots, x_n$ comes from a specified *continuous* distribution. This goodness of fit problem has been encountered already in Chapter 7 where the $\chi^2$ goodness of fit test was introduced. In a sense, that test served as an introduction to distribution-free methods.

Basic concepts for the tests introduced in this section are (i) the *order statistic(s)* of the sample, and (ii) the *empirical distribution function*.

DEFINITION 14.2.1.   *Order statistic* [see II, Chapter 15].   When the sample values $x_1, x_2, \ldots, x_n$ are arranged in increasing order of magnitude, denote the smallest by $x_{(1)}$, the next smallest by $x_{(2)}, \ldots$, and the largest by $x_{(n)}$. The set

$$x_{(1)}, x_{(2)}, \ldots, x_{(n)}$$

with

$$x_{(1)} < x_{(2)} < \ldots < x_{(n)}$$

is the order statistic, or the set of order statistics, of the sample. (If the underlying random variable had been discrete we should have had to take into account that samples with coincident observations may occur, a feature which materially complicates the theory of order statistics. With a continuous random variable, however, the possibility of coincident observations may be ignored, provided the observations are recorded with sufficient accuracy. [See II, § 15.1.]

Familiar statistics that are functions of the order statistic are the *percentiles* (including as important special cases the *lower quartile*, the *median*, and the *upper quartile*) and the *range*. These may be defined as follows:

*Range.*   The range is simply $x_{(n)} - x_{(1)}$.

*Median.*   If the number of observations is odd, with $n = 2m + 1$, the median is $x_{(m+1)}$, that is the middle one when the observations are arranged in order of magnitude. If the number of observations is even, with $n = 2m$, the median is conventionally taken as $\frac{1}{2}(x_{(m)} + x_{(m+1)})$.

*Quartiles, Percentiles.*   The lower quartile, the median and the upper quartile are intended to provide a partition of the order statistics into four subsets of equal size. Strictly speaking, this requires that the sample size $n$ be of the form $n = 4k + 3$, in which case the lower quartile is $x_{(k+1)}$, *the median* $x_{(2k+2)}$, and the upper quartile $x_{(3k+3)}$.

Deciles (tenths) and percentiles (hundredths) are similarly defined.

DEFINITION 14.2.2.   *Empirical distribution function.*   The *empirical distribution function $F_n(x)$* is defined as follows:

$$F_n(x) = \begin{cases} 0 & x < x_{(1)}, \\ k/n & x_{(k)} \le x < x_{(k+1)} \\ 1 & x \ge x_{(n)}. \end{cases} \qquad k = 1, 2, \ldots, n-1,$$

In other words $F_n(x)$ is a step function with jumps of $1/n$ at each of the values $x_{(1)}, x_{(2)}, \ldots, x_{(n)}$.

EXAMPLE 14.2.1.   *Empirical distribution function.*   Suppose a random sample of size $n = 6$ from a continuous distribution yields observations

$$2\cdot1, \quad -0\cdot6, \quad 0\cdot2, \quad 3\cdot0, \quad -1\cdot0, \quad 1\cdot3.$$

The order statistic is then the (increasing) ordered set

$$-1\cdot0, \quad -0\cdot6, \quad 0\cdot2, \quad 1\cdot3, \quad 2\cdot1, \quad 3\cdot0.$$

The empirical distribution function starts at 0 and has steps of $1/6$ at $-1\cdot0$, $-0\cdot6, 0\cdot2, 1\cdot3, 2\cdot1, 3\cdot0$ as illustrated in Figure 14.2.1.

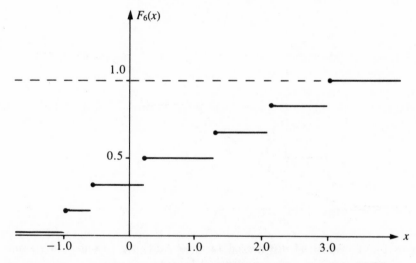

Figure 14.2.1: Empirical distribution function $F_6(x)$ for Example 14.2.1.

If we obtained a second random sample of size 6 from this continuous random variable, the values would vary, and hence the empirical distribution function $F_6(x)$ would be different. It is the nature of this variation of $F_6(x)$ from the underlying distribution function $F(x)$ of the random variable $X$ which we investigate in a goodness of fit context.

### 14.2.1.   Kolmogorov–Smirnov Test: One Sample

Suppose that we wish to test the simple null hypothesis that a specified continuous function $F_0(x)$ is the distribution function from which the random sample $x_1, x_2, \ldots, x_n$ arose. Thus we wish to test

$$H_1: F(x) = F_0(x) \quad \text{for all } x$$

against

$$H_2: F(x) \neq F_0(x) \text{ for some } x.$$

The test statistic used in the Kolmogorov–Smirnov one sample test is

$$D_n(\mathbf{x}) = \sup_{-\infty < z < \infty} |F_n(z) - F_0(z)|,$$

that is, the largest difference between the empirical distribution function and the specified distribution function.

For example, Figure 14.2.2 shows the value of $D_6(\mathbf{x})$ for the data in Example 14.2.1 if $F_0(x)$ is as illustrated.

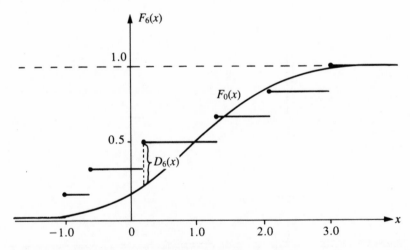

Figure 14.2.2: Empirical distribution function $F_6(x)$ for Example 14.2.1 and the specified $F_0(x)$.

The test statistic is a value of a random variable $D_n(\mathbf{X})$, which depends on $X_1, X_2, \ldots, X_n$ through $F_n(z)$, and it provides a measure of how far $F_n(z)$ is from $F_0(z)$. Clearly large values of $D_n(\mathbf{x})$ will make us doubt $H_1$, for then $F_n(z)$ must be some way from $F_0(z)$, at least for some values of $z$. We therefore take as our critical region [see § 5.12.2] the set

$$A_2 = \{\text{observations: } D_n(\mathbf{x}) > c_\alpha\},$$

where $c_\alpha$ is a constant chosen so that $P(A_2|H_1) = \alpha$. The determination of $c_\alpha$ appears at first glance to be a hopeless task for this general non-parametric approach. In fact by some simple probability theory involving the probability integral transformation [see II, Theorem 10.7.2] one can derive the remarkable result that, if the null hypothesis $H_1$ is true, $D_n(\mathbf{x})$ comes from a distribution which does not depend on the actual form of the specified $F_0(x)$, but only on the size $n$ of the sample. In other words we have a distribution-free method, since to determine $A_2$ we merely need to specify $c_\alpha$ for a particular significance level $\alpha$ by setting $P(A_2|H_1) = \alpha$; for a particular $\alpha$ and $n$ we obtain the same value of $c_\alpha$ whatever the specified distribution, be it Uniform, Normal, Gamma, or whatever.

The critical values $c_\alpha$ of the exact distribution of $D_n(\mathbf{X})$ under hypothesis $H_1$ have been tabulated for various values of $n$; see, for example Siegel (1956, Table E, p. 251) or Owen (1962, Table 15.1, pp. 423–425). A set of approximations for the critical values which works well for $n > 35$ is as follows:

$$\alpha = 0\cdot20 \quad 0\cdot10 \quad 0\cdot05 \quad 0\cdot01$$

$$c_\alpha \qquad \frac{1\cdot07}{\sqrt{n}} \quad \frac{1\cdot22}{\sqrt{n}} \quad \frac{1\cdot36}{\sqrt{n}} \quad \frac{1\cdot63}{\sqrt{n}}.$$

EXAMPLE 14.2.2.   *The Kolmogorov–Smirnov test.*   Suppose that we wish to test the hypothesis that the six observations in Example 14.2.1 form a random sample from a Normal $(1, 1)$ distribution. In other words, $F_0(z)$ is the distribution function of a Normal $(1, 1)$ random variable, in fact as illustrated in Figure 14.2.2. We can then derive $D_6(\mathbf{x})$ either from the graph or mathematically as follows. We first recognize that $D_6(\mathbf{x})$ must occur at a $z$-value corresponding to one of the observed values. The pair of values $d_1$ and $d_2$ as illustrated in Figure 14.2.3 are then calculated at each such $z$-value, and are shown in

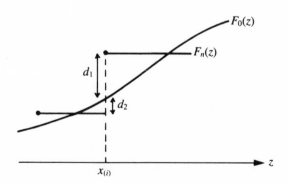

Figure 14.2.3: Illustration of differences $d_1$ and $d_2$.

Table 14.2.1. The third column shows the value of

$$F_0(z) = P(X \le z)$$
$$= \Phi(z-1),$$

| $z$ | $F_6(z)$ | $F_0(z)$ | $d_1$ | $d_2$ |
|------|----------|----------|---------|---------|
| $-1 \cdot 0$ | 0·1667 | 0·0228 | 0·1439 | 0·0228 |
| $-0 \cdot 6$ | 0·3333 | 0·0548 | 0·2785 | 0·1119 |
| $0 \cdot 2$ | 0·5000 | 0·2119 | 0·2881 | 0·1214 |
| $1 \cdot 3$ | 0·6667 | 0·6179 | 0·0488 | 0·1179 |
| $2 \cdot 1$ | 0·8333 | 0·8643 | 0·0310 | 0·1976 |
| $3 \cdot 0$ | 1·0000 | 0·9772 | 0·0228 | 0·1439 |

Table 14.2.1: Evaluation of $D_6(\mathbf{x})$ in Example 14.2.2.

where $\Phi(\cdot)$ is the cumulator distribution function of the standard Normal [see Appendix T3].

From the tables mentioned above we find that for a test at the 5% significance level for $n = 6$ the critical region is given by

$$A_2 = \{\text{observations: } D_6(\mathbf{x}) > 0 \cdot 52\}.$$

Here we observe $D_6(\mathbf{x}) = 0 \cdot 2881$ from Table 14.2.1, so that our set of observations does not fall in this critical region. Our significance level is therefore greater than 5% and we may regard the data as consistent with the hypothesis that they come from a Normal $(1, 1)$ distribution.

It should be noted that similar problems could have been tackled by the $\chi^2$ goodness of fit test of Chapter 7. [see § 7.4]. The potential advantage of the Kolmogorov–Smirnov test is that it does not group the data (with the subsequent loss of information) but considers the individual observed values. It can also be used successfully with small samples and is believed to be more powerful generally than the $\chi^2$ test.

The question of whether the test can be extended to the case of a composite null hypothesis which is not completely specified—for example, suppose in Example 14.2.2 that $H_1$ specifies a Normal $(\theta, 1)$ distribution with $\theta$ unknown—remains open.

### 14.2.2.  Kolmogorov–Smirnov Test: Two Samples

The Kolmogorov–Smirnov test for two samples again makes use of the concept of the empirical distribution function. Here though we address ourselves to the question of whether or not two independent samples of observations come from the same distribution. More specifically we have a random sample $x_1, x_2, \ldots, x_{n_1}$ from a population with continuous distribution function

$F(x)$ and an independent random sample $y_1, y_2, \ldots, y_{n_2}$ from a population with continuous distribution function $G(y)$. We wish to test

$$H_1: F(z) = G(z) \quad \text{for all } z$$

against

$$H_2: F(z) \neq G(z) \quad \text{for some } z.$$

Notice that we do not actually specify what the common form of $F(z)$ and $G(z)$ is in $H_1$.

From the two samples we can determine two empirical distribution functions $F_{n_1}(x)$ and $G_{n_2}(y)$ respectively. The test statistic used in the Kolmogorov-Smirnov two sample test is

$$D_{n_1,n_2}(\mathbf{x}, \mathbf{y}) = \sup_{-\infty < z < \infty} |F_{n_1}(z) - G_{n_2}(z)|,$$

which is the largest difference between the two empirical distribution functions. This is a value of a random variable $D_{n_1,n_2}(\mathbf{X}, \mathbf{Y})$ which depends on $X_1, X_2, \ldots, X_{n_1}$ through $F_{n_1}(z)$ and on $Y_1, Y_2, \ldots, Y_{n_2}$ through $G_{n_2}(z)$. If $H_2$ is true we would expect the empirical distribution functions to be 'far apart', so that we take as critical region of a test of size $\alpha$:

$$A_2 = \{\text{observations: } D_{n_1,n_2}(\mathbf{x}, \mathbf{y}) > c_\alpha\},$$

where again $c_\alpha$ is a constant chosen so that $P(A_2|H_1) = \alpha$. The exact sampling distribution of $D_{n_1,n_2}(\mathbf{X}, \mathbf{Y})$ under $H_1$ is known and has been tabulated; see, for example, Siegel (1956, Table $L$, p. 278) and Owen (1962, Table 15.4, pp. 434–436 for $n_1 = n_2 = n \leq 40$, and Massey (1952) for $n_1, n_2 \leq 10$.

Again the distribution does not depend on the common form of $F$ and $G$ in $H_1$. For large $n_1$ and $n_2$ (greater than about 40) the following approximations may be used to determine $c_\alpha$.

| $\alpha = 0.20$ | $0.10$ | $0.05$ | $0.01$ |
|---|---|---|---|
| $c_\alpha$:   $1.07\sqrt{\dfrac{n_1 n_2}{n_1 + n_2}}$ | $1.22\sqrt{\dfrac{n_1 n_2}{n_1 + n_2}}$ | $1.36\sqrt{\dfrac{n_1 n_2}{n_1 + n_2}}$ | $1.63\sqrt{\dfrac{n_1 n_2}{n_1 + n_2}}.$ |

EXAMPLE 14.2.3.   *The Kolmogorov–Smirnov two-sample test.*   Suppose that we obtain the following two independent random samples from two populations:

|  | | | | | | |
|---|---|---|---|---|---|---|
| Sample 1: | $2 \cdot 1$ | $-0 \cdot 6$ | $0 \cdot 2$ | $3 \cdot 0$ | $-1 \cdot 0$ | $1 \cdot 3$ |
| Sample 2: | $1 \cdot 0$ | $2 \cdot 6$ | $-0 \cdot 5$ | $0 \cdot 6$ | $1 \cdot 8$ | |

In Figure 14.2.4 we show the two empirical distribution functions $F_6(z)$ and $G_5(z)$, and see that

$$D_{6,5}(\mathbf{x}, \mathbf{y}) = 0 \cdot 33.$$

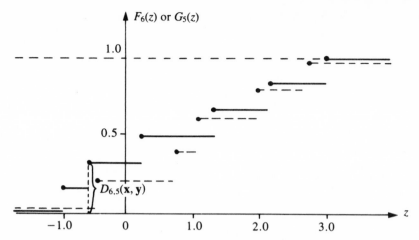

Figure 14.2.4: Empirical distribution functions $F_6(z)$ [————] and $G_5(z)$ [--------] for Example 14.2.3.

We find from the tables mentioned (Massey, 1952) that for a test at the 5% significance level for $n_1 = 6$, $n_2 = 5$ the critical region is given by

$$A_2 = \{\text{observations: } D_{6,5}(\mathbf{x}, \mathbf{y}) > 0 \cdot 67\}.$$

Here our observed value of $D_{6,5}(\mathbf{x}, \mathbf{y})$ does not fall in the critical region. Our significance level therefore exceeds 5% and so the data are regarded as being consistent with the null hypothesis $H_1$ that the two populations have the same underlying distribution.

## 14.3. TESTS BASED ON ORDER STATISTICS

The tests in section 14.2 are based on summaries of the observations through the empirical distribution function [see Definition 14.2.2]. Another major class of distribution-free tests uses the concept of order statistics [see Definition 14.2.1 and II, Chapter 15].

The *order statistic* of a sample arranges the (numerical) observations in increasing order of magnitude. The tests are then based on various summaries of this order statistic. For example, some involve comparisons of the values in the order statistic with the median [see § 14.2 and II, § 15.5] of the distribution; in others we require the further concept of *rank*, where the smallest observation is ranked 1, the next smallest rank 2, and so on.

In determining the ranks we sometimes encounter the problem of ties. Since we assume that the underlying distribution is continuous, in theory there should be no ties, that is no two observations should be the same. However, because of practical considerations in measuring techniques, for example, it will sometimes happen that some observations may coincide. When we consider the ranks in the order statistic the common practice is to give to each tied observation the average of the ranks they would have received if they had been slightly different. For example, if we have a sample of observations

$$-3, \quad -1, \quad 0, \quad 2, \quad 2, \quad -1, \quad 2, \quad 1, \quad 3,$$

the order statistic and ranks are given by

$$-3 \quad -1 \qquad -1 \qquad 0 \quad 1 \quad 2 \quad 2 \quad 2 \quad 3$$

$$\text{Rank:} \quad 1 \quad 2 \cdot 5 \qquad 2 \cdot 5 \quad 4 \quad 5 \quad 7 \quad 7 \quad 7 \quad 9.$$

In the next four Sections we illustrate such tests and give examples of procedures which are applicable to a single sample (§ 14.4), and to matched-pairs samples (§ 14.5), two samples (§ 14.6) and several samples (§ 14.7).

## 14.4. ONE SAMPLE TESTS

In the tests in this section we consider a random sample of observations $x_1, x_2, \ldots, x_n$ from the distribution of a continuous random variable $X$, which has an unspecified distribution function $F(x)$. In the corresponding parametric testing situation we looked for example at the $t$-test [see § 5.8.2] which tested hypotheses about the expected value. In the non-parametric context the expected value loses much of its relevance, and concentration is focused instead on the concept of the (population) *median*, $m$; that is, the value $m$ of $X$ for which $F(m) = \frac{1}{2}$.

### 14.4.1. Sign Test

On the basis of the random sample $x_1, x_2, \ldots, x_n$ we wish to test the null hypothesis $H_1$ that the median $m$ equals a specified value $m_0$. Consider then the test of

$$H_1: m = m_0$$

against

$$H_2: m \neq m_0.$$

The sign test uses as test statistic the number of observations $R$ greater than $m_0$. If an observation actually equals $m_0$ the convention is to ignore it and reduce $n$ by 1. Regardless of the underlying distribution of $X$ (provided it is continuous) we know that each observation has probability $\frac{1}{2}$ of being greater than the median $m$, independently of any other observations. Hence if $H_1$ is true and $m = m_0$, then $R$ will have a Binomial distribution with

$$P(R = r|H_1) = \binom{n}{r}(\tfrac{1}{2})^r(1 - \tfrac{1}{2})^{n-r}$$

$$= \binom{n}{r}(\tfrac{1}{2})^n \qquad (r = 0, 1, 2, \ldots, n). \tag{14.4.1}$$

Clearly relatively small or large values of $R$ would make us unhappy about $H_1$, and so we set the critical region of the form

$$A_2 = \{0, 1, \ldots, c_\alpha, n - c_\alpha, n - c_\alpha + 1, \ldots, n\},$$

where $c_\alpha$ is determined from $P(A_2|H_1) \leq \alpha$, the size of the test [c.f. § 5.2.1(f)]. The evaluation of $c_\alpha$ can be either by direct calculation, by the use of tables of the Binomial distribution or by the use of the Normal approximation to the Binomial [see II, § 11.4.7]. Under $H_1$, $R$ is approximately Normal $(\frac{1}{2}n, \frac{1}{2}\sqrt{n})$, and the approximation is reasonably good for $n \geq 10$.

The above test is two-sided [see § 5.2.1], and we could derive similar one-sided [see § 5.2.3] tests for the case where $H_2$ states that $m > m_0$ (or $m < m_0$). In such cases $A_2$ would be of the form $\{n - c_\alpha, n - c_\alpha + 1, \ldots, n\}$ (or $\{0, 1, \ldots, c_\alpha\}$) respectively.

EXAMPLE 14.4.1.    *The sign test.*    A random sample of 10 observations from a continuous distribution yields

$$6\cdot4, \quad 5\cdot9, \quad 4\cdot9, \quad 4\cdot8, \quad 6\cdot0, \quad 4\cdot7, \quad 7\cdot0, \quad 5\cdot5, \quad 7\cdot1, \quad 5\cdot6.$$

Suppose $H_1$ states that $m = m_0 = 5\cdot0$ and the alternative $H_2$ states that $m > 5\cdot0$. The critical region $A_2$ is of the form

$$A_2 = \{10 - c, 10 - c + 1, \ldots, 10\},$$

and we need to select the largest $c$ such that $P(A_2|H_1) \leq 0\cdot05$ for a significance test at the 5% level [see § 5.2.1(f)]. From (14.4.1)

$$P(R = 10|H_1) + P(R = 9|H_1) = (\tfrac{1}{2})^{10} + 10(\tfrac{1}{2})^{10} = 0\cdot011,$$

$$P(R = 10|H_1) + P(R = 9|H_1) + P(R = 8|H_1) = (\tfrac{1}{2})^{10} + 10(\tfrac{1}{2})^{10} + 45(\tfrac{1}{2})^{10} = 0\cdot055.$$

Thus for a test at the 5% significance level we take $c = 1$ and $A_2 = \{9, 10\}$. Here we observe $R = 7$, which does not fall within the critical region. At the 5% significance level therefore we have no evidence to reject the null hypothesis that $m' = 5\cdot0$.

An alternative approach is to derive the *significance probability* [see § 5.2.1(f)], that is the probability under the null hypothesis of obtaining the observed result or one at least as extreme (c.f. § 5.2.2, § 5.3). Here we observed $R = 7$ so that the significance probability for this one-sided test is given by

$$P(R \geq 7|H_1) = P(R = 7|H_1) + P(R = 8|H_1) + P(R = 9|H_1) + P(R = 10|H_1)$$

$$= 120(\tfrac{1}{2})^{10} + 45(\tfrac{1}{2})^{10} + 10(\tfrac{1}{2})^{10} + (\tfrac{1}{2})^{10}$$

$$= 0\cdot172.$$

This is a large probability [see Table 5.2.1] and the data must be judged as being in conformity with the null hypothesis.

### 14.4.2.  Wilcoxon Signed Ranks Test

Clearly the sign test throws away some information. A better non-parametric test would make use not only of the ordinal relation of whether the differences $x_i - m_0$ were positive or negative, but also of the relative sizes of these differences. Again let us consider a two-sided test of

$$H_1: m = m_0$$

against

$$H_2: m \neq m_0.$$

Suppose we define

$$z_i = x_i - m_0 \qquad (i = 1, 2, \ldots, n),$$

and then rank the absolute values $|z_i|$, i.e. we rank the $z_1, z_2, \ldots, z_n$ regardless of sign. As in the sign test we ignore any observations which yield $z_i = 0$.

EXAMPLE 14.4.2.   *Wilcoxon signed ranks.*   The $z_i$ values in Example 14.4.1 with $m_0 = 5\cdot0$ are

$$1\cdot4, \quad 0\cdot9, \quad -0\cdot1, \quad -0\cdot2, \quad 1\cdot0, \quad -0\cdot3, \quad 2\cdot0, \quad 0\cdot5, \quad 2\cdot1, \quad 0\cdot6$$

We rank these irrespective of sign to give

|        | $-0\cdot1$ | $-0\cdot2$ | $-0\cdot3$ | $0\cdot5$ | $0\cdot6$ | $0\cdot9$ | $1\cdot0$ | $1\cdot4$ | $2\cdot0$ | $2\cdot1$ |
|--------|------|------|------|-----|-----|-----|-----|-----|-----|-----|
| Rank:  | 1    | 2    | 3    | 4   | 5   | 6   | 7   | 8   | 9   | 10  |

The Wilcoxon signed ranks test uses these ranks through either of the two test statistics

$$V_+ = \text{sum of the ranks of positive differences, } z_i$$

or

$$V_- = \text{sum of the ranks of negative differences, } z_i.$$

If $H_1$ is true we would expect $V_+$ and $V_-$ to be roughly equal, whereas under $H_2$ we would expect one of these two sums to predominate.

In our example

$$V_+ = 4+5+6+7+8+9+10 = 49$$

and

$$V_- = 1+2+3 = 6.$$

Suppose we take the smaller, $V_-$, as the test statistic. Relatively small or large values of $V_-$ would make us reject $H_1$ and so observations which produce such values would go into the critical region. To determine the critical region or alternatively the significance probability we need to derive the distribution of $V_-$ under the null hypothesis $H_1$. Under $H_1$ each observation has probability

$\frac{1}{2}$ of being above or below the median $m_0$; so each of the $z_i$ values (and correspondingly each of the ranks) has probability $\frac{1}{2}$ of being $+$ or $-$. Each of the $2^n$ different possible arrangements of signs for the $z$ values thus has the same probability of $1/2^n$ under $H_1$. To obtain the observations which fall in the critical region we need to enumerate all possible cases which lead to extreme values of $V_-$. Alternatively we can derive the significance probability.

Thus in our example

$$P(V_- \leq 6|H_1) = \frac{\text{No. of ways } V_- \leq 6}{2^{10}}.$$

We evaluate the numerator as follows.

| Value of $V_-$ | Set of ranks |
|:---:|:---:|
| 0 | Empty set |
| 1 | $\{1\}$ |
| 2 | $\{2\}$ |
| 3 | $\{3\}, \{1, 2\}$ |
| 4 | $\{4\}, \{1, 3\}$ |
| 5 | $\{5\}, \{1, 4\}, \{2, 3\}$ |
| 6 | $\{6\}, \{1, 5\}, \{2, 4\}, \{1, 2, 3\}.$ |

Thus there are 14 ways in which $V_- \leq 6$, and so

$$P(V_- \leq 6|H_1) = \frac{14}{2^{10}} = 0 \cdot 0137.$$

For a two-sided test the significance probability is of the form

$$P(V_- \leq 6 \text{ or } V_- \geq 55 - 6 = 49|H_1).$$

Here the greatest value of $V_-$ is $1 + 2 + \ldots + 10 = 55$. Since under $H_1$ the distribution of $V_-$ is symmetric we obtain a significance probability of $2 \times 0 \cdot 0137 = 0 \cdot 027$. Hence at the 5% level of significance we would reject the null hypothesis that the median is $5 \cdot 0$ in favour of the hypothesis $H_2$. From the magnitude of $V_-$ we would conclude that the median was larger than $5 \cdot 0$.

This procedure is somewhat tedious in general. Fortunately tables are available which provide the critical values of $V_-$ (or $V_+$) in the critical region: see, for example, Siegel (1956, Table $G$, p. 254); or the distribution function of $V_-$ (or $V_+$): see for example, Owen (1962, Table 11.1, pp. 325–330).

For larger values of $n(>20)$ we may use the following Normal approximation for $V_-$, namely under $H_1$:

> $V_-$ is approximately Normal with expectation $n(n+1)/4$ and variance $n(n+1)(2n+1)/24$.            (14.4.2)

$V_+$ has an identical distribution under $H_1$.

In the event of ties if we assign average ranks where relevant, the effect on the distribution of $V_-$ (or $V_+$) is negligible unless the proportion of ties is very large. A correction to the approximation in (14.4.2) is to reduce the variance by

$$\frac{1}{48} \sum_{j=1}^{T} t_j(t_j^2 - 1),$$

where there are $T$ sets of ties and the $j$th set consists of $t_j$ observations $(j = 1, 2, \ldots, T)$.

## 14.5. MATCHED PAIRS

Because of the wide variability in experimental units it is often advantageous in the comparison of two treatments or methods to match the units in pairs, and apply one treatment to each unit [cf. § 9.3]. In this way a better comparison may be obtained in that any difference will be due (it is to be hoped) to differences in the treatments rather than differences in the units [see Example 5.8.1]. Alternatively in testing for the effect of a treatment an individual may act as his own control, and, for example, provide a measurement before treatment and one after treatment.

In either case we obtain pairs of related measurements $(x_1, y_1)$, $(x_2, y_2), \ldots, (x_n, y_n)$ as our observations and we are led to consider matched pairs samples. For such data the usual technique is to consider the differences

$$d_i = x_i - y_i \qquad (i = 1, 2, \ldots, n).$$

In a Normal parametric framework the paired sample $t$-test [see Example 5.8.1] may be appropriate for testing for equality of expected values or of zero expected value for the random variable $D = X - Y$. In the non-parametric framework we can employ the sign test and the Wilcoxon signed ranks test of section 14.4 applied to the random sample $d_1, d_2, \ldots, d_n$ of differences. The appropriate tests would specify the median value $m_0 = 0$ in the null hypothesis $H_1$.

## 14.6. TWO-SAMPLE TESTS

Suppose that $x_1, x_2, \ldots, x_{n_1}$ is a random sample from a population with continuous distribution function $F(x)$; and $y_1, y_2, \ldots, y_{n_2}$ is an independent random sample from a population with continuous distribution function $G(y)$. The *combined order statistic* of these two samples is simply the set of observations arranged in order of magnitude irrespective of sample.

EXAMPLE 14.6.1. *Combined order statistic of two samples.* Suppose that we observe the following random sample of size 7 from population I:

$$3 \cdot 7, \quad -1 \cdot 1, \quad 2 \cdot 6, \quad 2 \cdot 3, \quad 4 \cdot 1, \quad 0 \cdot 8, \quad 3 \cdot 9;$$

and a second independent random sample of size 5 from population II:

$$4·6, \quad 4·0, \quad 5·3, \quad 4·4, \quad 3·0.$$

We could represent the result diagrammatically as in Figure 14.6.1. The combined order statistic, together with the ranks of the observations, is then as shown below.

$$\begin{array}{ccccccccccccc}
& -1·1 & 0·8 & 2·3 & 2·6 & 3·0 & 3·7 & 3·9 & 4·0 & 4·1 & 4·4 & 4·6 & 5·3 \\
\text{Rank:} & 1 & 2 & 3 & 4 & 5 & 6 & 7 & 8 & 9 & 10 & 11 & 12
\end{array}$$

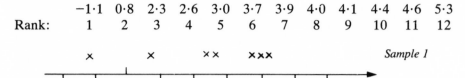

Figure 14.6.1: Observations in Example 14.6.1.

The three tests which we describe in this section tackle the question of whether the two populations have the same distribution centred around the same value. They are thus tests of the null hypothesis

$$H_1: F(z) = G(z) \text{ for all } z$$

against

$$H_2: F(z) \neq G(z) \text{ for at least some } z.$$

The tests are especially powerful against shift of location alternatives. In a Normal parametric framework the analogous test is the two sample $t$-test [see § 5.8.4].

### 14.6.1.  The Two-sample Median Test

The first test is based on the sample median of the combined order statistic, and may be considered as a generalization of the sign test (§ 14.4.1) to the case of two independent samples. We determine the number of values $m_1$ in the sample from population I which exceed the sample median.

EXAMPLE 14.6.2.  *The two-sample combined median test.*  The sample median of the combined order statistic of Example 14.6.1 is $\frac{1}{2}(3·7 + 3·9) = 3·8$. We may categorize the observations as follows.

|                  | Exceed median | Less than median | Totals |
|------------------|:---------:|:----------:|:------:|
| Sample from I    | 2         | 5          | 7      |
| Sample from II   | 4         | 1          | 5      |
| Totals           | 6         | 6          | 12     |

More generally for the case where $n_1 + n_2$ is even, we obtain a classification of the form

|  | Exceed median | Less than median | Totals |
|---|---|---|---|
| Sample from I | $m_1$ | $n_1 - m_1$ | $n_1$ |
| Sample from II | $m_2$ | $n_2 - m_2$ | $n_2$ |
| Totals | $\frac{1}{2}(n_1 + n_2)$ | $\frac{1}{2}(n_1 + n_2)$ | $n_1 + n_2$ |

For the case where $n_1 + n_2$ is odd, one of the observations will coincide with the sample median. It is common practice to ignore an observation which falls exactly on the median, and to reduce either $n_1$ or $n_2$ by 1 correspondingly. We will then be able to compile a table as above, and in fact here we proceed as if this has already been done.

If the two populations have the same medians we would expect the sample observations from each population to be similarly spread throughout the combined order statistic. The median test uses as test statistic the random variable $M_1$ which counts the number of observations in the sample from population I which exceed the sample median of the combined order statistic. If $H_1$ is true we would expect $M_1$ to have a distribution centred around $\frac{1}{2}n_1$. A simple combinatorial exercise reveals that

$$P(M_1 = m_1) = \frac{\binom{n_1}{m_1}\binom{n_2}{a - m_1}}{\binom{n_1 + n_2}{a}}, \qquad a = \tfrac{1}{2}(n_1 + n_2),$$

namely a hypergeometric distribution [see II, § 5.3], if $H_1$ is true. Values of $M_1$ remote from $\frac{1}{2}n_1$ will make us reject $H_1$ in favour of $H_2$, so that as critical region we would take

$$A_2 = \{m_1 : |m_1 - \tfrac{1}{2}n_1| \geq k\},$$

where $k$ is chosen so that the size of the test is at most $\alpha$.

Alternatively we may derive the significance probability of obtaining a result as extreme as the one we obtained, or more so, namely

$$P(M_1 \geq m_1 \text{ or } M_1 \leq n_1 - m_1) \quad \text{if } m_1 > \tfrac{1}{2}n_1,$$

or

$$P(M_1 \leq m_1 \text{ or } M_1 \geq n_1 - m_1) \quad \text{if } m_1 < \tfrac{1}{2}n_1.$$

This is an example of Fisher's exact test for a $2 \times 2$ table. (See § 5.4.2.)

If the sample sizes are large we may use instead the $\chi^2$ method (§ 7.2.1) to test the hypothesis $H_1$. This approximate test would be suitable if $n_1 + n_2 \geq 20$ and if the expected numbers in each cell are not too small (e.g. not less than 5; but see Cochran's criterion in § 7.5.1).

EXAMPLE 14.6.3 (continuation of Example 14.6.2).   We derive the significance probability of the sample quoted in Example 14.6.1. Sets of values as extreme as or more extreme than the observed one are, in similar tabular form, as follows.

| 1 | 6 |
|---|---|
| 5 | 0 |

| 2 | 5 |
|---|---|
| 4 | 1 |

| 5 | 2 |
|---|---|
| 1 | 4 |

| 6 | 1 |
|---|---|
| 0 | 5 |

The significance probability is then given by

$$\frac{\binom{7}{1}\binom{5}{5}}{\binom{12}{6}} + \frac{\binom{7}{2}\binom{5}{4}}{\binom{12}{6}} + \frac{\binom{7}{5}\binom{5}{1}}{\binom{12}{6}} + \frac{\binom{7}{5}\binom{5}{0}}{\binom{12}{6}}$$

$$= 0 \cdot 0076 + 0 \cdot 1136 + 0 \cdot 1136 + 0 \cdot 0076$$

$$= 0 \cdot 242.$$

This is a high probability. The data must be regarded as consistent with the null hypotheses $H_1$ that the underlying distributions for the two populations are the same. Alternatively we can see from the above probabilities that the critical region for a test at the 5% significance level is

$$A_2 = \{m_1 : m_1 = 1 \text{ or } 6\}.$$

Hence, since $m_1 = 2$, our observation does not fall in the critical region at the 5% level and we have the same conclusion as above.

### 14.6.2.   Wilcoxon–Mann–Whitney Test

The median test clearly throws away some of the information in the data. We can make more efficient use of the information in the combined order statistic by considering instead the ranks of observations. The following non-parametric test procedure was derived by Wilcoxon (1945) and Mann and Whitney (1947). We describe it in terms of the data of Example 14.6.3.

The combined order statistic with ranks, as we have seen, is

|       | −1·1 | 0·8 | 2·3 | 2·6 | 3·0 | 3·7 | 3·9 | 4·0 | 4·1 | 4·4 | 4·6 | 5·3 |
|-------|------|-----|-----|-----|-----|-----|-----|-----|-----|-----|-----|-----|
| Rank: | 1    | 2   | 3   | 4   | **5** | 6   | 7   | **8** | 9   | **10** | **11** | **12** |

We have indicated in bold type the ranks for the observations from population II (the smaller sample). The test statistic used in the Wilcoxon–Mann–Whitney test is the sum of the ranks from one sample. Here we could use

$$R_1 = 1+2+3+4+6+7+9 = 32$$

or

$$R_2 = 5+8+10+11+12 = 46$$

More generally, if $r_1, r_2, \ldots, r_{n_1}$ are the ranks of $x_1, x_2, \ldots, x_{n_1}$ and $s_1, s_2, \ldots, s_{n_2}$ are the ranks of $y_1, y_2, \ldots, y_{n_2}$ in the combined order statistic, then we can use either of the two test statistics

$$R_1 = r_1 + r_2 + \ldots + r_{n_1}$$

or

$$R_2 = s_1 + s_1 + \ldots + s_{n_2}.$$

(Notice that in general if we know the value of $R_1$ then we also know the value of $R_2$ since

$$R_1 + R_2 = 1 + 2 + \ldots + (n_1 + n_2)$$
$$= \tfrac{1}{2}(n_1 + n_2)(n_1 + n_2 + 1)$$

Since the two test statistics are equivalent it is perhaps simpler to use the one from the smaller population.)

If the null hypothesis $H_1$ that $F$ and $G$ are the same distribution functions is true, then we should not expect a preponderance of observations from one population in either end of the combined order statistic: the values should be dispersed throughout the combined order statistic. For the general alternative hypothesis $H_2$, relatively large or relatively small values of the test statistic ($R_2$, say,) would make us doubt the null hypothesis $H_1$. Hence we take as the critical region the set

$$A_2 = \{\text{observations: } R_2 \le c_1 \text{ or } R_2 \ge c_2\}, \qquad (14.6.1)$$

where $c_1$ and $c_2$ are constants chosen to make the size of $A_2$ at most $\alpha$. Alternatively we could evaluate the significance probability of observing a result as extreme as or more extreme than the one which we observed.

EXAMPLE 14.6.4.   *Significance probability in a Wilcoxon–Mann–Whitney test.*   Under $H_1$, $R_2$ has a symmetric distribution with expected value $\tfrac{1}{2}n_2(n_1 + n_2 + 1) = 32.5$ (see below). Thus the significance probability is

$$P(R_2 \ge 46 \text{ or } R_2 \le 19 | H_1).$$

If $H_1$ is true the $n_2 = 5$ ranks of the $y$'s correspond to a random sample of size $n_2 = 5$ from the integers $1, 2, \ldots, n_1 + n_2 = 12$. Since all possible combined order statistics are equally likely under $H_1$, the significance probability is

$$\frac{\text{No. of ways } R_2 \geq 46 \text{ or } \leq 19}{\binom{12}{5}}. \tag{14.6.2}$$

The evaluation of this probability can be tackled directly as follows by simply listing all the ways in which $R_2 \geq 46$.

$$12+11+10+9+8 = 50 \qquad\qquad 12+11+10+9+4 = 46$$

$$12+11+10+9+7 = 49 \qquad\qquad 12+11+10+8+5 = 46$$

$$12+11+10+9+6 = 48 \qquad\qquad 12+11+10+7+6 = 46$$

$$12+11+10+8+7 = 48 \qquad\qquad 12+11+ 9+8+6 = 46$$

$$\qquad\qquad\qquad\qquad\qquad\qquad\qquad 12+10+ 9+8+7 = 46$$

$$12+11+10+9+5 = 47$$

$$12+11+10+8+6 = 47$$

$$12+11+ 9+8+7 = 47$$

Hence we obtain the number of ways in which $R_2 \geq 46$ to be 12. Since $R_2$ is symmetric under $H_1$, the number of ways in which $R_2 \leq 19$ is the same, and so the significance probability (14.6.2) is given by

$$\frac{2 \times 12}{\binom{12}{5}} = 0 \cdot 030.$$

This is a sufficiently low probability to provide evidence of moderate strength against the null hypothesis that the two populations have the same underlying distribution. Notice the comparison with the sign test, where by summarizing the data too crudely we were unable to obtain evidence against $H_1$.

This tedious procedure may be avoided by consulting the tables that are available for the critical values $c_1$ and $c_2$ in (14.6.1); see, for example, Owen (1962, Table 11.5, pp. 354–361).

Some tables use the alternative test statistic $U$ defined by

$$U = n_1 n_2 + [\tfrac{1}{2} n_1 (n_1 + 1)] - R_1$$

or

$$U = n_1 n_2 + [\tfrac{1}{2} n_2 (n_2 + 1)] - R_2.$$

Here $U$ measures the number of times an observation in the larger sample precedes an observation in the smaller sample in the combined order statistic; for tables involving $U$, see, for example, Siegel (1956, Tables J, K, pp. 271–277) or Owen (1962, Tables 11.2–11.4, pp. 331–353).

For large samples we may use Normal approximations for $R_1$, $R_2$ (or $U$), and these approximations work well for $n_1$, $n_2$ as small as 7. Under the null hypothesis $H_1$ we expect the average rank of an observation from population I to be approximately equal to the average rank of an observation from population II. Since the total sum of ranks is $\frac{1}{2}(n_1 + n_2)(n_1 + n_2 + 1)$, then the average rank is $\frac{1}{2}(n_1 + n_2 + 1)$. Hence we expect that

$$E(R_1) = \tfrac{1}{2}n_1(n_1 + n_2 + 1) \quad \text{and} \quad E(R_2) = \tfrac{1}{2}n_2(n_1 + n_2 + 1).$$

It can be shown that

$$\text{var}\,(R_1) = \text{var}\,(R_2) = n_1 n_2(n_1 + n_2 + 1)/12. \tag{14.6.3}$$

In our example the sample sizes are somewhat small to expect the approximation to be very accurate; but as an illustration we see that $R_2$ is approximately Normal $(32.5, \sqrt{37 \cdot 9167})$ if $H_1$ is true. Hence approximately

$$P(R_2 \geq 46 | H_1) = 1 - \Phi\left(\frac{46 - 32 \cdot 5}{6 \cdot 1577}\right)$$

$$= 1 - \Phi(2 \cdot 192)$$

$$= 0 \cdot 0142.$$

The significance probability is therefore $2 \times 0 \cdot 0142 = 0 \cdot 028$ which compares with the exact value of $0 \cdot 030$ derived earlier.

It is possible that ties may occur among the observations, and we will assign average ranks where appropriate in the combined order statistic. If the ties occur between observations from within one sample then there will be no effect on $R_1$ or $R_2$. If the ties occur between observations from different samples the effect will in general be small, but a correction factor in the normal approximation is to multiply the variance in (14.6.3) by

$$1 - \frac{\sum_{j=1}^{T} t_j(t_j^2 - 1)}{(n_1 + n_2)\{(n_1 + n_2)^2 - 1\}},$$

where there are $T$ sets of ties and the $j$th set consists of $t_j$ observations $(j = 1, 2, \ldots, T)$.

One-sided tests could be derived for situations in which the alternative $H_2$ specifies a shift in location difference between $F$ and $G$ in a particular direction.

### 14.6.3. Runs

Another two-sample distribution-free test for the equality of two underlying distributions uses the concept of *runs* in the combined order statistic. This test again uses the data via the combined order statistic but summarizes it in yet another way.

EXAMPLE 14.6.5. *The data of Example 14.6.1 examined for runs.* In the combined order statistic, instead of recording the ranks we simply note the population from which each observation comes:

| | −1·1 | 0·8 | 2·3 | 2·6 | 3·0 | 3·7 | 3·9 | 4·0 | 4·1 | 4·4 | 4·6 | 5·3 |
|---|---|---|---|---|---|---|---|---|---|---|---|---|
| Population: | I | I | I | I | II | I | I | II | I | II | II | II |

We use as test statistic the number $W$ of runs of observations in the combined order statistic from the same population. In the Example, $W = 6$.

If the null hypothesis $H_1$ that the two underlying distributions are equal is true, the observations from I and II should be well mixed, and the total number of runs should be 'large'. On the other hand if the two distributions are different, for example if they are widely separated or if one is well dispersed and the other fairly compact, then $W$ is likely to be 'small'.

The exact distribution of $W$ if $H_1$ is true can be determined by some combinatorial analysis (see, for example Siegel (1956, p. 138)) and the critical region is of the form

$$A_2 = \{\text{observations: } w \le c_\alpha\}.$$

Tables of the critical $c_\alpha$ values are available; see, for example, Siegel (1956, Table F, pp. 252–253) or Owen (1962, Tables 12.4, 12.5, pp. 373–382).

EXAMPLE 14.6.6 (continuation of Example 14.6.5. *Significance of the sample*). From the table of critical values for a test at the 5% significance level for $n_1 = 7$, $n_2 = 5$, we obtain

$$A_2 = \{\text{observations: } w \le 3\}.$$

Since we observed $w = 6$, our value of $W$ lies outside the critical region and so we have no evidence to reject $H_1$ at the 5% level of significance.

Again a Normal approximation is available which uses the fact that if $H_1$ is true, then $W$ is approximately Normal, with

$$E(W) = 1 + 2n_1 n_2 / (n_1 + n_2)$$

and

$$\text{var}(W) = \frac{2n_1 n_2 (2n_1 n_2 - n_1 - n_2)}{(n_1 + n_2)^2 (n_1 + n_2 - 1)}.$$

This approximation requires each of $n_1$ and $n_2$ to equal at least 20 for reasonable accuracy.

The problem of ties in the runs test can cause difficulties if the ties occur between observations from different samples. The interested reader is referred to Siegel (1956, p. 143).

It is of interest to note that the three tests in sections 14.6.1, 14.6.2 and 14.6.3 do not all lead to the same conclusion. By summarizing the combined order statistics in different ways in the three tests we may reach different conclusions. In general it is believed that the Wilcoxon–Mann–Whitney test is the most sensitive, or, equivalently, is the most powerful [see § 5.3.1], at least for the case of an alternative hypothesis specifying a shift of location.

## 14.7. SEVERAL SAMPLES

We now extend the ideas to the case where we want to compare several samples. These tests study situations analogous to the one-way analysis of variance models in the Normal parametric framework [see § 5.8.7].

We have $n_i$ observations from the $i$th of $k$ populations ($i = 1, 2, \ldots, k; \sum_{i=1}^{k} n_i = n$) and wish to test the null hypothesis

$H_1$: all the populations have the same underlying distribution

against

$H_2$: they do not.

As with the median test (§ 14.6.1) and the Wilcoxon–Mann–Whitney test (§ 14.6.2), the extensions which we consider have particular relevance to situations where the alternative $H_2$ specifies shift of medians (or location) rather than shape.

### 14.7.1. Median Test

The two-sample median test of section 14.6.1 can be generalized in a straightforward manner to $k$ samples. The combined order statistic is again obtained and the sample median is determined. Then for each sample we classify each observation as being above or below this sample median. Again by convention we ignore any observations which actually equal the sample median and reduce the relevant sample sizes.

EXAMPLE 14.7.1. *A three-sample median test.* Suppose we have $k = 3$ populations with random samples as follows:

　　　From I:　　21, 50, 6, 69, 42, 34, 26, 57, 14, 31;

　　　From II　　10, 49, 22, 40, 24, 54, 12, 29, 25, 17, 32, 61;

　　　From III:　3, 15, 9, 18, 1, 33, 11, 5, 16, 30, 41.

Here $n_1 = 10$, $n_2 = 12$, $n_3 = 11$. The combined order statistic can be determined and the sample median, here the seventeenth value, is found to be 25. The classification of observations is given below. Notice that we ignore the observation 25 in sample II and reduce $n_2$ by 1.

|  |  | Exceed median | Less than median | Totals |
|---|---|---|---|---|
|  | I | 7 | 3 | 10 |
| Sample | II | 6 | 5 | 11 |
|  | III | 3 | 8 | 11 |
| Totals |  | 16 | 16 | 32 |

If $H_1$ is true we expect about half of each sample from each population to be less than the overall sample median and about half to be greater. Provided each sample size is greater than about 10 we can use the $\chi^2$ goodness of fit test as described in section 7.5.2. for a $k \times 2$ contingency table. With test statistic

$$X^2 = \sum_{\substack{\text{all} \\ \text{cells}}} \frac{(\text{observed})^2}{\text{expected}} - n,$$

we know that $X^2$ will have a $\chi^2(k-1)$ distribution if $H_1$ is true, and that large values of $X^2$ will imply departure from $H_1$.

EXAMPLE 14.7.2.   *The $\chi^2$ approximation to the procedure of Example 14.7.1.* In our example the expected values under $H_1$ are

|        |     | Exceed median | Less than median |
|--------|-----|---------------|------------------|
|        | I   | 5             | 5                |
| Sample | II  | 5·5           | 5·5              |
|        | III | 5·5           | 5·5              |

Hence

$$X^2 = \frac{7^2}{5} + \frac{3^2}{5} + \frac{6^2}{5\cdot5} + \frac{5^2}{5\cdot5} + \frac{3^2}{5\cdot5} + \frac{8^2}{5\cdot5} - 32$$

$$= 3\cdot96.$$

Since for $k = 3$ the critical region of size $0\cdot05$ is of the form

$$A_2 = \{\text{observations:} \ X^2 > \chi^2_{0.95}(2) = 5\cdot99\},$$

where $\chi^2_{0.95}(2)$ denotes the 95% percentile of the $\chi^2$ distribution on 2 d.f., we see that we have no evidence at the 5% level of significance to reject the null hypothesis $H_1$ of equality in the underlying distributions.

## 14.7.2.   Kruskal–Wallis Test

Again we can make better use of the information in the combined order statistic. As with the Wilcoxon–Mann–Whitney test in section 14.6.2. for two samples we use the rankings of the observations within the combined order statistic.

For each of the $k$ samples we evaluate the sums of the ranks

$$R_i = \text{sum of ranks of elements in the } i\text{th sample} \ (i = 1, 2, \ldots, k).$$

EXAMPLE 14.7.3. *The Kruskal–Wallis treatment of the data of Example 14.7.1.* The combined order statistic together with the ranks for the data in the Example are as follows:

| | 1 | 3 | 5 | 6 | 9 | 10 | 11 | 12 | 14 | 15 | 16 |
|---|---|---|---|---|---|---|---|---|---|---|---|
| Rank | 1 | 2 | 3 | ④ | 5 | 6 | 7 | 8 | ⑨ | 10 | 11 |

| | 17 | 18 | 21 | 22 | 24 | 25 | 26 | 29 | 30 | 31 | 32 |
|---|---|---|---|---|---|---|---|---|---|---|---|
| Rank | 12 | 13 | ⑭ | 15 | 16 | 17 | ⑱ | 19 | 20 | ㉑ | 22 |

| | 33 | 34 | 40 | 41 | 42 | 49 | 50 | 54 | 57 | 61 | 69 |
|---|---|---|---|---|---|---|---|---|---|---|---|
| Rank | 23 | ㉔ | 25 | 26 | ㉗ | 28 | ㉙ | 30 | ㉛ | 32 | ㉝ |

Here we have shown the ranks for population I as ④, for example, the ranks for population II as **6**, for example, and the ranks for population III as 1, for example. Then

$$R_1 = 4 + 9 + \ldots + 33 = 210,$$

$$R_2 = 6 + 8 + \ldots + 32 = 230,$$

$$R_3 = 1 + 2 + \ldots + 26 = 121.$$

For each individual in a particular sample we could evaluate the average rank $R_i/n_i$ $(i = 1, 2, \ldots, k)$. If $H_1$ is true and the underlying populations are identical we would expect all the average ranks to be about the same, and indeed we would expect them to be approximately equal to the overall average rank

$$\bar{R} = \frac{1 + 2 + \ldots + n}{n} = \tfrac{1}{2}(n + 1).$$

As a test statistic we therefore use a measure which is sensitive to the deviations of $R_i/n_i$ from $\bar{R}$, and the Kruskal–Wallis test statistic is

$$K = \sum_{i=1}^{k} n_i \left( \frac{R_i}{n_i} - \bar{R} \right)^2$$

$$= \sum_{i=1}^{k} \frac{R_i^2}{n_i} - \frac{n(n+1)^2}{4}.$$

Clearly large values of $K$ will be significant, so that the critical region is of the form

$$A_2 = \{\text{observations: } K > \text{constant}\}.$$

Again some tables are available for exact critical values of $K$ for small sample sizes; for example both Siegel (1956, Table 0, pp. 282–283) and Owen (1962, Table 14.2, pp. 420–422) consider the case for $k = 3$ and $n_i \le 5$ $(i = 1, 2, 3)$. If nearly all $n_i > 5$ we may use a convenient approximation which states that

$$\frac{12}{n(n+1)} K \quad \text{is } \chi^2(k-1) \quad \text{if } H_1 \text{ is true.}$$

If ties occur between the observations we again assign the average ranks to the corresponding observations in the combined order statistic. A small number of ties will have little effect on the test statistic $K$; however a satisfactory correction factor in the $\chi^2$ approximation is obtained by replacing $K$ by $K/a$, where

$$a = 1 - \frac{\sum_{j=1}^{T} t_j(t_j^2 - 1)}{n(n^2 - 1)}$$

where there are $T$ sets of ties and the $j$th set consists of $t_j$ observations $(j = 1, 2, \ldots, T)$.

EXAMPLE 14.7.4.   *The $\chi^2$ approximation to the procedure of Example 14.7.3.*   Here $n_1 = 10$, $n_2 = 12$, $n_3 = 11$, so we make use of the $\chi^2$ approximation. For a test of size 0·05 we have a critical region of the form

$$A_2 = \left\{ \text{observations: } K > \frac{33 \times 34}{12} \chi_{0.95}^2(2) = 560 \cdot 1 \right\}.$$

For our data

$$K = \frac{(210)^2}{10} + \frac{(230)^2}{12} + \frac{(121)^2}{11} - \frac{33(34)^2}{4}$$

$$= 612 \cdot 3.$$

Hence our observation falls in the critical region and so we reject the null hypothesis at the 5% level and decide that the underlying distributions in the three groups differ. Notice once again that by using the information from the combined order statistic in a more efficient way we have reached a different conclusion to that in section 14.7.1.

## 14.8.   RANDOMIZATION TESTS

In the tests introduced in sections 14.3 to 14.7, we condensed the information in the sample data by some summary of the combined order statistic, thus introducing the possibility of a loss of efficiency due to not using the actual observed values. Randomization tests [see § 5.7] are non-parametric techniques which make use of the actual observed (numerical) values and appeal to the random nature of the observations. They are useful for small sample sizes, but can become quite cumbersome as the sample sizes increase [see Example 5.7.1]. The underlying techniques used in the exact Wilcoxon signed ranks test (§ 14.4.2) and Wilcoxon–Mann–Whitney test (§ 14.6.2) are employed; indeed these two tests can be viewed as randomization tests applied to the ranks. We illustrate the underlying ideas for the two-sample situation. Details of applications in other situations can be found, for example, in Siegel (1956, pp. 88, 152), Silvey (1975, pp. 148–150).

EXAMPLE 14.8.1.   *Randomization test for two samples.*   Suppose that we wish to assess the effects of two treatments I and II with 8 experimental units. We allocate at random $n_1 = 4$ of the units to treatment I, the remaining $n_2 = 4$ being given treatment II. The responses are as shown.

$$\text{Treatment I:} \quad 3, \quad 8, \quad 1, \quad 5$$
$$\text{Treatment II:} \quad 10, \quad 4, \quad 6, \quad 14.$$

We wish to test the null hypothesis $H_1$ that the difference in the treatments has no effect, against the alternative $H_2$ that treatment II induces higher responses. If $H_1$ were true, we would have obtained the eight responses shown above, however the treatments had been allocated. In fact there are

$$\binom{n_1 + n_2}{n_1} = \binom{8}{4} = 70$$

different possible groupings into two sets of four. Under the random allocation scheme each possible grouping is equal likely, and so has probability

$$1 \bigg/ \binom{n_1 + n_2}{n_1} = \frac{1}{70}.$$

If $S_2$ is the sum of the responses with treatment II, then relatively large values of $S_2$ will make us doubt the null hypothesis. Hence we would use a critical region of the form

$$A_2 = \{\text{observations: } S_2 > c\},$$

or evaluate the significance probability of observing a result as extreme as or more extreme than the one we observed.

Here $S_2 = 10 + 4 + 6 + 14 = 34$, and the significance probability is given by

$$\frac{\text{No. of ways } S_2 \geq 34}{70}.$$

Below we see that there are six ways in which we could have observed $S_2 \geq 34$ for these data points, namely

$$14 + 10 + 8 + 6 = 38,$$
$$14 + 10 + 8 + 5 = 37,$$
$$14 + 10 + 8 + 4 = 36,$$
$$14 + 10 + 8 + 3 = 35,$$
$$14 + 10 + 6 + 5 = 35,$$
$$14 + 10 + 6 + 4 = 34.$$

Hence the significance probability is $6/70 = 0\cdot086$. This relatively high probability indicates that the data are consistent with the null hypothesis of no treatment difference.

For larger $n_1$, $n_2$ this procedure would obviously become somewhat tedious. Siegel (1956, p. 154) discusses an approximate solution. Alternatively the Wilcoxon–Mann–Whitney approximation may be used with satisfactory results.

## 14.9.   RANK CORRELATION MEASURES

In parametric models we have encountered the concepts of covariance and correlation [see (2.1.8)], which measure the degree of association between pairs of random variables from a bivariate distribution [see II, § 6.1]. We can derive estimates of these measures from a random sample of observations; for example, recall the sample product-moment correlation coefficient (2.1.9). We can derive analogous measures of association in the distribution-free framework by using the concept of rank from the orderings of the observations. As examples we derive two such measures, namely Spearman's rank correlation coefficient and Kendall's rank correlation coefficient. For more details of these and other measures the interested reader is referred to Kendall (1962).

Suppose we have $n$ pairs of observations $(x_1, y_1), (x_2, y_2), \ldots, (x_n, y_n)$ which form a random sample from some (unspecified) bivariate continuous distribution. We determine the order statistic of $x_1, x_2, \ldots, x_n$; and separately determine the order statistic of $y_1, y_2, \ldots, y_n$. If the two variables are highly dependent or have a high degree of association, we would expect the two rankings for the two elements in each pair to be roughly the same (or perhaps in the reverse order if the correlation is negative). If on the other hand there is no dependence we would not expect any such sort of trend.

### 14.9.1.   Spearman's Rank Correlation Coefficient

Spearman's rank correlation coefficient is a simple modification of Pearson's correlation coefficient (2.1.19), in which the $x_i$'s and $y_i$'s are replaced by their ranks. Using the knowledge that the ranks are $1, 2, \ldots, n$ in some order for each variable, we can show with some elementary manipulation that Spearman's rank correlation coefficient $r_s$ reduces to

$$r_s = 1 - \frac{6 \sum_{i=1}^{n} d_i^2}{n(n^2 - 1)}, \tag{14.9.1}$$

where

$$d_i = \text{rank } (x_i) - \text{rank } (y_1), \qquad (i = 1, 2, \ldots, n).$$

Strictly speaking a correction factor is necessary in (14.9.1) if there are any ties in the two ranking sets, but the correction has negligible effect unless the proportion of ties becomes relatively large.

The coefficient has the property that $-1 \le r_s \le 1$. We obtain values near $+1$ if larger $x$'s go with larger $y$'s; and values near $-1$ if larger $x$'s go with smaller $y$'s. Under the null hypothesis $H_1$ that the random variables are not associated, a property of the sampling distribution of $r_s$ is that its expectation is zero.

Relevant tables of the exact distribution of $r_s$ under $H_1$ are available; see, for example, Siegel (1956, Table $P$, p. 284) and Owen (1962, Table 13.2, pp. 400–406). For $n \geq 10$ a suitable test of $H_1$ uses the fact that

$$r_s \sqrt{\frac{n-2}{1-r_s^2}}$$

has approximately Student's distribution on $n-2$ degrees of freedom if $H_1$ is true, (see § 2.5.5). The critical region of size $\alpha$ for the usual test of $H_1$ against the alternative hypothesis $H_2$ that the variables are associated is then of the form

$$A_2 = \left\{ \text{observations: } |r_s| \sqrt{\frac{n-2}{1-r_s^2}} > t_\alpha(n-2) \right\}.$$

where $t_\alpha(n-2)$ is the $(1-\tfrac{1}{2}\alpha)$ percentile of Student's distribution on $n-2$ d.f. One could derive similar critical regions for one-sided tests in which $H_2$ specifies a positive (negative) association.

EXAMPLE 14.9.1.   *Spearman's rank correlation coefficient.*   Suppose that the scores of seven individuals in two tests are as follows:

| Individual | 1 | 2 | 3 | 4 | 5 | 6 | 7 |
|---|---|---|---|---|---|---|---|
| Test 1 | 31 | 82 | 25 | 26 | 53 | 30 | 29 |
| Test 2 | 21 | 55 | 8 | 27 | 32 | 42 | 26. |

The ranks are as follows:

| Individual | 1 | 2 | 3 | 4 | 5 | 6 | 7 |
|---|---|---|---|---|---|---|---|
| Test 1 | 5 | 7 | 1 | 2 | 6 | 4 | 3 |
| Test 2 | 2 | 7 | 1 | 4 | 5 | 6 | 3 |
| Difference, $d_i$ | 3 | 0 | 0 | −2 | 1 | −2 | 0. |

Hence

$$\Sigma d_i^2 = 18$$

and

$$r_s = 0 \cdot 68.$$

The significance probability $P(|r_s| \geq 0 \cdot 68)$ is, from the quoted tables, given by $2 \times 0 \cdot 055 = 0 \cdot 11$. This is a high probability and the test indicates that the data are consistent with the null nypothesis $H_1$, that is, the hypothesis that the scores from the two tests are not related to one another.

## 14.9.2.   Kendall's Rank Correlation Coefficient

Kendall's rank correlation coefficient, $\tau$, is an alternative measure of correlation which is derived as follows. Suppose that we rank the $x$-values and look at the corresponding ordering of the $y$'s.

EXAMPLE 14.9.2.   *Kendall's rank correlation coefficient.*   If we arrange the individuals in order of ranks of the *x*-variable we have

| Individual    | 3 | 4 | 7 | 6  | 1 | 5 | 2 |
|---------------|---|---|---|----|---|---|---|
| Rank: Test 1  | 1 | 2 | 3 | 4  | 5 | 6 | 7 |
| Rank: Test 2  | 1 | 4 | 3 | 6  | 2 | 5 | 7 |
| Score         | 6 | 1 | 2 | −1 | 2 | 1 | 0 |

We consider the corresponding *y* rankings as follows to derive the scores as shown. For each individual in turn we score +1 if any of the rankings to the right are larger, in other words if the pair of rankings is in the correct order; similarly we score −1 for any rankings to the right which are smaller. Thus, for example for individual 4 we score

$$(-1)+1+(-1)+1+1=1;$$

whilst for individual 6 we score

$$(-1)+(-1)+1=-1.$$

Kendall's $\tau$ is defined by

$$\tau = \frac{\text{sum of scores}}{\frac{1}{2}n(n-1)}, \tag{14.9.2}$$

where the denominator is in fact the maximum possible score which would occur if we had perfect agreement between the rankings. Notice that $\tau$ must satisfy the relation $-1 \le \tau \le 1$, and indeed $\tau$ has similar properties to $r_s$. For example, under the null hypothesis $H_1$ of no dependence we would expect about equal numbers of +1's and −1's, and so we would expect the sum of scores to be about 0. The critical region of size $\alpha$ of $H_1$ against the alternative hypothesis $H_2$ that the variables are associated is of the form

$$A_2 = \{\text{observations: } |\tau| > c_\alpha\}.$$

To determine the constant $c_\alpha$ relevant tables of the exact distribution of $\tau$ under the null hypothesis $H_1$ are available; see, for example, Siegel (1956, Table Q, p. 285) or Owen (1962, Table 13.1, pp. 396–399).

For $n \ge 10$ we may use the fact that, under $H_1$, $\tau$ is approximately normal, with expected value equal to zero and variance equal to

$$\frac{2(2n+5)}{9n(n-1)}.$$

In the event of ties in our rankings we need to redefine $\tau$ in (14.9.2). In our scoring system we allocate 0 to any pair of ranks which are the same, and define the rank correlation coefficient by

$$\tau = \frac{\text{sum of scores}}{\{\frac{1}{2}n(n-1)-\sum_{j=1}^{T} t_j(t_j-1)\}^{1/2}\{n(n-1)-\sum_{j=1}^{s} s_j(s_j-1)\}^{1/2}}$$

where there are $T$ sets of ties in the $x$-rankings and the $j$th set consists of $t_j$ observations ($j = 1, 2, \ldots, T$); and $S$ sets of ties in the $y$-rankings and the $j$th set consists of $s_j$ observations ($j = 1, 2, \ldots, s$).

EXAMPLE 14.9.3 (continuation of Example 14.9.2).  Here the sum of scores is 11, which gives a value of $\tau = 0 \cdot 52$. Using the tables quoted we find that the significance probability is $2 \times 0 \cdot 068 = 0 \cdot 14$, so that again at the 5% level of significance there is no reason to reject the null hypothesis $H_1$ of no dependence.

I.R.D.

## 14.10.   FURTHER READING AND REFERENCES

Texts dealing with inference often have a section on distribution-free ('nonparametric') methods; see for example Chapter 9 of Silvey (1975—See Bibliography C); but, in addition, as was pointed out in section 14.1, there are a large number of excellent specialist works, some of the better ones are listed below, together with references to particular methods and tables mentioned in the chapter.

A recent publication by Neave (1978) contains many of the tables relevant to this chapter, and so provides a very convenient handbook.

Kendall, M. G. (1962). *Rank Correlation Methods*, Griffin, London.

Hollander, M. and Wolfe, D. A. (1973). *Nonparametric Statistical Methods*, John Wiley & Sons, New York.

Lehmann, E. L. (1975). *Nonparametrics: Statistical Methods Based on Ranks*, Holden-Day, San Francisco.

Mann, H. B. and Whitney, D. R. (1947). On a Test of Whether One of Two Random Variables is Stochastically Larger than the Other, *Ann. Math. Statist.* **18**, 50–60.

Massey, F. J. (1952). Distribution Table for the Deviation Between Two Sample Cumulatives, *Ann. Math. Statist.* **23**, 435–441.

Neave, H. R. (1978). *Statistical Tables*. George Allen & Unwin, London.

Noether, G. E. (1967). *Elements of Nonparametric Statistics*, John Wiley & Sons, New York.

Siegel, S. (1956). *Nonparametric Statistics for the Behavioural Sciences*, McGraw-Hill, New York.

Walsh, J. E. (1962). *Handbook of Nonparametric Statistics*, Vol. I, Van Nostrand, Princeton, N.J.

Walsh, J. E. (1965). *Handbook of Nonparametric Statistics*, Vol. II. Van Nostrand, Princeton, N.J.

Walsh, J. E. (1968). *Handbook of Nonparametric Statistics*, Vol. III, Van Nostrand, Princeton, N.J.

Wilcoxon, F. (1945). Individual Comparisons by Ranking Methods, *Biometrics Bulletin* **1**, 80–83.

# CHAPTER 15

# Bayesian Statistics

## 15.1 INTRODUCTION

In this chapter, we shall attempt to provide basic answers to the following questions.

What are Bayesian methods, and what distinguishes them from other statistical methods?

Are Bayesian methods applicable to all types of statistical problem?

How do the results obtained using Bayesian methods differ from those obtained using other methods?

In fact, not even the first of these questions is entirely straightforward to answer. There is no official body to lay down precisely what should or should not be included under the heading of Bayesian methods, and there are considerable differences of opinion, both philosophical and practical, among those who might, collectively, be referred to as 'Bayesian' statisticians. However, some attempt at a general categorization is obviously required, and we shall proceed on the basis of the following:

*Bayesian methods are those which result from a systematic attempt to formulate and solve statistical problems using Bayes' Theorem.*

We therefore begin with a detailed examination of *Bayes' Theorem* in both the *discrete* case (§ 15.2) and the *continuous* case (§ 15.3).

The theorem itself expresses a relationship among various probabilities, and the specification of certain of these is a prerequisite for its use. Consideration of some simple examples reveals that a statistician's willingness to make systematic use of Bayes' Theorem depends, essentially, on what he regards as a legitimate interpretation of the concept of probability. Those for whom probability can legitimately be regarded as a measure of *degree of belief* [see de Finetti (1974, 1975)] capable of being assessed for any uncertain entity whatsoever, see no obstacle to the systematic use of Bayes' Theorem. Those, on the other hand, who regard the *frequentist* interpretation of probability as the sole legitimate one [see II, Chapter 2] regard Bayes' Theorem as of limited use, applicable only in those situations where the probabilities to be specified can be based upon observed frequencies. The answer to the second of the questions we posed is thus seen to depend, in part, on whether one can, in some sense, 'justify' the claims that degrees of belief can and should be thought of as probabilities and assigned numerical values. These problems will be examined in Chapter 19.

Irrespective of the methods employed for their solution, many statistical problems have, essentially, the following basic feature in common: prior to obtaining a particular set of data, a number of probability models are considered as possibly appropriate for the situation under study; given the data, some form of learning about the relative appropriateness of these models takes place.

What distinguishes the Bayesian paradigm (§§ 15.2–15.4) from other statistical approaches is that, *prior to obtaining the data, the statistician considers his degrees of belief for the possible models and represents them in the form of probabilities.* Once the data are obtained, Bayes' Theorem enables the statistician to calculate a new set of probabilities, which represent revised degrees of belief in the possible models, taking into account the new information provided by the data. For a given set of possible models, the basic process underlying Bayesian methodology is summarized schematically in Figure 15.1.1.

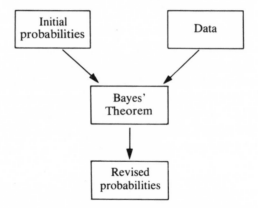

Figure 15.1.1: The basic Bayesian paradigm.

It is fairly standard terminology to refer to the initial probabilities as *prior probabilities* (since they are assessed before obtaining the data), and to refer to the revised probabilities as *posterior probabilities* (since they are calculated after obtaining the data). Of course, the terms prior and posterior are *relative* to a particular set of data; today's beliefs are posterior to yesterday's data, but prior to tomorrow's.

To answer the third of the questions that we posed initially, we shall need to consider the following topics:

(a) particular forms of inferential problems, such as point and interval estimation [see § 3.1], and more general ideas, such as those relating to sufficient statistics [see § 3.4];

(b) the need for technical simplifications and approximations in order that the paradigm can actually be used to obtain numerical results;

(c) detailed analysis, using the Bayesian paradigm, of situations involving standard probability models such as the Binomial, Poisson and Normal distributions [see II, §§ 5.2.2, 5.4 and 11.4].

Topic (a) will be treated in detail in section 15.4 and topic (b) will be considered in section 15.3.5. Section 15.5 will consider the Bayesian analysis of some standard univariate probability models; and section 15.6 will consider basic ideas of Bayesian statistics in multiparameter situations. A short bibliography is given in section 15.7.

## 15.2. BAYES' THEOREM: THE DISCRETE CASE

### 15.2.1. Learning from a Single Data Set

Let us suppose that a statistician considers the finite list of models

$$\{M_1, M_2, \ldots, M_k\}$$

to constitute an exclusive and exhaustive set of possible probability models for a particular situation under study.

Let us further suppose that, before data is obtained, the statistician assigns *prior probabilities* (see § 15.1)

$$\{P(M_1), P(M_2), \ldots, P(M_k)\}$$

to these models, where $0 \le P(M_i) \le 1$, $i = 1, 2, \ldots, k$, and

$$P(M_1) + P(M_2) + \ldots + P(M_k) = 1.$$

Each probability model defines a probability distribution over the possible data that might be obtained. In particular, if we denote by $D$ the data actually obtained, the probabilities of the data as defined by each of the alternative models are given by the conditional probabilities [see II, § 3.9.1]

$$\{P(D|M_1), P(D|M_2), \ldots, P(D|M_k)\}.$$

Considered in terms of the $\{M_1, M_2, \ldots, M_k\}$, for given $D$, these quantities are often referred to as the *likelihoods* [see §§3.5.4 and 4.13.1] of the $M_i$'s, given $D$.

Having considered an exclusive and exhaustive set of probability models and obtained data $D$, the statistician has specified a set of prior probabilities together with a set of likelihoods. He now wishes to revise his prior probabilities in the light of the information provided by the data. Expressed mathematically, he wishes to calculate

$$\{P(M_1|D), P(M_2|D), \ldots, P(M_k|D)\},$$

his probabilities for the alternative models, conditional now on having observed data $D$. The mathematical result which expresses these *posterior probabilities* (see § 15.1) in terms of the prior probabilities and likelihoods is *Bayes' Theorem* [see II, § 16.4], which, for the situation we are considering, can be stated as follows.

THEOREM 15.2.1.   *Bayes' Theorem (discrete form).   If* $\{M_1, M_2, \ldots, M_k\}$ *are an exclusive and exhaustive set of probability models, and the prior probabilities* $\{P(M_1), P(M_2), \ldots, P(M_k)\}$ *and likelihoods* $\{P(D|M_1), P(D|M_2), \ldots, P(D|M_k)\}$ *are specified such that* $P(D) > 0$, *then the posterior probabilities are given by*

$$P(M_i|D) = \frac{P(D|M_i)P(M_i)}{P(D)}, \qquad i = 1, 2, \ldots, k,$$

*where*

$$P(D) = P(D|M_1)P(M_1) + P(D|M_2)P(M_2) + \ldots + P(D|M_k)P(M_k).$$

*Proof.*   By the definition of conditional probability [see II, § 3.9.2],

$$P(M_i|D)P(D) = P(M_i \cap D) = P(D|M_i)P(M_i)$$

and the result is immediate.

The expression for $P(D)$ is simply a statement of the Theorem of Total Probability [see II, § 16.2].

EXAMPLE 15.2.1.   In a certain animal kingdon, genotypes $BB$ and $Bb$ are black, genotype $bb$ is brown, and all matings invariably result in a litter of seven offspring.

A black animal, which is known to have resulted from a $Bb \times Bb$ mating, is itself mated with a brown animal and the seven offspring are found to be all black.

What are the probabilities that the black parent is of genotype $BB$, $Bb$, respectively?

If we denote by $M_1$ the model which assumes type $BB$, and by $M_2$ the model which assumes type $Bb$, and if we let $D$ denote the observed data (seven black offspring), we require precisely $P(M_1|D)$ and $P(M_2|D)$. Since $\{M_1, M_2\}$ are an exclusive and exhaustive set of models in this case, Bayes' Theorem can be applied, provided we can specify $\{P(M_1), P(M_2)\}$ and $\{P(D|M_1), P(D|M_2)\}$.

Let us begin by considering these latter quantities.

Given mating $M_1$, that is $BB \times bb$, all offspring are necessarily of type $Bb$ (one gene being received from each parent) and so $P(D|M_1) = 1$.

Given mating $M_2$, that is $Bb \times bb$, Mendel's laws state that each offspring has, independently, probability $\frac{1}{2}$ of being of type $Bb$: it follows that $P(D|M_2) = (\frac{1}{2})^7 = \frac{1}{128}$.

Now let us consider the prior probabilities, $P(M_1)$ and $P(M_2)$. By Mendel's laws, the mating $Bb \times Bb$ of the parents of the black animal results in offspring

*BB*, *Bb* and *bb* with probabilities $\frac{1}{4}, \frac{1}{2}$ and $\frac{1}{4}$, respectively. The quantities $P(M_1)$, $P(M_2)$ denote the probabilities of *BB* and *Bb*, respectively, given that one or other of these obtains (recall that we are considering a black parent). It follows that $P(M_1) = \frac{1}{3}$, $P(M_2) = \frac{2}{3}$, since

$$P(M_1) = \frac{\frac{1}{4}}{\frac{1}{2} + \frac{1}{4}}, \quad P(M_2) = \frac{\frac{1}{2}}{\frac{1}{2} + \frac{1}{4}}.$$

Applying Bayes' Theorem, we obtain

$$P(M_1|D) = \frac{P(D|M_1)P(M_1)}{P(D|M_1)P(M_1) + P(D|M_2)P(M_2)}$$

$$= \frac{1 \times \frac{1}{3}}{1 \times \frac{1}{3} + \frac{1}{128} \times \frac{2}{3}}$$

$$= \frac{64}{65}.$$

Since $P(M_2|D) = 1 - P(M_1|D)$, we have

$$P(M_2|D) = \frac{1}{65}.$$

Of course, this latter could have been also obtained directly by using the theorem.

EXAMPLE 15.2.2.   Let us consider the same problem as in Example 15.2.1, except that the black parent is now known to have resulted from a $BB \times Bb$ mating.

Using the same notation, we still have $P(D|M_1) = 1$, $P(D|M_2) = \frac{1}{128}$, since these only require consideration of the final black-brown mating.

The quantities $P(M_1)$ and $P(M_2)$ are now changed, however. By Mendel's laws, the $BB \times Bb$ mating, which results in the black parent, will give type *BB* with probability $\frac{1}{2}$ and type *Bb* with probability $\frac{1}{2}$. We thus have $P(M_1) = P(M_2) = \frac{1}{2}$, and

$$P(M_1|D) = \frac{P(D|M_1)P(M_1)}{P(D|M_1)P(M_1) + P(D|M_2)P(M_2)}$$

$$= \frac{1 \times \frac{1}{2}}{1 \times \frac{1}{2} + \frac{1}{128} \times \frac{1}{2}}$$

$$= \frac{128}{129},$$

with $P(M_2|D) = 1 - P(M_1|D) = \frac{1}{129}$.

These two Examples illustrate well the kind of situation in which prior probabilities based on frequencies are available (Mendel's laws are backed by exhaustive experimentation), and also how a change in prior probabilities modifies the posterior probabilities, even though the data (and hence the likelihoods) remain the same.

EXAMPLE 15.2.3. Let us again consider the problem of the earlier examples, but this time suppose that *we do not know* which of the matings, $Bb \times Bb$ or $BB \times Bb$, gave rise to the black parent (although we know that it was one of the two).

Can we now assign values to $P(M_1)$ and $P(M_2)$? If we denote the mating $Bb \times Bb$ by $G_1$ and the mating $BB \times Bb$ by $G_2$, it might appear that we can use the Theorem of Total Probabilities [see II, § 16.2] to write

$$P(M_1) = P(M_1|G_1)P(G_1) + P(M_1|G_2)P(G_2),$$

$$P(M_2) = P(M_2|G_1)P(G_1) + P(M_2|G_2)P(G_2).$$

From Example 15.2.1, we have

$$P(M_1|G_1) = \tfrac{1}{3}, \qquad P(M_2|G_1) = \tfrac{2}{3},$$

and from Example 15.2.2, we have

$$P(M_1|G_2) = \tfrac{1}{2}, \qquad P(M_2|G_2) = \tfrac{1}{2}.$$

The problem therefore reduces to assigning values to $P(G_1)$ and $P(G_2)$, and this is the point at which controversy begins.

In section 15.1, we referred to differences of opinion regarding the legitimate usage of the concept of probability: in particular we drew attention to the *frequentist* and *degree of belief* interpretations. In the context of this example, statisticians committed to one or other of these interpretations might argue as follows.

### Frequentist interpretation

Without any knowledge of the mechanism by which the unknown mating, $G_1$ or $G_2$, came about, it is impossible to provide a frequency-based assignment of the probabilities $P(G_1)$ and $P(G_2)$. This makes it impossible to provide a frequency-based assignment of $P(M_1)$ and $P(M_2)$ and *thus renders Bayes' Theorem inapplicable so far as this inference problem is concerned.*

### Degree of belief interpretation

All probabilities are basically representations of degrees of belief, and so one can assign values to $P(G_1)$ and $P(G_2)$ to reflect one's beliefs about the relative plausibilities of the two types of mating.

For example, if one had no information at all, other than that one of $G_1$, $G_2$ took place, the specification

$$P(G_1) = \tfrac{1}{2}, \qquad P(G_2) = \tfrac{1}{2}$$

might appeal to some statisticians as a representation of 'ignorance', and would lead to

$$P(M_1) = \tfrac{1}{3} \times \tfrac{1}{2} + \tfrac{1}{2} \times \tfrac{1}{2} = \tfrac{5}{12},$$

$$P(M_2) = \tfrac{2}{3} \times \tfrac{1}{2} + \tfrac{1}{2} \times \tfrac{1}{2} = \tfrac{7}{12}.$$

Substituting into Bayes' Theorem, we then obtain

$$P(M_1|D) = \frac{P(D|M_1)P(M_1)}{P(D|M_1)P(M_1) + P(D|M_2)P(M_2)}$$

$$= \frac{1 \times \frac{5}{12}}{1 \times \frac{5}{12} + \frac{1}{128} \times \frac{7}{12}}$$

$$= \frac{640}{647},$$

with $P(M_2|D) = 1 - P(M_1|D) = \frac{7}{647}$.

Comparing this example with the two earlier ones, we see how restricted the applicability of Bayes' Theorem becomes if only frequency-based probabilities are regarded as legitimate. This serves to emphasize the point made in section 15.1; that systematic use of Bayes' Theorem as a statistical tool requires a willingness to express all forms of prior belief in the form of probabilities, whether or not there is a frequency basis to these probabilities.

A further important point to emerge from the comparison of Example 15.2.3 with Examples 15.2.1 and 15.2.2 is the following. Whereas in the first two examples the prior probabilities $P(M_1)$, $P(M_2)$ may be regarded as 'objective', in the sense that there is a consensus among scientists that the stated values are appropriate, this cannot be claimed in Example 15.2.3. Here, we are unavoidably involved in the 'subjective' assessment of $P(G_1)$ and $P(G_2)$; there is typically no data available, and it is unavoidably a matter of personal belief. In the example, we used $P(G_1) = P(G_2) = \frac{1}{2}$ for illustration. This choice certainly has some intuitive appeal as a statement of 'ignorance' about $G_1$, $G_2$, but if someone had a personal 'hunch' in favour of $G_1$, say, he might prefer not to use these values.

The point is that, in general, if we admit the use of probability as a representation of personal beliefs we can no longer talk about a 'correct choice' of prior probabilities. The choice made depends on the individual and on the information available to him at the time. In cases where individuals have a great deal of common information their 'prior' probabilities are often similar and 'objective' agreement is possible. If there is little 'hard' information available, subjective assessments may differ and an 'objective' consensus may not be possible.

### 15.2.2.  Learning from Several Data Sets

Suppose that, as in section 15.2.1, we have an exclusive and exhaustive set of models $\{M_1, M_2, \ldots, M_k\}$, together with specified prior probabilities $\{P(M_1), P(M_2), \ldots, P(M_k)\}$, but that data is now obtained in two stages, resulting in data sets $D_1$ and $D_2$.

If we simply combine these into one data set, denoted by $D = D_1 \cap D_2$, then, given the specification of likelihoods $\{P(D|M_1), P(D|M_2), \ldots, P(D|M_k)\}$, we can proceed as in the previous section to obtain the posterior probabilities $\{P(M_1|D), P(M_2|D), \ldots, P(M_k|D)\}$. In many practical situations, however, data arrives sequentially, $D_1$ being obtained before $D_2$, and we wish to first revise our probabilities in the light of $D_1$, and *then later* revise them further in the light of $D_2$. The process is shown schematically in Figure 15.2.2.

From a common-sense point of view, we should clearly hope to arrive at the same eventual posterior probabilities, $P(M_i|D_1 \cap D_2)$, $i = 1, 2, \ldots, k$, irrespective of whether the data $D = D_1 \cap D_2$ were processed altogether, using Bayes' Theorem once, or in two stages, as in Figure 15.2.1.

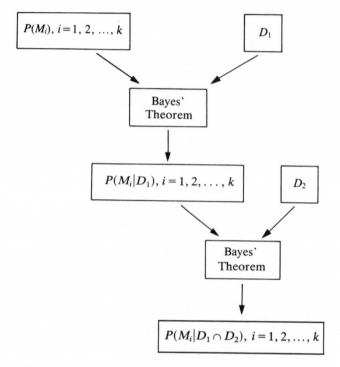

Figure 15.2.1: Learning in stages using Bayes' Theorem.

In fact, we can easily verify mathematically that this common-sense result is correct.

Suppose we have observed $D_1$, and have completed the first stage of Figure 15.2.1. Then, remembering that all inputs to the second stage of the figure are *conditional on $D_1$*, we have available for the second stage use of Bayes' Theorem the quantities

   $\{P(M_1|D_1), \ldots, P(M_k|D_1)\}$,   the model probabilities *prior to $D_2$*,

and

   $\{P(D_2|M_1 \cap D_1), \ldots, P(D_2|M_k \cap D_1)\}$,   the second stage *likelihoods.*

Using Bayes' Theorem, these combine to give

$$P(M_i|D_1 \cap D_2) = \frac{P(D_2|M_i \cap D_1)P(M_i|D_1)}{P(D_2|D_1)} \qquad (i = 1, 2, \ldots, k),$$

where

$$P(D_2|D_1) = P(D_2|M_1 \cap D_1)(P(M_1|D_1) + \ldots + P(D_2|M_k \cap D_1)P(M_k|D_1).$$

But, from the first stage use of Bayes' Theorem,

$$P(M_i|D_1) = \frac{P(D_1|M_i)P(M_i)}{P(D_1)} \qquad (i = 1, 2, \ldots, k).$$

Substituting this into the expression for $P(M_i|D_1 \cap D_2)$, we obtain

$$P(M_i|D_1 \cap D_2) = \frac{P(D_2|M_i \cap D_1)P(D_1|M_i)P(M_i)}{P(D_2|D_1)P(D_1)} \qquad (i = 1, 2, \ldots, k),$$

$$= \frac{P(D_1 \cap D_2|M_i)P(M_i)}{P(D_1 \cap D_2)} \qquad (i = 1, 2, \ldots, k).$$

But this, of course, is the result we obtain from a single direct use of Bayes' Theorem for the combined data set $D = D_1 \cap D_2$ and so the desired result is established.

Two further points are worth noting:

(a) Both Figure 15.2.1 and the mathematical development given above can be extended to *any number* of stages.
(b) It is often the case that, conditional on each $M_i$, data sets $D_1$ and $D_2$ provide *independent* information, so that $P(D_2|M_i \cap D_1) = P(D_2|M_i)$. This simplifies the calculations.

EXAMPLE 15.2.4. Preliminary medical examination of a patient reveals that he is suffering from one of a mutually exclusive set of medical conditions, $M_1$, $M_2$, $M_3$, each of which is initially assessed to be equally likely. Further investigation reveals the presence of symptom $X$ (data $D_1$) and symptom $Y$ (data $D_2$), the occurrences of which are known to be independent of each other, given any particular medical condition. Extensive past records are available, enabling doctors to calculate $P(D_j|M_i)$ for $j = 1, 2$, and $i = 1, 2, 3$. The results are summarized below.

|       | $M_1$ | $M_2$ | $M_3$ |
|-------|-------|-------|-------|
| $D_1$ | 0·5   | 0·7   | 0·8   |
| $D_2$ | 0·5   | 0·25  | 0·9   |

Using these figures, we can revise our probabilities for $M_1$, $M_2$, $M_3$, first using $D_1$ only, and then including $D_2$, also, at a second stage.

For the first stage, we have, from Bayes' Theorem,

$$P(M_i|D_1) = \frac{P(D_1|M_i)P(M_i)}{P(D_1)} \qquad (i = 1, 2, 3),$$

where

$$P(D_1) = P(D_1|M_1)P(M_1) + P(D_1|M_2)P(M_2) + P(D_1|M_3)P(M_3)$$
$$= \tfrac{5}{10} \times \tfrac{1}{3} + \tfrac{7}{10} \times \tfrac{1}{3} + \tfrac{8}{10} \times \tfrac{1}{3}$$
$$= \tfrac{2}{3}.$$

We therefore obtain:

$$P(M_1|D_1) = \frac{\tfrac{5}{10} \times \tfrac{1}{3}}{\tfrac{2}{3}} = \tfrac{5}{20} = 0 \cdot 25$$

$$P(M_2|D_1) = \frac{\tfrac{7}{10} \times \tfrac{1}{3}}{\tfrac{2}{3}} = \tfrac{7}{20} = 0 \cdot 35$$

$$P(M_3|D_1) = \frac{\tfrac{8}{10} \times \tfrac{1}{3}}{\tfrac{2}{3}} = \tfrac{8}{20} = 0 \cdot 40.$$

For the second stage, recalling the independence of $D_1$ and $D_2$, given $M_i$, Bayes' Theorem takes the form

$$P(M_i|D_1 \cap D_2) = \frac{P(D_2|M_i)P(M_i|D_1)}{P(D_2|D_1)} \qquad (i = 1, 2, 3),$$

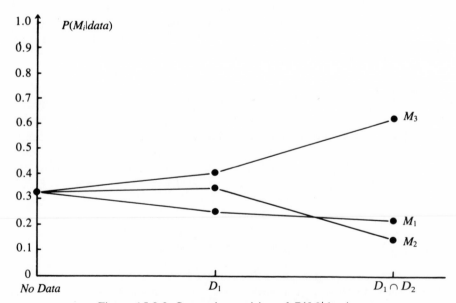

Figure 15.2.2: Successive revision of $P(M_i|\text{data})$.

where

$$P(D_2|D_1) = P(D_2|M_1)P(M_1|D_1) + P(D_2|M_2)P(M_2|D_1) + P(D_2|M_3)P(M_3|D_1)$$

$$= \tfrac{10}{20} \times \tfrac{5}{20} + \tfrac{5}{20} \times \tfrac{7}{20} + \tfrac{18}{20} \times \tfrac{8}{20}$$

$$= \tfrac{229}{400}.$$

We therefore obtain:

$$P(M_1|D_1 \cap D_2) = (\tfrac{10}{20} \times \tfrac{5}{20})/(\tfrac{229}{400}) = \tfrac{50}{229} = 0 \cdot 218,$$

$$P(M_2|D_1 \cap D_2) = (\tfrac{5}{20} \times \tfrac{7}{20})/(\tfrac{229}{400}) = \tfrac{35}{229} = 0 \cdot 153,$$

$$P(M_3|D_1 \cap D_2) = (\tfrac{18}{20} \times \tfrac{8}{20})/(\tfrac{229}{400}) = \tfrac{144}{229} = 0 \cdot 629.$$

The revision of the probabilities with increasing information is summarized in Figure 15.2.2.

### 15.2.3.  Bayes' Theorem Expressed in Terms of Odds

Using the notation of section 15.1, the simple version of Bayes' Theorem has the form

$$P(M_i|D) = \frac{P(D|M_i)P(M_i)}{P(D)},$$

which, since $P(D)$ does not involve explicitly the argument $M_i$, can be summarized in the form

$$P(M_i|D) \propto P(D|M_i)P(M_i).$$

In other words,

| *Posterior* *Probability* is proportional to *Likelihood* × *Prior* *Probability* |
|---|

This expression gives some insight into how the ingredients combine to influence posterior probabilities, but there is an alternative way of writing the theorem which gives perhaps greater insight.

If we consider just two alternative models, $M_1$ and $M_2$, say, then, from Bayes' Theorem,

$$\frac{P(M_1|D)}{P(M_2|D)} = \frac{P(D|M_1)}{P(D|M_2)} \cdot \frac{P(M_1)}{P(M_2)}.$$

Now, for any event $A$, the ratio $P(A)/(1 - P(A))$ is called the *odds on $A$* [c.f. II; § 2.3], and so the above expression can be summarized in the form

| *Posterior* *Odds* equal *Likelihood* *Ratio* × *Prior* *Odds*. |
|---|

In this way, we can see how the ratio of likelihoods plays a basic role in transforming relative prior beliefs for the two models into relative posterior beliefs.

In some ways, the *logarithm of the odds* is a more convenient way to measure relative beliefs. Applied to the above, we obtain the expression

$$
Log\left(\frac{Posterior}{Odds}\right) \quad equals \quad Log\left(\frac{Likelihood}{Ratio}\right) + Log\left(\frac{Prior}{Odds}\right).
$$

On the log-odds scale, the logarithm of the likelihood ratio acts as an *additive transformation* from prior to posterior beliefs.

### 15.2.4.   Extension to an Infinite List of Possible Models

In section 15.2.1, we assumed that the collection of possible probability models under consideration could be written as a *finite list* $\{M_1, M_2, \ldots, M_k\}$.

In fact, all the results obtained in the finite case can be extended easily to the case of an *infinite list* of possible models $\{M_1, M_2, \ldots\}$. Given prior probabilities $P(M_i)$, $i = 1, 2, \ldots$, such that $\sum_{i=1}^{\infty} P(M_i) = 1$, and with likelihoods $P(D|M_i)$, $i = 1, 2, \ldots$, for given data $D$ Bayes' Theorem has the form

$$
P(M_i|D) = \frac{P(D|M_i)P(M_i)}{P(D)} \qquad i = 1, 2, \ldots,
$$

where

$$
P(D) = \sum_{j=1}^{\infty} P(D|M_j)P(M_j).
$$

The proof follows straightforwardly from the same arguments as were used in the finite case.

EXAMPLE 15.2.5.   To see how such an infinite list of possible models might arise, consider the following. Suppose the number $X$ of telephone calls received per minute at a certain switchboard is known to be a Poisson random variable [see II, § 5.4] whose mean $\theta$ is an unknown, positive integer. Then, if data $D$ consists of the observation $X = x$, we have an infinite list of possible models $\{M_1, M_2, \ldots\}$, generating Poisson likelihoods $P(D|M_i)$, $i = 1, 2, \ldots$, where $M_i$ corresponds to the model $\theta = i$, $i = 1, 2, \ldots$.

## 15.3.   BAYES' THEOREM: THE CONTINUOUS CASE

### 15.3.1.   The Continuous Form of Bayes' Theorem

In the previous section, we dealt with the appropriate form of Bayes' Theorem when the set of possible models under consideration can be written as a *finite or infinite list*.

In many cases, however, there is a *continuum* of possible models, and a *discrete* representation of all possible models in the form of a list is impossible. As examples of this situation, we note the following:

(i) a random variable $X$ has a Binomial distribution [see II, § 5.2.2] corresponding to $n$ 'trials' with unknown chance of 'success', $\theta$, at each 'trial'; if the only constraint on the parameter $\theta$ is that $0 \le \theta \le 1$, then the set of possible models can be identified with the set $\{\theta; 0 \le \theta \le 1\}$ of all possible *parameter* values;

(ii) if we consider $X$ defined in Example 15.2.5, but now assume only that the parameter $\theta$ is positive, the set of all possible models again corresponds to the set of all possible parameter values; in this case, the set is the positive real line $\{\theta; \theta \in \mathbb{R}^+\}$ [see I, § 2.6].

(iii) a random variable $X$ is defined to have a Normal distribution [see II: § 11.4] with unknown mean, $\mu$, and unknown variance, $\sigma^2$; if there are no constraints on these parameters, the set of possible models corresponds to the set $\{\mu, \sigma^2; \mu \in \mathbb{R}, \sigma^2 \in \mathbb{R}^+\}$ [see I, § 2.6.1].

Each of these examples is a special case of the following general situation:

a random variable $X$ has a probability distribution defined in terms of an unknown parameter $\theta$, which belongs to a specified set of possible parameter values $\Theta$.

The distribution of $X$ may be discrete or continuous [see II, Chapters 5, 10], and either or both of $X$ and $\theta$ may be vector-valued [see II, Chapter 13]; in any case, we shall write $p(x|\theta)$ to denote the *likelihood* [see §§ 3.5.4, 6.2.1] for the particular parameter value $\theta$, given that we observe $X = x$; regarded as a function of $x$, for given $\theta$, $p(x|\theta)$ may be a probability mass function, that is, a discrete p.d.f., or a density, that is, a continuous p.d.f. [see II, § 10.1], depending on whether $X$ is discrete or continuous, respectively.

Prior probabilities over the set of possible models correspond, in this general situation, to a distribution of probability over the set $\Theta$ [see I, Chapter 1] of possible parameter values; since $\Theta$ is either an interval of the real line, or a region in the plane, or a region of some higher-dimensional space, whose dimension depends on that of the parameter, prior beliefs will need to be specified in the form of a *prior probability density*,

$$p(\theta), \theta \in \Theta, \quad \text{such that} \int_{\Theta} p(\theta)\, d\theta = 1$$

[see IV, Chapter 4].

In the light of the data, $X = x$, we wish to revise this prior distribution and obtain a *posterior probability density* $p(\theta|x)$, $\theta \in \Theta$; the result which links $p(\theta|x)$ to $p(x|\theta)$ and $p(\theta)$ is as follows.

THEOREM 15.3.1. *Bayes' Theorem (continuous form).* *If $\Theta$ is a set of possible parameter values with $p(\theta)$, $\theta \in \Theta$, a prior probability density over $\Theta$, and if $p(x|\theta)$ denotes the likelihood of $\theta$, given that we observe $X = x$, then the* *posterior probability density* $p(\theta|x)$ *is given by*

$$p(\theta|x) = \frac{p(x|\theta)p(\theta)}{p(x)}, \qquad \theta \in \Theta,$$

*where*

$$p(x) = \int_{\Theta} p(x|\theta)p(\theta) \, d\theta.$$

*Proof.*   By the definition of a conditional probability density [see II, § 13.1.4],

$$p(\theta|x) = \frac{p(x, \theta)}{p(x)} = \frac{p(x|\theta)p(\theta)}{p(x)},$$

where $p(x, \theta)$ is the joint density [see II, § 13.1.1]. The form of $p(x)$ follows from the definition of a marginal density [see II, § 13.1.2]

$$p(x) = \int_{\Theta} p(x, \theta) \, d\theta = \int_{\Theta} p(x|\theta)p(\theta) \, d\theta.$$

The content of the continuous form of Bayes' Theorem is most easily remembered from the summary:

| | | |
|---|---|---|
| *Posterior Density* | is proportional to   *Likelihood* × | *Prior Density* . |

Symbolically, we have

$$p(\theta|x) \propto p(x|\theta)p(\theta)$$

where here, as above, the notation $p(\cdot)$, $p(\cdot|\cdot)$ is used throughout to represent the appropriate marginal or conditional density, with the understanding that, for example, $p(\theta|x)$ and $p(x|\theta)$ are, in general, completely *different* densities.

The proportionality sign simply indicates that a multiplicative factor not involving $\theta$ is omitted from the right-hand side. The 'shape' of $p(\theta|x)$ is determined by the product $p(x|\theta)p(\theta)$, and the missing factor

$$[p(x)]^{-1} = \left[ \int_{\Theta} p(x|\theta)p(\theta) \, d\theta \right]^{-1}$$

serves merely to normalize $p(\theta|x)$ so that $\int_{\Theta} p(\theta|x) \, d\theta = 1$.

The specification of $p(\theta)$ clearly requires some discussion. First of all, let us emphasize again the point made in section 15.2.1. There is no such thing as 'the correct choice' of $p(\theta)$ in a given problem. The actual choice of $p(\theta)$

depends on individual judgement in the light of the information and experience available at the time.

EXAMPLE 15.3.1.    In the case of a Binomial distribution with unknown parameter $\theta$, $0 \leq \theta \leq 1$, Figure 15.3.1 displays some possible forms of $p(\theta)$ that might be specified.

Case (a) would represent the belief of someone who regarded all values of $\theta$, $0 \leq \theta \leq 1$, as 'equally plausible', in the sense that sub-intervals of equal length are all assigned the same probability, irrespective of where they are located in the interval from 0 to 1.

Case (b) represents the belief that $\theta = \frac{3}{4}$ is the 'most likely' value, but that $\theta < \frac{3}{4}$ is about five times as likely as $\theta > \frac{3}{4}$, and so on. The prior probability that $\theta$ lies in any particular interval is given by the area under the curve in that interval.

Case (c) represents a belief that values of $\theta$ close to 0 or 1 are much more plausible than values in the centre of the interval.

### 15.3.2.   The Assessment of Prior Densities

We have stated above that the specification of a prior density depends on individual experience and beliefs. This poses the very real problem of how to convert such experiences and beliefs into the form of a probability density function $p(\theta)$. This problem has been exhaustively studied by both statisticians and psychologists, and many theoretical and experimental papers have been published reporting the results of such studies. A convenient reference, which itself contains an extensive bibliography, is the article 'Subjective probability and its measurement' by Hampton, Moore and Thomas (1973).

Among the many possible approaches, the following are, perhaps, the two most important.

### (i)   *Smoothing of historical data*

Suppose that a manufacturer is uncertain about the proportion, $\theta$, of defective items that will result from a new production process that is to be introduced, but that he has available to him a histogram [see § 3.2.2] showing the relative frequencies with which the proportion of defectives fell in various intervals when a number of very similar previous production processes were introduced. Figure 15.3.2 shows such a histogram, together with a smooth curve, which, suitably normalized so as to contain total area 1, might well reasonably reflect, in the form of a probability density $p(\theta)$, the manufacturer's prior beliefs.

The technique is very straightforward; we simply seek to produce a smooth density which reflects the form of an historical frequency distribution. Even here, however, we cannot escape the subjective, judgemental nature of the specification, since we must *assess* the relevance and homogeneity of the historical data in relation to our current problem.

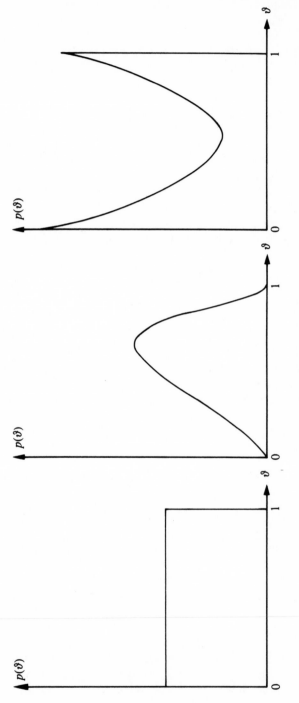

Figure 15.3.1: Possible forms of $p(\theta)$.

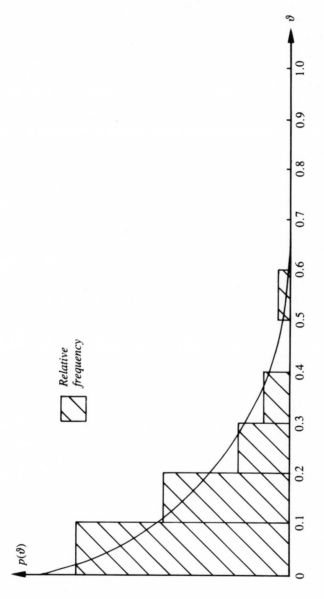

Figure 15.3.2: $p(\theta)$ as a smoothing of historical data.

(ii) *Judgemental curve fitting*

In the absence of sufficient, relevant past data to permit the use of (i), we are forced to try to elicit our beliefs directly by a process of self-interrogation.

Consider, for example, a manufacturer introducing an entirely new type of production process, who does not feel that historical data regarding the proportions of defectives resulting from other types of process is directly relevant to his new process.

The following kind of procedure can be used to elicit directly his beliefs about $\theta$, the proportion of defectives that will occur under the new process.

(a) The manufacturer is asked to give upper and lower limits between which, in his opinion, the proportion $\theta$ will be. In practice, we might ask him to give values such that there is only, according to him, a one in a hundred chance of each of these limits being exceeded. The values specified would then give, approximately, the first and ninety-ninth percentiles of his distribution [see § 5.2.2: Inverse Tables].

(b) We then ask the manufacturer to give the median (or fiftieth percentile) [see § 14.2] of his distribution. In other words, we ask him for that value of $\theta$ which would lead him to be equally willing to bet on the proportion being above or below this value.

(c) Having obtained the median, which divides his distribution in half, we ask the manufacturer to divide it further into quarters, by specifying the twenty-fifth and seventy-fifth percentiles. To do this for the seventy-

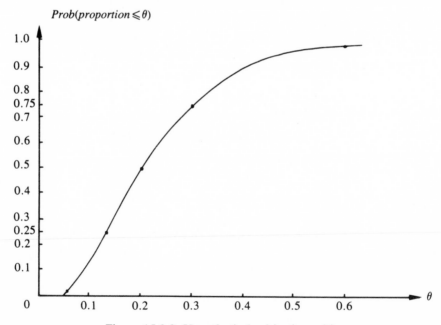

Figure 15.3.3: Hypothetical subjective c.d.f.

fifth percentile, for example, we ask the manufacturer to concentrate on those values of $\theta$ lying between the median and the upper limit, and then to choose a value in this range such that the probabilities of $\theta$ lying above or below this value are assessed to be approximately equal.

At the end of this interrogation process, we have five points on the cumulative distribution function [see II, § 10.1.1] of $\theta$; namely, the first, twenty-fifth, fiftieth, seventy-fifth and ninety-ninth percentiles. A smooth curve can be drawn through these to get a reasonable approximation to the manufacturer's subjective cumulative distribution function (c.d.f.).

Figure 15.3.3 shows such a function corresponding to the following (hypothetical) elicited values:

$$\text{(a)}\ \ 0{\cdot}05, 0{\cdot}6; \qquad \text{(b)}\ 0{\cdot}2; \qquad \text{(c)}\ 0{\cdot}125, 0{\cdot}3.$$

The corresponding histogram for subintervals $0{\cdot}05{-}0{\cdot}1, 0{\cdot}1{-}0{\cdot}15$, etc. is shown in Figure 15.3.4, together with a smoothed approximation to the probability density function (p.d.f.).

We have been considering the assessment of a prior distribution for a single parameter. The assessment of a joint distribution for two or more parameters is straightforward if, *a priori*, assessments can be considered independently (since we then reduce to a series of one-dimensional assessments), but can present considerable difficulties if there are complex dependencies among the parameters. One aspect of this problem will be considered further in section 15.6.

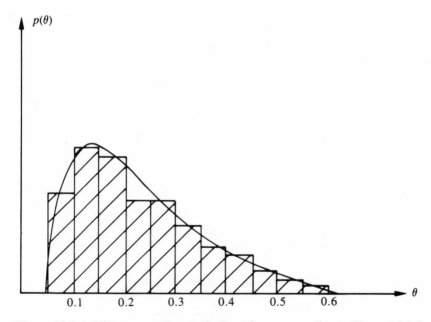

Figure 15.3.4: Histogram and smoothed p.d.f. corresponding to Figure 15.3.3.

### 15.3.3.   An Illustration of Bayes' Theorem for a Single Unknown Parameter

EXAMPLE 15.3.2.   Let us suppose that two engineers, $A$ and $B$, are interested in the breaking strength, $\theta$, in appropriate units, of a material which has so far not been systematically tested in the laboratory. Engineer $A$ has made extensive studies of similar materials, and, when the judgemental curve fitting procedure of the previous section is applied, his answers to questions (a), (b), (c) are as follows:

$$\text{(a) } 450, 550; \qquad \text{(b) } 500; \qquad \text{(c) } 485, 515.$$

Sketches of the c.d.f. and corresponding p.d.f. implied by these assessed percentiles reveal that, to a good approximation, the prior distribution of engineer $A$ for $\theta$ can be represented by a Normal distribution with mean 500 and standard deviation 20. We shall denote this symbolically by $p_A(\theta) = N(500, 20)$.

Engineer $B$ is not so familiar with materials of this kind, and when interrogated according to the judgemental curve fitting procedure answers as follows:

$$\text{(a) } 215, 585; \qquad \text{(b) } 400; \qquad \text{(c) } 345, 455.$$

This gives a prior distribution which is easily seen to be closely approximated by a Normal distribution with mean 400 and standard deviation 80. We shall denote this symbolically by $p_B(\theta) = N(400, 80)$.

Suppose now that both engineers are given the result of an experiment which gave rise to an observed breaking strength $x = 450$, and that both assume this to be the realization of a random variable $X$, Normally distributed with mean $\theta$ and standard deviation 40. The engineers therefore accept as the *likelihood* input to Bayes' Theorem the form

$$p(x|\theta) = \frac{1}{\sqrt{2\pi}\,40}\exp\left[-\frac{1}{2}\left(\frac{x-\theta}{40}\right)^2\right].$$

For engineer $A$, this is to be combined with

$$p_A(\theta) = \frac{1}{\sqrt{2\pi}\,20}\exp\left[-\frac{1}{2}\left(\frac{\theta-500}{20}\right)^2\right],$$

and, for engineer $B$, with

$$p_B(\theta) = \frac{1}{\sqrt{2\pi}\,80}\exp\left[-\frac{1}{2}\left(\frac{\theta-400}{80}\right)^2\right].$$

If we define as the *standardized likelihood* the ratio

$$\frac{p(x|\theta)}{p(x)} = \frac{p(x|\theta)}{\int_{\Theta} p(x|\theta)p(\theta)\,d\theta},$$

then, from section 15.3.1, we see that Bayes' Theorem has the form

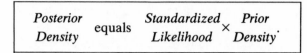

Figures 15.3.5 to 15.3.7 show the approximate forms of prior densities, standardized likelihood and posterior densities for $A$ and $B$. In the case of engineer $A$, the prior density is more concentrated than the standardized likelihood and so, when these functions are multiplied together to form the

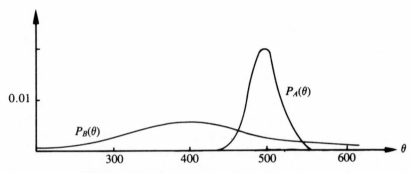

Figure 15.3.5: Prior densities for $A$ and $B$.

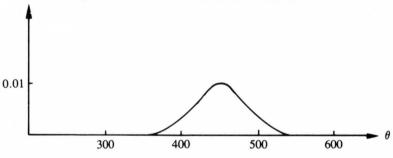

Figure 15.3.6: Standardized likelihood.

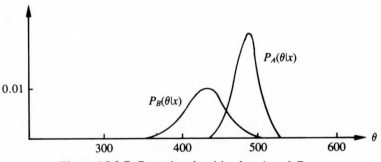

Figure 15.3.7: Posterior densities for $A$ and $B$.

posterior density, the prior density dominates and the resulting posterior density looks similar to the prior density. This reflects the fact that a single, rather inaccurate, observation cannot have much impact on relatively strongly held *a priori* beliefs.

In the case of engineer *B*, however, it is the standardized likelihood which is more concentrated than the prior density. This results in the posterior density having a form much closer to the standardized likelihood than to the prior density, and reflects the fact that if *a priori* beliefs are rather vague, then even a somewhat inaccurate observation will lead to a radical revision of prior beliefs.

In fact, it can be shown (see § 15.5.3) that $p_A(\theta|x) = N(490, 17\cdot9)$ and $p_B(\theta|x) = N(440, 35\cdot7)$. Comparing these forms with the corresponding prior densities, we see that engineer *A* has learnt very little from the experiment (in the sense that his posterior belief is not very different from his prior belief), whereas engineer *B* has learnt quite a lot (his posterior belief is considerably different from his prior belief). This illustrates a fundamental tenet of the Bayesian approach: *data does not create beliefs; rather, it modifies existing beliefs.*

On the other hand, we see from the Figures 15.3.8 and 15.3.9 that the data has had the effect of moving the two distributions of belief *closer together*. This movement towards 'consensus' of beliefs becomes even more marked as the quantity of data increases. Suppose that a total of 100 independent experiments are performed and result in a mean observed breaking strength of $\bar{x} = 470$. This can be regarded as a realization of a Normally distributed random variable with mean $\theta$ and standard deviation $40/\sqrt{100} = 4$. The likelihood resulting from the 100 experiments therefore has the form

$$p(\bar{x}|\theta) = \frac{1}{\sqrt{2\pi}4} \exp\left[-\frac{1}{2}\left(\frac{\bar{x}-\theta}{4}\right)^2\right],$$

the (approximate) standardized version of which is shown in Figure 15.3.8,

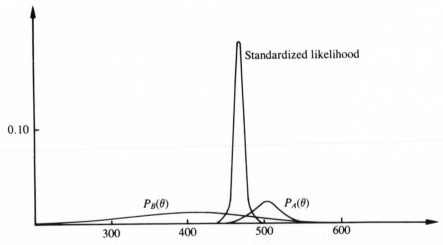

Figure 15.3.8: Prior densities and standardized likelihood for 100 observations.

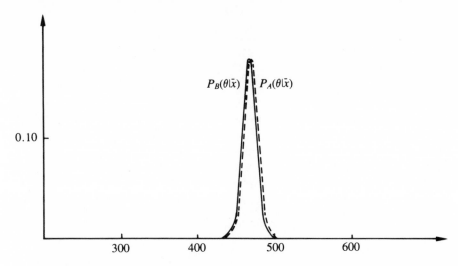

Figure 15.3.9: Posterior densities for *A* and *B* from 100 observations.

together with the prior densities $p_A(\theta)$, $p_B(\theta)$. The resulting posterior densities
are shown in Figure 15.3.9 and can be shown (see § 15.5.3) to be

$$p_A(\theta|\bar{x}) = \text{Normal }(471.2, 3.9)$$

$$p_B(\theta|\bar{x}) = \text{Normal }(469.8, 3.995),$$

which are practically indistinguishable.

Figures 15.3.5 to 15.3.9 clearly illustrate the way in which the form of the
posterior density depends on the relative flatness or peakedness of the prior
density in comparison with the flatness or peakedness displayed by the likelihood. When only small amounts of data are available, the prior density may
appear as peaked as the likelihood, and the posterior density will reflect a
compromise between prior beliefs and the information conveyed by the data.
When a large amount of data is available, however, the very sharply peaked
nature of the likelihood will make a typical prior density appear very flat in
comparison. Since the posterior density is given by the product of the standardized likelihood and the prior density, multiplying by the latter is rather like
multiplying by a constant function (with respect to the variable $\theta$) and the
shape and location of the posterior is almost entirely determined by the
likelihood.

### 15.3.4.   An Illustration of Bayes' Theorem for Two Unknown Parameters

EXAMPLE 15.3.3.   In a paper entitled 'The Numerical Computation of
Posterior Distributions in Bayesian Statistical Inference', Reilly (1976) considers the data given in Table 15.3.1, which were simulated using the model

$$\log(y_i) = \log(\alpha + \beta x_i) + \varepsilon_i, \qquad i = 0, 2, \ldots, 5,$$

| $x$ | 0 | 1 | 2 | 3 | 4 | 5 |
|---|---|---|---|---|---|---|
| $y$ | 4·11 | 6·32 | 8·21 | 10·43 | 14·29 | 16·78 |

Table 15.3.1

where the $\varepsilon_i$ are independently, Normally distributed with means zero and variances all equal to $(0·1398)^2$. In fact, the actual parameter values used in the simulation were $\alpha = 5$, $\beta = 2$, but Reilly analyses the data under the assumption that $\alpha$ and $\beta$ are *unknown* parameters, whose values are to be inferred using Bayes' Theorem. Denoting by **y** the vector of observations $(y_0, y_1, \ldots, y_5)$, and omitting specific reference to the fixed $x$ values, Bayes' Theorem in this example can be written, ignoring the normalizing constant, as

$$p(\alpha, \beta|\mathbf{y}) \propto p(\mathbf{y}|\alpha, \beta)p(\alpha, \beta)$$

where

$$p(\mathbf{y}|\alpha, \beta) = \prod_{i=0}^{5} \frac{1}{\sqrt{2\pi}\,(0·1398)} \exp\left[-\frac{1}{2(0·1398)^2}\left\{\log\left(\frac{y_i}{\alpha + \beta x_i}\right)\right\}^2\right]$$

$$\propto \exp\left[-25·58 \sum_{i=0}^{5} \left\{\log\left(\frac{y_i}{\alpha + \beta x_i}\right)\right\}^2\right],$$

and $p(\alpha, \beta)$ is a joint prior density over an appropriate range of values.

Figure 15.3.10 displays contours of the *joint posterior density* $p(\alpha, \beta|\mathbf{y})$, corresponding to the specification of $p(\alpha, \beta)$ as a uniform density over the range $2 \leq \alpha \leq 7$, $1 \leq \beta \leq 4$. The contours displayed contain (starting with the outside contour) 99·9, 98·6, 95·0, 74·6 and 49·5 percent, respectively, of the joint posterior probability.

Figure 15.3.11 displays the *marginal posterior density* functions for $\alpha$ and $\beta$, defined by

$$p(\alpha|\mathbf{y}) = \int_1^4 p(\alpha, \beta|\mathbf{y})\, d\beta, \qquad p(\beta|\mathbf{y}) = \int_2^7 p(\alpha, \beta|\mathbf{y})\, d\alpha.$$

In fact, in this example, both the calculation of the joint density contours and the integration of the joint density to obtain the marginals require careful numerical work, carried out on a computer. The article by Reilly outlines a form of crude numerical procedure for performing these calculations. More efficient numerical integration methods are described and illustrated in a paper by Naylor and Smith (1982).

We shall consider, in section 15.4, some of the possible approaches that might be adopted for summarizing the information conveyed by posterior densities. The basic message from Figure 15.3.11 is, however, easily understood. Starting from prior beliefs which viewed all values in the ranges $2 \leq \alpha \leq 7, 1 \leq \beta \leq 4$, as equally plausible, after having obtained the six observations recorded in Table 15.3.1 we are now fairly sure that $2·5 \leq \alpha \leq 6$, with a 'most likely' value at about $\alpha = 4$, and that $1·5 \leq \beta \leq 3·5$ with a 'most likely' value

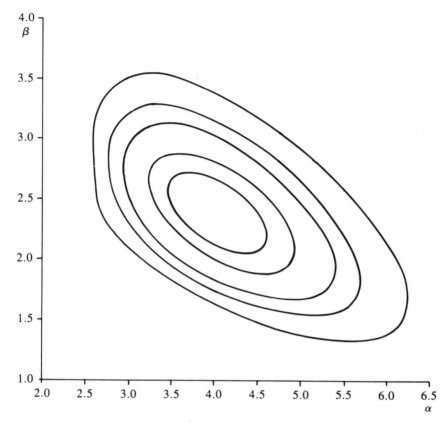

Figure 15.3.10: Contours of $p(\alpha, \beta|y)$.

at about $\beta = 2\cdot4$. The range of uncertainty is slightly reduced, but, with only a small number of observations, beliefs are not yet very concentrated.

When a model specification involves two, or more, parameters, it is often of practical interest to learn about a single function of these parameters. For example, in the case considered above, we may be interested in the parameter $\gamma = \alpha/\beta$. The marginal posterior density $p(\gamma|y)$ is found from the joint posterior density $p(\alpha, \beta|y)$, either by means of the usual transformation of variables technique [see II, § 10.7], where this is analytically tractable, or by direct numerical techniques such as those explained in the articles by Reilly or by Naylor and Smith. The approximate form, derived from $p(\alpha, \beta|y)$ by numerical methods, is shown in Figure 15.3.12.

### 15.3.5.  Approximate Analysis Under Great Prior Uncertainty

We have seen in section 15.3.3 that situations in which great prior uncertainty exists (relative to the information contained in the data) correspond,

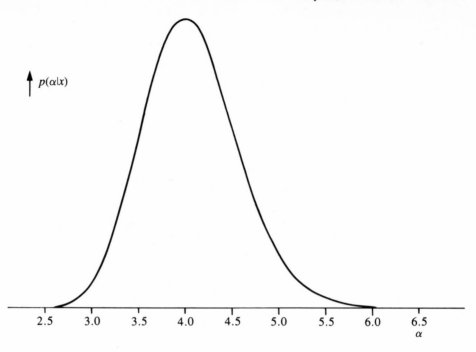

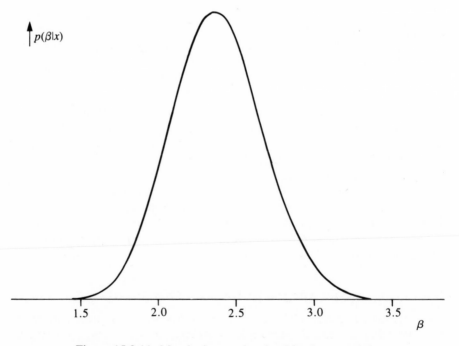

Figure 15.3.11: Marginal posterior densities for $\alpha$ and $\beta$.

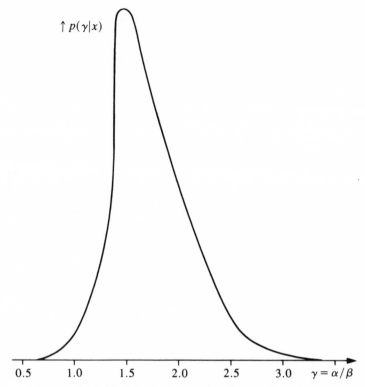

Figure 15.3.12: Marginal posterior density for $\gamma$.

mathematically, to a prior density function which is relatively flat in the region
where the (standardized) likelihood, corresponding to data $x$, is peaked. This
may arise from a moderate size experiment in conjunction with very diffuse
prior beliefs, or even from a very large experiment with moderately strong
prior beliefs, and leads to the kind of picture shown in Figure 15.3.13, for a
single parameter $\theta$.

Recalling that Bayes' Theorem may be summarized in the form

'posterior density = standardized likelihood × prior density',

we see that if the prior density is approximately constant, as a function of $\theta$,
over the range where the likelihood is concentrated, then, in this case of
relative great prior uncertainty, Bayes' Theorem gives the approximate result

$$p(\theta|x) \approx \frac{p(x|\theta)}{\int_\Theta p(x|\theta)\,d\theta} \propto p(x|\theta),$$

since $p(\theta) \approx$ constant appears in both numerator and denominator of Bayes'
Theorem and so cancels.

We see, therefore, that, when prior beliefs are relatively weak, posterior
beliefs are dictated by the location and shape of the likelihood. In particular,

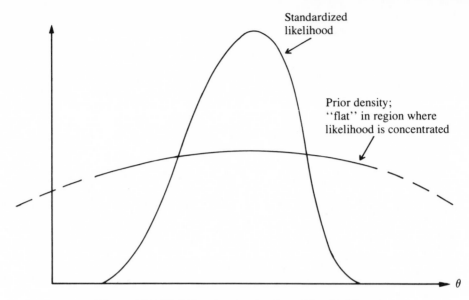

Figure 15.3.13: Prior density dominated by likelihood.

we see that the value, $\theta^*$, of $\theta$ which is considered 'most likely', given the data $x$, is the value which maximizes the likelihood $p(x|\theta)$: that is, the *maximum likelihood estimate* [see Chapter 6] much used by non-Bayesian statisticians.

The approximate argument given thus far tells us that, in the case of great prior uncertainty, posterior beliefs will peak around the maximum likelihood estimate, $\theta^*$, of $\theta$. In fact, if we are able to accept a few more approximating assumptions we can not only learn about the *location*, but also about the *shape* of the likelihood, and thus of the approximate posterior density.

To see this, let us recall from the argument given above that, approximately,

$$p(\theta|x) \propto \exp\{L(\theta)\},$$

where

$$L(\theta) = \log\{p(x|\theta)\}.$$

Now let us suppose that $L(\theta)$ can be well-approximated by considering a Taylor series expansion [see IV, § 3.6] about $\theta^*$, the series expansion being continued as far as the quadratic term. In other words, we are assuming that the logarithm of the likelihood function can be well approximated by a quadratic function in the vicinity of the maximum likelihood estimate, $\theta^*$. Assuming this form of approximation, we have

$$L(\theta) \approx L(\theta^*) + (\theta - \theta^*)L'(\theta^*) + \tfrac{1}{2}(\theta - \theta^*)^2 L''(\theta^*),$$

where $L'(\theta^*)$, $L''(\theta^*)$ denote the first and second derivatives of $L(\theta)$, with respect to $\theta$, evaluated at $\theta = \theta^*$. If we further assume that $p(x|\theta)$ has a unique maximum at $\theta^*$, then so does $L(\theta)$, since taking the logarithm of a function does not change the location of its turning point; in particular, it follows that $L'(\theta^*) = 0$ (the derivative at the maximum is zero).

Noting that, as a function of $\theta$, $L(\theta^*)$ is constant, we have the approximation

$$L(\theta) = \text{constant} - \tfrac{1}{2}(\theta - \theta^*)^2 / (-L''(\theta^*)).$$

The motivation for rewriting the quadratic term in this way becomes clear if we now note that

$$p(\theta|x) \propto \exp\left\{-\frac{1}{2\sigma_*^2}(\theta - \theta^*)^2\right\}$$

where $\sigma_*^2 = (-L''(\theta^*))^{-1}$ [cf. § 6.2.5].

It follows from the form of $p(\theta|x)$ that, provided the assumptions we have made are not unreasonable, *posterior beliefs about $\theta$ are well approximated by a Normal density with mean $\theta^*$ and variance $\sigma_*^2$*. The *location* of posterior beliefs is therefore determined by $\theta^*$, the maximum likelihood estimate, and the *spread* of posterior beliefs is inversely proportional to minus the second derivative of the log-likelihood (at $\theta^*$). This latter is really quite an intuitive measure of spread: the second derivative is measuring how quickly the gradient of the log-likelihood is changing (from positive to negative, hence the minus sign); if the gradient changes quickly, this indicates that the likelihood is sharply peaked and thus the spread is small.

EXAMPLE 15.3.4.   Suppose that data $x$ consists of the number of successes in $n$ independent trials, each with chance of success $\theta$, so that $x$ is the realization of a binomial random variable $X$ and

$$p(x|\theta) = \binom{n}{x} \theta^x (1-\theta)^{n-x}; \qquad x = 0, 1, \ldots, n; \qquad 0 \le \theta \le 1.$$

If $n$ is quite large, and $p(\theta)$ is relatively flat, we might expect the approximation discussed above to be reasonable, and so it suffices to calculate $\theta^*$ and $L''(\theta^*)$.

Noting that

$$L(\theta) = \log\{p(x|\theta)\} = \text{constant} + x \log\{\theta\} + (n-x) \log\{1-\theta\},$$

we easily obtain

$$L'(\theta) = \frac{x}{\theta} - \frac{(n-x)}{(1-\theta)},$$

and solving the equation

$$L'(\theta^*) = 0$$

leads to the maximum likelihood estimate

$$\theta^* = \frac{x}{n}.$$

Differentiating a second time gives

$$L''(\theta) = -\frac{x}{\theta^2} - \frac{(n-x)}{(1-\theta)^2}$$

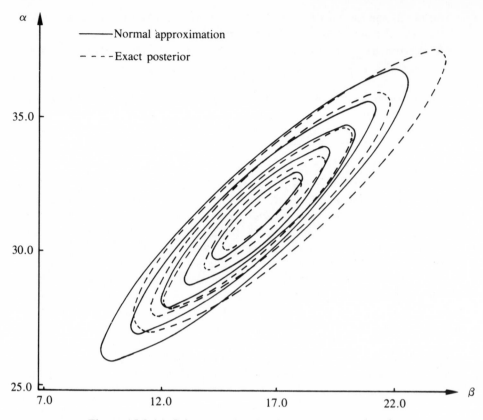

Figure 15.3.14: Joint posterior density contours for $(\alpha, \beta)$.

and, hence,

$$L''(\theta^*) = -n \left[ \frac{x}{n} \left( 1 - \frac{x}{n} \right) \right]^{-1}.$$

Applying the general result obtained above, we conclude that posterior beliefs for $\theta$ should be well approximated by

$$p(\theta|x) = N \left\{ \frac{x}{n}, \sqrt{\frac{1}{n} \left[ \frac{x}{n} \left( 1 - \frac{x}{n} \right) \right]} \right\}.$$

Recalling that the interval defined by 'mean $\pm 2$ standard deviations' contains approximately 95% of the probability in the case of a Normal distribution, we see that a Bayesian statistician using this approximation would quote posterior odds of about 19 to 1 on $\theta$ lying in the interval

$$\frac{x}{n} \pm \frac{2}{\sqrt{n}} \left[ \frac{x}{n} \left( 1 - \frac{x}{n} \right) \right]^{1/2}$$

[see Example 4.7.1].

It is interesting to compare this statement with the corresponding kind of inference that might be made without making use of Bayes' Theorem. The 'natural' estimator of $\theta$ is $X/n$, whose mean is $\theta$ and whose variance is $(1/n)[\theta(1-\theta)]$. If $n$ is large, the estimator $X/n$ should be 'close' to $\theta$ and the distribution of $X/n$ should be approximately Normal [see Example 4.7.1]. These facts make it 'plausible' to consider

$$\frac{X}{n} \pm \frac{2}{\sqrt{n}} \left[ \frac{X}{n} \left( 1 - \frac{X}{n} \right) \right]^{1/2}$$

as defining an approximate '95% *confidence interval*' for $\theta$. Having observed $X = x$, a non-Bayesian statistician might therefore end up, in this situation, making an identical numerical statement to that made by a Bayesian (even though the logic of the analysis is entirely different; see also section 15.4).

EXAMPLE 15.3.5.   Suppose that $x$ is the total number of occurrences observed from $n$ independent processes, in each of which occurrences follow a Poisson distribution with parameter $\theta$. It follows that $x$ may be regarded as the realization of a random variable $X$ having a Poisson distribution with parameter $n\theta$, and hence

$$p(x|\theta) = (n\theta)^x \, e^{-n\theta}/x!, \qquad \theta > 0.$$

If we follow through the steps required for the approximation in this case ($n$ large, $p(\theta)$ relatively flat), we have

$$L(\theta) = \log\{p(x|\theta)\} = \text{constant} + x \log\{\theta\} - n\theta,$$

and so

$$L'(\theta) = \frac{x}{\theta} - n,$$

and solving the equation

$$L'(\theta^*) = 0$$

leads to the maximum likelihood estimate

$$\theta^* = x/n.$$

Differentiating a second time gives

$$L''(\theta) = -x/\theta^2,$$

and, hence,

$$L''(\theta^*) = -n \left( \frac{x}{n} \right)^{-1}.$$

Posterior beliefs for $\theta$ in this case should therefore be reasonably approximated by

$$p(\theta|x) = N\left\{\frac{x}{n}, \sqrt{\frac{1}{n}\left(\frac{x}{n}\right)}\right\},$$

which implies approximate posterior odds of about 19 to 1 on $\theta$ lying in the interval

$$\frac{x}{n} \pm 2\left[\frac{1}{n}\left(\frac{x}{n}\right)\right]^{1/2}.$$

In this case, also, a non-Bayesian argument leads to the same approximation if we note that the estimator $X/n$ is, for large $n$, approximately Normally distributed with mean $\theta$ and variance $\theta/n$ [see Example 4.7.2].

Intervals based on posterior densities will be considered in more detail in section 15.4.2, and compared more systematically with non-Bayesian confidence intervals.

For moderate values of $n$, the approximation can often be improved by working with a function of $\theta$, rather than with $\theta$ itself. Further details can be found, for example, in section 7.2 of Lindley (1965).

In the case of more than one parameter, an extension of the above argument leads to a higher-dimensional (multivariate) form of Normal approximation. For illustration, let us consider two parameters, $\alpha$ and $\beta$, with likelihood $p(x|\alpha, \beta)$ and log-likelihood $L(\alpha, \beta) = \log\{p(x|\alpha, \beta)\}$. If $p(\alpha, \beta)$ is approximately constant, then the posterior density is given by

$$p(\alpha, \beta|x) \propto \exp\{L(\alpha, \beta)\}.$$

If $L(\alpha, \beta)$ is expanded as far as the quadratic terms in a two-dimensional Taylor series [see IV, § 5.8] about the joint maximum likelihood estimates, $\alpha^*, \beta^*$, we obtain

$$L(\alpha, \beta) \approx L(\alpha^*, \beta^*) + (\alpha - \alpha^*)L_\alpha'(\alpha^*, \beta^*) + (\beta - \beta^*)L_\beta'(\alpha^*, \beta^*)$$
$$+ \tfrac{1}{2}(\alpha - \alpha^*)^2 L_\alpha''(\alpha^*, \beta^*) + (\alpha - \alpha^*)(\beta - \beta^*)L_{\alpha\beta}''(\alpha^*, \beta^*)$$
$$+ \tfrac{1}{2}(\beta - \beta^*)^2 L_\beta''(\alpha^*, \beta^*),$$

where $L_\alpha'$, $L_\beta'$ denote the partial derivatives of $L$ with respect to $\alpha$, $\beta$, respectively, and $L_\alpha''$, $L_{\alpha\beta}''$, $L_\beta''$ denote the second order partial derivatives. Since $L_\alpha'(\alpha^*, \beta^*) = L_\beta'(\alpha^*, \beta^*) = 0$, we can write

$$p(\alpha, \beta|x) \propto \exp\left\{-\tfrac{1}{2}[\alpha - \alpha^*, \beta - \beta^*]' \, \Sigma_*^{-1}\, [\alpha - \alpha^*, \beta - \beta^*]\right\},$$

where

$$\Sigma_* = \begin{bmatrix} -L_\alpha''(\alpha^*, \beta^*) & -L_{\alpha\beta}''(\alpha^*, \beta^*) \\ -L_{\alpha\beta}''(\alpha^*, \beta^*) & -L_\beta''(\alpha^*, \beta^*) \end{bmatrix}^{-1}.$$

The approximate form of $p(\alpha, \beta|x)$ is thus that of a Bivariate Normal distribution [see II, § 13.4.6(i)] with mean $(\alpha^*, \beta^*)$, the maximum likelihood estimates, and covariance matrix $\Sigma_*$.

The method of approximation extends in an obvious way to more than two parameters.

## 15.4. BAYESIAN APPROACHES TO TYPICAL STATISTICAL QUESTIONS

### 15.4.1. Point Estimation

One of the most commonly posed statistical problems is that of providing, on the basis of observed data, an *estimate* [see § 3.1] of an unknown parameter.

As we have seen in section 15.3, given a model involving an unknown parameter $\theta$, and given data $x$, the Bayesian statistician can calculate the posterior density $p(\theta|x)$, corresponding to any particular prior density specification for $\theta$. In a sense, given $x$ and the specified likelihood and prior density, the description of the form of $p(\theta|x)$, either analytically as a mathematical function, or graphically, provides a 'complete picture' of what is now believed about the unknown parameter $\theta$. If we are asked to give a single (point) estimate of $\theta$, the question for a Bayesian statistician becomes

'What single number *best summarizes* beliefs, as represented by $p(\theta|x)$?'

Of course, the phrase 'best summary' is, as it stands, undefined: we must ask 'best with respect to what criterion?' But as soon as we begin to pose such questions, we are led towards the idea that in order to judge whether a method of choosing an estimate is sensible or not we must know something about the practical problem under study, and the actual consequences of the various discrepancies between 'estimate' and 'true value' that might arise. When viewed in this light, statistical estimation is seen to be a special kind of 'decision', whose potential consequences must be quantified before an 'optimal decision' can be made (or, in this case, 'best estimate' provided).

A detailed presentation of the decision-theoretic approach to estimation is given in section 19.2.1. In this section, we shall just consider three of the possible summaries of the posterior density that might be thought *intuitively* appealing (these are further discussed in section 19.2.1).

(i) *Mode.* The mode, $\theta^*$, of the posterior density is defined by

$$p(\theta^*|x) = \sup_{\theta \in \Theta} p(\theta|x)$$

[see II, § 10.1.3] and typically is unique. The estimate $\theta^*$ is the 'most likely' value of $\theta$ and, as we pointed out in section 15.3.5, is (approximately) equal to the maximum likelihood estimate in situations where the prior density is (approximately) specified to be constant.

(ii) *Median.* The median of the posterior density, $\tilde{\theta}$, is defined by

$$\int_{-\infty}^{\tilde{\theta}} p(\theta|x)\, d\theta = \int_{\tilde{\theta}}^{\infty} p(\theta|x)\, d\theta$$

and typically is unique. It is the point such that we have equal beliefs that the true $\theta$ lies above or below the value.

(iii) *Mean.* The mean of the posterior density, $\hat{\theta}$, is defined by

$$\hat{\theta} = \int_{-\infty}^{\infty} \theta p(\theta|x)\, d\theta$$

and is unique.

In section 15.5, we give some examples of point estimates and, for the purposes of illustration, we shall mostly use the posterior mean. It should be stressed once again, however, that a rationale can be given for each of these forms of estimate and also for other forms [see § 19.2.1], and the actual choice in any given problem must therefore depend on the relevant (decision-theoretic) considerations.

### 15.4.2.  Interval Estimation

It is clear that whatever form of point estimate is chosen it provides a poor summary of the complete posterior density, $p(\theta|x)$. A half-way house between the, perhaps complicated, description of the complete density and the over-simplification of the point estimate is the idea of a *credible interval.*

Given a posterior density function $p(\theta|x)$, we quote two values of $\theta$, $a$ and $b$ $(a < b)$, such that the posterior probability of $\theta$ lying in the interval from $a$ to $b$ is equal to some specified value (say, 90, 95, or 99%, or whatever is appropriate for the problem under study). *More precisely, we say that, given* $p(\theta|x)$, *the interval* $(a, b)$ *is a* $100(1-\alpha)\%$ *posterior credible interval for* $\theta$ *if*

$$\int_{a}^{b} p(\theta|x)\, d\theta = 1 - \alpha \qquad (0 \le \alpha \le 1)$$

(cf. Definition 4.1.1). When we use values $0\cdot1$, $0\cdot05$ or $0\cdot01$ for $\alpha$, we speak of 90, 95 or 99% credible intervals for $\theta$. The general idea is illustrated in Figure 15.4.1.

It will be seen that a 'credible interval' is formally identified with the 'probability interval' defined in Definition 4.1.1, when the probability interval is taken with respect to the posterior distribution $p(\theta|x)$.

In general, we can find many pairs of values $(a, b)$ which provide a $100(1-\alpha)\%$ credible interval, for a specified $\alpha$. This can readily be seen by considering Figure 15.4.1 and imagining $b$ displaced to the right of its present position. The shaded area would now contain *more* than $100(1-\alpha)\%$ of the probability, but this could clearly be reduced again to $100(1-\alpha)\%$ by shifting $a$ by some suitable amount to the right. Whichever particular interval is chosen, the

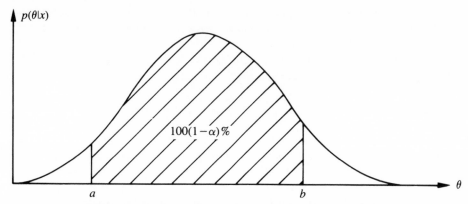

Figure 15.4.1: Illustration of a posterior credible interval.

interpretation of a credible interval is that, in terms of posterior beliefs about $\theta$, given $x$, one is

$100(1-\alpha)\%$ sure that $\theta$ lies in the interval $(a, b)$.

If we consider further the fact that many different credible intervals can be given for any particular choice of $\alpha$, we see that the notion of a credible interval as presented so far is not entirely satisfactory. For example, in Figure 15.4.2 both the intervals $(a_1, b_1)$ and $(a_2, b_2)$ are $100(1-\alpha)\%$ posterior

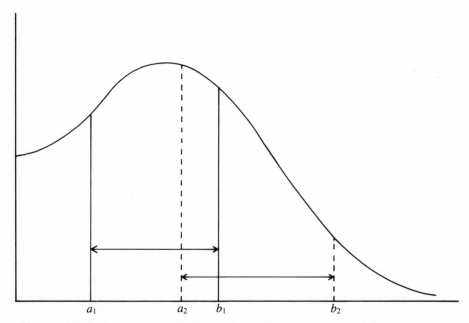

Figure 15.4.2: Two different $100(1-\alpha)\%$ credible intervals $(a_1, b_1)$ and $(a_2, b_2)$.

credible intervals for $\theta$, but, from the point of view of communicating information about $\theta$, $(a_1, b_1)$ is clearly preferable since, for a given $\alpha$, a shorter interval
represents a more unambiguous inference. This observation motivates a more
refined notion of credible interval. An interval $(a, b)$ is said to be a $100(1-\alpha)\%$
*highest posterior density interval* if

  (i) $(a, b)$ is a $100(1-\alpha)\%$ credible interval;
  (ii) for all $\theta' \in (a, b)$ and $\theta'' \notin (a, b)$, $p(\theta'|x) \geq p(\theta''|x)$.

The extra condition (ii) requires that no value of $\theta$ included in the interval
$(a, b)$ should have an ordinate, $p(\theta|x)$, of the posterior density lower than any
value of $\theta$ excluded from $(a, b)$. Clearly, this produces the shortest possible
credible interval, for a given $\alpha$, and hence the 'most informative' interval
estimate summary of $p(\theta|x)$. Examples will be given in section 15.5.

It is important to distinguish the credible interval approach from the super-
ficially similar approach to interval estimation based on confidence intervals
[see §§ 4.1.3, 4.2]. The former uses $p(\theta|x)$ and refers directly to *the probability
of $\theta$ lying in a particular interval.* The latter considers intervals with random
endpoints, $a(X)$ and $b(X)$, which have the property that, in terms of the
distribution $f(x|\theta)$, *the probability of the interval containing $\theta$ has the value*
$1-\alpha$, for suitably specified $\alpha$. The particular interval $[a(x), b(x)]$ which obtains
when $X = x$ is observed is then referred to as a $100(1-\alpha)\%$ confidence
interval.


### 15.4.3. Significance Testing

In many situations modelled by a family of densities, $p(x|\theta)$ with unknown
$\theta$, it is of interest to examine whether the data is or is not 'compatible' with
a *particular* value of the unknown parameter, $\theta_0$ say. If we consider $H_0$: $\theta = \theta_0$
as a basic hypothesis of interest (usually called the *null hypothesis*; [see § 5.2.1]),
we are led to seek a procedure for 'rejecting' or 'not rejecting' such an
hypothesis [see § 5.12].

A *significance testing* approach has been developed in a non-Bayesian
framework and details have been given in Chapter 5. One way of looking at
such non-Bayesian significance tests in the case of parametric models, $f(x|\theta)$,
is to note that we 'reject' $H_0$: $\theta = \theta_0$ at the $100\alpha\%$ level if $\theta_0$ does not lie in
a (sensible) $100(1-\alpha)\%$ confidence interval.

Proceeding analogously, a possible *Bayesian significance test* is provided by
rejecting $H_0$: $\theta = \theta_0$ 'at the $100\alpha\%$ level' if $\theta_0$ lies outside the $100(1-\alpha)\%$
*highest posterior density interval.*

A different Bayesian approach to significance testing has been developed
by Jeffreys (1967), and a brief introduction to the basic ideas is given by
Lindley (1972). We shall not give details here, but this approach to the problem
of comparing the plausibilities of *two* particular rival hypothesised values of
$\theta$ will be considered in section 19.2.2.

### 15.4.4.   Prediction

If we consider 'predicting' a future observation or observations, $x$ say, on the basis of previous data $y$, the Bayesian statistician will seek to derive his distribution of belief for $x$, given $y$. Probabilistically, we are led to consider $\pi(x|y)$, the so-called *predictive density* for $x$, given $y$.

Typically, we shall not have a model which specifies this density directly. Rather, we will usually have a probability model for $x$, $g(x|\theta)$, depending on an unknown parameter $\theta$, which itself appears in the model assumed for $y$. If $p(\theta|y)$ is the posterior density for $\theta$, given the previous data $y$, and if $x$ and $y$ are independent, given $\theta$, we may obtain the predictive density for $x$ from the formula

$$\pi(x|y) = \int_\Theta g(x|\theta)p(\theta|y)\, d\theta.$$

If a 'single-figure' prediction is required, we simply choose a point estimate summary of the density $\pi(x|y)$ using the ideas of section 15.4.1. If a predictive interval is required, we can use the idea of a credible interval or a 'highest predictive density interval', an obvious modification of the ideas of section 15.4.2 applied to the density $\pi(x|y)$.

A very full account of the theory and application of predictive densities is given in Aitchison and Dunsmore (1975). We shall provide some specific examples in section 15.5.

### 15.4.5.   Summarizing Data: Sufficient Statistics

In an earlier chapter, we considered the problem of summarizing a set of observations in the form of a small number of summary statistics without losing relevant information. This led to the notion of *sufficient statistics* [see § 3.4].

In order to re-examine this problem from a Bayesian point of view, we can argue as follows. Given a likelihood defined by $f(x|\theta)$ and a prior density $p(\theta)$, the Bayesian approach tells us to combine these, using Bayes' Theorem, and thus to determine $p(\theta|x)$, the posterior density given *all the data, x*. Suppose now that $t(x)$ is some summary of the data $x$; for example, we might have $x = (x_1, \ldots, x_n)$ and $t(x) = n^{-1} \sum x_i$, or $t(x) = \min\{x_1, \ldots, x_1\}$, etc. Since $p(\theta|x)$ is the 'complete' representation of current beliefs about $\theta$, given $x$, the summary $t(x)$ can only be said 'not to have lost any relevant information' if $p(\theta|t(x))$ is equal to $p(\theta|x)$. Moreover, for an agreed $f(x|\theta)$, if statisticians with different specifications of $p(\theta)$ are to agree that $t(x)$ constitutes an acceptable sufficient summary we should require that

$$p(\theta|t(x)) = p(\theta|x) \quad \text{for all } p(\theta).$$

If this condition holds, we say that $t(x)$ is a *Bayes' sufficient statistic.*

This definition of a sufficient statistic appears to be entirely different from the (non-Bayesian) definition given in section 3.4.1. In fact, however, the two definitions can be shown to be equivalent. A detailed demonstration of this fact can be found in Raiffa and Schlaifer (1961). Provided, therefore, that the form of probability model $f(x|\theta)$ is agreed, both Bayesian and non-Bayesian statisticians will base their analyses on the same data summary (the sufficient statistics).

### 15.4.6.   The Likelihood Principle

Suppose an investigation consists of a sequence of independent experiments, each having a chance $\theta$ of resulting in a success and $1-\theta$ of resulting in a failure. Suppose further that it is reported that a total of $n$ experiments were performed and $y$ successes were obtained.

If we denote the data by $x = (n, y)$, do we have sufficient information to write down a probability model relating $x$ to $\theta$?

The answer, of course, is *no*, since we are not told what mode of experimentation (or sampling) was employed. For example, if $n$ were fixed in advance and we simply observed $y$ then we would have the Standard Binomial probability distribution with [see II, § 5.2.2]

$$f(y|\theta, n) = \binom{n}{y} \theta^y (1 - \theta)^{n-y}, \qquad y = 0, 1, \ldots, n.$$

If, on the other hand, we had decided to fix $y$ in advance and then observe $n$—that is, to continue experimenting until $y$ successes were achieved and then to note how many experiments, $n$, had been required—we would instead have the Negative-Binomial distribution [see II, § 5.2.4] with

$$f(n|\theta, y) = \binom{n-1}{y-1} \theta^y (1 - \theta)^{n-y}, \qquad n = y, y+1, y+2, \ldots.$$

There are, of course, many other sampling rules that might have been employed—for example, 'continue experimenting until lunchtime, and then stop' (fixing neither $n$ or $y$)—but the point to which we wish to draw attention can be illustrated using just the Binomial and Negative-Binomial forms.

The basic question is as follows: given the data alone, that is, without being told the mode of sampling, can we make standard inference statements about $\theta$? For example, can we provide point or interval estimates?

The answer to this is that the Bayesian approach *can* provide inferences without knowing whether the Binomial or Negative-Binomial distribution is appropriate, whereas many non-Bayesian procedures *cannot*.

To see this, let us make the assumption that prior beliefs about $\theta$ are not influenced by the form of sampling employed, so that we can specify $p(\theta)$ independently of the form of likelihood.

For the *Binomial* assumption [see Example 3.4.6(a)], Bayes' Theorem gives

$$p(\theta|x) = p(\theta|n, y) \propto f(y|\theta, n) p(\theta)$$

$$\propto \binom{n}{y} \theta^y (1-\theta)^{n-y} p(\theta)$$

$$\propto \theta^y (1-\theta)^{n-y} p(\theta),$$

since $\binom{n}{y}$ does not involve $\theta$, and thus

$$p(\theta|x) = \frac{\theta^y (1-\theta)^{n-y} p(\theta)}{\int_0^1 \theta^y (1-\theta)^{n-y} p(\theta) \, d\theta}.$$

For the *Negative-Binomial* assumption [see Example 3.4.6(b)], Bayes' Theorem gives

$$p(\theta|x) = p(\theta|n, y) \propto f(n|\theta, y) p(\theta)$$

$$\propto \binom{n-1}{y-1} \theta^y (1-\theta)^{n-y} p(\theta)$$

$$\propto \theta^y (1-\theta)^{n-y} p(\theta),$$

since $\binom{n-1}{y-1}$ does not involve $\theta$, and thus

$$p(\theta|x) = \frac{\theta^y (1-\theta)^{n-y} p(\theta)}{\int_0^1 \theta^y (1-\theta)^{n-y} p(\theta) \, d\theta}.$$

*The posterior density $p(\theta|x)$ is thus seen to be identical in the two cases.*

However, if we consider non-Bayesian procedures, such as minimum variance unbiased estimation (MVUE), [see § 3.5.2] confidence intervals [see Chapter 4], or significance tests [see Chapter 5], the precise forms of these differ according as we make the Binomial or Negative-Binomial assumption. For example, under the Binomial assumption $Y/n$ is a MVUE, whereas under the Negative-Binomial assumption $(Y-1)/(n-1)$ is the MVUE [see Example 3.4.8].

The Binomial/Negative-Binomial example is a special case of the following situation. We have two different probability models $f_1(x|\theta)$, $f_2(x|\theta)$, satisfying

$$f_1(x|\theta) = c_1(x)g(x, \theta), \qquad f_2(x|\theta) = c_2(x)g(x, \theta)$$

with two different ranges of possible values of $x$. The part of the density depending on $\theta$ is the same in the two cases. The proportional form of Bayes' Theorem makes clear that, given $x$ and $p(\theta)$, the form of $p(\theta|x)$ will be the same in both cases, since it does not require the explicit form of $c_i(x)$, $i = 1, 2$.

Another way of interpreting this is to note that $g(x, \theta)$ takes into account only the *actual* $x$ observed, whereas the forms of $c_i(\cdot)$, $i = 1, 2$ reflects the range of *all possible* $x$ that might have been obtained.

The *likelihood principle* (see § 2.3 of Cox and Hinkley (1974)—see Bibliography C for further details) suggests that inferences about $\theta$ should be based only upon $g(x, \theta)$, not taking into account $c_i(x)$, $i = 1, 2$. Bayesian procedures (and others such as maximum likelihood) may be said to *obey* the likelihood principle; procedures like MVUE are said to *violate* the likelihood principle.

A discussion of the pros and cons of this, and other principles underlying debates on statistical inference, can be found in e.g. § 6.8.2 of Barnett (1982)—see Bibliography C.

## 15.5. BAYESIAN INFERENCE FOR SOME UNIVARIATE PROBABILITY MODELS

### 15.5.1.  Inferences for the Binomial and Related Distributions

If $y$ denotes the number of 'successes' obtained in $n$ independent 'trials', each with chance $\theta$ of a 'success', then if $n$ is fixed in advance we obtain the Binomial probability model [see II, § 5.2.2]:

$$f(y|\theta, n) = \binom{n}{y} \theta^y (1 - \theta)^{n-y}, \qquad y = 0, 1, \ldots, n.$$

Let us suppose that prior beliefs about $\theta$ $(0 \le \theta \le 1)$ are specified in the form of a density $p(\theta)$. The posterior density [see § 15.3.1] is then given by

$$p(\theta|y, n) = \frac{f(y|\theta, n)p(\theta)}{\int_0^1 f(y|\theta, n) p(\theta) \, d\theta} \qquad (0 \le \theta \le 1).$$

As we have remarked in our general discussion of the continuous form of Bayes' Theorem, the form of $p(\theta|y, n)$ can easily be computed numerically *for any choice of* $p(\theta)$. However, this is tiresome if we wish to explore how $p(\theta|y, n)$ varies for *different choices* of $p(\theta)$, since we require a separate numerical exercise for each specification. Moreover, simply producing a curve numerically fails to provide any analytic insight into the manner in which the data and prior beliefs interact to form posterior beliefs.

For these reasons, it is of interest to examine a particular *mathematical* representation of $p(\theta)$ (whilst bearing in mind the practical approaches to assessing *actual* forms of $p(\theta)$ discussed in section 15.3.2). In order to obtain mathematical representations which afford both theoretical and practical insights, it would be very convenient if we could discover a *family of probability densities* which, by varying the small number of parameters of a mathematical function, could be made to generate a range of 'shapes' of prior beliefs which would adequately represent many actual forms of belief occurring in practice.

In the Binomial case, we require a family of functions defined on the interval $0 \le \theta \le 1$, and specified in terms of just a few parameters, which can be varied to provide a number of flexible forms. Such a family of densities is the *beta family* [see II, § 11.6], defined, for $a > 0$, $b > 0$, by

$$p(\theta) = \frac{\Gamma(a+b)}{\Gamma(a)\Gamma(b)} \theta^{a-1}(1 - \theta)^{b-1} \qquad (0 \le \theta \le 1),$$

where $\Gamma(\cdot)$ is the gamma function [see IV, § 10.2], having the property $\Gamma(z) = z\Gamma(z-1)$, $z > 0$. Since $p(\theta)$ is a density, we have $\int p(\theta)\, d\theta = 1$, and so,

$$\int_0^1 \theta^{a-1}(1-\theta)^{b-1}\, d\theta = \frac{\Gamma(a)\Gamma(b)}{\Gamma(a+b)}.$$

By varying $a$ and $b$, a very wide family of shapes can be generated, some examples being shown in Figure 15.5.1. With the advent of interactive computing, it is possible to interrogate a subject in order to discover whether (and, if so, for which $a$, $b$) his beliefs about $\theta$ can be adequately represented by a beta density.

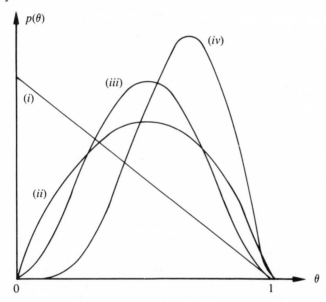

Figure 15.5.1: Examples of beta densities: (i) $a = 1$, $b = 2$; (ii) $a = 2$, $b = 2$; (iii) $a = 3$, $b = 3$; (iv) $a = 5$, $b = 3$.

For a particular choice of $a$, $b$, the proportional form of Bayes' Theorem gives

$$p(\theta|y, n) \propto f(y|\theta, n)\, p(\theta)$$
$$\propto \theta^y(1-\theta)^{n-y} \times \theta^{a-1}(1-\theta)^{b-1}$$
$$\propto \theta^{a+y-1}(1-\theta)^{b+n-y-1}.$$

Hence we have

$$p(\theta|y, n) = \frac{\theta^{a+y-1}(1-\theta)^{b+n-y-1}}{\int_0^1 \theta^{a+y-1}(1-\theta)^{b+n-y-1}\, d\theta}$$

$$= \frac{\Gamma(a+b+n)}{\Gamma(a+y)\Gamma(b+n-y)}\, \theta^{a+y-1}(1-\theta)^{b+n-y-1},$$

using the result noted above in order to evaluate the integral.

Now we have an interesting result. *The posterior density is also of the beta form*, but with parameters $a + y$ and $b + n - y$ in place of the prior parameters $a$, $b$. Schematically, we may write this as

$$\frac{\text{Prior}}{\text{Beta } (a, b)} \quad \text{with} \quad \frac{\text{Likelihood}}{\text{Binomial } [n, y]} \quad \text{implies} \quad \frac{\text{Posterior}}{\text{Beta } (a + y, b + n - y)}.$$

This result provides a very simple rule for updating beliefs in the case of a Binomial probability model and prior beliefs represented by a beta density. The above analysis then suffices as a once-and-for-all solution for any choice of $a$, $b$ and for any data $n$, $y$. Provision of credible intervals, for example, can straightforwardly be made by reference to suitable tables of the beta distribution, or by noting that the quantity

$$\left(\frac{b + n - y}{a + y}\right)\left(\frac{\theta}{1 - \theta}\right)$$

has an $F_{2(a+y),2(b+n-y)}$ distribution [see § 2.5.6(b)], so that tables of the $F$-distribution may be utilized.

For example, if $F$, $\bar{F}$ denote the lower and upper $100(\alpha/2)\%$ points of this $F$-distribution [see § 2.5.6(a)], a $100(1 - \alpha)\%$ posterior credible interval for $\theta$ is given by

$$\left[\frac{(a + y)F/(b + n - y)}{1 + (a + y)F/(b + n - y)}, \frac{(a + y)\bar{F}/(b + n - y)}{1 + (a + y)\bar{F}/(b + n - y)}\right].$$

Since lower percentage points of the $F$-distribution are not usually tabulated, it is more convenient to note that the lower $100p\%$ point of an $F_{u,v}$-distribution is equal to the reciprocal of the upper $100p\%$ point of an $F_{v,u}$-distribution [see § 2.5.6(a)]. If we denote the latter by $F^*$, in the case $u = 2(a + y)$, $v = 2(b + n - y)$ the credible interval for $\theta$ can be rewritten in the form

$$\left[\frac{1}{1 + (b + n - y)F^*/(a + y)}, \frac{1}{1 + (b + n - y)/(a + y)\bar{F}}\right].$$

EXAMPLE 15.5.1.   If $a = 1$, $b = 1$, $n = 15$, $y = 4$, the quantity $(12/5) \times [\theta/(1 - \theta)]$ has an $F_{10,24}$ posterior distribution. With $\alpha = 0 \cdot 05$, for example, we have $\bar{F} = 2 \cdot 64$, $F^* = 3 \cdot 37$ and so a 95% posterior credible interval for $\theta$ is given by

$$\left[\frac{1}{1 + (12)(3 \cdot 37)/5}, \frac{1}{1 + (12)/(5)(2 \cdot 64)}\right] = (0 \cdot 11, 0 \cdot 52).$$

(Note that this is *not* a highest posterior density interval, since standard tables do not—except in the case of symmetric densities—provide upper and lower values calculated on this basis. Results which are calculated with a view to providing highest posterior density intervals are given in *Tables for Bayesian Statisticians* by Isaacs, Christ, Novick and Jackson (1974). [See Bibliography G.] In most cases, however, the results are very similar.)

Some insight into the way in which posterior inferences combine prior information with that contained in the data is obtained by examining the form of the point estimate of $\theta$ provided by the mean of the posterior distribution. The posterior mean is given by

$$\hat{\theta} = \int_0^1 \theta p(\theta|y, n) \, d\theta$$

$$= \frac{\Gamma(a+b+n)}{\Gamma(a+y)\Gamma(b+n-y)} \int_0^1 \theta^{(a+y+1)-1} (1-\theta)^{b+n-y-1} \, d\theta$$

$$= \frac{\Gamma(a+b+n)}{\Gamma(a+y)\Gamma(b+n-y)} \frac{\Gamma(a+y+1)\Gamma(b+n-y)}{\Gamma(a+b+n+1)}$$

$$= \frac{a+y}{a+b+n},$$

using results stated above for the integral and for the gamma function.

When we examine $\hat{\theta}$ more closely, we see that it can be rewritten in the form

$$\hat{\theta} = \frac{(a+b)\left(\dfrac{a}{a+b}\right) + n\left(\dfrac{y}{n}\right)}{a+b+n}$$

$$= (1-w)\left(\frac{a}{a+b}\right) + w\left(\frac{y}{n}\right),$$

where $w = n/(a+b+n)$.

This reveals that the mean of the posterior distribution is a *weighted average* of two quantities, $a/(a+b)$ and $y/n$. The former is, in fact, *the mean of the prior distribution* (which can be shown by direct calculation, or can be deduced from the form of the posterior mean with $n = 0$, $y = 0$); the latter is *the 'natural' estimate of $\theta$ using the data alone* (and is also the maximum likelihood estimate and MVUE).

The posterior estimate thus combines what the data tells us, $y/n$, with our 'best guess' before seeing the data, $a/(a+b)$. As the amount of data increases, that is as $n$ becomes larger and larger, the weight $w$ which attaches to the data-based estimate, $y/n$, becomes larger:

$$\frac{n}{a+b+n} = \frac{1}{1+(a+b)/n} \to 1 \quad \text{as } n \to \infty.$$

Conversely, if we have no data, that is $n = 0$, we use the prior estimate $a/(a+b)$ (since $w = 0$). The form of the posterior estimate therefore adapts itself, in an intuitively sensible way, taking account of the amount of data available for modifying prior beliefs.

In addition to the summary of the posterior density provided by a point estimate, we could also look at what happens to the 'spread' of the posterior as $n$ increases. If we are to assess the spread by looking at the variance of the posterior distribution, we shall require the form of the variance of a beta distribution with parameters $(a+y, b+n-y)$. Using standard methods for finding variances [see II, § 10.4.1], this is easily shown to be equal to

$$[(a+y)(b+n-y)]/[(a+b+n)^2(a+b+n+1)].$$

As $n \to \infty$, this variance clearly tends to 0. Thus, whatever the particular prior choice of $a$ and $b$, as the amount of data increases beliefs become more and more concentrated around the posterior mean. The latter itself comes more and more to resemble $y/n$ (as we have just seen) and so, eventually, no matter which particular beta prior we adopt, we would all come to believe more and more strongly that the true chance of success $\theta$ is 'very close' to the observed frequency of successes $y/n$.

More specifically, if $y$ and $n-y$ become very large compared with $a$ and $b$, the posterior density will be (approximately) beta with parameters $(y, n-y)$. Thus, the effect of a large amount of data will have been to force a range of different prior beliefs (represented by many different choices of $a$ and $b$) into a posterior *consensus*. This is the key to the Bayesian answer to accusations of the lack of 'objectivity' in Bayesian methods as a result of the 'subjective' intrusion of prior beliefs. For the Bayesian, it is natural to regard subjective beliefs as primary, and 'objective consensus' as a special case, arising when the amount of data available is large enough to overwhelm and dominate everyone's prior beliefs, forcing them into the same posterior shape.

In cases where the amount of data is *not* large enough to force this kind of consensus, a Bayesian statistician regards it as right and proper that prior beliefs will have an effect on posterior inferences. If conditions for achieving 'consensus' are not satisfied, there is no merit in pretending that there is an 'objective' answer. All that a statistician can do in such cases is to display a *range* of posterior inferences corresponding to a *range* of different kinds of prior belief. The reader (of a scientific report) or client (of a statistical consultant) can then assess his own reaction to the data by identifying the particular prior to posterior analyses that correspond most closely to his own prior beliefs. Such a display of analyses is greatly facilitated—as we have just seen—by exploiting a *mathematical* (rather than a purely numerical) approach using a flexible, tractable family of prior distributions. If a family can be found which is (a) rich enough to represent most 'shapes' of prior belief occurring in practice, and (b) 'fits nicely' with the likelihood, so that the mathematical form of the posterior density is easily identified, the inference process is then completely summarized by noting how the parameters of the prior density transform to those of the posterior. In the case of the beta density, we saw that the prior parameters $(a, b)$ are transformed into posterior parameters $(a+y, b+n-y)$, the data entering into this transformation via the sufficient statistics, $n$ and $y$.

Suppose now that having observed $y$ successes in $n$ trials, we are interested in 'predicting' the number of successes that will result in a further $m$ independent trials. If $X$ denotes this future number of successes, a Bayesian statistician will wish to calculate

$$\pi(x|m, y, n) = P(X = x|m, y, n), \qquad (x = 0, 1, \ldots, m).$$

Using straightforward probability calculus [see II, Theorem 16.2.2], these probabilities are given by

$$\pi(x|m, y, n) = \int_0^1 f(x|m, \theta)p(\theta|y, n) \, d\theta,$$

an example of a *predictive* distribution (see § 15.4.4).

In the case we are considering, assuming a prior beta $(a, b)$ distribution for $\theta$, we obtain, for $x = 0, 1, \ldots, m$,

$$\pi(x|m, y, n) = \binom{m}{x} \frac{\Gamma(a+b+n)}{\Gamma(a+y)\Gamma(b+n-y)}$$

$$\times \int_0^1 \theta^x(1-\theta)^{m-x}\theta^{a+y-1}(1-\theta)^{b+n-y-1} \, d\theta$$

$$= \binom{m}{x} \frac{\Gamma(a+b+n)}{\Gamma(a+y)\Gamma(b+n-y)}$$

$$\times \int_0^1 \theta^{a+y+x-1}(1-\theta)^{b+n-y+m-x-1} \, d\theta$$

$$= \binom{m}{x} \frac{\Gamma(a+b+n)}{\Gamma(a+y)\Gamma(b+n-y)} \frac{\Gamma(a+y+x)\Gamma(b+n-y+m-x)}{\Gamma(a+b+n+m)}.$$

EXAMPLE 15.5.2.   If $a = b = 1$, so that $p(\theta)$ is taken to be a uniform density over the interval $(0, 1)$, and if $m = x = 1$, so that we are considering the probability that a single further trial will turn out to be a success, $\pi(x|m, y, n)$ simplifies to give

$$P_r(X = 1|m = 1, y, n) = \frac{\Gamma(n+2)}{\Gamma(y+1)\Gamma(n-y+1)} \frac{\Gamma(y+2)\Gamma(n-y+1)}{\Gamma(n+3)}$$

$$= \frac{y+1}{n+2},$$

using the fact that $\Gamma(z+1) = z\Gamma(z)$.

The results discussed in this chapter have been developed on the basis of the Binomial probability model. However, as we indicated in section 15.4.6, models such as the Negative-Binomial will lead to identical forms of $p(\theta|y, n)$, for any choice of $p(\theta)$.

### 15.5.2.   Inferences for the Poisson Distribution

If $x = (x_1, x_2, \ldots, x_n)$ is a random sample from a Poisson distribution with parameter $\theta$ [see II, § 5.4], we have the probability model

$$f(x|\theta) = \prod_{i=1}^{n} \frac{\theta^{x_i} e^{-\theta}}{x_i!} \qquad (x_i \geq 0, \, i = 1, 2, \ldots, n)$$

$$= \frac{\theta^{n\bar{x}} e^{-n\theta}}{\prod_i (x_i)!},$$

where $n\bar{x} = (x_1 + x_2 + \ldots + x_n)$.

To carry out the Bayesian analysis we must specify a prior probability density $p(\theta)$ for $\theta$, the parameter appearing in the Poisson distribution. In this case, $\theta$ can be any positive real number and so

$$p(\theta|x) = \frac{f(x|\theta)p(\theta)}{\int_0^\infty f(x|\theta)p(\theta) \, d\theta}, \qquad 0 \leq \theta < \infty.$$

As we argued in the previous section, it would be very convenient if we could find a family of probability density functions, generating a wide range of shapes of possible prior beliefs as we vary the parameters of the family, and 'fitting together' in a tractable way with the likelihood defined by the Poisson model.

Such a family is provided by the gamma densities [see II, § 11.3], which have the form

$$p(\theta) = \frac{b^a \theta^{a-1} e^{-b\theta}}{\Gamma(a)}, \qquad (0 \leq \theta < \infty),$$

for any choices of $a > 0$, $b > 0$. As we vary $a$ and $b$, a wide variety of shapes can be generated, some examples being shown in Figure 15.5.2. Since $p(\theta)$ is a density, we have $\int_0^\infty p(\theta) \, d\theta = 1$, and hence

$$\int_0^\infty \theta^{a-1} e^{-b\theta} \, d\theta = \Gamma(a)/b^a.$$

Using this result, it is easy to show that the mean of the gamma distribution is equal to $a/b$, and the variance is equal to $a/b^2$ [c 11.3.2]. By appropriately choosing the parameters $a$, $b$, a shape can be chosen to reflect the location and spread of a wide range of actual prior beliefs.

For a particular choice of $a$, $b$, the proportional form of Bayes' Theorem gives

$$p(\theta|x) \propto f(x|\theta)p(\theta)$$

$$\propto \theta^{n\bar{x}} e^{-n\theta} \theta^{a-1} e^{-b\theta}$$

$$\propto \theta^{a+n\bar{x}-1} e^{-(b+n)\theta}.$$

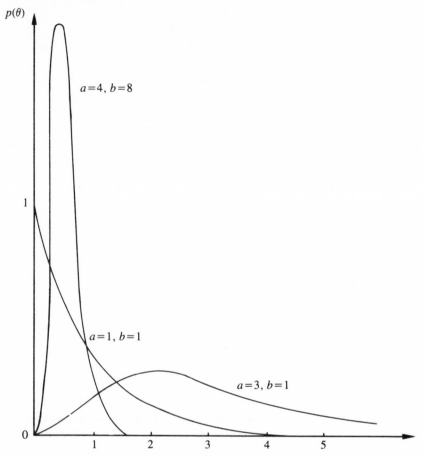

Figure 15.5.2: Examples of gamma densities: $a = 1$, $b = 1$; $a = 3$, $b = 1$; $a = 4$, $b = 8$.

Hence we have

$$p(\theta|x) = \frac{\theta^{a+n\bar{x}-1} e^{-(b+n)\theta}}{\int_0^\infty \theta^{a+n\bar{x}-1} e^{-(b+n)\theta} \, d\theta}$$

$$= \frac{(b+n)^{a+n\bar{x}} \theta^{a+n\bar{x}-1} e^{-(b+n)\theta}}{\Gamma(a+n\bar{x})},$$

the expression for the integral being obtained from the form noted above.

Comparing the forms of $p(\theta)$ and $p(\theta|x)$, we note that the latter is also a gamma density, and that we have demonstrated a general result which can be expressed schematically in the form

| | | | | |
|---|---|---|---|---|
| $\dfrac{\text{Prior}}{\text{Gamma } (a, b)}$ | with | $\dfrac{\text{Likelihood}}{\text{Poisson } [n\bar{x}]}$ | implies | $\dfrac{\text{Posterior}}{\text{Gamma } (a + n\bar{x}, b + n)}$ . |

This result provides a simple rule for updating beliefs in the case of a Poisson probability model and prior beliefs represented by a gamma density. The parameters of the latter are simply transformed using the sufficient statistics $n$ and $\bar{x}$.

Posterior credible intervals can easily be derived using tables of the $\chi^2$-distribution by noting that the quantity $2(b+n)\theta$ has a $\chi^2$-distribution with $2(a+n\bar{x})$ degrees of freedom, a fact which is easily demonstrated by the standard transformation technique [c.f. § 2.5.4(a)].

The posterior mean, a possible choice of point estimate, is given by

$$\hat{\theta} = \int_0^\infty \theta p(\theta|x) \, d\theta$$

$$= \frac{(b+n)^{a+n\bar{x}}}{\Gamma(a+n\bar{x})} \int_0^\infty \theta^{(a+n\bar{x}+1)-1} e^{-(b+n)\theta} \, d\theta$$

$$= \frac{(b+n)^{a+n\bar{x}}}{\Gamma(a+n\bar{x})} \cdot \frac{\Gamma(a+n\bar{x}+1)}{(b+n)^{a+n\bar{x}+1}}$$

$$= \frac{a+n\bar{x}}{b+n},$$

using the fact that $\Gamma(z+1) = z\Gamma(z)$.

Noting that we can write

$$\frac{a+n\bar{x}}{b+n} = \frac{b(a/b) + n(\bar{x})}{b+n},$$

we see again that the posterior mean is a weighted average of the *prior estimate* of $\theta (a/b)$ and the *data-based estimate* $(\bar{x})$. As $n$ becomes large, the weight attached to $\bar{x}$ becomes larger and approaches 1. Moreover, the posterior variance $(a+n\bar{x})/(b+n)^2$ tends to 0 and so beliefs become more and more concentrated around $\bar{x}$, no matter what the precise initial choice of $a$, $b$ was, provided the latter are small compared with $n\bar{x}$ and $n$. In this case, a possibly widely different set of prior beliefs will be forced to a posterior consensus of beliefs, well represented by a gamma distribution with parameters $n\bar{x}$ and $n$. The mean and variance of this posterior distribution are given by $\bar{x}$ and $\bar{x}/n$, respectively, and if we take the mean $\pm 2$ s.d. as an approximate 95% posterior credible interval, we have a verification of a result obtained in Example 15.3.5 of section 15.3.5 using a different approach.

If we are interested in a predictive distribution for the next observation to be made, $y = x_{n+1}$, we calculate

$$\pi(y|x) = \int_0^\infty f(y|\theta) p(\theta|x) \, d\theta$$

$$= \frac{(b+n)^{a+n\bar{x}}}{y!\,\Gamma(a+n\bar{x})} \int_0^\infty \theta^{a+n\bar{x}+y-1} e^{-(b+n+1)\theta} \, d\theta$$

$$= \frac{(b+n)^{a+n\bar{x}}}{y!\,\Gamma(a+n\bar{x})} \cdot \frac{\Gamma(a+n\bar{x}+y)}{(b+n+1)^{a+n\bar{x}+y}}.$$

EXAMPLE 15.5.3. Suppose $y = 0$, so that we are interested in the probability of the next observation being zero, then

$$P_r(Y = 0|x) = \left(\frac{b+n}{b+n+1}\right)^{a+n\bar{x}}.$$

This is seen to be an intuitively sensible form. All other things being equal, small values of $n\bar{x}$ (i.e. a small total number of observations on the $n$ previous occasions) will lead to values close to 1 (particularly if $n$ is large). On the other hand, if $n\bar{x}$ is large the quantity in brackets (which is less than 1) is raised to a high power and a small probability is obtained.

### 15.5.3. Inferences for the Normal Distribution

*Unknown mean, known variance*

If $x = (x_1, x_2, \ldots, x_n)$ is a random sample from a Normal distribution with *unknown* mean $\mu$, and *known* variance $\sigma^2$, we have from (2.5.1) the probability model

$$f(x|\mu, \sigma^2) = \prod_{i=1}^{n} \left[\left(\frac{1}{2\pi\sigma^2}\right)^{1/2} \exp\left(-\frac{1}{2\sigma^2}(x_i - \mu)^2\right)\right]$$

$$= \left(\frac{1}{2\pi\sigma^2}\right)^{n/2} \exp\left(-\frac{1}{2\sigma^2} \sum_{i=1}^{n} (x_i - \mu)^2\right)$$

$$= \left(\frac{1}{2\pi\sigma^2}\right)^{n/2} \exp\left(-\frac{1}{2\sigma^2}\left[\sum_{i=1}^{n} (x_i - \bar{x})^2 + n(\bar{x} - \mu)^2\right]\right),$$

and to perform a Bayesian analysis we must specify $p(\mu)$ and calculate

$$p(\mu|x, \sigma^2) = \frac{f(x|\mu, \sigma^2)p(\mu)}{\int_{-\infty}^{\infty} f(x|\mu, \sigma^2)p(\mu)\, d\mu}, \qquad -\infty < \mu < \infty.$$

Although the posterior density can be calculated numerically for *any* choice of $p(\mu)$, as in sections 15.5.1 and 15.5.2 it is of some interest to examine the form of $p(\mu|x, \sigma^2)$ resulting from the specification of a particular mathematical form for $p(\mu)$.

In this case, the choice

$$p(\mu) = \left(\frac{1}{2\pi\beta}\right)^{1/2} \exp\left(-\frac{1}{2\beta}(\mu - \alpha)^2\right)$$

represents beliefs about $\mu$ which are centred at the value $\alpha$ and which fall away symmetrically either side of $\alpha$ in the manner of a Normal curve. The variance parameter $\beta$ reflects the strength of these beliefs. If $\beta$ is small, the curve is narrow and peaked; if $\beta$ is large, the curve is very flat and spread out (i.e. beliefs are rather vague).

To derive the posterior density, we note that

$$p(\mu|x, \sigma^2) \propto f(x|\mu, \sigma^2)p(\mu)$$

$$\propto \exp\left(-\frac{n}{2\sigma^2}(\bar{x}-\mu)^2\right) \times \exp\left(-\frac{1}{2\beta}(\mu-\alpha)^2\right).$$

Noting that

$$\frac{n}{\sigma^2}(\mu-\bar{x})^2 + \frac{1}{\beta}(\mu-\alpha)^2 = \left(\frac{n}{\sigma^2}+\frac{1}{\beta}\right)\mu^2 - 2\left(\frac{n}{\sigma^2}\bar{x}+\frac{\alpha}{\beta}\right)\mu + \ldots,$$

where the remaining terms do not involve $\mu$, we see, on completing the square, that

$$p(\mu|x, \sigma^2) \propto \exp\left\{-\frac{1}{2((n/\sigma^2)+(1/\beta))^{-1}}\left[\mu - \frac{(n/\sigma^2)\bar{x}+(\alpha/\beta)}{(n/\sigma^2)+(1/\beta)}\right]^2\right\}.$$

We recognize from this that the posterior distribution for $\mu$ is Normal with mean

$$\alpha^* = \left(\frac{n}{\sigma^2}\bar{x}+\frac{\alpha}{\beta}\right)\Big/\left(\frac{n}{\sigma^2}+\frac{1}{\beta}\right)$$

and variance

$$\beta^* = \left(\frac{n}{\sigma^2}+\frac{1}{\beta}\right)^{-1}.$$

Schematically, we can summarize the above in the form

| Prior | with | Likelihood | implies | Posterior |
|-------|------|-----------|---------|-----------|
| Normal $(\alpha, \beta)$ | | Normal $[n\bar{x}, \sigma^2]$ | | Normal $(\alpha^*, \beta^*)$ |

We see that $\alpha^*$ again has the form of a weighted average between $\bar{x}$ (the data-based estimate of $\mu$) and $\alpha$ (the prior mean for $\mu$). If $\sigma^2/n$ is small compared with $\beta$, the weight concentrates on $\bar{x}$ and $\beta^* \to \sigma^2/n$. In this latter case of relatively vague prior beliefs, we can say that the posterior distribution for $\mu$ is approximately Normal with mean $\bar{x}$ and variance $\sigma^2/n$. Alternatively, this implies that, in terms of posterior beliefs about $\mu$, given $\bar{x}$, the quantity

$$\frac{\sqrt{n}(\mu-\bar{x})}{\sigma}$$

has a Standard Normal distribution [see II, § 11.4.1]. Using tables of the Normal distribution [see Appendix (T4)] to identify upper and lower percentage points, credible intervals for $\mu$ can be straightforwardly obtained.

It is important not to confuse the development given above with the statement that, in terms of the distribution of $\bar{X}$, given $\mu$, the quantity

$$\frac{\sqrt{n}\,(\bar{X}-\mu)}{\sigma}$$

has a Standard Normal distribution. The probability distributions on the basis of which the statements are made (i.e. that of $\bar{X}$ given $\mu$, and $\mu$ given $\bar{x}$, respectively) are of a different kind; one is determined by $f(x|\mu, \sigma^2)$, the probability model, the other by $p(\mu|x, \sigma^2)$, a distribution of belief obtained from Bayes' Theorem. On the other hand, the *numerical* forms of (a) credible intervals and (b) confidence intervals (for a specified content $100(1-\alpha)\%$) will be identical for any given $\bar{X} = \bar{x}$.

Thus, in this case, when prior beliefs are relatively vague, the numerical answers provided by Bayesian statisticians will not differ from those provided by non-Bayesian statisticians, even though the methods employed are derived in an entirely different way. Note, however, that there will *not* be a unique Bayesian answer—and thus no such agreement with non-Bayesian results—if prior beliefs are not vague compared with the information provided by the data (i.e. if $\sigma^2/n$ is not small compared with $\beta$).

*Unknown mean, unknown variance*

In the more commonly occurring situation where both $\mu$ and $\sigma^2$ are unknown, we must assign a *joint* prior density function $p(\mu, \sigma^2)$ to the two unknown parameters, deriving a joint posterior density function

$$p(\mu, \sigma^2|x) \propto f(x|\mu, \sigma^2)p(\mu, \sigma^2).$$

A very general discussion of the assignment of $p(\mu, \sigma^2)$ would have to consider many possible cases. For example: are prior beliefs for $\mu$ and $\sigma^2$ independent, so that $p(\mu, \sigma^2)$ can be written in the form $p(\mu)p(\sigma^2)$, the product of separately assessed marginal densities for $\mu$ and $\sigma^2$? The specific choice to be made will clearly depend on considerations of this kind, which are not present in the one-parameter case.

We shall not provide here a catalogue of possible prior to posterior analysis. Rather, we shall just consider briefly the special case of *independent* and *vague* prior beliefs for $\mu$ and $\sigma^2$. The reason for this will become clear shortly.

To obtain an analysis of this special case we proceed as follows. We recall first that in the previous section ($\mu$ unknown, $\sigma^2$ known) a 'vague' prior representation of beliefs about $\mu$ was obtained by considering a Normal prior density whose variance, $\beta$, was taken to be 'large'. This, in effect, results in a very 'flat' form for $p(\mu)$, so that we can think of $p(\mu)$ as (approximately) constant. Similarly, we saw in section 15.5.2 that a vague prior specification for a positive-valued parameter could be obtained by considering a gamma $(a, b)$ density with $a$ and $b$ 'small'. Applying this latter idea to $\sigma^2$, and combining it with the above suggestion of a 'constant' approximation for $p(\mu)$, we find the resultant 'vague prior approximation' $p(\mu, \sigma^2) \propto \sigma^{-2}$.

Of course, this does *not* represent a proper probability density and should *not* be thought about seriously as a 'genuine' representation of prior beliefs. The interpretation is rather the following. If we were to assess and represent genuine beliefs which were 'very vague', in the sense that the posterior density would be dominated by 'what the data said' (i.e. the likelihood will be very peaked, compared with the prior), then the resulting posterior, obtained from the properly assessed prior, would differ very little from that obtained by using this form of approximation. The choice $p(\mu, \sigma^2) \propto \sigma^{-2}$ therefore simply provides a convenient *approximate* 'short-cut' for illustrating the kind of inferences that arise in this case.

Using Bayes' Theorem, we obtain

$$p(\mu, \sigma^2 | x) \propto f(x | \mu, \sigma^2) p(\mu, \sigma^2)$$

$$\propto (\sigma^2)^{-[(n/2)+1]} \exp \left\{ -\frac{1}{2\sigma^2} \left[ \sum_{i=1}^{n} (x_i - \bar{x})^2 + n(\bar{x} - \mu)^2 \right] \right\}.$$

To obtain marginal posterior densities for $\mu$ and $\sigma^2$, respectively, we must integrate this joint density, first with respect to $\sigma^2$ and then with respect to $\mu$. The details of this are straightforward, but not particularly interesting in themselves, and so we shall just summarize the results of the integration.

So far as $\mu$ is concerned, it turns out that the quantity

$$\frac{\sqrt{n}\,(\theta - \bar{x})}{\sqrt{\sum(x_i - \bar{x})^2 / n - 1}}$$

has Student's $t$-distribution with $n-1$ degrees of freedom [see § 2.5.5], whereas for $\sigma^2$, the quantity

$$\frac{\sum(x_i - \bar{x})^2}{\sigma^2}$$

has a $\chi^2$-distribution with $n-1$ degrees of freedom [see § 2.5.4(a)].

From Standard tables of the $t$- and $\chi^2$-distributions, interval estimates for $\mu$ and $\sigma^2$ can easily be obtained.

We should stress here that the distributions of the above quantities are based on $p(\mu, \sigma^2 | x)$, the approximate posterior distribution of beliefs for $\mu$ and $\sigma^2$, given data $x$.

These results may call to mind the similar, but conceptually distinct, frequency-based results given in section 2.5. Those results are derived from the frequency distributions of $\bar{X}$ and $\sum(X_i - \bar{X})^2$, *given* $\mu$ and $\sigma^2$. However, we see again (as we also saw in earlier sections) that when a Bayesian analysis is based upon vague prior beliefs, the reported inferences from the analysis are *numerically* identical to those given by non-Bayesian methods.

## 15.6  BAYES METHODS WHEN MODELS CONTAIN MANY PARAMETERS

### 15.6.1.  Inappropriateness of 'Completely Vague' Prior Specifications

One of the general conclusions we can draw from the kinds of examples we have considered is that although the Bayesian approach is very different conceptually from Standard statistical methods, in practice the 'answers' obtained will often not differ much from those given by Standard procedures. In particular, we have seen this to be the case for many Standard probability models containing one or two unknown parameters when initial beliefs are not strong compared with the amount of information contained in the data.

Of course, it is *not* true, even with models containing just one or two parameters, that Bayesian 'answers' will *always* be similar to non-Bayesian answers. If few data are available and prior information is substantial this will not be the case.

However, when models contain *many* parameters, even a seemingly 'large' sample may, in effect, not really contain overwhelming evidence about the unknown aspects of the model, since the information is 'spread' over many parameters. On the other hand, when a model contains many parameters it will typically be the case that we have substantial information about the *relationships* that exist among such parameters. After all, parameters usually 'represent' something 'real', and the individual 'real' elements are likely to be highly interconnected or we should not be considering them altogether in one model.

In such situations, specification of independent, vague priors to each parameter would usually *not* be a genuine representation of prior information. And yet typical non-Bayesian procedures correspond to Bayesian procedures derived using just such a prior specification. It follows that in many-parameter situations there is a great deal of scope for finding Bayesian forms of inferences that will differ considerably from standard forms.

### 15.6.2.  Simple Examples

Suppose that data consist of $k$ groups of observations, $n$ in each group, and that all observations may be assumed Normally and independently distributed with equal variances (known or unknown; it does not matter) and unknown group expectations $\theta_1, \theta_2, \ldots, \theta_k$. Given these assumptions, how should the parameters be estimated?

Standard procedures (for example, least squares (§ 3.5.2) or maximum likelihood (§ 3.5.4), would lead to the use of the sample means $\bar{x}_1, \bar{x}_2, \ldots, \bar{x}_k$, where $\bar{x}_i = (x_{i1} + x_{i2} + \ldots + x_{in})/n$, as estimates of $\theta_1, \theta_2, \ldots, \theta_k$. This would also be the solution from a Bayesian point of view if a joint prior specification for $(\theta_1, \theta_2, \ldots, \theta_k)$ had the following form:

(a) $p(\theta_1, \theta_2, \ldots, \theta_k) = \prod_i p(\theta_i)$, so that beliefs about any individual $\theta_i$ are *independent* of beliefs about other $\theta_j$'s; this means that all the $\theta_j$'s are regarded as *unrelated* parameters;

(b) each $p(\theta_i)$ corresponds to a *vague prior specification*.

But is this a *realistic* form of prior specification? Consideration of some concrete examples falling into the general structure of $k$ groups with $n$ observations in each suggests not.

For example, suppose the observations are yields of a particular crop, the groups corresponding to *slightly* different growing conditions. Suppose further that there are twenty groups $(k = 20)$ with two plants in each group $(n = 2)$. In this case, it seems intuitively much more sensible to consider estimating $\theta_i$ by an estimate of the form $w\bar{x}_i + (1 - w)\bar{x}$, where $0 < w < 1$ and $\bar{x}$ is the overall mean, $(x_{11} + x_{12} + \ldots + x_{kn})/kn$. This is a *weighted average* of information drawn from just the $i$th group $(\bar{x}_i)$ and overall information $(\bar{x})$, and reflects a feeling that $n = 2$ is small compared with $kn = 40$ and that it might be inefficient to use only information from two observations if we really believed that the effects of the different growing conditions should only be *slightly* different. The weight $w$ should reflect the relative sizes of $k$ and $n$, as well as some measure of how similar we feel the groups to be.

As a second example, suppose, instead, that observations within each group were revealed to be replicate, observed responses to a given level of a stimulus, the stimulus level being fixed for each group, but increasing as we proceed from the first to the $k$th group (with, say, levels $s_1 < s_2 < \ldots < s_k$). The 'stimulus-response' framework might refer to fertilizers and crop yields, drug doses and patient recovery rates, sensory stimulus and physiological response, and so on [cf. § 6.6].

In many such contexts, the stimulus levels span the range over which it is known that responses tend to increase, flatten out, and finally decrease as a result of excess stimulus. If we again take $n = 2$, $k = 20$, many people would be unhappy intuitively with using $\bar{x}_i$ to estimate the true response $\theta_i$ corresponding to stimulus $s_i$. Some might suggest fitting, say, a quadratic curve through the plot of the $\bar{x}_i$ against $s_i$, and then using the fitted value corresponding to $s_i$ as an estimate. Others might favour a weighted average between this fitted value and the group mean $\bar{x}_i$, the relative weights depending in the general case on $k$ and $n$.

The choice of these particular examples is not really the issue here. The point to note is that background details of the real situation provides information very different from that encapsulated in the 'independent, vague prior' form discussed above. The Bayesian conclusion is that knowledge of the relationships that hold between parameters by virtue of their *meanings* should be incorporated into the model through a prior distribution. Mathematically, this suggests using a *hierarchical* model of the form

$$\begin{cases} p(x|\theta_1, \theta_2, \ldots, \theta_k) \\ p(\theta_1, \theta_2, \ldots, \theta_k|\phi) \\ p(\phi), \end{cases}$$

where the first stage relates parameters to observations, the second specifies the nature of the relationship existing among the parameters, and the third stage incorporates numerical information (if any) about the general form of relationship specified in the second stage.

As an example, the stimulus-response situation may be modelled by considering [see § 1.4.2(i)]

$$\begin{cases} x_{ij} \sim N(\theta_i, \sigma^2) & i=1,\ldots,k, j=1,\ldots,n, \\ \theta_i \sim N(\phi_0 + \phi_1 s_i + \phi_2 s_i^2, \tau^2) & i=1,\ldots,k \\ p(\phi_0, \phi_1, \phi_2) \approx \text{constant}. \end{cases}$$

The second stage here would represent the information that true response means lie approximately on a quadratic curve, while the third stage would indicate vagueness about the precise numerical form of the quadratic.

It can be shown that the posterior mean for $\theta_i$ from such a model is of the form

$$w\bar{x}_i + (1-w)(\hat{\phi}_0 + \hat{\phi}_1 s_i + \hat{\phi}_2 s_i^2),$$

where $w = n\tau^2/(n\tau^2 + \sigma^2)$ and $\hat{\phi}_0 + \hat{\phi}_1 s + \hat{\phi}_2 s^2$ denotes the quadratic curve fitted by least squares through the means $\bar{x}_i$.

An introduction to the ideas of Bayesian hierarchical models can be found in articles by Lindley and Smith (1972), and in Smith (1973). A good introduction to the extension of such ideas to time-series models can be found in Harrison and Stevens (1976).

A.F.M.S.

## 15.7. FURTHER READING AND REFERENCES

The following short list includes some of the books that might usefully be consulted for further study of Bayesian methods and philosophy, and gives details of references to specific books and papers mentioned in the chapter.

Aitchison, J. and Dunsmore, I. R. (1975). *Statistical Prediction Analysis*, Cambridge University Press.

Box, G. E. P. and Tiao, G. C. (1973). *Bayesian Inference in Statistical Analysis*, Addison-Wesley.

de Finetti, B. (1974, 1975). *Theory of Probability*, Vol. I, Vol. II, Wiley.

Hampton, J. M., Moore, P. G. and Thomas, H. (1973). Subjective Probability and Its Measurement, *J. Roy. Statist. Soc.* (A) **136**, 21.

Harrison, P. J. and Stevens, C. F. (1976). Bayesian Forecasting, *J. Roy. Statist. Soc.* (*B*) **38**, 205.

Jeffreys, H. (1967). *Theory of Probability*, Third edition, Clarendon Press, Oxford.

Lindley, D. V. (1965). *Introduction to Probability and Statistics from a Bayesian Viewpoint*, Part 2, *Inference*, Cambridge University Press.

Lindley, D. V. (1972). *Bayesian Statistics: A Review*, S.I.A.M., Philadelphia.

Lindley, D. V. and Smith, A. F. M. (1972). Bayes Estimates for the Linear Model, *J. Roy. Statist. Soc.* (*B*) **34**, 1.

Naylor, J. C. and Smith, A. F. M. (1982). Application of a Method for the Efficient Computation of Posterior Distributions, *Applied Statistics* **31**.

Raiffa, H. and Schlaifer, R. (1961). *Applied Statistical Decision Theory*, Harvard Business School, Boston.

Reilly, P. M. (1976). The Numerical Computation of Posterior Distributions in Bayesian Statistical Inference, *Applied Statistics*, **25**, 201.

Smith, A. F. M. (1973). Bayes Estimates in One-way and Two-way Models, *Biometrica* **60**, 319.

CHAPTER 16

# Multivariate Analysis: Classical Methods

## 16.1. INTRODUCTION

Statistics is concerned with populations of entities, and samples from those populations. When each entity in the samples has but one qualitative or quantitative aspect of interest to the statistician, the population and the sample are *univariate*. The entities might be adult human beings, and the variate might be their height, (a quantitative variate) or their hair colour (a qualitative variate). When for each entity in the sample two or more variates are recorded we are dealing with a *multivariate* population: *bi*variate if the number of variates per individual is two, *tri*variate if three, and so on. With a population of adult human beings the variates might be: $X_1$, the height; $X_2$, the weight; $X_3$, the age; and $X_4$, the blood-pressure. We should then be concerned with a 4-variate population, and a *vector* random variable $\mathbf{X}$ with four *components* $X_1, X_2, X_3, X_4$:

$$\mathbf{X} = \begin{pmatrix} X_1 \\ X_2 \\ X_3 \\ X_4 \end{pmatrix}.$$

It is convenient for mathematical reasons to arrange the $X_r$ as entries in a column vector as shown, but this is typographically inconvenient. We therefore normally rewrite the vector in the form of its vector/matrix *transpose*, [see I, § 6.5] indicating this by a prime:

$$\mathbf{X} = (X_1, X_2, X_3, X_4)'.$$

In general, for a $p$-variate population, we need a random vector $\mathbf{X}$ with $p$ components $X_i$: $\mathbf{X} = (X_1, X_2, \ldots, X_p)'$, the r.v. $X_i$ referring to the $i$th component of a member of the population, we shall call $X_i$ the $i$th component random variable.

689

A 'vector observation' on the $i$th member of the sample will provide an ordinary scalar (i.e. numerically valued) observation on each of the scalar random variables $X_1$, $X_2$, $X_3$, $X_4$ for that member. Denote these scalar observations by $x_{1j}$, $x_{2j}$, $x_{3j}$, $x_{4j}$ respectively, the second subscript ('$j$') serving to identify the $j$th member of the sample. The *vector observation* on that member of the sample will be the vector

$$\mathbf{x}_j = (x_{1j}, x_{2j}, x_{3j}, x_{4j})'.$$

A bivariate example of this would be

$$\mathbf{x}_j = \begin{pmatrix} x_{1j} \\ x_{2j} \end{pmatrix} = \begin{pmatrix} 65 \cdot 2 \\ 110 \cdot 5 \end{pmatrix}$$

say; here $X_1$ represents the height in inches and $X_2$ the weight in pounds of a randomly chosen adult member of a human community; for the $j$th person the height was $65 \cdot 2$ inches and the weight $110 \cdot 5$ pounds. A *sample of size k* from a $p$-variate population will provide $k$ observation vectors, one from each sample member:

$$\mathbf{x}_1, \mathbf{x}_2, \dots, \mathbf{x}_k,$$

each being of order $(p \times 1)$, with

$$\mathbf{x}_j = (x_{1j}, x_{2j}, \dots, x_{pj})', \qquad j = 1, 2, \dots, k.$$

The $k$ sample vectors $\mathbf{x}_j$ are $k$ realizations of the vector random variable $\mathbf{X}$ whose $p$-variate distribution is under investigation.

It is often convenient to assemble the observation vectors to form a sample matrix $\mathbf{S}$ [see I, § 6.2]:

$$\mathbf{S} = (\mathbf{x}_1 \vdots \mathbf{x}_2 \vdots \dots \vdots \mathbf{x}_k) \qquad (16.1.1)$$

of order $(p \times k)$. For a sample of size $k = 4$ on a bivariate ($p = 2$) distribution of human heights ($X_1$) and weights ($X_2$) this would take the form

$$\mathbf{S} = \begin{pmatrix} x_{11} & \vdots & x_{12} & \vdots & x_{13} & \vdots & x_{14} \\ x_{21} & \vdots & x_{22} & \vdots & x_{23} & \vdots & x_{24} \end{pmatrix} \begin{matrix} \leftarrow \text{height} \\ \leftarrow \text{weight} \end{matrix} \qquad (16.1.2)$$

$$\begin{matrix} \uparrow & \uparrow & \uparrow & \uparrow \\ \text{1st person} & \text{2nd person} & \text{3rd person} & \text{4th person} \\ \text{in} & \text{in} & \text{in} & \text{in} \\ \text{sample} & \text{sample} & \text{sample} & \text{sample} \end{matrix}$$

In general the $(i, j)$ entry in $\mathbf{S}$ ($i = 1, 2, \dots, p$; $j = 1, 2, \dots, k$) is $x_{ij}$, that is the observation of the $i$th component $X_i$ on the $j$th member of the sample.

We note for future reference that, defining $\mathbf{1}$ as the column vector $(1, 1, \dots, 1)'$,

(i) $\mathbf{S1}$ is the $(p \times 1)$ column vector given by

$$\mathbf{S1} = (\textstyle\sum x_{1j}, \sum x_{2j}, \dots, \sum x_{pj}),$$

(ii) $\mathbf{1'S}$ is the $(1 \times k)$ row vector given by

$$\mathbf{1'S} = (\sum x_{i1}, \sum x_{i2}, \ldots, \sum x_{ik}),$$

(iii) $\mathbf{SS'} = \sum \mathbf{x}_j \mathbf{x}_j'$ is the symmetric $(p \times p)$ matrix whose $(i, j)$ entry is $\sum x_{ir} y_{jr}$.

In our sample of a human population, in which $X_2$ denoted weight, the mean weight in a sample of size $k$ would be

$$\bar{x}_2 = \sum_{j=1}^{k} x_{2j} / k.$$

Similarly, we define the sample mean of the $i$th component in general as

$$\bar{x}_i = \sum_{j=1}^{k} x_{ij} / k, \qquad i = 1, 2, \ldots, p; \tag{16.1.3}$$

that is, $k\bar{x}_r$ is the sum of the entries in the $r$th row of the sample matrix $\mathbf{S}$ of (16.1.1.).

For the sample matrix (16.1.2) this would give

$$\bar{x}_1 = (x_{11} + x_{12} + x_{13} + x_{14})/4$$

and

$$\bar{x}_2 = (x_{21} + x_{22} + x_{23} + x_{24})/4$$

for the mean height $\bar{x}_1$ and the mean weight $\bar{x}_2$. Using the rule for vector addition [see II, § 6.2] it will be seen that this is equivalent to writing

$$\begin{pmatrix} \bar{x}_1 \\ \bar{x}_2 \end{pmatrix} = \frac{1}{4} \left\{ \begin{pmatrix} x_{11} \\ x_{21} \end{pmatrix} + \begin{pmatrix} x_{12} \\ x_{22} \end{pmatrix} + \begin{pmatrix} x_{13} \\ x_{23} \end{pmatrix} + \begin{pmatrix} x_{14} \\ x_{24} \end{pmatrix} \right\}$$

$$= \tfrac{1}{4}(\mathbf{x}_1 + \mathbf{x}_2 + \mathbf{x}_3 + \mathbf{x}_4)$$

$$= \bar{\mathbf{x}}, \text{ say.}$$

Similarly in general, the *sample mean vector*

$$\bar{\mathbf{x}} = \begin{pmatrix} \bar{x}_1 \\ \vdots \\ \bar{x}_p \end{pmatrix} = (\bar{x}_1, \bar{x}_2, \ldots, \bar{x}_p)',$$

which records the sample means of the 1st, 2nd, $\ldots$, and $p$th characteristics, may be alternatively defined as

$$\bar{\mathbf{x}} = \frac{1}{k}(\mathbf{x}_1 + \mathbf{x}_2 + \ldots + \mathbf{x}_k)$$

$$= \sum_{i=1}^{k} \mathbf{x}_i / k. \tag{16.1.4}$$

Equivalently

$$\bar{x} = \frac{1}{k} S1$$

where $1 = (1, 1, \ldots, 1)'$ [see I, § 6.2], since $S1$ is a vector whose components are the row-sums of $S$.

For each of the $p$ 'characteristics' (height, weight, etc.) we define the *sample sum of squares*

$$\left.\begin{aligned} a_{rr} &= \sum_{i=1}^{k} (x_{ri} - \bar{x}_r)^2 \\ &= \sum_{i=1}^{k} x_{ri}^2 - k\bar{x}_r^2, \end{aligned}\right\} \qquad r = 1, 2, \ldots, p, \qquad (16.1.5)$$

and the *sample sum of products*

$$\left.\begin{aligned} a_{rs} &= \sum_{i=1}^{k} (x_{ri} - \bar{x}_r)(x_{si} - \bar{x}_s) \\ &= \sum x_{ri} x_{si} - k\bar{x}_r \bar{x}_s, \end{aligned}\right\} \qquad r, s = 1, 2, \ldots, p, (r \neq s) \quad (16.1.6)$$

where, by definition, $a_{rs} = a_{sr}$. (Note the conventional use of 'sum of squares'. It really means 'sum of squared deviations from the appropriate sample mean'. Similarly for products.) Thus for the data matrix (16.1.2), we have

$$a_{11} = (x_{11} - \bar{x}_1)^2 + (x_{12} - \bar{x}_1)^2 + (x_{13} - \bar{x}_1)^2 + (x_{14} - \bar{x}_1)^2$$

as the sample sum of squares for heights, and similarly for $a_{22}$, the sample sum of squares for weights. The sample sum of products is:

$$(x_{11} - \bar{x}_1)(x_{21} - \bar{x}_2) + (x_{12} - \bar{x}_1)(x_{22} - \bar{x}_2) + (x_{13} - \bar{x}_1)(x_{23} - \bar{x}_2)$$
$$+ (x_{14} - \bar{x}_1)(x_{24} - \bar{x}_2).$$

The sums of squares and products may be assembled in the form of matrix $A$ having $a_{rs}$ as its $(r, s)$ entry. This is the 'sample sum of squares and products matrix',

$$A = \begin{pmatrix} a_{11} & \cdots & a_{1p} \\ \vdots & & \vdots \\ a_{p1} & \cdots & a_{pp} \end{pmatrix} \qquad (16.1.7)$$

a symmetric [see I, § 6.2(v)] $(p \times p)$ matrix.

It follows from the definition that the sum of squares and products matrix $A$ can be expressed in terms of the sample matrix $S$ and the mean vector $\bar{x}$ as [see (16.1.2')]

$$A = SS' - k\bar{x}\bar{x}' \qquad (16.1.8)$$

whence

$$A = \sum_{i=1}^{k} (\mathbf{x}_i - \bar{\mathbf{x}})(\mathbf{x}_i - \bar{\mathbf{x}})'$$

$$= \sum_{r=1}^{k} \mathbf{x}_i \mathbf{x}_i' - k\bar{\mathbf{x}}\bar{\mathbf{x}}'. \qquad (16.1.9)$$

The *sample variances* are the quantities

$$a_{rr}/(k-1), \qquad r = 1, 2, \ldots, p$$

and the *sample covariances* are the quantities

$$a_{rs}/(k-1), \qquad r, s = 1, 2, \ldots, p (r \neq s),$$

[see § 2.1.2(b)] the divisors $k-1$ being chosen so as to ensure that these are unbiased estimates [see § 3.3.2] of the corresponding population variances and covariances [see § 2.1.2(a)]. The *sample covariance matrix* (sample variance-covariance matrix, sample dispersion matrix) is the symmetric ($p \times p$) matrix $\mathbf{C}$ given by

$$\left.\begin{aligned} \mathbf{C} &= \mathbf{A}/(k-1) \\ &= (\mathbf{SS}' - k\bar{\mathbf{x}}\bar{\mathbf{x}}')/(k-1) \\ &= \sum_{i=1}^{k} (\mathbf{x}_i - \bar{\mathbf{x}})(\mathbf{x}_i - \bar{\mathbf{x}})'/(k-1). \end{aligned}\right\} \qquad (16.1.9)$$

The $(r, r)$ entry in $\mathbf{C}$ is the $r$th sample variance, and the $(r, s)$ entry the $(r, s)$ sample covariance.

We now define the *expectation of a vector variable*

$$\mathbf{X} = (X_1, X_2, \ldots, X_p)'$$

as the vector

$$E(\mathbf{X}) = \{E(X_1), E(X_2), \ldots, E(X_p)\}', \qquad (16.1.10)$$

and the expectation of a matrix random variable

$$\mathbf{Z} = \begin{pmatrix} Z_{11}, & Z_{12}, & \cdots & Z_{1n} \\ Z_{21}, & Z_{22}, & \cdots & Z_{2n} \\ Z_{m1}, & Z_{m2}, & \cdots & Z_{mn} \end{pmatrix} \qquad (16.1.11)$$

$$= [Z_{rs}]$$

as

$$E(\mathbf{Z}) = [E(Z_{rs})], \qquad (16.1.12)$$

that is, as the matrix obtained from (16.1.11) on replacing each entry $Z_{rs}$ by its expectation $E(Z_{rs})$:

It follows from the unbiasedness of the sample variances and covariances as estimates of the corresponding population values that

$$E(\bar{\mathbf{X}}) = \boldsymbol{\mu}$$

and

$$\left.\begin{array}{c} \\ \\ \\ \end{array}\right\} \quad\quad (16.1.13)$$

$$E(\mathbf{C}) = \mathbf{V}$$

where $\boldsymbol{\mu}$ denotes the expectation vector and $\mathbf{V}$ the dispersion matrix of the population.

The reader is reminded of the following definition:

DEFINITION 16.1.1.    The *dispersion matrix* (also called the variance matrix, the variance-covariance matrix, or the covariance matrix) of a vector random variable $\mathbf{X} = (X_1, X_2, \ldots, X_p)'$ is a $p \times p$ matrix $\mathbf{V} = (v_{i,j})$, where

$$v_{ii} = \text{var}(x_i)$$
$$v_{ij} = \text{cov}(x_i, x_j)$$
$$i, j = 1, 2, \ldots, p$$

The *correlation matrix* is obtained from this on replacing $v_{ij}$ by $v_{ij}/\sqrt{(v_{ii}v_{jj})}$,    $i, j = 1, 2, \ldots, p$.

DEFINITION 16.1.2.    The *cross-covariance marix* of two vector random variables, $X = (X_1, X_2, \ldots, X_p)'$ and $\mathbf{Y} = (Y_1, Y_2, \ldots, Y_q)'$ is a rectangular $(p \times q)$ matrix $\mathbf{C} = (c_{ij})$ where

$$c_{ij} = \text{cov}(X_i, Y_j), \quad\quad i = 1, 2, \ldots, p, j = 1, 2, \ldots, q.$$

The *cross-correlation matrix* is obtained from this on replacing $c_{ij}$ by $c_{ij}/\sqrt{(v_{ii}w_{jj})}$,   $i = 1, 2, \ldots, p$,   $j = 1, 2, \ldots, q$,   where   $v_{ii} = \text{var}(X_i)$   and   $w_{jj} = \text{var}(X_j)$.

It is convenient to introduce here some terminology that will be used in the sequel.

DEFINITION 16.1.3.    *Positive-definite and non-negative definite matrices.* A square matrix $\mathbf{V}$ with real entries is called *positive-definite* if

$$\mathbf{a}'\mathbf{V}\mathbf{a} > 0$$

for every real vector $\mathbf{a}$ (*other than the vector consisting solely if zero entries*): $V$ *is called non-negative-definite* if

$$\mathbf{a}'\mathbf{V}\mathbf{a} \geq 0$$

under the same circumstances.

In particular, if $\mathbf{V}$ is the dispersion matrix of a random vector $\mathbf{X} = (X_1, X_2, \ldots, X_p)$ in which the components are linearly independent in the sense that every non-trivial linear combination $\sum a_j X_j$ is a non-degenerate random variable (i.e. is not a constant, and so has strictly positive variance) then, if we define **a** as $(a_1, a_2, \ldots, a_p)$,

$$\mathbf{a'Va} = \text{var} \left( \sum a_j X_j \right) > 0$$

[see I, 19.6.14)], so that **V** must then be positive-definite.

(The reader is warned that a positive matrix, that is a matrix all of whose entries are positive, is not necessarily positive-definite. A similar warning applies to non-negative matrices.)

The results quoted in (16.1.13) point to a notational difficulty, which we propose to deal with as follows.

*Notation*

We have hitherto distinguished between a random variable $X$ and a realization $x$ of that random variable; in this section we have therefore distinguished between a random variable $X_j$ and a realization $x_{ji}$ of that random variable. However, we now also need to distinguish between a matrix **G** and an entry $g_{ij}$ in that matrix. The capital letter—small letter conventions of probability theory and of those matrix theory are in collision. We could get round this of course by using, say, Greek letters to denote vector or matrix random variables, and Latin letters for vectors and matrices, but the gain in clarity from an unambiguous notation would be outweighed by the increase in complexity. We therefore adopt the following:

*Notational convention*

The standard notation of vector and matrix algebra will be retained: **x** is a column vector; **x'**, its transpose, is a row-vector; **A** is a matrix; etc. In the present chapter, when it is convenient to do so, we abandon the former convention by which we invoked any induced random variable $X$ [see Definition 2.2.1] to describe the sampling distribution of an observation or statistic $x$, using $E(X)$ for the sampling expectation of $x$ [see Definition 2.3.1], and so forth. We shall permit the use of '$E(x)$' to denote the sampling expectation of $x$; similarly for variances, covariances, etc.; similarly for vector observations **x**, and for a matrix **G** of observations, etc., relying on the context to avoid ambiguities.

## 16.2. SAMPLES FROM MULTIVARIATE NORMAL (MVN) DISTRIBUTIONS

Just as in the analysis of univariate samples, it is possible to obtain summary measures from the data and use them to estimate population parameters, as indicated in section 16.1. However, not all methods of multivariate analysis are simply analogous to analyses used for univariate samples. Most of the methods presented in this chapter have no equivalent in univariate analysis.

### 16.2.1.  Maximum Likelihood Estimation

Given a random sample $(\mathbf{x}_1, \mathbf{x}_2, \ldots, \mathbf{x}_k)$ of size $k$ from a multivariate Normal (MVN) distribution [see II, § 13.4] with mean vector $\boldsymbol{\mu}$ and covariance matrix $\mathbf{V}$, a distribution whose description we abbreviate to MVN$(\boldsymbol{\mu}, \mathbf{V})$, the likelihood as a function of the parameters involved in $\boldsymbol{\mu}$ and $\mathbf{V}$ is

$$l(\boldsymbol{\mu}, \mathbf{V}) = \frac{1}{(2\pi)^{pk/2}|\mathbf{V}|^{k/2}} \exp -\tfrac{1}{2} \sum_{i=1}^{k} (\mathbf{x}_i - \boldsymbol{\mu})'\mathbf{V}^{-1}(\mathbf{x}_i - \boldsymbol{\mu}) \qquad (16.2.1)$$

since the observation vectors $\mathbf{x}_i$ are independent. (Note that the *elements* of the vector random variable are not mutually independent unless $\mathbf{V} = \mathbf{I}$, the unit matrix.) The log likelihood is

$$\log l(\boldsymbol{\mu}, \mathbf{V}) = -\tfrac{1}{2}kp \log 2\pi - \tfrac{1}{2}k \log |\mathbf{V}| - \tfrac{1}{2} \sum_{i=1}^{k} (\mathbf{x}_i - \boldsymbol{\mu})'\mathbf{V}^{-1}(\mathbf{x}_i - \boldsymbol{\mu}).$$

The last, which is the only one involving $\boldsymbol{\mu}$, term may be rewritten as

$$-\tfrac{1}{2} \sum_{i=1}^{k} (\mathbf{x}_i - \bar{\mathbf{x}})'\mathbf{V}^{-1}(\mathbf{x}_i - \bar{\mathbf{x}}) - \tfrac{1}{2}k(\bar{\mathbf{x}} - \boldsymbol{\mu})'\mathbf{V}^{-1}(\bar{\mathbf{x}} - \boldsymbol{\mu}), \qquad (16.2.2)$$

from which it may be seen that the log likelihood is maximized with respect to variations of $\boldsymbol{\mu}$ when $\boldsymbol{\mu} = \bar{\mathbf{x}}$; hence $\bar{\mathbf{x}}$ is the maximum likelihood estimate of $\boldsymbol{\mu}$.

This problem of estimating the elements of $\mathbf{V}$ is not so easy. There are not $p^2$ but $\tfrac{1}{2}p(p+1)$ elements to estimate, since $\mathbf{V}$ is symmetric [see I, § 6.7(v)], this number being made up of $p$ variances and $\tfrac{1}{2}p(p-1)$ covariances.

The procedure is to work in terms of $\mathbf{W} = \mathbf{V}^{-1}$, replacing the determinant $|\mathbf{V}|$ by the equivalent $1/|\mathbf{W}|$. [See I, § 6.9ff]. By estimating the elements of $\mathbf{W}$ and using a multivariate analogue of the invariance property of maximum likelihood estimators [see § 6.2.6], it may be shown that the maximum likelihood estimate of $\mathbf{V}$ is

$$\hat{\mathbf{V}} = \mathbf{A}/k, \qquad (16.2.3)$$

where $\mathbf{A}$ is defined as in (16.1.9). This estimate may be adjusted for unbiasedness by muliplying by $k/(k-1)$. The details of the proof may be found in Anderson (1958, pp. 47–48) or Morrison (1976, pp. 99–100). These are similar to the results obtained for maximum likelihood estimation from a univariate Normal distribution [cf. Example 6.4.1]

EXAMPLE 16.2.1.   *Estimation of the disperion matrix of a bivariate Normal distribution.*   Consider a random sample $(\mathbf{x}_1, \mathbf{x}_2, \ldots, \mathbf{x}_k)$ of size $k$ from the bivariate Normal distribution with mean vector

$$\boldsymbol{\mu} = (\mu_1, \mu_2)'$$

and covariance matrix

$$V = \begin{pmatrix} v_{11} & v_{12} \\ v_{12} & v_{22} \end{pmatrix} = \begin{pmatrix} \sigma_1^2 & \rho\sigma_1\sigma_2 \\ \rho\sigma_1\sigma_2 & \sigma_2^2 \end{pmatrix},$$

with inverse

$$V^{-1} = W = \begin{pmatrix} w_{11} & w_{12} \\ w_{12} & w_{22} \end{pmatrix}.$$

The term in the log-likelihood corresponding to the last term in (16.2.2) in

$$-\frac{1}{2} \sum_{i=1}^{k} [w_{11}(x_{1i}-\mu_1)^2 + 2w_{12}(x_{1i}-\mu_1)(x_{2i}-\mu_2) + w_{22}(x_{2i}-\mu_2)^2].$$

Differentiation, with respect first to $\mu_1$ and then to $\mu_2$, shows that the maximum occurs when

$$w_{11} \sum_i (x_{1i}-\mu_1) + w_{12} \sum_i (x_{2i}-\mu_2) = 0$$

and

$$w_{12} \sum_i (x_{1i}-\mu_1) + w_{22} \sum_i (x_{2i}-\mu_2) = 0,$$

i.e. when

$$w_{11}\mu_1 + w_{12}\mu_2 = w_{11}\bar{x}_1 + w_{12}\bar{x}_2$$

and

$$w_{12}\mu_1 + w_{22}\mu_2 = w_{12}\bar{x}_1 + w_{22}\bar{x}_2,$$

whence the MLE of $\mu_1$ is

$$\hat{\mu}_1 = \bar{x}_1$$

and that of $\mu_2$ is

$$\hat{\mu}_2 = \bar{x}_2$$

whence

$$\hat{\mu} = \bar{x}.$$

The terms in the log-likelihood involving $W$ $(= V^{-1})$ then become

$$\tfrac{1}{2}k \log |W| - \tfrac{1}{2}k \sum_{i=1}^{k} (x_i - \bar{x})' W(x_i - \bar{x})$$

$$= \tfrac{1}{2}k \log (w_{11}w_{22} - w_{12}^2)$$

$$-\frac{1}{2} \sum_{i=1}^{k} [w_{11}(x_{1i}-\bar{x}_1)^2 + 2w_{12}(x_{1i}-\bar{x}_1)(x_{2i}-\bar{x}_2) + w_{22}(x_{2i}-\bar{x}_2)^2].$$

If we differentiate in turn with respect to $w_{11}$, $w_{12}$ and $w_{13}$ we obtain the following likelihood equations for the MLE of the $w_{rs}$:

$$\frac{kw_{22}}{w_{11}w_{22}-w_{12}^2}-\sum_i(x_{1i}-\bar{x}_1)^2=0,$$

$$\frac{-2kw_{12}}{w_{11}w_{22}-w_{12}^2}-\sum_i(x_{1i}-\bar{x}_1)(x_{2i}-\bar{x}_2)=0,$$

$$\frac{kw_{11}}{w_{11}w_{22}-w_{12}^2}-\sum_i(x_{2i}-\bar{x}_2)^2=0.$$

It follows from the invariant property of maximum likelihood estimates that the MLE of the correlation coefficient

$$\rho=v_{12}/\sqrt{(v_{11}v_{22})}$$

is

$$\hat{\rho}=\hat{v}_{12}/\sqrt{(\hat{v}_{11}\hat{v}_{12})},$$

the sign of $\hat{\rho}$ being determined by the sign of $\hat{v}_{12}$.

Since

$$V=W^{-1}$$

$$=\frac{1}{w_{11}w_{22}-w_{12}^2}\begin{pmatrix}w_{22}&-w_{12}\\-w_{12}&w_{11}\end{pmatrix},$$

the above equations reduce to

$$\hat{v}_{11}=\frac{1}{k}\sum_i(x_{1i}-\bar{x}_1)^2/k=\frac{k}{k-1}c_{11},$$

$$\hat{v}_{12}=\frac{1}{k}\sum_i(x_{1i}-\bar{x}_1)(x_{2i}-\bar{x}_2)=\frac{k}{k-1}c_{12},$$

and

$$\hat{v}_{12}=\frac{1}{k}\sum_i(x_{2i}-\bar{x}_2)^2=\frac{k}{k-1}c_{22},$$

where $c_{11}$ and $c_{22}$ are the sample variances and $c_{12}$ the sample covariance (see (16.1.9)).

### 16.2.2.   Some Sampling Distributions

In the present section we consider a vector random variable $\mathbf{X}$ which has the MVN distribution with expectation vector $\mathbf{p}$ and dispersion matrix $\mathbf{V}$ [see II, § 13.4]. $\mathbf{X}_1, \mathbf{X}_2, \ldots, \mathbf{X}_2$ are statistical copies of $\mathbf{X}_1$ i.e. they are independent vector random variables each with the same distribution as $\mathbf{X}$. They may be regarded as the random vectors induced by the observation vectors $\mathbf{x}_1, \mathbf{x}_2, \ldots, \mathbf{x}_k$.

It is easy to see that the sampling distribution [see § 2.2] of the sample mean $\bar{x} = \sum_1^k x_j/k$ is MVN $(\mu, V/k)$.

We next consider the sampling distribution of the sample sum of squares and products matrix (16.1.9), namely the matrix

$$A = \sum_{i=1}^k (x_i - \bar{x})(x_i - \bar{x})'.$$

The $k$ vectors $x_i - \bar{x}$, $i = 1, 2, \ldots, k$, are not independent of each other as they have the term $\bar{x}$ in common.

When one speaks of the sampling distribution of $A$ one means of course the joint sampling distribution of the $\frac{1}{2}p(p+1)$ algebraically distinct statistics

$$a_{rr} \ (r = 1, 2, \ldots, p) \quad \text{and} \quad a_{rs} \ (r, s = 1, 2, \ldots, p) \quad (r < s).$$

This generalization of the univariate problem of finding the distribution of $\sum_1^k (x_r - \bar{x})^2$ is therefore not a trivial one. In the univariate case, with independent Normal $(\mu, \sigma)$ observations, an orthogonal transformation enables one to write

$$\sum_1^k (x_r - \bar{x})^2 = \sum_1^{k-1} z_r^2$$

where the $z_r$ are realizations of independent Normals each with zero expectation and variance $\sigma^2$; it follows that the sampling distribution of $\sum_1^k (x_r - \bar{x})^2/\sigma^2$ is chi-squared on $k-1$ degrees of freedom [see § 2.5.4(a)]. In the present case a generalization of that technique enables one to write

$$A = \sum_{j=1}^k z_j z',$$

where the $z_j$ are mutually independent random vectors each with the same MVN$(0, V)$ distribution. It then turns out that the sampling distribution of $A$ is the Wishart distribution with parameter $V$ and $n = k-1$ degrees of freedom, the density function being

$$f(A, V, n) = \frac{|A|^{(n-p-1)/2} \exp{-\frac{1}{2}\text{trace } AV^{-1}}}{2^{(np)/2} \pi^{p(p-1)/4} |V|^{n/2} \prod_{i=1}^p \Gamma[\frac{1}{2}(n+1-i)]}, \quad A \text{ positive definite.}$$

(16.2.4)

When $p = 1$,

$$A = \sum (x_i - \bar{x})^2,$$

and if $V = 1$ then the sample is from a univariate Normal distribution with expectation $\mu$ and variance 1. The density above becomes that of a chi-squared variate on $(k-1)$ degrees of freedom, confirming the well-known univariate result. In the case where $V = \sigma^2$, the density of

$$AV^{-1} = \sum (X_i - \bar{X})^2/\sigma^2$$

has a chi-squared distribution on $(k-1)$ degrees of freedom. Corresponding to the results for univariate Normal samples [see § 2.5.4(c)], $\bar{\mathbf{X}}$ and $\mathbf{A}$ are independently distributed and they are jointly sufficient statistics for $\boldsymbol{\mu}$ and $\mathbf{V}$ [see Example 3.4.8].

The results in this section may be extended to deal with several independent samples from different MVN populations. For example suppose that $\bar{\mathbf{x}}_1$ and $\bar{\mathbf{x}}_2$ are the sample means from samples of size $k_1$ and $k_2$ from MVN distributions with expectation vectors $\boldsymbol{\mu}_1, \boldsymbol{\mu}_2$ and common covariance matrix $\mathbf{V}$; then $\bar{\mathbf{X}}_1 - \bar{\mathbf{X}}_2$ has the MVN distribution with expectation vector $\boldsymbol{\mu}_1 - \boldsymbol{\mu}_2$ and covariance matrix $\mathbf{V}(1/k_1 + 1/k_2)$. Further, if $\mathbf{A}_1, \mathbf{A}_2, \ldots, \mathbf{A}_m$ are independent and Wishart-distributed with parameter $\mathbf{V}$ and degrees of freedom $n_1, n_2, \ldots, n_k$ then $\sum_{i=1}^{m} \mathbf{A}_i$ has a Wishart distribution with $\sum_{i=1}^{m} n_i$ degrees of freedom. When $p = 1$ and $\mathbf{V} = \mathbf{I}$ this becomes the reproductive property of independent chi squared variates [see Proposition 2.5.1].

### 16.2.3. Tests and Confidence Regions for the Expectation Vector

The methods used to construct confidence intervals for the mean $\mu$ or to test hypotheses such as $H_0 : \mu = \mu_0$ say, in a univariate Normal distribution, with known variance $\sigma^2$, [see Examples 4.2.1 and 4.5.2] are based on the percentage points of the standard Normal variate, $z$, where

$$z = \sqrt{k}\,(\bar{x} - \mu)/\sigma.$$

When the value of $\sigma$ is not known corresponding procedures may be followed using Student's variate on $(k-1)$ degrees of freedom given by

$$t = \sqrt{k}\,(\bar{x} - \mu)/s$$

$s^2$ being the usual unbiased estimator of $\sigma^2$. Alternatively the $z$, $t$ variates may both be squared giving variates which have a $\chi^2$ distribution on 1 degree of freedom and an $F$ variate on 1 and $(k-1)$ degrees of freedom respectively.

The difficulty of using the univariate sample as a pattern for obtaining the relevant statistics in the multivariate case is obvious. It turns out that the best procedure is not to work with the known sampling distribution of

$$\sqrt{k}\,(\bar{\mathbf{x}} - \boldsymbol{\mu}),$$

namely the MVN$(0, \mathbf{V})$, but to work with quadratic functions which are generalizations of the squares of the $z$ and $t$ variates mentioned above.

The sampling distribution of the quadratic form

$$k(\bar{\mathbf{x}} - \boldsymbol{\mu})' \mathbf{V}^{-1} (\bar{\mathbf{x}} - \boldsymbol{\mu})$$

(a scalar quantity), is a $\chi^2$ distribution on $p$ degrees of freedom. To test the hypothesis $H_0 : \mu = \mu_0$ when $\mathbf{V}$ is known (this being the analogue of the univariate Normal with known variance), the critical region at significance level $\alpha$ [see § 5.12] is the set of vectors **x** for which

$$k(\bar{\mathbf{x}}-\boldsymbol{\mu}_0)'\mathbf{V}^{-1}(\bar{\mathbf{x}}-\boldsymbol{\mu}_0) \geq c(p, \alpha) \qquad (16.2.5)$$

where $c(p, \alpha) = \chi^2_{(p)}(1-\alpha)$ is the $(1-\alpha)\times 100\%$ point of the $\chi^2$ distribution on $p$ degrees of freedom.

To obtain a confidence region for $\boldsymbol{\mu}$, the inequality

$$k(\bar{\mathbf{x}}-\boldsymbol{\mu})'\mathbf{V}^{-1}(\bar{\mathbf{x}}-\boldsymbol{\mu}) \leq c(p, \alpha) \qquad (16.2.6)$$

may be inverted to give a region in the $p$-dimensional space of $\boldsymbol{\mu}$. This region will be an ellipsoid with its centre at $\bar{\mathbf{x}}$.

For the bivariate sample in Example 16.2.1 (16.2.3) becomes

$$k[w_{11}(\bar{x}_1-\mu_1)^2 + 2w_{12}(\bar{x}_1-\mu_1)(\bar{x}_2-\mu_2) + w_{22}(\bar{x}_2-\mu_2)^2] \leq c(p, \alpha).$$

The region is an ellipse in the two dimensional space of $\mu_1, \mu_2$, with its centre at $(\bar{x}_1, \bar{x}_2)$.

When **V** is unknown the univariate generalization of the square of Student's variate becomes Hotelling's $T^2$ statistic

$$T^2 = k(\bar{\mathbf{x}}-\boldsymbol{\mu})'\mathbf{C}^{-1}(\bar{\mathbf{x}}-\boldsymbol{\mu}), \qquad (16.2.7)$$

where **C** is the sample covariance matrix (16.1.9). To test the hypothesis $H_0: \boldsymbol{\mu} = \boldsymbol{\mu}_0$ the critical region at significance level $\alpha$ is the set of vectors **x** for which

$$k(\bar{\mathbf{x}}-\boldsymbol{\mu}_0)'C^{-1}(\bar{\mathbf{x}}-\boldsymbol{\mu}_0) \geq d(p, \alpha) \qquad (16.2.8)$$

where

$$d(p, \alpha) = \frac{p(k-1)}{k-p} F_{(p,k-p)}(1-\alpha),$$

and $F_{(p,k-p)}(1-\alpha)$ is the $100(1-\alpha)\%$ point of the $F$ distribution on $p$ and $k-p$ degrees of freedom [see § 2.5.6].

A confidence region for $\boldsymbol{\mu}$ in $p$-dimensional space may be obtained by using the percentage points of the $F$ distribution above. This will again be an ellipsoid.

### 16.2.4.    Two-sample Problems

Suppose that $\bar{\mathbf{x}}_1, \bar{\mathbf{x}}_2$ are the sample means and $\mathbf{C}_1, \mathbf{C}_2$ are the sample covariance matrices of samples of sizes $k_1$ and $k_2$ from $p$-variate MVN($\boldsymbol{\mu}_1, \mathbf{V}$) and MVN($\boldsymbol{\mu}_2, \mathbf{V}$) distributions respectively. Then, if **V** is a known matrix, the critical region for testing the hypothesis $H_0: \boldsymbol{\mu}_1 = \boldsymbol{\mu}_2$, at confidence level $\alpha$ is

$$\left(\frac{k_1 k_2}{k_1 + k_2}\right)(\bar{\mathbf{x}}_1 - \bar{\mathbf{x}}_2)'\mathbf{V}^{-1}(\bar{\mathbf{x}}_1 - \bar{\mathbf{x}}_2) \geq c(p, \alpha),$$

if **V** is known, where $c(p, \alpha)$ is as defined in connection with (16.2.5).

If $V$ is unknown, the unbiased estimates $C_1$ and $C_2$ of $V$ may be pooled to give

$$C = [(k_1 - 1)C_1 + (k_2 - 1)C_2]/(k_1 + k_2 - 2), \qquad (16.2.9)$$

and the critical region for testing $H_0$: $\mu_1 = \mu_2$ at significance level $\alpha$ is based on the statistic

$$T^2 = \left(\frac{k_1 k_2}{k_1 + k_2}\right)(\bar{x}_1 - \bar{x}_2)' C^{-1}(\bar{x}_1 - \bar{x}_2).$$

The critical region in the set of vectors for which

$$T^2 \geq \left(\frac{k_1 + k_2 - 2p}{k_1 + k_2 - p - 1}\right) F_{p, k_1 + k_2 - p - 1}(1 - \alpha).$$

### 16.2.5.   Inferences about the Correlation Coefficient

Consider a random sample of size $k$ from the bivariate Normal distribution with mean vector $(\mu_1, \mu_2)'$ and covariance matrix

$$V = \begin{pmatrix} \sigma_1^2 & \rho\sigma_1\sigma_2 \\ \rho\sigma_1\sigma_2 & \sigma_2^2 \end{pmatrix}.$$

It was shown in Example 16.2.1 that the maximum likelihood estimate, $\hat{\rho} = r$, of the correlation coefficient $\rho$ is

$$r = \frac{\displaystyle\sum_{i=1}^{k} (x_{1i} - \bar{x}_1)(x_{2i} - \bar{x}_2)}{\sqrt{\left\{\displaystyle\sum_{i=1}^{k} (x_{1i} - \bar{x}_1)^2 \sum_{i=1}^{k} (x_{2i} - \bar{x}_2)^2\right\}}}. \qquad (16.2.10)$$

It can be shown (see Anderson (1959) pp. 62–64) that $r$ has the following sampling density when $\rho = 0$:

$$\frac{\Gamma[\frac{1}{2}(k-1)]}{\Gamma[\frac{1}{2}(k-2)]\sqrt{\pi}}(1 - r^2)^{(k-4)/2}, \qquad -1 \leq r \leq 1,$$

or, equivalently, that the sampling distribution of

$$(k-2)^{1/2}\frac{r}{\sqrt{(1-r^2)}} \qquad (16.2.11)$$

is Student's $t$-distribution with $(k-2)$ degrees of freedom.

The most important application of this result is to the testing of the hypothesis $H_0$: $\rho = 0$ against $H_1$: $\rho \neq 0$. For a given significance level $\alpha$, $H_0$ is rejected if

$$(k-2)r^2/(1-r^2) \geq \{t_{(k-2)}(1 - \tfrac{1}{2}\alpha)\}^2,$$

where $t_{(k-2)}(1 - \tfrac{1}{2}\alpha)$ is the $(1 - \tfrac{1}{2}\alpha) \times 100\%$ point of Student's $t$-distribution on $k-2$ degrees of freedom.

Suppose that the regression $E(X_2|X_1=x_1)$ of $X_2$ on $X_1$ [see Example 4.5.3] is given by

$$E(X_2|X_1) = \beta_0 + \beta_1 x_1;$$

then

$$\beta_1 = \rho \sigma_1 / \sigma_2.$$

It follows that, when $\sigma_1 \sigma_2 > 0$, the propositions

$$\text{`}p = 0\text{'}, \qquad \text{`}\beta_1 = 0\text{'}$$

are equivalent. A test for the significance of $\beta_1$ is therefore a test for the significance of $\rho$. In conventional testing, [see Example 5.5.2] the null hypothesis

$$H_0: \beta_1 = 0$$

is 'rejected' at level $\alpha$ if

$$|b_1|/\text{s.d.}(b_1) \geq t_{k-2}(1-\tfrac{1}{2}\alpha), \qquad (16.2.12)$$

where $b_1$ is the least squares estimate of $\beta_1$ [see § 6.5.3(i)] and 's.d.$(b_1)$' denotes its estimated sampling standard deviation. On inserting the appropriate formulae in (16.2.12) it will be seen that the left-hand member of that inequality reduces to (16.2.4)—as it should, of course since the latter has Student's distribution.

In testing whether $\rho$ is zero or not one is of course testing whether $X_1$ and $X_2$ are independent.

For large or moderate $k$, the transformation of $r$ known as 'Fisher's $z$' may be made, [see § 2.7.3(b)] viz.:

$$z = \tfrac{1}{2} \log \left( \frac{1+r}{1-r} \right).$$

$z$ is asymptotically Normally distributed with mean

$$\tfrac{1}{2} \log \left( \frac{1+\rho}{1-\rho} \right)$$

and variance $1/(k-1)$. (Here logs are taken to base $e$).

The hypothesis $H_0: \rho = \rho_0$ may therefore be tested by using tables of the standard Normal distribution [see Example 5.2.2].

The tests above may be modified to deal with partial correlation coefficients [see II, (13.4.25)]. For example, consider a random sample of size $k$ from a trivariate Normal. The partial correlation coefficient between $X_1$ and $X_2$ given $X_3 = x_3$ is

$$\rho_{12.3} = \frac{\rho_{12} - \rho_{12}\rho_{23}}{\sqrt{(1-\rho_{13}^2)(1-\rho_{23}^2)}} \qquad (16.2.13)$$

where $\rho_{ij}$ is the correlation between $X_i$ and $X_j$. (This is, by definition, the correlation coefficient of the conditional bivariate Normal distribution of $(X_1, X_2)$ given $X_3 = x_3$.

Under the hypothesis $\rho_{12.3} = 0$ the maximum likelihood estimate $r_{12.3}$ obtained from (16.2.13) on replacing the $\rho_{ij}$ by their corresponding maximum likelihood estimates is such that

$$\sqrt{k-3}\, r_{12.3}/\sqrt{1-r_{12.3}^2} \qquad (16.2.14)$$

has Student's distribution on $k-3$ degrees of freedom. The test of $\rho_{12.3}$ follows. Tests for individual partial correlations in $p$-variate Normals where the conditioning is on $q$ of the variates are based on a similar statistic to (16.2.14), which has Student's distribution on $(k-q-2)$ degrees of freedom.

### 16.2.6.   Non-central Distributions

In sections (16.2.3) and (16.2.4) tests for mean vectors were discussed and their distributions were given under the null hypothesis, but the distribution of the test statistics for values of the parameter specified by the alternative hypothesis were not mentioned.

For example, suppose that $H_0$: $\boldsymbol{\mu} = \boldsymbol{\mu}_0$ and $\mathbf{V}$ is known. The test statistic

$$k(\bar{\mathbf{x}} - \boldsymbol{\mu}_0)' \mathbf{V}^{-1}(\bar{\mathbf{x}} - \boldsymbol{\mu}_0)$$

has a (central) $\chi^2_{(r)}$ distribution provided $\boldsymbol{\mu} = \boldsymbol{\mu}_0$. If the power function of the test or probability of type II error is required for $\boldsymbol{\mu} \neq \boldsymbol{\mu}_0$ then the test statistic no longer has this distribution but has instead a *non-central* $\chi^2_{(p)}$ distribution [see § 2.8.1] with *noncentrality parameter*

$$k(\boldsymbol{\mu} - \boldsymbol{\mu}_0)' \mathbf{V}^{-1}(\boldsymbol{\mu} - \boldsymbol{\mu}_0). \qquad (16.2.6)$$

Note that this is zero when $\boldsymbol{\mu} = \boldsymbol{\mu}_0$.

For the case where $\mathbf{V}$ is unknown the test statistic has an $F_{(p, k-p)}$ distribution with the non-centrality parameter (16.2.6), which contains the unknown $\mathbf{V}$. Tables of non-central distributions are available, [see, e.g. Hartor and Owen, Vol. 1 (1970) for non-central $\chi^2$, Owen (1962) and Resnikoff and Lieberman (1957) for non-central—all in Bibliography G; and see Graybill (1976). (Bibliography C) for non-central F.]

## 16.3.   PRINCIPAL COMPONENTS

### 16.3.1.   Introduction

In this section it will be assumed, for ease of exposition, that the $p$-component random variable $\mathbf{X}$ has mean $\mathbf{0}$ and positive-definite [see Definition 16.1.1] variance-covariance matrix $\mathbf{V}$ [see (16.1.13), ff.]. In other words, the vector random variable $\mathbf{X} = \mathbf{X}^* - \boldsymbol{\mu}$ where $\mathbf{X}^*$ is the original random variable with mean $\mathbf{\mu}$. Note that at this point it is not necessary to make any assumptions about distributional form.

Principal components are orthogonal linear transformations (i.e. uncorrelated random variables) [see § 2.5.3(f)] of $\mathbf{X}$ with the first having the largest variance, down to the $p$th which has the smallest variance. Under certain scaling assumptions, it may be shown that the variances of the principal components are the eigenvalues [see I, Definition 7.1.1] of $\mathbf{V}$ and the coefficients of $\mathbf{X}$ in the transformation are the elements of the corresponding eigenvector [see I, Definition 7.1.1′].

Principal components analysis is concerned with reducing the number of variables under consideration by using only the first few principal components and discarding those linear combinations which have small variances.

### 16.3.2.  Population Principal Components

Let the first principal component $Y_1$ of the random vector $\mathbf{X}$ be

$$Y_1 = \sum_{i=1}^{p} c_{1i} X_i = \mathbf{c}_1' \mathbf{X}.$$

Clearly

$$E(Y_1) = 0$$

and [see II, § 13.3.2]

$$\text{var}(Y_1) = E(Y_1^2) = \sum_{i=1}^{p} \sum_{j=1}^{p} c_{1i} c_{1j} E(X_i X_j) = \mathbf{c}_1' \mathbf{V} \mathbf{c}_1.$$

The coefficient vector $\mathbf{c}_1$ is chosen such that var $(Y_1)$ is a maximum subject to the condition

$$\sum_{i=1}^{p} c_{1i}^2 = \mathbf{c}_1' \mathbf{c}_1 = 1. \tag{16.3.1}$$

This is a problem of maximizing under constraint and can be solved by using Lagrange's multipliers. [See IV, § 5.15.] The problem becomes that of finding, find $\mathbf{c}_1$ to maximize

$$\mathbf{c}_1' \mathbf{V} \mathbf{c}_1 - \lambda_1 (\mathbf{c}_1' \mathbf{c}_1 - 1)$$

where $\lambda_1$ is an undetermined Lagrange multiplier. On taking the derivative of this with respect to $\mathbf{c}_1$ and equating to $\mathbf{0}$, we obtain the equation

$$(\mathbf{V} - \lambda_1 \mathbf{I}) \mathbf{c}_1 = \mathbf{0} \tag{16.3.2}$$

where $\mathbf{I}$ is the unit matrix. As we are interested only in the condition where $\mathbf{c}_1 \neq \mathbf{0}$, we must have the determinantal condition [see I, § 5.9]:

$$|\mathbf{V} - \lambda_1 \mathbf{I}| = 0.$$

Thus $\lambda_1$ is an eigenvalue of $\mathbf{V}$, and $\mathbf{c}_1$ is the corresponding eigenvector.

The expression (16.3.2) may be rewritten as

$$\mathbf{Vc}_1 = \lambda_1 \mathbf{c}_1. \tag{16.3.3}$$

Premultiplying this by $\mathbf{c}_1$ gives

$$\mathbf{c}_1' \mathbf{Vc}_1 = \lambda_1 \mathbf{c}_1' \mathbf{c}_1$$

$$= \lambda_1, \text{ by } (16.3.1). \tag{16.3.4}$$

The left-hand side of (16.3.4) is var $(Y_1)$. Now the whole object of the exercise has been to maximize var $(Y_1)$. It follows that $\lambda_1$ is the largest eigenvalue of $\mathbf{V}$.

To find the second principal component

$$Y_2 = \mathbf{c}_2' X$$

two conditions are imposed on $\mathbf{c}_2$, namely the normalizing condition

$$\mathbf{c}_2' \mathbf{c}_2 = 1, \tag{16.3.5}$$

and the orthogonality condition

$$\mathbf{c}_1' \mathbf{c}_2 = 0. \tag{16.3.6}$$

$\mathbf{c}_2$ is now found such that var $(Y_2)$ is maximized under the two constraints, requiring the use of two Lagrangian multipliers $\lambda_2$ and $\beta$. We have to maximize the expression

$$\mathbf{c}_2' \mathbf{Vc}_2 - \lambda_2 (\mathbf{c}_2' \mathbf{c}_2 - 1) - \beta (\mathbf{c}_2' \mathbf{c}_2 - 0). \tag{16.3.7}$$

On taking the derivative of (16.3.6) and equating to $\mathbf{0}$, it will be found by using the orthogonality condition (16.3.6) that $\beta = 0$, and, by using the condition (16.3.5), that $\lambda_2$ is the second largest eigenvalue of $\mathbf{V}$, that $\lambda_2$ is equal to var $(Y_2)$, and that $\mathbf{c}_2$ is its corresponding eigenvector.

The process is repeated until all the eigenvectors and eigenvalues have been shown to be the coefficients of the linear combination and the variances of the principal components respectively. Note that to prove the result for the $k$th principal component we have to maximize var $(Y_k)$ under the set of $k$ constraints consisting of the normalizing condition

$$\mathbf{c}_k' \mathbf{c}_k = 1$$

and the $(k-1)$ orthogonality conditions:

$$\mathbf{c}_k' \mathbf{c}_r = 0, \qquad r = 1, 2, \ldots, k-1.$$

An unfortunate property of principal components is that they depend on the scales on which the original variables are measured, that is they are not scale invariant. For example changes of measurements of dimensions from feet to inches and of time from hours to seconds will in general produce different eigenvalues and eigenvectors. For this reason it is perhaps better to work with the standardized variates

$$X_i / \sigma_{ii} = Z_i$$

which have expected value 0 and unit variance. In this case the variance-covariance matrix of $\mathbf{Z}$ is the correlation matrix of $\mathbf{X}$, $\mathbf{R}$ say. The principal components using this transformation may now be obtained from the eigenvectors of $\mathbf{R}$ and their 'variances' from the corresponding eigenvalues.

Nothing has been assumed about the rank [see I, § 5.6] of $\mathbf{V}$ in the foregoing discussion. If the matrix is not of full rank then some of the smaller eigenvalues are zero and $\mathbf{X}$ may be transformed into fewer than $p$ components. The only firm requirement on $\mathbf{V}$ is that it is non-negative definite [see Definition 16.1.3].

Since

$$\sum_{i=1}^{p} \lambda_i = \text{trace } \mathbf{V},$$

[see I, § 6.2(vii)] the sum of the eigenvalues may be thought of as the total variation of the population, and the first $m$ principal components with the $m$ largest variances may be said to 'account for' the proportion of the total variation given by

$$\sum_{i=1}^{m} \lambda_i \bigg/ \sum_{i=1}^{p} \lambda_i.$$

When this proportion is large enough, the components with variances $\lambda_{m+1}, \ldots, \lambda_p$ may be discarded, the population being adequately represented by $m$ components.

If the correlation matrix $\mathbf{R}$ is used instead then (assuming $\mathbf{V}$ to be of full rank)

$$\sum_{i=1}^{p} \lambda_i = p.$$

A choice has to be made between extracting components from the covariance matrix or from the correlation matrix. The amount of variation explained by corresponding components using each method will be different. Using the covariance matrix may of course produce components with large variances simply because of the scale on which one of its $x$-values has been measured.

### 16.3.3. Sample Principal Components

The preceding sections have discussed the extraction of principal components from populations using the population parameters. In practice the parameters have to be estimated from a random sample, of size $k$, say.

The data matrix may be regarded as being made up of the original observations measured from their means, and the sample principal components may be extracted from the sample covariance matrix $\mathbf{C}$ or from the sample correlation matrix, using the same method as in the last section. Care must be taken in computing values of principal components or scores for specific $p$-variate observations under the two methods. Using the covariance matrix these scores are simply a linear function of the original observations measured from their mean. For the correlation matrix, the score is a linear function of the standardized variates.

### 16.3.4.　Numerical Example

Principal components evaluated from measurements of nine characters of six botanical clones.

Jeffers (1965) used principal components in an attempt to identify linear combinations of nine variables, comprising measurements on leaves of six clones of poplar, which best discriminated between the six clones.

The nine variables were:

$x_1$ = length of the petiole
$x_2$ = length of the blade of the leaf
$x_3$ = width of the leaf at its widest point
$x_4$ = width of the leaf halfway along the blade
$x_5$ = width of the leaf one third of the way along the blade
$x_6$ = width of the leaf two thirds of the way along the blade
$x_7$ = distance of the base of the leaf from the point at which the petiole joins
　　　the blade
$x_8$ = angle of the first major vein with mid-rib
$x_9$ = angle of the first minor vein with mid-rib.

Five observations on each of the six clones were taken ($k = 30$). The sample correlation matrix for the nine variables is given in Table 16.3.1. All the eigenvalues of the correlation matrix and their corresponding eigenvectors are given in Table 16.3.2.

From Table 16.3.2 it is seen that most of the variation is explained by the first two components.

| | | | | | | | | |
|---|---|---|---|---|---|---|---|---|
| 1·000 | | | | | | | | |
| −0·409 | 1·000 | | | | | | | |
| 0·731 | 0·156 | 1·000 | | | | | | |
| 0·624 | −0·121 | 0·781 | 1·000 | | | | | |
| 0·699 | 0·190 | 0·982 | 0·820 | 1·000 | | | | |
| 0·674 | 0·012 | 0·907 | 0·940 | 0·935 | 1·000 | | | |
| 0·767 | −0·570 | 0·567 | 0·583 | 0·515 | 0·597 | 1·000 | | |
| 0·364 | −0·569 | 0·268 | 0·427 | 0·239 | 0·401 | 0·743 | 1·000 | |
| 0·564 | −0·428 | 0·535 | 0·652 | 0·496 | 0·635 | 0·822 | 0·833 | 1·000 |

Table 16.3.1: Correlations between nine variables (from Jeffers (1965)).

| Variable | Component | | | | | | | | |
|---|---|---|---|---|---|---|---|---|---|
| | 1 | 2 | 3 | 4 | 5 | 6 | 7 | 8 | 9 |
| $x_1$ | 0·349 | −0·029 | 0·644 | −0·174 | −0·036 | 0·382 | 0·524 | 0·081 | 0·071 |
| $x_2$ | −0·121 | 0·600 | −0·393 | −0·423 | −0·151 | −0·098 | 0·504 | 0·028 | 0·066 |
| $x_3$ | 0·362 | 0·321 | 0·066 | −0·271 | 0·145 | 0·089 | −0·458 | −0·407 | 0·532 |
| $x_4$ | 0·370 | 0·149 | −0·134 | 0·679 | −0·146 | −0·158 | 0·325 | −0·459 | 0·016 |
| $x_5$ | 0·357 | 0·356 | 0·034 | −0·103 | 0·236 | 0·061 | −0·210 | −0·030 | −0·793 |
| $x_6$ | 0·384 | 0·235 | −0·112 | 0·287 | 0·103 | −0·085 | −0·097 | 0·780 | 0·256 |
| $x_7$ | 0·356 | −0·286 | 0·130 | −0·340 | −0·155 | −0·795 | 0·047 | 0·007 | −0·038 |
| $x_8$ | 0·268 | −0·428 | −0·486 | −0·155 | 0·615 | 0·155 | 0·276 | 0·063 | 0·052 |
| $x_9$ | 0·349 | −0·257 | −0·378 | −0·155 | −0·682 | 0·383 | −0·156 | 0·050 | −0·089 |
| Eigen-value $\lambda_i$ | 5·648 | 2·051 | 0·663 | 0·338 | 0·128 | 0·097 | 0·042 | 0·023 | 0·009 |
| % variation | 62·8 | 22·8 | 7·4 | | | | | | 100 |
| Cumu-lative % variation | 62·8 | 85·6 | 93·0 | | | | | | 100 |

(Components 4–9 are grouped together, annotated with the value 7.)

Table 16.3.2: Eigenvectors and their associated eigenvalues of the correlation matrix in Table 16.3.1.

In attempting to give a physical interpretation to the components it may be helpful to divide through by the largest entry in the eigenvector and to retain those components for interpretation which have absolute values of 0·7 or greater (Jeffers (1967)).

For example in Table 16.3.2, components 1 and 2 together give a summary of the components of the leaf, and component 5 contrasts the vein angles $(x_8, x_9)$. There is no importance to be attached to the sign of the coefficients since they may all be reversed without changing the analysis; they may be used only to indicate contrasting effects on the component. In the example the standardized scores of the observations represented by the first two principal components may be plotted and then used to discriminate between the clones.

### 16.3.5.  Some Sampling Distributions

If it is assumed that **X** has a multivariate Normal distribution with expectation vector **0** and covariance matrix **V** the population principal components will be Normally distributed, since they are linear functions of $X_1, \ldots, X_p$, [see II, § 13.4.7] and independent. It does not matter whether the components have been extracted from the covariance or correlation matrix.

For sample principal components the sampling distributions of the eigenvalues and eigenvectors from the covariance matrix only are generally considered. The exact small sample distributions are complicated, and the situation is exacerbated when several of the eigenvectors of **V** coincide. Large sample distributional results are available under the MVN assumption and these may be used to test hypotheses concerning the eigenvalues and eigenvectors of **V**. The reader is referred to Morrison (1976), pp. 292–299 for details.

## 16.4.  FACTOR ANALYSIS

### 16.4.1.  Introduction

Factor analysis is a technique related to principal components analysis in that it again considers the interrelationship between the $p$ scalar random variables in which make up the random vector **X**, analysing the covariance matrix or correlation matrix of the variables. Instead of subjecting **X** to a linear transformation, factor analysis suggests that the vector random variable may be expressed as a linear model involving random variables known as factors, fewer than $p$ in number; in addition there is an error term. In consequence, the correlation between $X_1, X_2, \ldots, X_p$ may be accounted for by the correlations between the linear functions of fewer than $p$ factors.

Since the factors are unobservable, the analysis concentrates on the decomposition of the covariance or correlation matrix suggested by the linear model.

### 16.4.2. Factor Model

It is assumed that the *p*-variate random variable $\mathbf{X}$ has mean $\mathbf{0}$ and covariance matrix $\mathbf{V}$ of full rank $p$. Alternatively $\mathbf{X}$ is obtained by subtracting the mean vector $\boldsymbol{\mu}$ from the original random variables $\mathbf{X}^*$.

Suppose that each $X_i$ may be expressed in terms of $k$ unobservable factors $f_1, f_2, \ldots, f_p$, with $(f_1, \ldots, f_k)' = \mathbf{f}$, $k < p$, such that

$$X_i = \sum_{j=1}^{k} \lambda_{ij} f_j + e_i, \qquad i = 1, 2, \ldots, p, \qquad (16.4.1)$$

where the parameters $\lambda_{ij}$ in the linear model are known as the *factor loadings* and $e_i$ is a random error term associated only with $X_i$.

It is further assumed that the $f_j$ are uncorrelated random variables with mean 0 and variance 1, that the $e_i$ are uncorrelated random variables with mean 0 and unknown variance $\psi_{ii}^2$, and that the $f_j$ are uncorrelated with $e_i$.

Under these conditions the variance of $X_i$ from (16.4.1) is

$$\text{var} (X_i) = \sigma_{ii}^2 = \sum_{i=1}^{k} \lambda_{ij}^2 + \psi_{ii}^2.$$

$\sum \lambda_{ij}^2$ is known as the *communality*. It represents that part of the variance of $X_i$ due to the 'factors'. $\psi_{ii}^2$ is that part of the variance of $X_i$ due to the error term.

The covariance of $X_r$, $X_s$ is given by

$$\text{cov} (X_r, X_s) = E(X_r X_s) = \sum_{j=1}^{k} \lambda_{rj} \lambda_{sj}, \qquad r \neq s.$$

Likewise

$$\text{cov} (X_i, f_j) = \text{cov} \left( \sum_{j=1}^{k} \lambda_{ij} f_j + e_i, f_j \right) = \lambda_{ij}.$$

If we represent the equations (16.4.1) in the concise form

$$\mathbf{X} = \mathbf{\Lambda} \mathbf{f} + \mathbf{e}$$

where $\mathbf{\Lambda}$ is the $(p \times k)$ matrix of loading parameters, the factor model implies

$$\mathbf{V} = \mathbf{\Lambda} \mathbf{\Lambda}' + \mathbf{\Psi}$$

where $\mathbf{\Psi}$ is the diagonal matrix [see I, § 6.7(iv)] of order $p$ containing the variances of the (uncorrelated) error terms.

### 16.4.3. Some Properties

Unlike principal component analysis, factor analysis (or the decomposition of $\mathbf{V}$) is unaffected by changes in scale. This is easily seen by multiplying each $X_i$ by a constant $c_i$ and rewriting the factor model and the corresponding decomposition of $\mathbf{V}$. In particular, suppose that $c_i = (\sigma_{ii})^{-1}$; then the conditions on the expectations, variances and covariances of the factors and error terms are unchanged although the factor model now includes factor loadings of $\lambda_{ij}/\sigma_{ii} = \lambda_{ij}^*$ say, and error variance $\psi_{ii}^2/\sigma_{ii}^2 = \psi_{ii}^{*2}$ say. The covariance matrix of $\mathbf{X}$ becomes the correlation matrix $\mathbf{R}$ under the transformation. This gives

$$\sum_{j=1}^{k} \lambda_{ij}^* + \psi_{ii}^{*2} = 1 \qquad (16.4.1)$$

and

$$\sum_{j=1}^{k} \lambda_{ij}^* \lambda_{lj}^* = \text{corr}\ (X_i, X_l).$$

Equation (16.4.1) gives the relationship between the variances of the transformed error term and the factor loadings. Care must be taken with the choice of $k$ to ensure that $\psi_{ii}^2$ is non-negative.

Despite being invariant under scale changes, the factor loadings are not unique. If $\mathbf{B}$ is an orthogonal $(k \times k)$ matrix [see I, § 6.7(vii)], the general factor model

$$\mathbf{X} = \Lambda \mathbf{f} + \mathbf{e} \quad \text{with } \mathbf{e} = (e_1, e_2, \ldots, e_n)'$$

may be rewritten as

$$\mathbf{X} = \Lambda(\mathbf{BB}')\mathbf{f} + \mathbf{e}$$

(since $\mathbf{BB}' = \mathbf{I}$) which is a model with factor loadings $\Lambda\mathbf{B}$ and factors $\mathbf{B}'\mathbf{f}$. The assumptions about the random variables making up the original model are not violated by this transformation. The corresponding decomposition of $\mathbf{V}$ is

$$\mathbf{V} = \Lambda\mathbf{BB}'\Lambda' + \Psi$$

$$= \Lambda\Lambda' + \Psi$$

as before. This means that there are an infinite number of factor loadings satisfying the original assumptions of the model. This difficulty may be overcome (see Lawley and Maxwell (1971), pp. 7–11) by introducing a constraint on $\Lambda$ such as that

$$\Lambda'\Psi^{-1}\Lambda \quad \text{is diagonal.} \qquad (16.4.2)$$

(This of course assumes that $\Psi$ is unique.)

In factor analysis an attempt is made to decompose the covariance matrix $\mathbf{V}$ containing $\frac{1}{2}p(p+1)$ parameters into two matrices, one $(\Lambda\Lambda')$ containing $pk$ parameters, and one $(\Psi)$ containing $p$ parameters. If a constraint such as (16.4.2) is introduced, this implies a further $\frac{1}{2}k(k-1)$ relationships between the parameters of $\Lambda$ and $\Psi$. A parsimonious representation of the interrelationships between $\mathbf{X}$ will therefore be obtained if the difference, $t$, between the number of parameters in $\mathbf{V}$ and the effective number in the model, taking account of the constraint, is positive, giving

$$t = \frac{1}{2}[(p-k)^2 - (p+k)].$$

For the model based on the correlation matrix, this measurement of efficiency will take the same value despite the relationship (16.4.1) since the number of parameters of $\mathbf{R}$ is $\frac{1}{2}p(p+1)-p$.

If $t=0$, a solution for $\Lambda$ and $\Psi$ may be obtained. However this only offers a new way of representing the interelationships between elements of $\mathbf{X}$, which uses the same number of parameters. Unfortunately an exact solution does not exist when there is a parsimonious representation.

### 16.4.4. Estimation

In practice, the population covariance or correlation matrices are unknown and the factor loadings and the variances $\Psi_{ii}^2$ have to be estimated from the sample covariance matrix $\mathbf{C}$ obtained from a random sample of size $n$. Before this procedure is carried out, the value of $k$ in the factor model has to be fixed. This value is chosen so that the factor model has at most the same effective number of parameters as the original covariance matrix ($t \ge 0$). It is usually assumed that $\mathbf{X}$, $\mathbf{f}$, and $\mathbf{e}$ all have MVN distributions. This enables a goodness of fit test to be carried out on the model for particular values of $k$.

There are two common estimation methods used in factor analysis. One uses maximum likelihood [see Chapter 6], which obviously requires the MVN distributional assumption [see II, § 13.4], and the other exploits the implied factorization of the covariance matrix brought about by principal components analysis. The second method requires only the distributional assumption for the derivation of goodness of fit tests.

Maximum likelihood factor analysis attempts to maximize the likelihood with respect to $\Lambda$, $\Psi$ subject to the diagonality assumption, where the expectation vector of the original observations is replaced by its maximum likelihood estimator $\bar{\mathbf{x}}$.

From section 16.2.1, the log likelihood with $\mu = \bar{\mathbf{x}}$ is

$$\log l(\mathbf{V}) = -\tfrac{1}{2}np \log 2\pi - \tfrac{1}{2}n \log |\mathbf{V}| - \tfrac{1}{2}\operatorname{trace} \mathbf{AV}^{-1}.$$

Assuming the factor model, $\mathbf{V}$ is replaced by $\Lambda\Lambda' + \Psi$, and $\log l(\Lambda, \Psi)$ is maximized to find the maximum likelihood estimators $\hat{\Lambda}$, $\hat{\Psi}$. This is not a simple problem and cannot in general be solved analytically. Details may be found in Lawley and Maxwell (1971).

The other method is known as principal factor analysis and uses ideas from principal components analysis. Recall that principal components analysis finds an orthogonal transformation of $\mathbf{X}$

$$\mathbf{Y} = \mathbf{\Gamma}\mathbf{X}$$

where the $i$th row of the orthogonal matrix $\mathbf{\Gamma}$ is the transpose of the eigenvector $\mathbf{c}_i$ corresponding to the $i$th largest eigenvalue $\alpha_i$. The covariance matrix $\mathbf{V}_y$ of $\mathbf{Y}$, is diagonal, with diagonal elements $\alpha_i$, where

$$\mathbf{V}_y = \mathbf{\Gamma}\mathbf{V}\mathbf{\Gamma}'$$

and

$$V = \Gamma' V_y \Gamma. \tag{16.4.3}$$

Press [(1972), pp. 318–319] suggests a solution based on (16.4.3) which assumes that the errors are small or $\Psi \approx 0$, in which case in the factor model becomes

$$V \approx \Lambda \Lambda'$$

giving

$$\Lambda \approx \Gamma'(V_y)^{1/2},$$

where $V_y^{1/2}$ is a diagonal matrix with each diagonal element equal to the positive square root of the corresponding diagonal element of $V_y$.

A more parsimonious representation of $V$ may now be obtained by choosing $\Lambda$ so that only the $k$ largest eigenvalues of $V$ are retained and the remaining $(p-k)$ are omitted, together with their corresponding eigenvectors. This then gives the $(p \times k)$ matrix of factor loadings.

If it cannot be assumed that $\Psi \approx 0$, a similar approach to that above is used to resolve the matrix

$$V - \Psi = \Lambda \Lambda',$$

or, more conveniently,

$$R - \Psi^* = \Lambda^* \Lambda^{*'}, \tag{16.4.4}$$

in terms of eigenvalues and eigenvectors. Note that in both cases $\Lambda \Lambda'$ is diagonal.

In practice, the sample covariance matrix $C$ or the sample correlation matrix $\hat{R}$ is used. In the second method where $\Psi^* \neq 0$ the diagonal elements of $\hat{R} - \hat{\Psi}^*$, which are the estimated communalities, have to be guessed. Details may be found in Mardia, Kent and Bibby [(1979), pp. 261–263].

The decompositions of the symmetric matrices in principal factor analysis are simply a restatement of well-known results in matrix algebra [see I, Chapters 6 and 7]. It is possible to relax the condition that the factors are uncorrelated. If the covariance matrix of $f$ is the $(k \times k)$ positive definite matrix $\Phi$ then the original covariance matrix becomes

$$V = \Lambda \Phi \Lambda' + \Psi.$$

### 16.4.5.  Discussion

There are obvious difficulties in the interpretation of factor loadings. By considering only the larger factor loadings for a particular factor and looking at the random variables corresponding to these loadings an overall description of the factor may be attempted. To make this easier it is possible to make an orthogonal transformation of the factors without changing the decomposition of the covariance matrix. This is known as rotating the factors. The factors are usually transformed so that as many of the factor loadings as possible are near zero, leaving larger ones to help in interpretation. Details of one method known as the varimax may be found in Kaiser (1958).

## 16.5.   CANONICAL CORRELATION

### 16.5.1.   Introduction

Consider two vector random variables $\mathbf{X} = (X_1, X_2, \ldots, X_p)'$ and $\mathbf{Y} = (Y_1, Y_2, \ldots, Y_q)'$ with $q \le p$, both of which have expectation $\mathbf{0}$. (This is not a restrictive assumption since it is only the correlations between the variables that are of interest.) Let the covariance matrices [see Definition 16.1.1] of $\mathbf{X}$ and $\mathbf{Y}$ be

$$\mathbf{V}_{11} = E(\mathbf{XX}'), \qquad \mathbf{V}_{22} = E(\mathbf{YY}'),$$

respectively, and let the cross-covariance matrix [see Definition 16.1.2] between $\mathbf{X}$ and $\mathbf{Y}$ be

$$\mathbf{V}_{12} = E(\mathbf{XY}').$$

It is assumed that $\mathbf{V}_{11}, \mathbf{V}_{22}$ are positive definite [see Definition 16.1.3] and $\mathbf{V}_{12}$ is of full rank $(q)$. Canonical correlation attempts to simplify the relationships between $\mathbf{X}$ and $\mathbf{Y}$, given by the covariances assembled in the matrix $\mathbf{V}_{12}$, by considering correlations between linear combinations of the two sets of random variables.

The method produces uncorrelated pairs of scalar or univariate random variables in which the two members of a given pair are correlated with each other. The pair with the largest pair correlation (in absolute value) is called the first pair, and this proceeds down to the $q$th pair which have the smallest (non-zero) correlation. In this way a more parsimonious representation of the relationship between $\mathbf{X}$, $\mathbf{Y}$ is obtained than $\mathbf{V}_{12}$.

The linear combinations are known as canonical variates. By analogy with principal components analysis, the first few canonical variates, with the largest canonical correlations, might be used to describe approximately the relationship between $\mathbf{X}$ and $\mathbf{Y}$.

### 16.5.2.   Population Canonical Correlations

Consider a linear combination of the components of $\mathbf{X}$ and another of the components of $\mathbf{Y}$, given by

$$U = \mathbf{a}'\mathbf{X},$$

and

$$W = \mathbf{b}'\mathbf{Y};$$

then $U$, $W$ have expectation zero, and variances and covariance

$$\text{var}\,(U) = \mathbf{a'V}_{11}\mathbf{a},$$

$$\text{var}\,(W) = \mathbf{b'V}_{22}\mathbf{b},$$

$$\text{cov}\,(U, W) = \mathbf{a'V}_{12}\mathbf{b}.$$

The correlation coefficient between $U$, $W$ is therefore

$$\rho = \frac{(\mathbf{a'V}_{12}\mathbf{b})}{[(\mathbf{a'V}_{11}\mathbf{a})(\mathbf{b'V}_{22}\mathbf{b})]^{1/2}}.$$

It is required to maximize (or minimize) the dimensionless quantity $\rho$ [see II, § 9.8.2]. Since this is independent of the scale of measurements of $U$ and $W$, there is no loss of generality in restricting $\mathbf{a}$ and $\mathbf{b}$ to values giving a unit variance for $U$ and $W$.

The problem now becomes that of finding the stationary value of

$$\rho = \mathbf{a'V}_{12}\mathbf{b}$$

with respect to $\mathbf{a}$, $\mathbf{b}$ subject to the constraints

$$\text{var}\,(U) = \mathbf{a'V}_{11}\mathbf{a} = 1,$$

$$\text{var}\,(W) = \mathbf{b'V}_{22}\mathbf{b} = 1.$$

Using Lagrange's multipliers $\alpha$, $\beta$, we have to find the unconditioned stationary value of

$$C = \mathbf{a'V}_{12}\mathbf{b} - \alpha(\mathbf{a'V}_{11}\mathbf{a} - 1) - \beta(\mathbf{b'V}_{22}\mathbf{b} - 1).$$

Thus we have to solve the equations

$$\frac{\partial C}{\partial \mathbf{a}} = \mathbf{V}_{12}\mathbf{b} - 2\alpha\mathbf{V}_{11}\mathbf{a} = \mathbf{0}, \tag{16.5.1}$$

$$\frac{\partial C}{\partial \mathbf{b}} = \mathbf{V}'_{12}\mathbf{a} - 2\beta\mathbf{V}_{22}\mathbf{b} = \mathbf{0}. \tag{16.5.2}$$

Premultiplying both sides of (16.5.1) by $\mathbf{a'}$, and both sides of (16.5.2) by $\mathbf{b'}$, and rearranging, gives

$$2\alpha\mathbf{a'V}_{11}\mathbf{a} = \mathbf{a'V}_{12}\mathbf{b}$$

and

$$2\beta\mathbf{b'V}_{22}\mathbf{b} = \mathbf{b'V}'_{12}\mathbf{a}.$$

From the restrictions on $a$, $b$,

$$\mathbf{a'V}_{11}\mathbf{a} = 1 \quad \text{and} \quad \mathbf{b'V}_{22}\mathbf{b} = 1,$$

where $\mathbf{b'V}'_{12}\mathbf{b} = \mathbf{b V}_{12}\mathbf{b'} = \rho$, these imply that $2\alpha = 2\beta = \lambda$, which is the stationary value of $\rho$.

From (16.5.1), noting that $\mathbf{V}_{11}$ is non-singular, we have

$$\mathbf{a} = \frac{1}{\lambda} \mathbf{V}_{11}^{-1} \mathbf{V}_{12} \mathbf{b}. \tag{16.5.3}$$

On substituting for $\mathbf{a}$ in (16.5.2) we find

$$(\mathbf{V}_{12}' \mathbf{V}_{11}^{-1} \mathbf{V}_{12} - \lambda^2 \mathbf{V}_{22}) \mathbf{b} = \mathbf{0},$$

whence

$$(\mathbf{V}_{22}^{-1} \mathbf{V}_{12}' \mathbf{V}_{11}^{-1} \mathbf{V}_{12} - \lambda^2 \mathbf{I}) \mathbf{b} = \mathbf{0}. \tag{16.5.4}$$

A non-zero solution to (16.5.4) exists if and only if the matrix on the left is singular [see I, § 5.9], i.e. if

$$|\mathbf{V}_{22}^{-1} \mathbf{V}_{12}' \mathbf{V}_{11}^{-1} \mathbf{V}_{12} - \lambda^2 \mathbf{I}| = 0, \tag{16.5.5}$$

or, equivalently, if $\lambda^2$ is an eigenvalue of the matrix $\mathbf{V}_{22}^{-1} \mathbf{V}_{12}' \mathbf{V}_{11}^{-1} \mathbf{V}_{12}$. The largest such eigenvalue is the maximum value of $\rho^2$, and $\mathbf{b}$ is its associated eigenvector. The corresponding vector $\mathbf{a}$ comes from (16.5.3).

An alternative method of solution to the problem is to substitute for $\mathbf{b}$ and solve for $\mathbf{a}$, replacing (16.5.4) by

$$(\mathbf{V}_{11}^{-1} \mathbf{V}_{12} \mathbf{V}_{22}^{-1} \mathbf{V}_{12}' - \lambda^2 \mathbf{I}) \mathbf{a} = \mathbf{0}.$$

Anderson (1958), pp. 292–295 shows that, not surprisingly, this second method produces a solution identical to the first. Note that the two matrices involved have the same rank $q$ (the rank of $\mathbf{V}_{12}$) and hence have $q$ non-zero eigenvalues.

The canonical variates obtained above are called the first canonical variates $U_1$, $W_1$. Their correlation $\rho = \text{corr}\,(U_1, W_1)$ is equal to $\lambda_1$ (in absolute value). The corresponding coefficients are $\mathbf{a}_1$ and $\mathbf{b}_1$.

The second pair of canonical variates

$$U_2 = \mathbf{a}_2 \mathbf{X}$$

$$W_2 = \mathbf{b}_2 \mathbf{Y}$$

are chosen to maximize the absolute value of their correlation $\mathbf{a}_2' \mathbf{V}_{12} \mathbf{b}_2$ subject to their having unit variance and to the orthogonality constraints

$$\text{cov}\,(U_1, U_2) = \text{cov}\,(W_1, W_2) = \text{cov}\,(W_1, U_2) = 0;$$

that is,

$$\mathbf{a}_1 \mathbf{V}_{11} \mathbf{a}_2 = \mathbf{b}_1' \mathbf{V}_{22} \mathbf{b}_2 = \mathbf{b}_1' \mathbf{V}_{12} \mathbf{a}_2 = 0.$$

The values of $\mathbf{a}_2$, $\mathbf{b}_2$ giving the stationary value of $\rho$ are obtained by finding the stationary value of

$$C = \mathbf{a}_2 \mathbf{V}_{12} \mathbf{b}_2 - \alpha(\mathbf{a}_2' \mathbf{V}_{11} \mathbf{a}_2 - 1) - \beta(\mathbf{b}_2' \mathbf{V}_{22} \mathbf{b}_2 - 1)$$

$$+ \gamma \mathbf{a}_1' \mathbf{V}_{11} \mathbf{a}_2 + \delta \mathbf{b}_1' \mathbf{V}_{22} \mathbf{b}_2 + \varepsilon \mathbf{b}_1' \mathbf{V}_{12} \mathbf{a}_2$$

where $\alpha$, $\beta$, $\gamma$, $\delta$, $\varepsilon$ are Lagrange multipliers. By using the orthogonality constraints after taking the derivative of $C$ with respect to $\mathbf{a}_2$, $\mathbf{b}_2$, it may be shown that $\gamma = \delta = \varepsilon = 0$; and, by using the variance constraint, that $2\alpha = 2\beta = \lambda_2$. The second largest value of $\rho^2$ is $\lambda_2^2$, the second largest eigenvalue of $\mathbf{V}_{22}^{-1}\mathbf{V}_{12}'\mathbf{V}_{11}^{-1}\mathbf{V}_{12}$, and $\mathbf{b}_2$ is its associated eigenvector. $\mathbf{a}_2$ may then be obtained from an equation similar to (16.5.3).

The procedure is then repeated to find the remaining $(q-2)$ canonical variates and their associated squared correlations (see Anderson (1958), p. 291 for details). Strictly the eigenvalues have to be distinct to produce $q$ pairs of canonical variates.

The relationship between $X$, $Y$ will then have been summarized by the correlations between $q$ pairs of canonical variates instead of by the $pq$ covariances in $\mathbf{V}_{12}$ (and strictly speaking the covariances in $\mathbf{V}_{11}$ and $\mathbf{V}_{22}$).

As in previous sections, the procedure may be carried out using the correlation matrices $\mathbf{R}_{11}$, $\mathbf{R}_{22}$ [see Definition 16.1.1] and the cross correlation matrix $\mathbf{R}_{12}$ [see Definition 16.1.2]. This produces the same canonical correlations as the previous method but the corresponding coefficients of the canonical variates are different.

### 16.5.2. Sample Canonical Correlations

In practice the population covariance or correlation matrices are unknown and have to be estimated by their corresponding sample matrices from a random sample of $n$ observations on $(\mathbf{X}, \mathbf{Y})$.

By analogy with principal components analysis only those canonical variates with large correlations need be retained, giving a representation of the interrelationships between $X$ and $Y$ with fewer than $q$ correlations. Furthermore if $\mathbf{C}_{12}$ is not of full rank then there are fewer than $q$ non-zero eigenvalues or correlations.

To obtain a formal test of the hypothesis that some of the population canonical correlations are zero it is assumed that the random variables $(\mathbf{X}, \mathbf{Y})$ have a multivariate normal (MVN) distribution, with multiplicity $(p+q)$. A test of the hypothesis that $\mathbf{V}_{12} = 0$, which implies that $\mathbf{X}$ and $\mathbf{Y}$ are independent, may be based on the statistic

$$-\{n - \tfrac{1}{2}(p+q+3)\} \log \prod_{i=1}^{q} (1 - \hat{\lambda}_i^2),$$

(where the $\hat{\lambda}_i^2$ are the ordered eigenvalues of $\mathbf{C}_{22}^{-1}\mathbf{C}_{12}'\mathbf{C}_{11}^{-1}\mathbf{C}_{12}$) which has an asymptotic $\chi^2_{pq}$ distribution.

The test may be modified to deal with the partial hypothesis that the final $t$ canonical correlations are zero $(t < q)$, by using the statistic

$$-\{n - \tfrac{1}{2}(p+q+3)\} \log \prod_{i=q-t+1}^{q} (1 - \hat{\lambda}_i^2),$$

which has an asymptotic $\chi^2_{(p-q+t)t}$ distribution [see § 2.5.4(a)].

### 16.5.4. Numerical Example

Canonical variates related to growth rate of a variety of apple tree.

In Pearce and Holland (1960) the growth of a certain variety of apple tree was studied at maturity, and, after four years, the following four measurements were considered.

$X_1 = \log$ (weight of mature tree above ground)
$X_2 = \log$ (basal trunk girth of mature tree)
$Y_1 = \log$ (total shoot growth during the first four years)
$Y_2 = \log$ (basal trunk girth at four years of age).

The following correlation matrices were obtained:

$$\hat{\mathbf{R}}_{11} = \begin{pmatrix} 1 & 0 \cdot 951 \\ 0 \cdot 951 & 1 \end{pmatrix}$$

$$\hat{\mathbf{R}}_{22} = \begin{pmatrix} 1 & 0 \cdot 898 \\ 0 \cdot 898 & 1 \end{pmatrix}$$

$$\hat{\mathbf{R}}_{12} = \begin{pmatrix} 0 \cdot 596 & 0 \cdot 517 \\ 0 \cdot 694 & 0 \cdot 619 \end{pmatrix}.$$

This gives

$$\hat{\mathbf{R}}_{22}^{-1} \hat{\mathbf{R}}_{12}' \hat{\mathbf{R}}_{11}^{-1} \hat{\mathbf{R}}_{12} = \begin{pmatrix} 0 \cdot 4939 & 0 \cdot 4403 \\ 0 \cdot 0340 & 0 \cdot 0414 \end{pmatrix}.$$

This matrix is of rank nearly one, since $\hat{\mathbf{R}}_{12}$ has rank nearly one. The second eigenvalue will therefore be near zero and may be ignored.

The first eigenvalue $\hat{\lambda}_1^2 = 0 \cdot 5249$ (the other is of the order of $0 \cdot 02$), corresponding to a canonical correlation of $\hat{\rho}_1 = 0 \cdot 725$, which is greater than the correlations in $\hat{\mathbf{R}}_{12}$.

The canonical coefficients $b_1$, $b_2$ may be obtained from (16.5.4) and are given by

$$-0 \cdot 0310 b_1 + 0 \cdot 4403 b_2 = 0$$

$$0 \cdot 0340 b_1 - 0 \cdot 4835 b_2 = 0.$$

Setting $b_1 = 1$ gives $b_2 = 0 \cdot 0704$. From (16.5.3), substituting for $b_1, b_2$,

$$a_1 = -0 \cdot 994$$

and

$$a_2 = 1 \cdot 966.$$

The values above are all calculated relative to $b_1 = 1$; however they should take values such that the canonical variates have unit variance. (This exercise need only be carried out for $b_1, b_2$, since it follows immediately from (16.5.3) that $a_1, a_2$ will then have unit variance.)

The modified coefficients become

$$a_1 = -0 \cdot 877, \qquad a_2 = 1 \cdot 734, \qquad b_1 = 0 \cdot 882, \qquad b_2 = 0 \cdot 0621.$$

The first canonical variates are

$$U_1 = -0 \cdot 887 X_1 + 1 \cdot 734 X_2$$
$$W_1 = 0 \cdot 882 Y_1 + 0 \cdot 0621 Y_2$$

where $X$, $Y$ are standardized variates.

$U_1$ represents a contrast between $X_1$ (log weight of mature tree) and $X_2$ (log basal trunk girth) with $X_2$ having about twice as much weight attached to it. $W_1$ is almost entirely $Y_1$ (log total shoot growth during the first four years). As pointed out in Pearce and Holland (1960) interpretation of the meaning of these variates should be left to the biologist.

## 16.6.  DISCRIMINANT ANALYSIS

### 16.6.1.  Introduction

Discriminant analysis is concerned with obtaining rules for the classification of multivariate observations into one of several categories or populations. In medicine, for example, it helps with diagnosis or prognosis, where, given observations on a patient, it is required to assign the patient to one of several disease types or medical conditions. Titterington *et al.* (1981) give examples of discriminant techniques applied to a prognosis problem concerning head-injured patients. There are many other applications of discriminant analysis in the literature (see for example the bibliography in Press (1972)).

The analysis presented here assumes that the number of populations or categories are known in advance. In cluster analysis, by contrast, the objective is to identify clusters or categories from the data [see Chapter 17.].

### 16.6.2.  Discrimination in Two Known Populations

Consider the problem of classifying a single multivariate observation $\mathbf{x} = (x_1, x_2, \ldots, x_p)'$ into one of two populations which have known $p$-variate densities $f_1(\mathbf{x})$, $f_2(\mathbf{x})$ (i.e. the parameters and form of the density are known). Suppose that the prior probabilities that an observation comes from population 1 or 2 are $p(1)$, $p(2) = 1 - p(1)$ respectively. Then, by Bayes' Theorem, the posterior probability that the observation comes from population 1 is [see Chapter 15]

$$p(1|\mathbf{x}) = \frac{p(1)f_1(\mathbf{x})}{p(1)f_1(\mathbf{x}) + p(2)f_2(\mathbf{x})}$$

and the posterior probability it comes from population 2 is

$$p(2|\mathbf{x}) = \frac{p(2)f_2(\mathbf{x})}{p(1)f_1(\mathbf{x}) + p(2)f_2(\mathbf{x})} = 1 - p(1|\mathbf{x}).$$

Classification of the observation may now be carried out by using the posterior odds that the observation comes from population 1

$$p(1|\mathbf{x})/[1-p(1|\mathbf{x})] = p(1)f_1(\mathbf{x})/p(2)f_2(\mathbf{x}). \qquad (16.6.1)$$

A sensible procedure would be to allocate the observation to population 1 if the odds are greater than 1, that is if, $p(1|x) > \frac{1}{2}$, and to population 2 if the odds are less than 1. Anderson (1958) pp. 130–131 shows that this procedure minimizes the probability of misclassification.

If one postulates a cost (or loss) function, [see Chapter 19] $C(2|1)$ and $C(1|2)$ representing the cost of misclassifying an observation from 1 and 2 respectively it may be shown [see, Anderson (1958), pp. 130–131] that, comparing (posterior) expected loss is due to misclassification, the modification to (16.6.1) gives the rule as follows: allocate to population 1 if

$$\frac{C(2|1)p(1)f_1(x)}{C(1|2)p(2)f_2(x)} \geq 1$$

and to population 2 otherwise.

In both cases the test may be reduced to one based on the likelihood ratio $f_1(\mathbf{x})/f_2(\mathbf{x})$, thus:

$$\text{if } f_1(x)/f_2(x) \geq k, \text{classify as '1';}$$

otherwise, classify as '2'; here the constant $k$ depends on the prior probabilities and the costs of misclassification. In the case where the prior expected losses are such that $C(2|1)p(1) = C(1|2)p(2)$, it turns out that $k = 1$.

### 16.6.3. Discrimination in Two Multivariate Populations

Suppose that the observation comes from one of two $p$-variate MVN populations with known means $\boldsymbol{\mu}_1$ and $\boldsymbol{\mu}_2$ and known common covariance matrix $\mathbf{V}$.

The likelihood ratio is

$$\frac{f_1(\mathbf{x})}{f_2(\mathbf{x})} = \exp\left[\mathbf{x}'\mathbf{V}^{-1}(\boldsymbol{\mu}_1 - \boldsymbol{\mu}_2) - \tfrac{1}{2}(\boldsymbol{\mu}_1 - \boldsymbol{\mu}_2)\,\mathbf{V}^{-1}(\boldsymbol{\mu}_1 + \boldsymbol{\mu}_2)\right]. \qquad (16.6.2)$$

Using this, the classification rule becomes the following:

$$\text{let } u = \mathbf{x}'\mathbf{V}^{-1}(\boldsymbol{\mu}_1 - \boldsymbol{\mu}_2) - \tfrac{1}{2}(\boldsymbol{\mu}_1 - \boldsymbol{\mu}_2)'\mathbf{V}^{-1}(\boldsymbol{\mu}_1 + \boldsymbol{\mu}_2);$$

$$\text{if } u \geq c, \text{classify as '1'; otherwise as '2'.} \qquad (16.6.3)$$

Here $c = \log_e k$.

The rule therefore depends on the scalar random variable $U$ which is a linear function of $\mathbf{X}$. $U$ has some interesting distributional properties:

If $X$ came from population 1, which is MVN $(\boldsymbol{\mu}_1, \mathbf{V})$, then $U = U_1$ has the Normal distribution with expectation $\frac{1}{2}\delta$, and variance $\delta$, while if $\mathbf{X}$ is from population 2, which is MVN $(\boldsymbol{\mu}_2, \mathbf{V})$, $U = U_2$ has the Normal $\delta$ distribution with expectation $-\frac{1}{2}\delta$, and variance $\delta$, where

$$\delta = (\boldsymbol{\mu}_1 - \boldsymbol{\mu}_2)'\mathbf{V}^{-1}(\boldsymbol{\mu}_1 - \boldsymbol{\mu}_2).$$

This quantity $\delta$ is known as the Mahalanobis distance between the two populations. It is now possible to obtain the probabilities of misclassification of the observation. For example, suppose that the observation is from population 1 but is allocated to population 2; then the probability of misclassification is

$$P(U_1 < c) = \Phi\left(\frac{c - \frac{1}{2}\delta}{\sqrt{\delta}}\right)$$

where $\Phi$ denotes the standard Normal integral.

The performance of the rule may be judged by calculating expected loss due to misclassification

$$C(2|1)P(U_1 < c)p(1) + C(1|2)P(U_2 \geq c)p(2).$$

Suppose now that the parameters of the $p$-variate Normal distribution are unknown and that samples of sizes $N_1$ and $N_2$ are taken from the two $p$-variate Normal populations. $\mu_1, \mu_2$ may be estimated as $\bar{x}_1, \bar{x}_2$ and the common covariance matrix $V$ from the pooled estimate $C$ given by

$$C = [(n-1)C_1 + (n_1-1)C_2]/(n_1+n_2-2),$$

where $C_1$, $C_2$ are the usual unbiased estimates of $V$ from the two separate samples.

A discriminant function may now be obtained by substituting for the parameters in (16.6.3) their corresponding unbiased estimates, giving the following rule:

$$\text{let } w = x'C^{-1}(\bar{x}_1 - \bar{x}_2) - \tfrac{1}{2}(\bar{x}_1 - \bar{x}_2)C^{-1}(\bar{x}_2 + \bar{x}_2); \qquad (16.6.4)$$

if $w \geq c$, classify the new observation $x$ into population 1; otherwise into 2.

The first term in $w$ is known as Fisher's linear discriminant function. Unfortunately, the distribution of $w$ is very complex, and hence the calculation of the probabilities of misclassification using this criterion becomes difficult. The exact distribution of $w$ is briefly discussed in Anderson (1958), pp. 138–139. Asymptotically, as $n_1$ and $n_2 \to \infty$, the probabilities of misclassification may be obtained by approximating $\delta$ above by

$$(\bar{x}_1 - \bar{x}_2)'C^{-1}(\bar{x}_1 - \bar{x}_2).$$

A further method of obtaining estimates of the probabilities of misclassification would be to use the discrimination rule for a set of data where it is known which observations are from which populations, sometimes called the training set in medical applications. An estimate of the probability of misclassifying into population 2 is then the ratio of the number misclassified into 2 to the total number of the observations from population 1. However as Morrison (1972) points out this is liable to produce a badly biased estimate.

A Bayesian formulation has been adopted for this section in that $p(1)$, $p(2)$ are the experimenter's assessment of the probability an observation comes from populations 1 and 2 based on prior knowledge of the populations.

Anderson (1958) pp. 133–136 also discusses a minimax solution which chooses the constant $c$ in the absence of prior probabilities to satisfy

$$C(1|2)\left[\Phi\left(\frac{-c-\frac{1}{2}\delta}{\sqrt{\sigma}}\right)\right] = C(2|1)\Phi\left(\frac{c-\frac{1}{2}\delta}{\sqrt{\sigma}}\right).$$

In many applications the discriminant rules are used with $c = 0$.

If the covariance matrices of the two populations are not assumed equal then the discriminant functions become quadratics in $\mathbf{X}$.

### 16.6.4. Discrimination in Several Populations

If there are more than two populations, there are several methods for classifying an observation into one of them. The discriminant rules (16.6.3) or (16.6.4) may be used for all pairs of populations and the values of the random variables $U$ or $W$ will then classify an individual observation into one of the populations (see Morrison (1972), pp. 239–245 for examples). Suppose that there are $m$ ($>2$) populations and that $w_{ij}$ is the value of the discriminant function for classifying an individual into population $i$ or $j$; then in the case where $c = 0$ for all pairs of populations, the rule would be: classify into $i$ if $w_{ij} \geq 0$ for all $j \neq i$.

A modification of the distance function may also be used to classify an individual observation; define

$$\delta_i = (\mathbf{x} - \bar{\mathbf{x}}_i)\mathbf{C}^{-1}(\mathbf{x} - \bar{\mathbf{x}}_i),$$

and allocate $\mathbf{x}$ to the population with the smallest $\delta_i$, $i = 1, 2, \ldots, m$. Morrison (1972) p. 241 shows that this rule is equivalent to that based on $w_{ij}$.

Rao (1973) suggests a rule based on the overall expected loss of classifying an observation in a particular population. This is defined as

$$\sum_{j=1}^{m} C(i|j)p(j)P(i|j)$$

for population $i$, where $P(i|j)$ is the probability of misclassifying an observation in population $j$ and $C(i|j)$ is defined as above ($C(i|i) = 0$). The population giving the smallest expected loss is then chosen. It is assumed that the distribution in each population is known.

Aitchinson and Dunsmore (1975) Chapter 11 present the full Bayesian approach to the problem when all the parameters are unknown but the form of the density function of the observations are known. Prior distributions are assigned to the unknown parameters and given the random sample of observations from each population, the posterior distributions are calculated.

The new observation $\mathbf{x}$ is classified by calculating the predictive probability [cf. Chapter 15]

$$p(i|\mathbf{x}, \text{Data}) \propto p(\mathbf{x}|i, \text{Data})p(i|\text{Data}),$$

and choosing that population with the largest such probability. The quantity $p(\mathbf{x}|i, \text{Data})$ is the marginal predictive distribution of $\mathbf{x}$ given that the observation comes from population $i$, and $p(i|\text{Data})$ is the predictive probability that the observation comes from population $i$.

To illustrate this approach consider a univariate random variable $X$ with density $f(x; \theta)$ where the single parameter $\theta$ is assigned a prior distribution $\pi(\theta)$; then given a random sample $\mathbf{x} = (x_1, x_2, \ldots, x_n)'$ as Data, the posterior distribution of $\theta$ is

$$\pi(\theta|\text{Data}) \propto f(\mathbf{x}; \theta)\pi(\theta).$$

The predictive distribution of a new observation $\xi$ is

$$p(\xi|\text{Data}) = \int_\theta f(\mathbf{x}; \theta)\pi(\theta|\text{Data})\, d\theta.$$

In section 16.3.5 an application of principal components to some forestry data was discussed. It was seen that most of the variation could be explained by the first two principal components. A plot of the scores or values of the principal components is therefore possible in this case where a cluster of points represent a variety of poplar. The first two principal components of the new observation may be calculated and it may be classified by observing whether it falls within an existing cluster.

To avoid overlapping clusters, the means of the principal components scores for the existing variates could be calculated, and the new observation could be classified on its proximity to a variety mean. Another method based on reducing the complexity of the data using scores on canonical variates is presented in Maxwell (1977) Chapter 9.

*Acknowledgement*: I should like to thank Dr. D. A. Quinney for help with the computations in Table 16.3.2.

P.J.

## 16.7.  FURTHER READING AND REFERENCES

Aitchison, J. and Dunsmore, I. R. (1975). *Statistical Prediction Analysis*, Cambridge University Press.

Anderson, T. W. (1958). *An Introduction to Multivariate Statistical Analysis*, Wiley.

Bishop, Y. M. M., Fienberg, S. E. and Holland, P. W. (1975). *Discrete Multivariate Analysis*, M.I.T. Press.

Giri, N. C. (1977). *Multivariate Statistical Inference*, Academic Press.

Goldstein, M. and Dillon, W. R. (1978). *Discrete Discriminant Analysis*, Wiley.

Jeffers, J. N. R. (1965). Principal components Analysis in Taxonomic Research, Forestry Commission Statistics Section Paper no. 83.

Jeffers, J. N. R. (1967). Two Case Studies in the Application of Principal Components, *Appl. Statist.* **16**, 225–236.

Kaiser, H. F. (1958). The Varimax Criterion for Analytic Rotation in Factor Analysis, *Psychometrika* **23**, 187–200.

Kendall, M. G. (1957). *A Course in Multivariate Analysis*, Griffin.

Lawley, D. N. and Maxwell, A. E. (1971). *Factor Analysis as a Statistical Method* (2nd ed.), Butterworths.

Lochenbruch, P. A. (1975). *Discriminant Analysis*, Hafner.

Mardia, K. V., Kent, J. T. and Bibby, J. M. (1979). *Multivariate Analysis*, Academic Press.

Maxwell, A. E. (1977). *Multivariate Analysis in Behavioural Research*, Chapman & Hall.

Morrison, D. F. (1976). *Multivariate Statistical Methods* (2nd ed.), McGraw-Hill.

Pearce, S. C. and Holland, D. A. (1960). Some Applications of Multivariate Methods in Botany, *Appl. Statist.* **9**, 1–7.

Press, S. J. (1972). *Applied Multivariate Analysis*, Holt, Rinehart & Winston.

Rao, C. R. (1973). *Linear Statistical Inference and its Applications*, Wiley.

Titterington, D. M., Murray, G. D., Murray, L. S., Spiegelhalter, D. J., Skene, A. M., Habbema, J. D. F. and Gepke, G. J. (1981). Comparison of Discrimination Techniques Applied to a Complex Data Set of Head Injured Patients (with discussion), *Journ. Roy. Statist. Soc.*, Series A **144** (to appear).

# CHAPTER 17

# Multivariate Analysis: Ordination, Multidimensional Scaling and Allied Topics

## 17.1. INTRODUCTION

In multivariate analysis, an ordination method is one in which members of multivariate samples are represented by points in some geometric space—usually Euclidean. Thus a confusing array of numbers becomes what is hoped to be a less confusing scatter of points revealing structures of interest such as clusters of points, collinearities, trends or other salient features of the sample. Of course the eye cannot hope to detect structure in a many-dimensional space, so there is a strong emphasis on methods which give interpretable two, or perhaps, three-dimensional approximations to the scatter. When interpreting diagrams of this kind, close pairs of points usually represent similar samples and distant pairs of points represent very dissimilar samples. Thus distance is the basic concept underlying the interpretation of most ordination diagrams but, as will be seen in section 17.13, below, other types of representation also occur.

The term ordination derives from ecology, where there is interest in representing different ecological communities of plants as a set of points on a line. This is a one-dimensional ordination or ordering. It was soon discovered that one dimension was often insufficient to give a realistic representation, so two or more dimensions were admitted. Although the concept or ordering is then invalid, the term ordination has persisted. Rather similarly, psychologists have long been concerned with representing the intensity of some indirectly observed stimulus as a point on a scale, analogous to a scale for an ordinary physical measurement. Again this gives a one-dimensional configuration and again further dimensions were found necessary to give a realistic representation, leading to the term *multidimensional scaling* which has now gained acceptance amongst statisticians and data-analysts. Thus ordination and multidimensional scaling are synonyms, borrowed from different sciences, for which there is no neutral statistical term.

727

The most simple form of ordination is the bivariate scatter diagram in which one variable is plotted against another (e.g. height against weight) each point representing one sample observation. Such diagrams are scrutinized for evidence of pattern in the same way as are more general ordinations. They are also used to indicate possible outliers which may be regarded as departures from pattern. Scatter diagrams suffer in a mild way from the major defect of many ordination methods, in that they are scale-dependent. Measuring height in inches and weight in pounds gives a different diagram from that derived from height measured in centimetres and weight in kilograms. This difficulty becomes more serious in general ordinations based on linear combinations of variable values.

When a sample of size $n$ has variable values $x_1, x_2, \ldots, x_n$ measured in some unspecified units, the two most usual forms of standardization are

(a) $$y_i = \log x_i$$

(b) $$z_i = x_i / s$$

where $s$ is the standard deviation of the sample. Note that the distance $y_i - y_j$ between the $i$th and $j$th samples is invariant to a multiplicative scale change (e.g. inches to centimetres) but not to an additive, or combination of additive with multiplicative, change (e.g. Fahrenheit to Centigrade). The distance $z_i - z_j$ is invariant to both additive and multiplicative scale changes. Such considerations govern the type of standardization to be used but much remains arbitrary. When all variables of a multivariate sample are measured on the same scale the need for standardization is less clear, but it may happen that the variation for some variables is very much greater than for others and can be usefully eliminated by using (b).

The use of standard error for standardization highlights a fundamental issue that should always be borne in mind when interpreting ordinations. All $s^2$ [see § 2.12(c)] that estimates a population variance $\sigma^2$ is clearly interpretable but when $s^2$ is estimated from a heterogeneous sample drawn from some unknown mixture of basic populations its use is less justified. In many ordinations the values $x_i$ of a variable cannot be regarded either as a random sample or even as realizations of a random variable. It may be realistic to regard them as an exhaustive enumeration of all the values of a variable in a finite population disociated from any probabilistic considerations. For example in taxonomy one is usually interested in the ordination of a finite number of species described by a set of characters. Of course, the individual representatives of a species could be sampled and, for some characters at least, there will be within-species variation. Thus it is possible to do ordinations of samples for each species separately and these may show features of interest. However, it is more usual to focus attention on the differences between species for which average, or some kind of typical, values are required for the variables for each species. Occasionally qualitative characters can be found, that do not vary within a species, which are ideally suitable for describing differences between species in terms of ordination.

The above remarks draw attention to some of the problems encountered when ordinating groups (e.g. species). Within-group and between-group ordination are both important and can, to some extent, be combined (see Gower and Digby (1980)). Between-group ordination is often disguised so that it appears like a within-group ordination, possibly because within-group variation either does not exist or is ignored as being unimportant.

## 17.2. PRINCIPAL COMPONENTS ANALYSIS

Suppose $_n\mathbf{X}_p$ is a data-matrix giving the values of $p$ variables for $n$ samples. When the $n$ samples can be regarded as a random sample of a multinormal (multivariate Normal) distribution [see II, § 13.4], principal components analysis can be developed as in section 16.3. This is the approach normally found in textbooks. In this chapter a different approach is taken which is useful in its own right and serves as an introduction to more general ordination problems. In this approach, which is related to Karl Pearson's paper of 1901, it is convenient, though not essential, to regard the $n$ samples as representing different groups and the associated $p$ variables as giving typical or average values of quantitative variables for each group.

Geometrically we can regard the values of the $i$th sample $(x_{i1}, x_{i2}, \ldots, x_{ip})$ as representing a point $P_i$ in a $p$-dimensional space with rectangular co-ordinate axes. Because we are considering between-group variation, there is no reason to expect an ellipsoidal scatter of points in multidimensional space, as would occur with a multinormal distribution. Any kind of scatter is acceptable and no distributional assumptions are required for the development given here. A simple scatter-diagram of, say, the $r$th and $s$th variables represents an orthogonal projection of the samples onto the plane defined by the $r$th and $s$th axes. It is natural to ask whether some other plane might not, in some sense, give a more representative projection. In principal components analysis the best fitting plane is defined as the one which minimizes the sum-of-squares of projections onto it; i.e. it is the plane that minimizes the sum-of-squares of the distances between itself and the samples. Instead of a plane we may wish to consider projections onto any $k$-dimensional linear sub-space of the full $p$-dimensional space.

Suppose the $k$-dimensional sub-space [see I, § 5.5] is defined as that spanned by any $k$ linearly independent vectors given as the columns of a matrix $_n\mathbf{H}_k$ then it is a standard result of linear algebra that the coordinates of the points $_n\mathbf{X}_p$ projected onto the sub-space are given by $\mathbf{XH(H}^T\mathbf{H)}^{-1}\mathbf{H}^T$. [Here, and throughout the chapter, $_n\mathbf{H}_k$ denotes a matrix $\mathbf{H}$ with $n$ rows and $k$ columns, and $\mathbf{H}^T$ denotes the transpose of $\mathbf{H}$.] Nothing is lost by requiring the linearly independent vectors to be orthogonal [see I, § 10.2] so that the projected coordinates simplify to $\mathbf{XHH}^T$ where $\mathbf{H}^T\mathbf{H} = \mathbf{I}$. The residuals [see § 8.2.4] are given by the coordinates orthogonal to the subspace and are $\mathbf{X(I} - \mathbf{HH}^T)$, so that the residual sum of squares is

$$\text{Trace } \mathbf{X(I} - \mathbf{HH}^T)\mathbf{X}^T$$

[see I, (6.2.4)] and this is to be minimized over $\mathbf{H}$. This is the same as maximizing

$$\text{Trace }(\mathbf{XHH}^T\mathbf{X}^T)$$

which may be written Trace $(\mathbf{H}^T\mathbf{SH})$ where $\mathbf{S}=\mathbf{X}^T\mathbf{X}$, a sums of squares and products matrix. Writing $\mathbf{S}=\mathbf{U\Lambda U}^T$, the spectral form [see I: § 7.10] of $\mathbf{S}$, where $\mathbf{U}$ is an orthogonal matrix and $\mathbf{\Lambda}$ is diag $(\lambda_1, \lambda_2, \ldots, \lambda_p)$, we have

$$\text{Trace }(\mathbf{H}^T\mathbf{SH}) = \sum_{i=1}^{p} \lambda_i(g_{i1}^2 + g_{i2}^2 + \ldots + g_{ik}^2)$$

where $g_{ij}$ are the elements of the orthonormal matrix $\mathbf{G}=\mathbf{U}^T\mathbf{H}$. Writing

$$f_i = g_{i1}^2 + g_{i2}^2 + \ldots + g_{ik}^2,$$

we have

$$0 \leq f_i \leq 1 \qquad (17.2.1)$$

because the sum of squares of the elements of a row of the orthonormal matrix $\mathbf{G}$ cannot exceed unity, the value obtained when $k=p$ and $\mathbf{G}$ becomes orthogonal. Also

$$\sum_{i=1}^{p} f_i = k \qquad (17.2.2)$$

because this quantity is the sum of squares of all the elements of $\mathbf{G}$ each of whose $k$ columns has sums-of-squares equal to unity. Thus we have to maximize the quantity

$$\sum_{i=1}^{p} \lambda_i f_i$$

subject to the constraints (17.2.1) and (17.2.2). This is a linear programming problem [see I, Chapter 11], so the maximum must occur when $(f_1, f_2, \ldots, f_p)$ is a vertex of the feasible region defined by (17.2.1) and (17.2.2). These vertices all occur when $k$ of the quantities $f_i$ are equal to unity and the remaining $p-k$ quantities are zero. When the eigenvalues [see I, Chapter 7] are ordered such that $\lambda_1 \geq \lambda_2 \geq \ldots \geq \lambda_p$ it is clear that the maximum is $\sum_{i=1}^{k} \lambda_i$ attained when $f_1 = f_2 = \ldots = f_k = 1$ and $f_i = 0$ for $i > k$. When $\lambda_{k+1} = \lambda_k$ this maximum is attained in many different ways. So far the additional sets of constraints that the columns of $\mathbf{G}$ are orthogonal has not been used. However the solution obtained without imposing this constraint has given a matrix $\mathbf{G}$ that is in fact orthonormal and so must maximize the trace over all orthonormal matrices. Thus at the maximum we have shown that the orthonormal matrix $\mathbf{G}$ becomes one whose rows as well as columns have unit length. This in turn implies that $\mathbf{G}$ may be partitioned in the form:

$$G = \begin{pmatrix} \mathbf{G}_k \\ \mathbf{0} \end{pmatrix}$$

where $\mathbf{G}_k$ is orthogonal of order $k$. Thus $\mathbf{H} = \mathbf{UG}$ is a matrix with orthogonal columns that are linear combinations of the first $\mathbf{k}$ eigenvectors of $\mathbf{S}$.

The above shows that the best fitting $k$-dimensional linear subspace that contains the origin is given by the space spanned by the $k$ (orthogonal) unit vectors that are the eigenvectors of $\mathbf{S}$ corresponding to the $k$-largest eigenvalues. The residual sum-of-squares that has been minimized has value

$$\text{Trace } (S) - \sum_{i=1}^{k} \lambda_i = \sum_{i=k+1}^{p} \lambda_i,$$

the sum of the $p-k$ smallest eigenvalues of $\mathbf{S}$.

It is of course possible that an even smaller residual sum-of-squares might be obtained if the $k$-dimensional subspace did not contain the origin. It is well known, and simple to prove, that sums-of-squares about the centroid (i.e. mean) are smaller than sums-of-squares about any other point. It follows that the best fitting $k$-dimensional sub-space should contain the mean. This implies that in the above, $\mathbf{X}$ should be replaced by deviations from the mean $(\mathbf{I}-\mathbf{N})\mathbf{X}$ where the elements of $\mathbf{N}$ all equal $1/n$. The definition of $\mathbf{S}$ is then replaced by $\mathbf{X}^T(\mathbf{I}-\mathbf{N})\mathbf{X}$, the corrected sums-of-squares and products matrix for the $n$ observations on $p$ variates. In the following we shall assume that the columns of $\mathbf{X}$ are expressed as deviations from their means so that $\mathbf{X}$ has zero column totals, and we can then dispense with the matrix $\mathbf{N}$.

The above derivation of the basic result of principal components analysis differs from that usually given in textbooks, which is as follows. It is routine to show by differentiation with Lagrange multipliers, that, when $k = 1$, $\mathbf{H}$ (which is then a vector) is given by the eigenvector of $\mathbf{S}$ corresponding to the largest eigenvalue $\lambda_1$. Similarly if $r$ directions are determined by the first $r$ eigenvectors of $\mathbf{S}$ then the direction *orthogonal* to this space that minimizes the residual sum-of-squares is that given by the $(r+1)$th eigenvector of $\mathbf{S}$. This argument establishes optima that are conditional on the dimensions already fitted [see § 16.3]. Our approach above shows that these conditional optima are globally optimal.

For ordination, the coordinates $\mathbf{XHH}^T$ of the projected points are required. This form is useless for plotting, as even when $k = 1$ the one-dimensional set of coordinates are expressed in $p$-dimensional form. More convenient is to choose a set of $k$ orthogonal axes in the sub-space; the coordinates $\mathbf{XH}$ are the simplest way of doing this. Any other set of orthogonal axes in the sub-space would do as well, but $\mathbf{XH}$ has the advantage that its first $c$ columns give coordinates for the best fit in $c$ dimensions.

In the terminology of components analysis, the eigenvectors given in the columns of $\mathbf{H}$ are *principal components*, or *principal axes*, and the coordinates $\mathbf{XH}$ are *component scores*. The coefficients $h_{ij}$ of the matrix $\mathbf{H}$ are termed the *loadings* of the $i$th sample on the $j$th variate. By setting $k = p$ all principal axes are obtained and $\mathbf{H}$ becomes the orthogonal matrix of all the eigenvectors of $\mathbf{S}$. As orthogonal transformations leave Euclidean distance invariant, this shows that the principal axes merely represent a change from the original coordinate axes to new axes which have the optimal properties established above: the relative disposition of the points is unchanged. Equivalently one may view the transformation as one in which the original variables become principal variables whose values are the scores.

Not infrequently the new variables are identified with underlying properties of the samples being studied; a process termed *reification*. Sometimes these identifications are quite convincing but it seems that the human mind is adept at forming interpretations that do not stand up to close inspection. The process of reification like the related identification of factors in factor analysis [see § 16.4] should be viewed with caution. Just as in factor analysis, a $k$-dimensional space is found by components analysis and any set of linearly independent directions in this space suffices for coordinate axes. there seems to be no reason why the mathematically determined principal axes should be more interpretable than any other set of coordinate axes in the same space. Thus the full battery of orthogonal and oblique factor rotation techniques, together with their dangers, are available in components analysis to help find interpretable axes. Fortunately reification is less important when components analysis is used for ordination than when analyzing samples from a single multivariate distribution. In ordination the emphasis is on plotting the points representing the samples.

Reduction from $p$-dimensions to give a good approximation in $k$-dimensions, $k < p$, is a key factor of all ordinations. A simple measure of the goodness-of-fit is $\sum_{i=1}^{k} \lambda_i / \sum_{i=1}^{p} \lambda_i$, usually expressed as a percentage. This quantity represents the ratio of the sum-of-squares about the origin (usually the centroid) amongst the projected points to the total sum-of-squares before projection. This is the same thing as the ratio of the sum-of-squares of all the pairwise distances amongst the projected points to the same quantity for the unprojected points. A high value of this ratio for a small value of $k$ indicates a good fit in few dimensions. Thus implicit in a components analysis is the hope that the $p$-dimensional cloud of points contains (at least approximately) a $k$-dimensional linear manifold with $k \ll p$. If the points have some simple structure but lie on a non-linear manifold, components analysis is unlikely to be successful. For example points lying on the surface of a sphere have no simple representation in less than three dimensions, unless sparsely sampled polar regions allow one of the map-projections to be used.

Very commonly the loadings of the first component are all positive, and not infrequently they have similar magnitudes. When this occurs the first component is often identified as a *size* component. In biological problems when measurements are made on growing organisms it is usual for different parts to grow at similar rates (referred to as *allometric* growth). In these circumstances correlations between all pairs of characters will be positive and the matrix **S** will have non-negative terms. The Frobenius–Perron theorem [see I, Theorem 7.11.1] then tells us that the maximum eigenvalue of **S** is associated with an eigenvector of non-negative loadings. Thus size effects automatically lead to the kind of phenomenon that is observed in practice. When size is not of primary interest it is common to plot coordinates only for the second and subsequent components, which are broadly speaking supposed to express differences in shape. A somewhat more formal approach is to treat the variable **X1** as a size variable. Eliminating size by projection leads to the shape variables $X(I-11^T/p)$ and the modified sums-of-squares-and-products matrix $S_{\text{SHAPE}} = (I-P)X^T(I-N)X(I-P)$ where $P = 11^T/p$. This has a zero eigenvalue corresponding to the vector $1$ and its non-zero eigenvalues lead to a principal components analysis of shape.

The full force of the warnings given in section 17.1 of this chapter, about the sensitivity of some forms of multivariate analysis to changes of scale are relevant to principal components analysis. When scales of measurement change for the variables it is usual first to standardize **X** so that **S** becomes a correlation matrix. In size/shape studies the log-transformation of **X** is more common. This is because shape is often thought of as a ratio of two variables and this ratio is unaffected by allometric growth. Further, eliminating size as above gives shape variables $(\log X)(I-11^T/p)$ which gives an $i$th size variable $\log\{x_i/(x_1 x_2 \dots x_p)^{1/p}\}$ with desired ratio form. Note that two samples, in one of which all the variable values are a multiple of the values for the other sample, have the same shape after transformation.

## 17.3. MULTIPLICATIVE MODELS AND THE ECKART–YOUNG THEOREM

Linear models underly many of the standard types of statistical analyses and have been studied for some 200 years (Chapters 8, 10, 11, 12). The theory of multiplicative models is much less well understood and is relatively little-known, although the beginnings of its study date back more than 50 years. Their place in this chapter is justified because the least squares analysis of a simple multiplicative model leads naturally to the Eckart–Young Theorem which is fundamental to several of the methods to be discussed later.

Consider an $m \times n$ table of observations $y_{ij}$ and suppose we wish to fit the multiplicative model

$$y_{ij} = \mu + \alpha_i + \beta_j + \gamma_i \gamma_j' + \varepsilon_{ij} \qquad (i = 1, 2, \dots, m; j = 1, 2, \dots, n)$$

where $\varepsilon_{ij}$ are i.i.d [see § 1.4.21i] error terms. As with linear models, this model is over-determined and only differences in parameter values are estimable. Unique estimates can be obtained only by specifying an origin for each set of estimates and, as usual, we set the origin at the centroid to give $\sum_i \alpha_i = \sum_j \beta_j = \sum_i \gamma_i = \sum_j \gamma_j' = 0$. These constraints are not essential but they have the advantage of treating all parameters equally and so lead to tidy algebra. The least-squares estimates are:

$$\left. \begin{aligned} \hat{\mu} &= y \\ \hat{\alpha}_i &= y_{i\cdot} - y_{\cdot\cdot} \\ \hat{\beta}_j &= y_{\cdot j} - y_{\cdot\cdot} \\ Z\hat{\gamma}' &= \Gamma'\hat{\gamma} \\ Z^T\hat{\gamma} &= \Gamma\hat{\gamma}' \end{aligned} \right\}$$

where the dot-notation indicates the mean value over the specified suffix, $\mathbf{Z}$ is a matrix of residuals with elements $z_{ij} = y_{ij} - y_{i.} - y_{.j} + y_{..}$, and

$$\Gamma = \sum_{i=1}^{m} \hat{\gamma}_i^2, \qquad \Gamma' = \sum_{j=1}^{n} \hat{\gamma}_j'^2.$$

The estimates of the linear parameters $\mu$, $\alpha_i$, $\beta_j$ are exactly as for the linear model. The two equations for the multiplicative parameter estimates may be written:

$$\begin{aligned}(\mathbf{Z}^T\mathbf{Z})\hat{\boldsymbol{\gamma}}' = \Gamma\Gamma'\hat{\boldsymbol{\gamma}}' \\ (\mathbf{Z}\mathbf{Z}^T)\hat{\boldsymbol{\gamma}} = \Gamma\Gamma'\hat{\boldsymbol{\gamma}}\end{aligned}\Bigg\}$$

so that $\hat{\boldsymbol{\gamma}}$ and $\hat{\boldsymbol{\gamma}}'$ are eigenvectors of $\mathbf{Z}^T\mathbf{Z}$ and $\mathbf{Z}\mathbf{Z}^T$ respectively, scaled so that the corresponding eigenvalue $\lambda = \Gamma\Gamma'$. With these estimates the residual sum of squares is $\Sigma(y_{ij} - y_{..})^2 - m\Sigma(y_{i.} - y_{..})^2 - n\Sigma(y_{.j} - y_{..})^2 - \Gamma\Gamma'$, and this is minimized by choosing $\lambda$ to be the biggest eigenvalue of $\mathbf{Z}^T\mathbf{Z}$. Now the eigenvectors of $\mathbf{Z}^T\mathbf{Z}$ and $\mathbf{Z}\mathbf{Z}^T$ may be expressed simultaneously as the singular vectors of $\mathbf{Z}$ in the form $\mathbf{Z} = \mathbf{U}\boldsymbol{\Sigma}\mathbf{V}^T$ where $\mathbf{U}$ is an orthogonal $m \times m$ matrix, $\mathbf{V}$ is an orthogonal $n \times n$ matrix and $\boldsymbol{\Sigma}$ is $m \times n$ with non-zero (and positive) values $\sigma_{ij}$ only when $i = j$. We shall therefore write these 'diagonal' values as $\sigma_i$ $(i = 1, 2, \dots, l)$ where $l = \min(m, n)$, and assume that these have been ordered so that $\sigma_1 \geq \sigma_2 \geq \dots \sigma_l \geq 0$. The quantities $\sigma_i$ are the singular values of $\mathbf{Z}$. We have that

$$\mathbf{Z}^T\mathbf{Z} = \mathbf{V}\boldsymbol{\Sigma}^T\boldsymbol{\Sigma}\mathbf{V}^T$$

and

$$\mathbf{Z}\mathbf{Z}^T = \mathbf{U}\boldsymbol{\Sigma}\boldsymbol{\Sigma}^T\mathbf{U}^T$$

so that $\mathbf{V}$ and $\mathbf{U}$ are eigenvectors corresponding to eigenvalues $\sigma_1^2, \sigma_2^2, \dots, \sigma_l^2$, and therefore $\boldsymbol{\gamma}$ is the first column of $\mathbf{U}$ and $\boldsymbol{\gamma}'$ is the first column of $\mathbf{V}$. Just as the additive parameters are undetermined up to an additive constant, the multiplicative parameters are undetermined up to a multiplicative constant. Thus $\hat{\boldsymbol{\gamma}}$ and $\hat{\boldsymbol{\gamma}}'$ can be replaced by $\rho\hat{\boldsymbol{\gamma}}$ and $\rho^{-1}\hat{\boldsymbol{\gamma}}'$ for any non-zero multiplicative constant $\rho$.

When $r = \text{rank}\,(Z) < l$ there are precisely $r$ non-zero singular values. It turns out that if we wish to fit an additional pair of multiplicative terms $\delta_i\delta_j'$ then $\hat{\boldsymbol{\delta}}$ and $\hat{\boldsymbol{\delta}}'$ are given by the second pair of singular vectors of $\mathbf{Z}$, corresponding to the singular value $\sigma_2$. Because the matrices $\mathbf{U}$ and $\mathbf{V}$ are orthogonal, $\hat{\boldsymbol{\delta}}$ is orthogonal to $\hat{\boldsymbol{\gamma}}$ and $\hat{\boldsymbol{\delta}}'$ is orthogonal to $\hat{\boldsymbol{\gamma}}'$. Third and subsequent pairs of multiplicative terms may be fitted by taking the third and subsequent pairs of singular vectors. This result is equivalent to stating that $\mathbf{Z}_s$, the best least-squares rank-$s$ fit to $\mathbf{Z}$, is obtained by setting $\mathbf{Z}_s = \mathbf{U}\boldsymbol{\Sigma}_s\mathbf{V}^T$ where $\boldsymbol{\Sigma}_s$ is $\boldsymbol{\Sigma}$ but with $\sigma = 0$ for $i > s$. The result is true for any rectangular matrix $\mathbf{Z}$, not just for residual matrices as defined above, and was first proved by Eckart and Young (1936). The Eckart–Young theorem is a basic result that underlies much of what follows.

Further developments of the multiplicative model will not be pursued here, apart from some brief comments.

In the above development, any of the additive parameters $\mu$, $\alpha_i$, $\beta_j$ may be dropped but those remaining have the same least-squares estimates as previously, and these estimates may be used to define a residual matrix $\mathbf{Z}$. The Eckart–Young theorem then shows that the multiplicative constants are estimated from the singular-value decomposition of the new form of $\mathbf{Z}$. Now, however, the row and column sums of $\mathbf{Z}$ are not both zero so not both of $\Sigma\gamma_i = 0$ and $\Sigma\gamma'_j = 0$ will be true. All these variants of the simple multiplicative model lead to orthogonal terms whose analysis can be exhibited in an analysis of variance. This analysis of variance differs from the analysis for the corresponding linear model mainly in the number of degrees of freedom to be associated with the multiplicative term(s) and associated significance tests.

With three or more subscripts the models may be extended in two ways. Either additional products of *pairs* of parameters can be allowed or products of more than two parameters may be accepted. Extensions of the former kind are straightforward but the latter type of extension introduces many difficulties, especially of uniqueness and in problems of estimation (see Gower (1977) for a fuller discussion and further references). One simple three-way model is discussed further in section 17.12 below.

## 17.4. BIPLOTS

The basic idea of a biplot is to represent the units of a data matrix by a set of points, as in components analysis, and the variables by a set of vectors *in the same space* as the units. The bi- refers to this duality and not to the usual use of two dimensions for the representation, which is a practical rather than a theoretical limitation.

Suppose we have a data matrix $\mathbf{X}$, assumed to be expressed in deviations from the variate means, and that the $n$ samples have been exhibited in an ordination given by their projections onto the plane determined by the first two principal axes. We might be interested to know how these plotted points relate to the original variates. One way of proceeding is to project each of the $p$ original axes, each of which relates to one of the variates, onto the same plane as for the samples. Each axis will become a vector through the origin. The simplest way of doing this is to plot the projection of one point on each axis and it is convenient to take the point unit distance from the origin. Thus we require the component scores for pseudo-samples with values given in the unit matrix $\mathbf{I}_p$ of order $p$. These are merely the rows of the matrix of loadings $\mathbf{H}$, whose first two columns give the relevant coordinates representing the $p$ variates in two dimensions. Joining each of these points to the origin gives one form of bi-plot. Generally speaking, points lying near a vector and far from the origin might be expected to score unusually well (either positively or negatively) on the associated variate.

The method is probably most easily understood from an examination of the singular value decomposition of **X** which we shall write in this section as **X** = **LΣH**$^T$. It follows that the variance-covariance matrix [see Definition 16.1.1] is **S** = **X′X** ≡ **HΣ**$^2$**H**$^T$, so that the columns of **H** are identified as the component loadings of section 2. Hence the scores for the samples are the rows of **XH** = **LΣ**. Similarly the scores for the pseudo-samples representing the variates are given as the rows of **IH** = **H**. It is the rows of **LΣ** and **H** which give the biplots and the inner product of the two matrices reproduces the data **X**, at least when all dimensions are plotted. When only $r$ dimensions are plotted, it follows from the Eckart–Young theorem that these represent the best, in a least-squares sense, rank-$r$ approximation to **X**. Thus if $P_i$ represents the $i$th sample, $Q_j$ the $j$th variable and $O$ the origin, then $\Delta(OP_i)\Delta(OQ_j)\cos(P_iOQ_j)$ approximates $x_{ij}$.

Instead of plotting unit pseudo-samples it might be interesting to project points on the original axes that are, say, one standard deviation distant from the origin. With good approximations the lengths of the resulting plotted vectors would give useful information on the relative variability of the original variates. When the matrix **X** is first normalized so that all variates have unit standard deviation, the projected vectors should have equal lengths, but few approximations exhibit this feature. Variances and covariances may be better approximated by using a different form of biplot. This plots the rows of **L** as coordinates for the samples and **HΣ** as vectors for the variates, so that the inner-product [see I, § 10.2] of the two sets again reproduces **X**. The lengths of the vectors are given by **HΣ**$^2$**H**$^T$ i.e. **S**. Thus the lengths are equal to the standard deviations and the inner-products of pairs of vectors gives rise to the covariances. The cosine of the angle between two vectors is the correlation of the two variates so that completely correlated variables give coincident vectors and poorly correlated pairs of variables will be nearly orthogonal. The approximation obtained by using two (say) dimensions is, of course, again the Eckart–Young approximation to **S** so that the proportion of sum-of-squares accounted for by $k$ dimensions is

$$\sum_{i=1}^{k} \sigma_i^4 \bigg/ \sum_{i=1}^{p} \sigma_i^4 = \sum_{i=1}^{k} \lambda_i^2 \bigg/ \sum_{i=1}^{p} \lambda_i^2,$$

where $\lambda_i$ is the $i$th eigenvalue of **S**. This form of biplot gives an improved plot of the variates compared to the former. However the distances between the units are modified. In the former plot, the distances approximate the Euclidean distances between pairs of rows of **X** as given by Pythagoras's theorem. Now the distances are obtained from **LL**$^T$ = **XS**$^{-1}$**X**$^T$. That is the distances are a kind of Mahalanobis distance [see e.g., Rao (1965)—Bibliog c], but it must be remembered that **S** = **XX**$^T$ and is not an independently derived within-population dispersion matrix. That **LL**$^T$ = **X**(**X**$^T$**X**)$^{-1}$**X**$^T$ is idempotent [see I, § 6.7(vii)] implies that the points with this 'Mahalanobis' distance have the property that their sum-of-squares is constant in all directions, and that any

$k$-dimensional projection will have sum-of-squares equal to $k$. The proportion of sum-of-squares accounted for is therefore $k/p$, which becomes $2/p$ for two-dimensional plots. Broadly speaking it is my opinion that the first form of biplot deals adequately with the points representing the samples but poorly with the vectors representing the variates, while the opposite is true for the second form of biplot. the best of both methods can be retained by plotting $\mathbf{L\Sigma}$ for the samples and $\mathbf{H\Sigma}$ for the variates but the inner-product relationship with $\mathbf{X}$ is then lost.

A different form of biplot arises when $\mathbf{X}$ is an $m \times n$ two-way table [see § 7.5.2] rather than a data matrix. There is now no special reason for working in terms of deviations from column means. Also, as rows and columns have equal status, the weighting by singular values should be balanced so that it is usual to plot $\mathbf{L\Sigma}^{1/2}$ and $\mathbf{H\Sigma}^{1/2}$. Otherwise the biplot procedure is as before but the variate/sample interpretation previously given to the two sets of points must now be discarded.

Of special interest is the case when $\mathbf{X}$ has rank 2 and hence only two terms in its singular value decomposition, giving

$$\mathbf{X} = \sigma_1 \mathbf{u}_1 \mathbf{v}_1 + \sigma_2 \mathbf{u}_2 \mathbf{v}_2'.$$

This form includes the simple additive model

$$\mathbf{X} = \mu \mathbf{1}\mathbf{1}^T + \alpha \mathbf{1}^T + \mathbf{1}\beta^T$$

which can always be reparameterised so that $\Sigma \alpha_i = \Sigma \beta_j = 0$. By writing

$$\mathbf{X} = (p\mathbf{1} + \alpha)\mathbf{1}^T + \mathbf{1}(q\mathbf{1} + \beta)^T$$

where $p + q = \mu$ it is easy to see that $\mathbf{X}$ has the required rank 2 form. Writing $\mathbf{N} = \mathbf{1}\mathbf{1}^T/n$ and $\mathbf{M} = \mathbf{1}\mathbf{1}^T/m$ we have that

$$(\mathbf{I} - \mathbf{N})\mathbf{X}(\mathbf{I} - \mathbf{M}) = \alpha_1 \bar{\mathbf{u}}_1 \bar{\mathbf{v}}_1^T + \sigma_2 \bar{\mathbf{u}}_2 \bar{\mathbf{v}}_2^T$$

where $\bar{\mathbf{u}}_1$ is the vector with elements $u_{1i} - \sum_{i=1}^n u_{1i}/n$ of deviations from the mean of $\mathbf{u}_1$, and so on for the other vectors. The geometrical interpretation of this transformation is that the origin for the points referring to rows is placed at their mean, and similarly for the column points. Thus the two configurations are unaltered except that one set is translated relative to the other and a new common origin is chosen. The old and new axes are parallel, so that the transformation has not changed any angles and any collinearities there may be will remain. Now because of the special additive form of $\mathbf{X}$ the left-hand side is zero, so that the $(i, j)$th element of the right-hand side gives

$$\frac{\sigma_1 \bar{u}_{1i} \bar{v}_{1j}}{\sigma_2 \bar{u}_{2i} \bar{v}_{2j}} = -1.$$

Thus the line joining $(\sigma_1^{1/2} \bar{u}_{1i}, \sigma_2^{1/2} \bar{u}_{2i})$ to the means of $\mathbf{u}_1$ and $\mathbf{u}_2$ is orthogonal to the line joining $(\sigma_1^{1/2} \bar{v}_{1j}, \sigma_2^{1/2} \bar{v}_{2j})$ to the means of $\mathbf{v}_1$ and $\mathbf{v}_2$. This is true for all values of $i$ and $j$ implying that the points plotted for the rows of $\mathbf{X}$ are collinear as are the points plotted for the columns. The two lines are orthogonal. This result does not depend on the singular values being distributed equally over rows and columns—any partition will suffice.

A more general rank-2 form of two-way table is:

$$\mathbf{X} = \mu \mathbf{1}\mathbf{1}^T + \boldsymbol{\alpha}\mathbf{1}^T + \mathbf{1}\boldsymbol{\beta}^T + \lambda \boldsymbol{\alpha}\boldsymbol{\beta}^T.$$

This is the model proposed by Tukey (1949) for his analysis of non-additivity. The rank-2 structure is evident by writing $\mathbf{X}$ in the form: $\mathbf{X} = (\mu - \lambda^{-1})\mathbf{1}\mathbf{1}^T + (\lambda^{-1/2}\mathbf{1} + \lambda^{1/2}\boldsymbol{\alpha})(\lambda^{-1/2}\mathbf{1} + \lambda^{1/2}\boldsymbol{\beta})^T.$

Proceeding as for the linear model above, we find that

$$(\mathbf{I} - \mathbf{N})\mathbf{X}(\mathbf{I} - \mathbf{M}) = \lambda \bar{\boldsymbol{\alpha}}\boldsymbol{\beta}^T$$

which is not, in general, zero so that in this case the biplots do not lead to orthogonal pairs of lines. That they lead to lines at all follows from noting that the rank-2 form of $\mathbf{X}$ implies that the vectors $\mathbf{u}_1$ and $\mathbf{u}_2$ both lie in the plane containing the vectors $\mathbf{1}$ and $\boldsymbol{\alpha}$. Thus there are scalars $p_1, q_1, p_2$ and $q_2$ such that:

$$\mathbf{u}_1 = p_1\mathbf{1} + q_1\boldsymbol{\alpha}$$

$$\mathbf{u}_2 = p_2\mathbf{1} + q_2\boldsymbol{\alpha}.$$

Eliminating $\boldsymbol{\alpha}$ gives $q_2\mathbf{u}_1 - q_1\mathbf{u}_2 = (p_1q_2 - p_2q_1)\mathbf{1}$, establishing the collinearity of the bipilot for the row points. Similarly the column-points are collinear.

The most general rank-2 model for a two-way table gives no collinearities, although sub-sets of points may be collinear suggesting that although a general model may be required for the whole table, sub-tables may be fitted by more simple models.

A model intermediate between the simple additive model and .the most general form of rank-2 model is the *columns regression* model of Mandel (1961)

$$\mathbf{X} = \mu \mathbf{1}\mathbf{1}^T + \boldsymbol{\alpha}\mathbf{1}^T + \mathbf{1}\boldsymbol{\beta}^T + \lambda \boldsymbol{\alpha}\boldsymbol{\gamma}^T.$$

and the corresponding *rows regression model*. Since

$$\mathbf{X} = \mathbf{1}(\mu\mathbf{1} + \boldsymbol{\beta})^T + \boldsymbol{\alpha}(\mathbf{1} + \lambda\boldsymbol{\gamma})^T$$

we have that there exist scalars $p_1, q_1, p_2, q_2$ such that

$$\mathbf{u}_1 = p_1\mathbf{1} + q_1\boldsymbol{\alpha}$$

$$\mathbf{u}_2 = p_2\mathbf{1} + q_2\boldsymbol{\alpha}$$

showing as before that the plot of row points is collinear. However $\mathbf{v}_1$ and $\mathbf{v}_2$ lie in the plane of $\mu\mathbf{1} + \boldsymbol{\beta}$ and $\mathbf{1} + \lambda\boldsymbol{\gamma}$ and do not give a linear plot. The forms for $\mathbf{u}_1$ and $\mathbf{u}_2$ show that the distance between the pair of points $P_i$ and $P_j$ representing the $i$th and $j$th rows is proportional to $\alpha_i - \alpha_j$ giving a visual appreciation of the estimated values of these parameters, a result that is also true for the simpler linear model obtained by setting $\lambda = 0$.

In the above it has been shown how the model-form leads to distinctive forms of biplots. The converse that the distinctive forms can only arise from the models considered is also true. These biplots can therefore be used to diagnose models of the linear, Tukey and rows/columns regression forms; in particular they indicate whether or not a simple linear model is adequate for describing a two-way table, (see Bradu and Gabriel (1978) for further details and a lead-in to the extensive literature on biplots which owes much to the enthusiastic work of K. R. Gabriel).

## 17.5. CORRESPONDENCE ANALYSIS

Correspondence analysis, like the biplot is a variant of the general multiplicative model discussed in section 17.3. Of special relevance to two-way contingency tables, it has also found applications with other types of data. The method (in French: analyse factorielle des correspondences) has found special favour among French statisticians led by Professor J. P. Benzecri. Recently Nishisato (1980) has aptly suggested the term *dual-scaling* for this whole area of data-analysis; he also gives an excellent historical review of the growth of interest in this subject. The starting point is a two-way table $\mathbf{X}$, which we shall think of as a table of counts. Suppose the row and column totals of $\mathbf{X}$ are arranged in the diagonals of matrices $\mathbf{R} = \text{diag}\,(\mathbf{X1})$, $\mathbf{C} = \text{diag}\,(\mathbf{1}^T\mathbf{X})$ then correspondence analysis operates on the matrix $\mathbf{Y}$ which is a specially standardized form of $\mathbf{X}$ defined by

$$\mathbf{Y} = \mathbf{R}^{-1/2}\mathbf{X}\mathbf{C}^{-1/2}.$$

(Here $\mathbf{R}^{-1/2}$ is the diagonal matrix whose $i$th diagonal element is $r_i^{-1/2}$, where $r_i$ is the $i$th diagonal element of $\mathbf{R}$.) We have that $\mathbf{Y}\mathbf{C}^{1/2}\mathbf{1} = \mathbf{R}^{-1/2}\mathbf{X1} = \mathbf{R}^{-1/2}\mathbf{R1} = \mathbf{R}^{1/2}\mathbf{1}$. Similarly $\mathbf{1}^T\mathbf{R}^{1/2}\mathbf{Y} = \mathbf{1}^T\mathbf{C}^{1/2}$. Hence $\mathbf{R}^{1/2}\mathbf{1}$, $\mathbf{C}^{1/2}\mathbf{1}$ is a singular-vector pair (when properly normalized) corresponding to a unit singular value. Thus the singular value decomposition of $\mathbf{Y}$ may be written

$$\mathbf{Y} = \mathbf{R}^{1/2}\mathbf{1}\mathbf{1}^T\mathbf{C}^{1/2}/x.. + \sum_{i=2} \sigma_i\mathbf{u}_i\mathbf{v}_i^T,$$

where $x..$ is the normalizer that arises from noting that the sum-of-squares of the elements of both $\mathbf{R}^{1/2}\mathbf{1}$ and $\mathbf{C}^{1/2}\mathbf{1}$ is merely the total of the matrix $\mathbf{X}$. When $\mathbf{X}$ is a non-negative matrix, the unit singular value is the biggest. This follows from noting that the singular values of $\mathbf{Y}$ are the square-roots of the eigenvalues of $\mathbf{Y}^T\mathbf{Y}$ which is itself a non-negative matrix. As we have shown that the unit singular value corresponds to a postiive vector, it follows from the Frobenius Theorem [see I: Theorem 7.11.1] that it must be the maximum. When $\mathbf{X}$ is not positive its row and column totals may not be positive so that $\mathbf{R}^{1/2}$ and $\mathbf{C}^{1/2}$ need not exist as real matrices. Even when the row and column totals are positive $\mathbf{Y}^T\mathbf{Y}$ need not be positive and then the maximum singular value of $\mathbf{Y}$ need not be unity. Rearranging the above singular value decomposition of $\mathbf{Y}$ gives $\mathbf{R}^{-1/2}\mathbf{X}\mathbf{C}^{-1/2} - \mathbf{R}^{1/2}\mathbf{1}\mathbf{1}^T\mathbf{C}^{1/2}/x.. = \sum_{i=2} \sigma_i\mathbf{u}_i\mathbf{v}_i^T$ showing that the right-hand side is the singular value decomposition of the matrix on the left-hand side, which has elements:

$$z_{ij} = \frac{x_{ij}}{\sqrt{(x_i.x_{.j})}} - \frac{\sqrt{(x_i.x_{.j})}}{x_{..}} = \left( x_{ij} - \frac{x_i.x_{.j}}{x_{..}} \right) \Big/ \sqrt{(x_i.x_{.j})}.$$

This latter form is the square root of an element of the Pearson chi-square criterion for independence of the row and column classifications of the contingency table $\mathbf{X}$ [see § 7.5.1]. It follows that $\sum_{i=2} \sigma_i^2$ gives a decomposition of chi-square with corresponding model terms $\sigma_i \mathbf{u}_i \mathbf{v}_i$. It is, perhaps, simpler to regard the method as fitting simple multiplicative models (including biplots) to the derived matrix $\mathbf{Z}$ and much depends on whether the transformation from $\mathbf{X}$ to $\mathbf{Z}$ is relevant and interpretable.

One situation when $\mathbf{Z}$ is useful arises in ecological studies where the rows of $\mathbf{X}$ correspond to different sites and the columns to different species of plant. Thus $x_{ij}$ gives the number of species $j$ found at site $i$. It is sometimes of interest to find scores for each site and, to a lesser extent, scores for each species so that sites can be ordered (hence *ordination*) according to supposed ecological gradients. However because some sites are richer in plant species than are others, and because some species occur in much greater numbers than do others, some form of adjustment is necessary. Suppose that the unknown site (rows) scores are in a vector $\mathbf{p}$ and the unknown species (columns) scores in a vector $\mathbf{q}$. Then the average score for site $i$, based on the species scores is $\sum_{j=1}^{m} x_{ij} q_j / x_i.$ and this is required to be proportional to the site score $p_i$. In matrix terms this becomes:

$$\mathbf{R}^{-1}\mathbf{X}\mathbf{q} = \sigma\mathbf{p}.$$

Similarly deriving the species scores from the site scores gives:

$$\mathbf{p}^T\mathbf{X}\mathbf{C}^{-1} = \sigma\mathbf{q}^T.$$

These equations imply that $\mathbf{R}^{1/2}\mathbf{p}$ and $\mathbf{C}^{1/2}\mathbf{q}$ are the singular vectors of $\mathbf{Y}$ which correspond to the singular value $\sigma$. The greatest value $\sigma = 1$, found above, corresponds to the uninteresting vectors

$$\mathbf{R}^{1/2}\mathbf{p} = \mathbf{R}^{1/2}\mathbf{1} \quad \text{and} \quad \mathbf{C}^{1/2}\mathbf{q} = \mathbf{C}^{1/2}\mathbf{1}$$

giving the constant scores $\mathbf{p} = \mathbf{1}$ and $\mathbf{q} = \mathbf{1}$.

Consequently the scores are obtained from the second singular vector pair of $\mathbf{Y}$ to give:

$$\mathbf{R}^{1/2}\mathbf{p} = \mathbf{u}_2 \quad \text{and} \quad \mathbf{C}^{1/2} = \mathbf{q} = \mathbf{v}_2,$$

leading to scores $\mathbf{p} = \mathbf{R}^{-1/2}\mathbf{u}_2$ and $\mathbf{q} = \mathbf{C}^{-1/2}\mathbf{v}_2$. Further pairs of singular vectors may be selected to give secondary, tertiary, etc. sets of scores. The second and third pairs of scaled singular vectors may be simultaneously plotted in a similar manner to that used with biplots, or we may wish to plot the scores themselves. Thus, as in the biplot technique, we have the freedom of distributing the singular value $\sigma_i$ over $\mathbf{u}_i$ and $\mathbf{v}_i$ in an arbitrary manner, but in correspondence analysis there is also the choice of plotting $\mathbf{R}^{-1/2}\mathbf{u}_i$ and $\mathbf{C}^{-1/2}\mathbf{v}_i$. One combination of the two forms of scaling that is commonly used, is to plot $\mathbf{R}^{-1/2}\mathbf{U}\boldsymbol{\Sigma}$ and $\mathbf{C}^{-1/2}\mathbf{V}\boldsymbol{\Sigma}$. The (squared) distance between pairs of points representing rows is then given by

$$\mathbf{R}^{-1/2}\mathbf{U}\boldsymbol{\Sigma}^2\mathbf{U}^T\mathbf{R}^{-1/2} = \mathbf{R}^{-1/2}\mathbf{Z}\mathbf{Z}^T\mathbf{R}^{-1/2} = \mathbf{R}^{-1}\mathbf{X}\mathbf{C}^{-1}\mathbf{X}^T\mathbf{R}^{-1}$$

plus terms that do not affect distance. The squared distance between the $i$th and $j$th rows (in the full space) derived from this expression is given by:

$$\left(\frac{\mathbf{x}_i}{r_i} - \frac{\mathbf{x}_j}{r_j}\right)^T \mathbf{C}^{-1} \left(\frac{\mathbf{x}_i}{r_i} - \frac{\mathbf{x}_j}{r_j}\right)$$

which has been termed the *chi-squared distance*. Thus when the $i$th and $j$th rows of $\mathbf{X}$ are proportionate their corresponding points coincide. Similar expressions apply to distances between column points. The low-dimensional representations are *not* least-squares approximations with this analysis.

It will be clear that many variants of singular value and rows-and-columns scaling are possible. In applications it is often difficult to determine exactly what form of scaling has been used.

## 17.6.   METRIC SCALING: PRINCIPAL COORDINATES ANALYSIS AND CLASSICAL SCALING

The starting point of all multidimensional scaling is a symmetric matrix $\mathbf{M}$ of order $n$, whose values $m_{ij}$ give some measure of association (e.g. similarity, dissimilarity, distance) between samples $i$ and $j$. $\mathbf{M}$ may be observed directly or it may be derived from more basic data such as a data-matrix as described in section 17.1, above. $\mathbf{M}$ is to be analysed to give an ordination with a set of $n$ coordinates in $k$ dimensions such that the distance between the $i$th and $j$th points approximates $m_{ij}$, or at least some function of $m_{ij}$. In metric scaling, discussed in this and the following section, the criteria used to judge the goodness-of-fit of the approximations $\hat{m}_{ij}$ are simple functions $f(m_{ij}, \hat{m}_{ij})$. In non-metric scaling, discussed in section 17.8, more general goodness-of-fit criteria are used.

Principal coordinates analysis and classical scaling are synonyms for a metric scaling method that is based on the principal components idea [see § 17.2]. It is assumed that $n$ points $P_i$ ($i = 1, 2, \ldots, n$) exist in at most $n-1$ dimensions whose inter-point distances $\Delta(P_i, P_j)$ are exactly $m_{ij}$. We shall return shortly to the validity of this assumption and how it can be relaxed. Given the points $P_i$, principal components analysis can be used to obtain an approximation in $k$ dimensions by projecting onto the $k$-dimensional sub-space that minimizes the sum-of-squares of the distances from the space.

The assumption that the values $m_{ij}$ are distances implies

(a) $m_{ii} = 0$      $i = 1, 2, \ldots, n$

(b) a real set of coordinates $P_i$ exist.

Neither may be true. When $m_{ii} \neq 0$ we may attempt to fit distances to the main body of the table, ignoring the diagonal terms, and/or we may seek a transformation to rectify matters. A common transformation is $m'_{ij} \leftarrow 1 - m_{ij}$ which is suitable for measures of similarity for which $m_{ii} = 1$. A good property of principal-coordinates analysis is that it establishes when a real configuration does not exist and indicates how serious a problem this may be.

The basic algebraic result that underlies principal-coordinates analysis is that if $\mathbf{M}$ is a symmetric matrix with any decomposition $\mathbf{YY}^T$ then the rows of $\mathbf{Y}$ may be taken as coordinates. The squared distance between points whose coordinates are given by the $i$th and $j$th rows of $\mathbf{Y}$ is $m_{ii} + m_{jj} - 2m_{ij}$. For a matrix with zero diagonal values this simplifies to $-2m_{ij}$. Thus to find coordinates that reproduce a set of distances $d_{ij}$, it is sufficient to set $m_{ij} = -\frac{1}{2}d_{ij}^2$. For a similarity matrix with unit diagonal values, the matrix $\mathbf{Y}$ immediately gives a set of squared distances $2(1 - m_{ij})$ which has the property that high similarities ($m_{ij}$ close to unity) map into short distances and low similarities ($m_{ij}$ close to zero) map into long distances. Having found the coordinates $\mathbf{Y}$, the $k$-dimensional approximation to the coordinates may be completed by a components analysis. Before this can be done a major difficulty has to be overcome. To appreciate this difficulty consider the spectral decomposition of $\mathbf{M} = \mathbf{X\Lambda X}^T$. This is of the required form with $\mathbf{Y} = \mathbf{X\Lambda}^{1/2}$. The trace of a distance matrix is zero and hence at least one eigenvalue in $\mathbf{\Lambda}$ must be negative, leading to an imaginary term in $\mathbf{\Lambda}^{1/2}$ and an imaginary set of coordinates. That is the coordinates $\mathbf{Y}$ found as above can never be real when $\mathbf{M}$ is a distance matrix. The same applies to any other decomposition of $\mathbf{M}$.

It turns out that a real set of coordinates, when they exist, can be found by transforming the elements of $\mathbf{M}$, into $\mathbf{M}^*$ in a manner to be described below, that ensures that $\mathbf{M}^*$ has a zero eigenvalue (strictly speaking one more zero eigenvalue than $\mathbf{M}$ if the points lie on a hypersphere, and two more if they do not. Consider the transformation $m_{ij}^* = m_{ij} - g_i - g_j$ where the terms $g_i (i = 1, 2, \ldots, n)$ are arbitrary, then

$$m_{ii}^* + m_{jj}^* - 2m_{ij}^* = m_{ii} + m_{jj} - 2m_{ij}$$

showing that any decompositon of $\mathbf{M}^*$ generates coordinates with the same inter-distances as those of $\mathbf{M}$. The matrix $\mathbf{M}^*$ is the most general form with this property for if $\mathbf{M} - \mathbf{G}$ preserves distance, then $g_{ii} + g_{jj} - 2g_{ij} = 0$, so that $\mathbf{G} = \mathbf{g1}^T + \mathbf{1g}^T$ where $\mathbf{1}$ is a vector of units and the vector $\mathbf{g}$ has elements $g_i = \frac{1}{2}g_{ii}$. The values $g_i$ may be chosen in many ways to ensure that $\mathbf{M}^*$ has a zero eigenvalue. For example we can make the $k$th row of $\mathbf{M}^*$ zero, by choosing

$$g_j = m_{kj} - \tfrac{1}{2}m_{kk}$$

so that $m_{ij}^* = m_{ij} - m_{ik} - m_{jk} + m_{kk}$, defining a matrix $\mathbf{M}_k^*$.

Any decompositon $\mathbf{M}^* = \mathbf{YY}^T$ will now have the $k$th row of $\mathbf{Y}$ zero and so the $k$th sample is placed at the origin. Alternatively we may choose $\mathbf{g}$ so that every row (and column) of $\mathbf{M}^*$ sums to zero. This requires

$$g_j = m_{j.} - \tfrac{1}{2}m_{..}$$

to give $m_{ij}^* = m_{ij} - m_{i\cdot} - m_{j\cdot} + m_{\cdot\cdot}$, defining a matrix $\mathbf{M}_0^*$, where the dot-notation implies that means have been taken over the relevant suffix. For obvious reasons $\mathbf{M}_0^*$ is often referred to as the doubly-centered form of $\mathbf{M}$. With this transformation we have

$$\mathbf{M}_0^* \mathbf{1} = 0$$

showing that $\mathbf{1}$ is an eigenvector that corresponds to an induced zero root. For any decomposition $\mathbf{M}_0^* = \mathbf{Y}\mathbf{Y}^T$ we have that $(\mathbf{1}^T\mathbf{Y})(\mathbf{Y}^T\mathbf{1}) = 0$. Now $(\mathbf{1}^T\mathbf{Y})/n = \mathbf{y}^T$ (say) is the row-vector of the centroid of the configuration; as we have shown that $\Sigma y_i^2 = 0$, it follows that each $y_i = 0$. The geometrical interpretation is that the columns of coordinates $\mathbf{Y}$ sum to zero, showing that the origin is at the centroid of the configuration of points. When $\mathbf{Y} = \mathbf{X}\mathbf{\Lambda}^{1/2}$, with $\mathbf{X}\mathbf{\Lambda}\mathbf{X}^T$ the spectral decomposition of $\mathbf{M}_0^*$, we have that $\mathbf{Y}^T\mathbf{Y} = \mathbf{\Lambda}$, diagonal, so that the coordinate axes are the principal axes of the configuration of points. It follows that the eigenvalues $\lambda_1, \lambda_2, \ldots, \lambda_{n-1}$, expressed in decreasing order, give the sums-of-squares accounted for by successive dimensions and can be used, as in components analysis, to choose the number of dimensions needed to give an acceptable approximation. Note that the transformation has ensured that $\lambda_n$ is zero. This is the principal coordinates method, which for any symmetric matrix $\mathbf{M}$ of order $n$ gives coordinates of $n$ points, such that $\Delta^2(P_iP_j) = m_{ii} + m_{jj} - 2m_{ij}$, centred at their centroid and referred to principal axes for coordinate axes. When the distances are Euclidean, coordinates found in this way are real. A necessary and sufficient condition for a real solution is that $\mathbf{M}^*$ be positive semi-definite for some vector $\mathbf{g}$. Sufficiency is obvious but necessity is harder to prove (see e.g. Blumenthal (1970)). In stating this result $\mathbf{M}^*$ is derived from $\mathbf{M}$ by choosing $\mathbf{g}$ in any way that gives an extra zero eigenvalue. The two choices discussed above may be written in matrix notation as:

$$\mathbf{M}_k^* = (\mathbf{I} - \mathbf{E}_k)\mathbf{M}(\mathbf{I} - \mathbf{E}_k^T)$$

and

$$\mathbf{M}_0^* = (\mathbf{I} - \mathbf{N})\mathbf{M}(\mathbf{I} - \mathbf{N})$$

where $\mathbf{E}_k = \mathbf{e}_k\mathbf{1}^T$, $\mathbf{N} = \mathbf{1}\mathbf{1}^T/n$ and $\mathbf{e}_k$ is a vector with unity in the $k$th position and zero elsewhere.

The general condition on the choice of $\mathbf{g}$ to induce a zero eigenvalue in $\mathbf{M}^*$ is that it satisfy

$$(\mathbf{1}^T\mathbf{M}^{-1}\mathbf{1})(\mathbf{g}^T\mathbf{M}^{-1}\mathbf{g}) = (1 - \mathbf{1}^T\mathbf{M}^{-1}\mathbf{g})^2$$

and it is a straightforward matter to show that the choices of $\mathbf{g}$ leading to $\mathbf{M}_k^*$ and $\mathbf{M}_0^*$ satisfy this relationship. The conditions for a Euclidean representation of the distances $(m_{ii} + m_{jj} - 2m_{ij})^{1/2}$ can be put more conveniently by constructing the symmetric matrix

$$\mathbf{F} = (\mathbf{I} - \mathbf{1}\mathbf{s}^T)\mathbf{M}(\mathbf{I} - \mathbf{s}\mathbf{1}^T)$$

where $\mathbf{s}$ is any vector that sums to unity that is not a null-vector of $\mathbf{M}$. Such a choice of $\mathbf{s}$ is equivalent to setting $\mathbf{g} = \mathbf{Ms} - \frac{1}{2}(\mathbf{s}^T \mathbf{Ms})\mathbf{1}$. Gower (1982) shows that the configuration is Euclidean if and only if $\mathbf{F}$ is positive semi-definite. The decomposition $\mathbf{F} = \mathbf{YY}^T$ gives the coordinates $\mathbf{Y}$ which have origin such that $\mathbf{s}^T \mathbf{Y} = 0$. The multiplicative forms given above show that $\mathbf{s} = 1/n\mathbf{1}$ for $\mathbf{M}_0$ and $\mathbf{s} = \mathbf{e}_k$ for $\mathbf{M}_k$. The squared distance of each point from the chosen origin is given by diag $\mathbf{F}$. This diagonal matrix may be written as the column-vector $(\mathrm{diag}\,\mathbf{M})\mathbf{1} - 2\mathbf{Ms} + (\mathbf{s}^T\mathbf{Ms})\mathbf{1}$ where the first term vanishes when $\mathbf{M}$ is itself a distance matrix. These results may be used to derive settings of $\mathbf{s}$ with useful geometric properties. For example if one requires the origin to be at the centre of the circum-hypersphere of the configuration then

$$(\mathrm{diag}\,\mathbf{M})\mathbf{1} - 2\mathbf{Ms} + (\mathbf{s}^T\mathbf{Ms})\mathbf{1} = R^2\mathbf{1}$$

where $R$ is the radius. The solution to this equation is straightforward but tedious. However when $\mathbf{M}$ is a distance matrix with diag $\mathbf{M} = 0$ we have

$$\mathbf{Ms} = k\mathbf{1}$$

for some $k$. Hence for non-singular $\mathbf{M}$

$$\mathbf{s} = k\mathbf{M}^{-1}\mathbf{1}$$

and because $\mathbf{s}^T\mathbf{1} = 1$ then $k = (\mathbf{1}^T\mathbf{M}^{-1}\mathbf{1})^{-1}$ thus identifying $\mathbf{s}$ and giving $R^2 = -(\mathbf{1}^T\mathbf{M}^{-1}\mathbf{1})^{-1}$. This setting of $\mathbf{s}$ defines $\mathbf{M}_c^* = \mathbf{M} - (\mathbf{11}^T/\mathbf{1}^T\mathbf{M}^{-1}\mathbf{1})$.

These results are mainly of geometric rather than statistical interest. The coordinates found all represent the same geometrical configurations but referred to different origins and different axes, the latter depending on the decomposition $\mathbf{YY}^T$ that is chosen. The principal-coordinates choice given by the spectral decomposition of $\mathbf{M}_0^*$ is of special importance because of the least-squares properties of principal components.

It may happen that $\mathbf{F}$ is not positive-semi-definite. The assumption that there is a real Euclidean configuration in $n-1$ dimensions with the given interpoint distances then breaks down, throwing doubts on the validity of the geometrical argument based on projections. When only the smaller eigenvalues are negative this will not be a serious problem. In a study of the effects of perturbing a matrix of a general Euclidean distances, Sibson (1979) has shown that several small negative eigenvalues may be generated. For determining dimensionality, he suggests the useful rule that the sum of the retained positive eigenvalues should approximate the sum of all the eigenvalues. That is the small negative eigenvalues may be regarded as cancelling out small inessential positive eigenvalues.

When large negative eigenvalues occur Euclidean ideas have to be rejected. The classical scaling process can still be justified through the Eckart–Young theorem (see section 3, above). We have that $\mathbf{M}_0^* = \mathbf{X}\boldsymbol{\Lambda}\mathbf{X}^T$ and the Eckart–Young theorem when applied to symmetric matrices says that the best rank-$r$ fit to $\mathbf{M}_0^*$ is obtained by setting to zero the $n-r$ eigenvalues of smallest modulus. Thus negative eigenvalues may be retained and the Euclidean representation of the ordination lost. Even when the *r* eigenvalues with largest modulus are positive, giving a Euclidean representation, the least-squares criterion that has been optimized is not the same as the original criterion. The two coincide only when *all* eigenvalues are non-negative. Although fits that include negative eigenvalues do not have distance interpretations they nevertheless may give useful simplifications of the data. Of all metric scaling methods, only classical scaling/principal coordinates analysis reveals good non-Euclidean approximations when they exist. The other methods, discussed in section 17.7, fit only Euclidean configurations and cannot be conveniently modified to include sufficiently flexible non-Euclidean models. The difficulty is that in an *r*-dimensional approximation each of the *r* axes may be imaginary, giving $2^r$ possibilities, only one of which is Euclidean. With principal-coordinates analysis the negative eigenvalues indicate how many imaginary axes to include, but other methods would require all $2^r$ possibilities to be explored independently. This is quite impracticable, especially when it has to be done for each value of *r* of potential interest.

This account would be incomplete without mentioning the additive constant problem, which is often regarded as an integral part of classical scaling. We are now concerned with the case when diag $(\mathbf{M}) = 0$ so that the non-diagonal values of $\mathbf{M}$ are putative distances. The question arises of whether a constant *k* can be found such that $m_{ij} + k$ are real Euclidean distances and hence can lead to real ordination even when the original $m_{ij}$ do not. Until very recently no closed-form solution was known to this problem although Torgerson (1958) described an iterative numerical algorithm. Before outlining the new solution to the additive constant problem, the solution first given by Lingoes (1971) to a simpler related problem is given. This requires the smallest constant *k* such that $m_{ij}^2 + k$ are real sqared Euclidean distances. Recalling that to reproduce distances $m_{ij}$ we require the matrix with entries $-\frac{1}{2}m_{ij}^2$, we shall assume in the remainder of this section that $\mathbf{M}$ is so defined. Then the matrix with equivalent adjusted squared distances is

$$\mathbf{L} = \mathbf{M} - \tfrac{1}{2}k(n\mathbf{N} - \mathbf{I}).$$

For classical scaling we have that the doubly-centred matrix

$$\mathbf{M}_0^* = (\mathbf{I} - \mathbf{N})\mathbf{M}(\mathbf{I} - \mathbf{N}) = \mathbf{X}\mathbf{\Lambda}\mathbf{X}^T.$$

The doubly-centred form of $\mathbf{L}$ is:

$$\mathbf{L}_0^* = \mathbf{M}_0^* + \tfrac{1}{2}k(\mathbf{I} - \mathbf{N}) = \mathbf{X}(\mathbf{\Lambda} + \tfrac{1}{2}k\mathbf{I})\mathbf{X}^T - \tfrac{1}{2}k(\mathbf{1}\mathbf{1}^T/n).$$

It has been shown that $n^{-1/2}\mathbf{1}$ is eigenvector of $\mathbf{M}_0^*$ corresponding to the zero eigenvalue. It follows that $\mathbf{1}$ is also an eigenvector of $\mathbf{L}_0^*$, again with zero eigenvalue and that the other eigenvectors also remain unchanged but the eigenvalues becomes $\lambda_i + \frac{1}{2}k$. To ensure a real configuration we merely have to choose *k* so that all $\lambda_i + \frac{1}{2}k$ are non-negative. If $\lambda_0$, the smallest eigenvalue, is negative it is sufficient to choose $k \geq -2\lambda_0$. When $k = 2\lambda_0$ a second zero eigenvalue is induced and the distances $(m_{ij}^2 - 2\lambda_0)^{1/2}$ have a real configuration in, at most, $n-2$ dimensions. If $n-1$ of the eigenvalues of $\mathbf{M}_0^*$ are equal, all the eigenvalues of $\mathbf{L}_0^*$ vanish leading to a degenerate solution, but this possibility is excluded by requiring $\lambda_0$ to be negative.

Returning to the basic additive-constant problem, Cailliez (1983) shows that $k$ is the largest eigenvalue of

$$\begin{pmatrix} \mathbf{0} & 2\mathbf{M}_0^* \\ -\mathbf{I} & 2\mathbf{P}_0^* \end{pmatrix}$$

where $\mathbf{P}_0^* = (\mathbf{I}-\mathbf{N})\mathbf{P}(\mathbf{I}-\mathbf{N})$ with $p_{ij}=m_{ij}$. With this value of $k$ the distances $m_{ij}+k$ have a real configuration in, at most, $n-2$ dimensions.

## 17.7.  METRIC SCALING: OTHER METHODS

This section continues the discussion of the previous one, giving an account of other criteria that have been used to derive an ordination from a sample distance-matrix. It is appropriate now to replace the general symmetric matrix $\mathbf{M}$ by a symmetric matrix $\mathbf{D}$ with elements $-\frac{1}{2}d_{ij}^2$ and zero diagonal values. The data are the positive elements $d_{ij}$ and although it is convenient to think of them in a general way as distances, there is no implication that they necessarily can be generated as Euclidean (or any other) distances between a set of points with real coordinates.

There are few known analytical results in this field but algorithmic solutions are available. This difference between analytic and algorithmic solutions is largely illusory because the application of the algebraic eigenvalue solution to classical metric scaling itself rests on the availability of suitable numerical algorithms. With the criteria about to be discussed, the numerical algorithms do not have a corresponding well-understood algebraic theory and so are less open to analysis. To this extent the methods are less well-founded than those discussed in section 17.6.

In the following we shall be concerned to find a set of coordinates $\mathbf{X}$ in the *specified* number $k$ of dimensions, which give rise to Euclidean distances $\delta_{ij}$, with values $-\frac{1}{2}\delta_{ij}^2$ contained in a matrix $\mathbf{\Delta}$ that is supposed to approximate the data $\mathbf{D}$. One criterion suggested for this is to find $\mathbf{X}$ such that $C_1 = \Sigma(d_{ij}-\delta_{ij})^2$ is minimized. This is termed *least-squares-scaling* and the criterion is sometimes referred to as *STRESS* (see section 17.8 for the corresponding non-metric scaling definition). The criterion is that given by maximum likelihood when the observations $d_{ij}$ are independently and identically normally-distributed. Differentating the criterion w.r.t. the elements of $\mathbf{X}$ leads to the Normal equations:

$$\mathbf{FX} = 0$$

where $\mathbf{F}$ is a symmetric matrix whose rows (and columns) sum to zero and where $f_{ij} = (d_{ij}-\delta_{ij})/\delta_{ij}$ when $i \neq j$. The elements of $\mathbf{F}$ are functions of $\mathbf{\Delta}$, and hence of $\mathbf{X}$, so the Normal equations are non-linear. Before going on to make some comments on their numerical solution we shall derive a few simple results.

The first thing to notice is that when $\mathbf{X}$ is a solution so is $\mathbf{X}(\mathbf{H}+\mathbf{1m})$ where $\mathbf{H}$ is an orthogonal matrix and $\mathbf{m}$ is an abitrary vector. These are the conditions (rotation and translation) to ensure that the distances between pairs of rows of $\mathbf{X}$ are unaltered and hence that $\mathbf{F}$ is invariant. Thus although no constraint has been put on $\mathbf{X}$ when deriving the Normal equations, we may assume that $\mathbf{X}$ has any convenient form of centring and orientation. For example if the columns of $\mathbf{X}$ sum to zero it follows that $\mathbf{XX}^T = \mathbf{\Delta}_0^*$, the centred distance matrix corresponding to $\mathbf{\Delta}$ (see $\mathbf{M}_0$ of section 17.6). Multiplying the Normal equation by $\mathbf{X}^T$ then gives:

$$\mathbf{F\Delta}_0^* = 0.$$

This expresses the Normal equations entirely in terms of the observed and fitted distances, giving $n^2$ equations for the $nk$ coordinates. These equations may be interpreted as identities between the observed and fitted distances but do not, in general seem to be of any particular interest. An important result follows from expanding Trace $(\mathbf{F\Delta}_0^*) = 0$. This gives:

$$\text{Trace } (\mathbf{F\Delta}_0^*) = \text{Trace } \{\mathbf{F}(\mathbf{I}-\mathbf{N})\mathbf{\Delta}(\mathbf{I}-\mathbf{N})\}$$

$$= \text{Trace } \{(\mathbf{I}-\mathbf{N})\mathbf{F}(\mathbf{I}-\mathbf{N})\mathbf{\Delta}\}$$

$$= \text{Trace } (\mathbf{F\Delta}).$$

Thus $\Sigma f_{ij}\delta_{ij}^2 = 0$ or $\Sigma(d_{ij}-\delta_{ij})\delta_{ij}=0$, showing that at the minimum of the criterion we have that

$$\Sigma(d_{ij}-\delta_{ij})^2 = \Sigma d_{ij}^2 - \Sigma \delta_{ij}^2,$$

which may be used as the basis of an analysis of variance in which the total sum of squares (of the observed distances) is equal to the sum of squares of the fitted distances plus the sum of squares of the residual distances. Since the residual sum of squares must be non-negative the average of the fitted distances is never greater than (nearly always less than) the average value of the observed distances. Centering at a particular sample point may also be adopted to give

$$\mathbf{F\Delta}_k^* = 0$$

to give further identities.

Two basic methods have been suggested for solving $\mathbf{FX} = 0$ for $\mathbf{X}$. The first is to use one of the many function optimization processes that have been developed by numerical analysts (see e.g. Murray (1972)). The second approach gives more insight and is based on work by Guttman (1968). We have that $\mathbf{F} = \mathbf{F}^* + (n\mathbf{I}-\mathbf{11}^T)$ where $\mathbf{F}^*$ is a symmetric matrix whose rows (and columns) sum to zero and where $f_{ij}^* = d_{ij}/\delta_{ij}$, $i \neq j$. Thus when $\mathbf{X}$ is in centred form, the normal equations may be written:

$$\frac{1}{n}\mathbf{F}^*\mathbf{X} = \mathbf{X}.$$

This may be used as the basis of a iterative sequence:

$$\frac{1}{n}\mathbf{F}_i^*\mathbf{X}_i = \mathbf{X}_{i+1}$$

in which $\mathbf{X}_1$ is some initial trial estimate perhaps given by the methods described in section 17.6. Note that if $\mathbf{X}_i$ is centred then so is $\mathbf{X}_{i+1}$. It has been shown that this sequence never increases the value of the criterion $C_1$ and under normal circumstances will converge to a solution. Convergence can be slow, but may be improved by choosing a constant $\alpha$ in the sequence:

$$\mathbf{X}_{i+1} = \frac{1}{n}\mathbf{F}_i^*\mathbf{X}_i + \alpha\left(\mathbf{X}_i - \frac{1}{n}\mathbf{F}_{i-1}^*\mathbf{X}_{i-1}\right).$$

Values of $\alpha = \varepsilon - 1$, with $\varepsilon$ a small positive number, square the convergence rate and reduce the number of iterations required. The whole process is very close to the old method for calculating eigenvectors by iterating multiplicatively on a vector. The acceleration technique [see III: § 2.4.1] is then seen as a variant of Aitken's acceleration technique obtained by fitting a parabola to three successive vectors in the iteration. Of course the main difference is that the matrix $\mathbf{F}_i^*$ itself changes at each iteration. The connection with eigenvectors is underlined by noting from the Normal equations that the columns of $\mathbf{X}$ are eigenvectors of $(1/n)\mathbf{F}^*$ all corresponding to unit eigenvalues.

Another possible iterative solution arises from adapting the distance-form of the normal equations to give:

$$\frac{1}{n}\mathbf{F}_i^*(\mathbf{\Delta}_0^*)_i = (\mathbf{\Delta}_0^*)_{i+1}.$$

This has some essential properties. For example if $(\mathbf{\Delta}_0^*)_i$ is centred by rows and columns then so is $(\mathbf{\Delta}_0^*)_{i+1}$ and for an exact fit $(\mathbf{\Delta}_0^*)_{i+1} = (\mathbf{\Delta}_0^*)_i$. Also provided $(\mathbf{\Delta}_0^*)_i$ has rank $k$, then so do the successive centred matrices in the sequence. It is satisfying to work entirely in terms of distances, avoiding any difficulties there may be in the arbitrary rotation and location of the coordinates $\mathbf{X}$. Note that there is no guarantee that $(\mathbf{\Delta}_0^*)_i$ remains symmetric, though if the process converges to a solution, $(\mathbf{\Delta}_0^*)_i$ must converge to symmetric form. However the convergence properties of the method remain unknown.

Another criterion that has been used for metric scaling is to find the coordinates $\mathbf{X}$ which minimize $C_2 = \Sigma(d_{ij}^2 - \delta_{ij}^2)^2$. This is termed *least-squares-squared scaling* and the criterion is sometimes referred to as *STRESS*. The normal equations, which resemble those for least-squares-scaling, are:

$$\mathbf{GX} = 0$$

where $\mathbf{G}$ is a symmetric matrix whose rows (and columns) sum to zero and where $g_{ij} = d_{ij}^2 - \delta_{ij}^2$. As before we have

$$\mathbf{G\Delta}_0^* = 0 \quad \text{and} \quad \mathbf{G\Delta}_k^* = 0.$$

Now

$$\text{Trace } (\mathbf{G}\Delta_0^*) = \Sigma(d_{ij}^2 - \delta_{ij}^2)\delta_{ij}^2 = 0.$$

Leading to an analysis of variance of squared distances

$$\Sigma(d_{ij}^2 - \delta_{ij}^2)^2 = \Sigma d_{ij}^4 - \Sigma \delta_{ij}^4.$$

Although the Normal equations in the form $\mathbf{G}\Delta_0^* = 0$ are only quadratic n $\delta_{ij}^2$ direct solutions are not known and the iterative sequences appropriate for $\mathbf{F}$ do not extend, so that conventional function optimization methods have to be used to obtain numerical solutions.

The criteria $C_1$ and $C_2$ are usually formulated in a weighted form, becoming

$$C_1 = \Sigma w_{ij}(d_{ij} - \delta_{ij})^2$$

and

$$C_2 = \Sigma w_{ij}(d_{ij}^2 - \delta_{ij}^2)^2.$$

The weights may be given or may be chosen as functions of the distances. Although it might be desirable to express weights in terms of the unknown fitted distances $\delta_{ij}$ it is more usual, and certainly more convenient, to use the observed distances. Common choices are $w_{ij} = d_{ij}^{-1}$ which gives greater weight to short distances and hence local accuracy, and $w_{ij} = d_{ij}$ which emphasises accurate representation of the longer distances. The particular choice $w_{ij} = d_{ij}^{-1}$ with $C_1$ has been termed *non-linear-mapping*. The Normal equations for $C_1$ and $C_2$ can be modified easily to take account of the weights, and the iterative sequence for minimizing $C_1$ modified.

Another characterization of these criteria is obtained as follows: Suppose $C_1$ has been evaluated for some configuraion (not necessarily giving a minimum) then the coordinates may be scaled to give new distances $\lambda\delta_{ij}$, in $k$ dimensional space. Thus

$$C_1(\lambda) = \Sigma w_{ij}(d_{ij} - \lambda\delta_{ij})^2,$$

which is minimized by choosing

$$\hat{\lambda} = \Sigma w_{ij}d_{ij}\delta_{ij}/\Sigma w_{ij}\delta_{ij}^2$$

giving,

$$C_1(\hat{\lambda}) = \Sigma w_{ij}d_{ij}^2(1 - \rho_1^2)$$

where

$$\rho_1^2 = (\Sigma w_{ij}d_{ij}\delta_{ij})^2(/\Sigma w_{ij}d_{ij}^2)(\Sigma w_{ij}\delta_{ij}^2).$$

Now $\rho$ is in correlation form, so that $0 \le \rho^2 \le 1$ and hence $C_1$ will be reduced whenever $\hat{\lambda} \ne 1$. At the minimum of $C_1$ no further reduction is possible and so $\hat{\lambda} = 1$. This is the same condition, now in weighted form, derived above from Trace $(\mathbf{F}\Delta_0^*) = 0$. The present approach shows that minimizing $C_1$ is the same as maximizing $\rho_1^2$ which has maximum value $\rho_1^2(\max) = \Sigma w_{ij}\delta_{ij}^2/\Sigma w_{ij}d_{ij}^2$.

Similar arguments show that

$$C_2(\hat{\lambda}) = \Sigma w_{ij} d_{ij}^4 (1 - \rho_2^2)$$

where

$$\rho_2^2 = (\Sigma w_{ij} d_{ij}^2 \delta_{ij}^2)/(\Sigma w_{ij} d_{ij}^4)(\Sigma w_{ij} \delta_{ij}^4)$$

so that $C_2$ is minimized by maximizing $\rho_2^2$ which has maximum value $\rho_2^2(\text{max}) = \Sigma w_{ij} \delta_{ij}^4 / \Sigma w_{ij} d_{ij}^4$ attained when $\hat{\lambda} = \Sigma w_{ij} d_{ij}^2 \delta_{ij}^2 / \Sigma w_{ij} \delta_{ij}^4 = 1$.

Another form of metric scaling termed *parametric mapping* has been proposed. The criterion to be minimized is

$$C_3 = (\Sigma d_{ij}^2 / \delta_{ij}^4)/\Sigma 1/ \delta_{ij}^2)^2.$$

Now a weighted uncentred correlation between $d_{ij}^{-2}$ and $\delta_{ij}^{-2}$ may be defined by

$$\rho_3^2 = (\Sigma w_{ij} d_{ij}^{-2} \delta_{ij}^{-2})^2/(\Sigma w_{ij} d_{ij}^{-4})(\Sigma w_{ij} \delta_{ij}^{-4}).$$

By setting $w_{ij} = d_{ij}^2$ we have:

$$\rho_3^2 = (C_3 \Sigma d_{ij}^{-2})^{-1},$$

which shows that minimizing $C_3$ maximizes $\rho_3^2$. An argument similar to that given above for least-squares-scaling with scaling-factor $\lambda$ shows that maximizing $\rho_3^2$ is equivalent to minimizing the weighted least-squares criterion for inverse squared distances, i.e.

$$\Sigma w_{ij} (d_{ij}^{-2} - \delta_{ij}^{-2})^2$$

with $w_{ij} = d_{ij}^2$. At the minimum we have that

$$\Sigma w_{ij} (d_{ij}^2 - \delta_{ij}^{-2})^2 = \Sigma w_{ij} d_{ij}^{-4} - \Sigma w_{ij} \delta_{ij}^{-4}$$

which may be compared with the other relationships underlying analyses of variance. Clearly parametric mapping is identical to least-squares-squared-scaling with weight $w_{ij} d_{ij}^{-4} \delta_{ij}^{-4} \sim d_{ij}^{-6}$ provided $d_{ij}$ and $\delta_{ij}$ are of similar magnitudes. Thus the method gives very little weight to the longer distances and exceptionally heavy weight to shorter distances. This corresponds to the original rationale of the method which was based on an *index of continuity* where local distances are important.

It was shown in section 17.6 that classical scaling of $\mathbf{M}$ gives a maximum number $r$ of real dimensions, where $r$ is the number of positive eigenvalues of $\mathbf{M}_0^*$. In this section $\mathbf{M}$ has been replaced by the matrix $\mathbf{D}$. The question arises as to how many real dimensions can be fitted to $\mathbf{D}$ when the metric scaling methods discused above are used. The answer is not known but I have conjectured that no more than $r$ dimensions, possibly less, can be fitted. By this I mean that if an attempt is made to fit more than $r$ dimensions then the criterion value will not be decreased and rank $(\mathbf{X}) \le r$. There are some pointers in support of this conjecture. Firstly when $r = n - 1$, all methods give an exact fit in $n-1$ dimensions. Secondly when $r < n-1$, then it can be proved by elementary methods that no mre than $n-2$ dimensions can be fitted. The conjecture implies that this bound can be attained only when $r = n-2$, and not necessarily then. Recent, unpublished, work has given what seems to be a formal proof of the truth of the conjecture.

## 17.8. NON-METRIC MULTIDIMENSIONAL SCALING

The problem remains that of approximating distance-like data given in a matrix with elements $d_{ij}$ by fitting quantities $\delta_{ij}$ (usually Euclidean) arising as the distances between all pairs of $n$ points in some specified number $k$ of dimensions. Now however no longer does $\delta_{ij}$ approximate $d_{ij}$ itself but only some function $f(d_{ij})$ of observed distance. Of course when the function $f(\cdot)$ is unknown in advance it is only necessary to use the methods of sections 17.6 and 17.7 on the transformed data. Usually $f(\cdot)$ is not known and has to be estimated; however its general form may be specified. For example we may require $f(\cdot)$ to be a polynomial, or a smooth monotone function [see IV, § 27] expressible in terms of $B$-splines [see II, § 6.5], or we may settle for a general monotone function. The latter choice is the classical problem of non-metric multidimensional scaling. The rationale of seeking a general monotonic transformation of $d_{ij}$ is a belief that even when little confidence can be placed in the absolute values of the observations, as is common in many applications, they may be ranked fairly reliably. That is more confidence can be put in the ordinal values of the data than in their absolute or cardinal values. Consequently a method of ordination based solely on ordinal information is worth investigation. The rather remarkable thing is that such methods are feasible and generally lead to well-defined stable configurations of points.

There are two basic problems to be solved:

(i) How to define a goodness of fit criterion that is invariant to monotonic transformations of the data $d_{ij}$.

(ii) Having decided on a suitable criterion, how to find the configuration of coordinates $\mathbf{X}$ that optimizes it.

These problems do not have unique solutions and in the following we review the main approaches that are in current use.

For any ordination $\mathbf{X}$, the scatter diagram of $d_{ij}$ against $\delta_{ij}$ may be plotted. The fit would be exact if $d_{ij}$ increases monotonically with $\delta_{ij}$, that is the line joining two successive values of $\delta_{ij}$ never decreases. The plotted monotonic function could then be taken as our function $f(\cdot)$ and it would satisfy $f(d_{ij}) = \delta_{ij}$. Although exact fits cannot be expected, it is always possible to fit a monotonic (or *isotonic*) regression of $\delta_{ij}$ against the rank order of the values $d_{ij}$. Quite straightforward algorithms are available for this (see Barlow *et al.* (1978)). There is, then, a value $f(d_{ij})$ that depends only on the ranking of $d_{ij}$ and this value can be compared with the fitted value $\delta_{ij}$. As a measure of goodness-of-fit it is natural to define a criterion $\Sigma(\delta_{ij} - f(\delta_{ij}))^2$ similar to that used in least-squares scaling. In practice, it is usual to normalize this criterion to facilitate the comparison of two or more non-metric scaling solutions. Two forms of normalization are used to give the criteria:

$$S_1^2 = \Sigma(\delta_{ij} - f d_{ij}))^2 / \Sigma \delta_{ij}^2$$
$$S_2^2 = \Sigma(\delta_{ij} - f(d_{ij}))^2 / \Sigma(\delta_{ij} - \delta..)^2$$

where $\delta..$ is the average value of $\delta_{ij}$. $S_1$ is termed *stress-formula* 1 and $S_2$ *stress-formula* 2. With ranked data there is always the problem of how to deal with any ties there may be. In the present context there are two possibilities. Either ties in values of $d_{ij}$ are made to correspond with ties in $\delta_{ij}$ or they are not—usually the latter gives the more satisfactory results.

The stress criteria may be used flexibly. When certain values of $d_{ij}$ are unknown, the corresponding term is merely omitted from the summation. This leads to a form of experimental design, whose develpment is as yet at a rudimentary stage, where observations $d_{ij}$ are omitted in a systematic manner, thus reducing experimental work needed to collect the initial data. Recent studies suggest that up to about one-third of the data may be missing and still enough remains to give a satisfactory ordination. Another possibility is to split the calculation of stress into separate components each using a different monotone regression, and even possibly a different type of regression. The most common form of this is to *split by rows* (or columns), sometimes termed *local order scaling*, where stress is calculated separately for each row (or column) of the data, and the sum of the separate values of $S^2$ is minimized. This approach is useful for example when the data are collected in ranked form within rows (or columns) as happens with some forms of multidimensional unfolding (see section 17.9). In previous sections we have had the constraint $d_{ij} = d_{ji}$ but this is no longer required, although the fitted values are always symmetrically related, i.e. $\delta_{ij} = \delta_{ji}$. Nevertheless it is usual to symmetricise the data by forming $\frac{1}{2}(d_{ij} + d_{ji})$. Note that even when this is not done, non-metric scaling gives no analysis of any asymmetric features there may be in the data. To do this requires the kind of methodology discussed in section 13.

Criteria other than stress have been proposed. Clearly most metric scaling criteria can be extended to non-metric form by replacing $d_{ij}$ by $f(d_{ij})$, where $f(d_{ij})$ is invariant to monotonic transformations of $d_{ij}$. One approach is to work in terms of the correlation between $\delta_{ij}$ and $f(d_{ij})$ to give a criterion

$$\mu^2 = (\Sigma \delta_{ij} f(d_{ij}))^2 / \Sigma \delta_{ij}^2 \Sigma (f(d_{ij}))^2$$

which looks very like the correlational forms of the metric scaling criteria discussed in section 7. Presumably when $f(d_{ij})$ is defined by the fitted values of a monotonic regression, as above, maximizing $\mu$ would be equivalent to minimizing stress were it not for the minor effects of the normalizing divisors used in the definitions of stress. This straightforward position has been complicated because those who prefer to use $\mu$ also prefer a different definition of $f(\cdot)$, which incidentally could have been used in the definition of stress. The function $f(\cdot)$ chosen is termed the *rank-image* transformation. This transformation is very simple and may be described as follows: if $d_{ij}$ is missing in the data then $f(d_{ij}) = \delta_{ij}$; if $d_{ij}$ is ranked $r$ in the data then $f(d_{ij})$ is the $r$th ranking value of $\delta_{ij}$, with suitable adjustments for ties. This ensures that when there is a monotonic relationship between $d_{ij}$ and $\delta_{ij}$ (possibly allowing for ties in either or both sequences) then $f(d_{ij}) = \delta_{ij}$ so that $\mu = 1$. When the monotonic relationship is not exact then $\mu < 1$.

Ramsay (1977) uses a maximum likelihood estimation process. Because the observations $d_{ij}$ are supposed to give only ordinal information, it is convenient rather to make distributional assumptions about the transformed values $f(d_{ij})$; e.g. that they have a Normal or log-normal distribution. Such assumptions may seem unrealistic but when they can be accepted a consequence of the likelihood approach is that confidence regions may be estimated for each of the fitted sets of co-ordinates.

Whether stress with a monotonic transformation or correlation with a rank image transformation is used, the computational problems are similar and standard function optimization procedures (see e.g. Murray (1972)) may be used to optimize the chosen criterion. Of course there are $nk$ elements of **X** to estimate so the computational problem is a major one and much work has been put into obtaining good and reliable algorithms that include convenient facilities for data manipulation. The outcome has been that to an extent that is much greater than with most other techniques, the methodology is embodied in a few widely-available computer programs. To produce these is a major project, not to be undertaken casually. The first of these programs was written around 1962 but, building on the earlier experience, a second generation is now available. Important programs that might be encountered are KYST (Kruskal, Young, Shepard, Torgerson), ALSCAL (de Leeuw, Takane, Young), MULTISCALE (Ramsay) and MINISSA (Guttman, Lingoes and Roskam). A good short simple introduction to non-metric scaling, including more details of the programs mentioned above, is given by Kruskal and Wish (1978).

Using these programs requires some skill. They may, and often do, converge on local optima rather than the true optimum of the chosen criterion; they may not converge at all. The user has to recognize such situations and take remedial action. He has to be able to judge when an acceptable number of dimensions has been fitted, a deision that is related to an interpretable ordination coupled with a satisfactory criterion value. An important product of this kind of analysis is that the values of $f(d_{ij})$ are available. These may be plotted against $d_{ij}$ to exhibit the form of transformation found. With monotonic regression this may be a function with major discontinuities, which may throw light on the interpretation of the data-values $d_{ij}$ or it may be fairly smooth, suggesting that a mathematical functional form may be fitted. At the beginning of this section mention was made of using polynomial regressions, or smooth monotonic functions expressed as $B$-splines. Facilities for these options are sometimes available in computer programs and may be regarded as stepping stones linking the basic forms of metric scaling to the general forms of non-metric scaling.

Section 17.7 closed with some remarks about the maximum numbers of dimensions that might be fitted by metric scaling methods. It has already been seen that there exist constants $c_1$ and $c_2$ such that the simple monotonic transformations $d_{ij} + c_1$ and $(d_{ij}^2 + c_2)^{1/2}$ always give exact Euclidean fits in $n - 2$ dimensions. Can better be done with a general monotonic transformation?

## 17.9.  MULTIDIMENSIONAL UNFOLDING

This problem and its name originated in psychology but applications are now widespread. Suppose we have a two-way table with rows representing different people and columns representing some activity or experience of these people. For illustrative purposes suppose the columns represent the newspapers they read. To complete the two-way table each person is asked to rank the newspapers in order of his preferences for them. Multidimensional unfolding gives a joint ordination of the people and the newspapers, in such a way that those newspapers given a high ranking by a person occur as points near the point representing the person. Thus the method is allied to techniques like biplots (section 17.4) and correspondence analysis (section 17.5) but, unlike those methods, multidimensional unfolding gives distances between the two sets of points that are directly interpretable. In its original form the joint ordination was one-dimensional and one could imagine the line folded over (hinged) at the point representing an individual of interest. The points representing the newspapers would then all be on the same side of the hinge and ranked correctly for that individual—at least they would be if an exact ordination were possible. The points representing individuals are often termed *ideal points*. Exact joint ordination in one dimension is rarely possible and higher dimensional ordinations improve the representation.

One way to do multidimensional unfolding is to use standard non-metric scaling techniques. We have already noted in section 17.8 that the non-metric scaling algorithms readily cope with missing data. The $n \times m$ table **A** of preference data may be regarded as the corner of an $(n + m) \times (n + m)$ symmetric matrix with the two symmetric matrices of order $n \times n$ and $m \times m$ missing. These missing matrices represent the unknown associations between the individuals and between the newspapers. Operating on this matrix of order $n + m$, non-metric multidimensional scaling gives an ordination with $n + m$ points, $n$ of these representing the rows of **A** and $m$ representing the columns of **A**. With ranked preference data we have no direct information on any ranking of individuals for each newspaper, so there is a clear case for using the split-by-rows method for computing the stress criterion of section 17.8. Further, stress-formula 2 must be used to avoid the degenerate solution with zero stress (formula 1) in which all the newspapers are represented by a single point. Because the data are weakly structured, degenerate solutions should be guarded against and particular attention paid to the possibility of inadmissible local optima.

Although the unfolding problem originated with preference data, it is now often generalized to allow quantitative values in $\mathbf{A}$. Rows and columns are then on an equal basis and there is no call to use the split-by-rows method; the non-metric multidimensional scaling method then performs better. With quantitative data of this kind it is instructive to investigate the possibilities of a fully metric approach. The case when $\mathbf{A}$ is part of a matrix arising as distances between points in $k$ dimensions will be investigated. We shall work in terms of a matrix $\mathbf{D}$ of squared distances. $\mathbf{D}$ has the same size as $\mathbf{A}$ but its elements may be a function of the elements of $\mathbf{A}$. Thus the unfolding problem is to find coordinates $\mathbf{X}(n \times k)$ and $\mathbf{Y}(m \times k)$ which generate the matrix $\mathbf{D}$. For the present we assume an exact fit is possible so that Pythagoras's theorem gives

$$\mathbf{D} = \mathbf{EU} + \mathbf{UF} - 2\mathbf{XY}^T$$

where $\dot{\mathbf{E}} = \text{diag}\,(\mathbf{XX}^T)$, $\mathbf{F} = \text{diag}\,(\mathbf{YY}^T)$ and $\mathbf{U}(n \times m)$ is a matrix of units. Let $\mathbf{N} = \mathbf{11}^T/n$ and $\mathbf{M} = \mathbf{11}^T/m$ so that

$$(\mathbf{I} - \mathbf{N})(-\tfrac{1}{2}\mathbf{D})(\mathbf{I} - \mathbf{M}) = \mathbf{XY}^T.$$

The joint translation of $\mathbf{X}$ and $\mathbf{Y}$ cannot affect the solution, so we may assume the centroid of $\mathbf{X}$ is at the origin, i.e. $\mathbf{NX} = 0$. Suppose the singular value decomposition of the left-hand side is $\mathbf{GH}^T$ where the $k$ non-zero singular vectors are absorbed into $\mathbf{H}$. Thus $\mathbf{G}^T\mathbf{G} = \mathbf{I}$, $\mathbf{X} = \mathbf{GT}$ and $(\mathbf{I} - \mathbf{N})\mathbf{Y} = \mathbf{H}(\mathbf{T}^T)^{-1}$ where $\mathbf{T}$ is an arbitrary non-singular matrix of order $k$. Thus it is not difficult to evaluate $\mathbf{X}$ and $\mathbf{Y}$, up to an arbitrary transformation, with respect to their own centroids and ignoring the translation of one with respect to the other. To progress, premultiply the first equation by $\mathbf{I} - \mathbf{N}$, postmultiply it by $\mathbf{1}$ and replace $\mathbf{X}$ by $\mathbf{GT}$ to give:

$$(\mathbf{I} - \mathbf{N})\mathbf{D1} + \mathbf{G}(2\mathbf{TY}^T\mathbf{1}) = m(\mathbf{I} - \mathbf{N})\,\text{diag}\,(\mathbf{GTT}^T\mathbf{G}^T)\mathbf{1}.$$

Let $\mathbf{TT}^T = \mathbf{S}$, a symmetric positive-definite-matrix, $\mathbf{d} = \mathbf{D1}/m$ the row means of $\mathbf{D}$, and $\mathbf{Y}^T\mathbf{1}/m = \mathbf{y}$ the displacement of the centroid of $\mathbf{Y}$ from the centroid of $\mathbf{X}$. Then the above equation becomes:

$$(\mathbf{I} - \mathbf{N})\mathbf{d} + 2\mathbf{GTy} = (\mathbf{I} - \mathbf{N})\mathbf{Cs}$$

where $\text{diag}\,(\mathbf{GSG}^T)\mathbf{1} = \mathbf{Cs}$, formed from a vector $\mathbf{s}$ of length $\tfrac{1}{2}k(k+1)$ containing the distinct elements of $\mathbf{S}(s_{11}, s_{21}, s_{22}, s_{31}, s_{32}, s_{33}, s_{41}, \ldots, s_{kk})$ and $\mathbf{C}$ a matrix $(n \times \tfrac{1}{2}k(k+1))$ containing functions of the elements of $\mathbf{T}$ calculated for the $i$th row of $\mathbf{C}$ by taking $\mathbf{g}_i$, the $i$th row of $\mathbf{G}$, and deriving the symmetric matrix $\mathbf{g}_i\mathbf{g}_i^T = \mathbf{A}_i$ (say). The elements of the lower triangular part of $2\mathbf{A}_i - \text{diag}\,(\mathbf{A}_i)$ strung out in row order (as in $\mathbf{s}$) are the required quantities.

Because $\mathbf{G}^T\mathbf{1} = 0$ and $\mathbf{G}^T\mathbf{G} = \mathbf{I}$ this equation for $\mathbf{y}$ may be simplied to give:

$$\mathbf{Y} = \tfrac{1}{2}\mathbf{T}^{-1}\mathbf{G}^T(\mathbf{Cs} - \mathbf{d})$$

which on back-substitution yields

$$\mathbf{BCs} = \mathbf{Bd}$$

where $\mathbf{B} = \mathbf{I} - \mathbf{N} - \mathbf{GG}^T$ which is idempotent. This equation can be solved for $\mathbf{s}$. It contains $n$ linear relationships to determine $\frac{1}{2}k(k+1)$ elements of $\mathbf{s}$, but $\mathbf{G}^T\mathbf{B} = \mathbf{0}$ and $\mathbf{1}^T\mathbf{B} = \mathbf{0}$ so, at most, there are $n - k - 1$ independent equations. For a solution we must have $n - k - 1 \geq \frac{1}{2}k(k+1)$, i.e. $n \geq \frac{1}{2}k(k+3)$. If this inequality fails and $m > n$, a solution may possibly be found merely by transposing $\mathbf{D}$. When the inequality fails with both $m$ and $n$ a unique solution might still exist but it cannot be found with this approach. When the system of equations is over-determined, the least-squares solution is

$$\mathbf{s} = (\mathbf{C}^T\mathbf{BC})^{-1}\mathbf{C}^T\mathbf{Bd}$$

which is exact, when there is a unique solution, as is being assumed. Having estimated the vector $\mathbf{s}$ we can reconstruct the matrix $\mathbf{S}$ and hence $\mathbf{T}$ may be taken to be a solution to any decomposition $\mathbf{S} = \mathbf{TT}^T$, each decomposition merely referring to a different joint orientation of $\mathbf{X}$ and $\mathbf{Y}$. The eigenvector decomposition $\mathbf{S} = \mathbf{L\Lambda L}^T$ gives $\mathbf{T} = \mathbf{L\Lambda}^{1/2}$ and

$$\mathbf{y} = \tfrac{1}{2}\mathbf{\Lambda}^{-1/2}\mathbf{L}^T\mathbf{G}^T(\mathbf{Cs} - \mathbf{d})$$

where $\mathbf{s}$ is given by the previous equation. Finally, the coordinates are given by

$$\mathbf{X} = \mathbf{GL\Lambda}^{1/2}, \qquad \mathbf{Y} = \mathbf{HL\Lambda}^{-1/2} + \mathbf{1}\mathbf{y}^T.$$

The above method will work with exact data and in principle can be expected to work even when an exact fit does not exist. There are however many difficulties. Firstly, different results, which are not simply related, will be obtained from $\mathbf{D}$ and $\mathbf{D}^T$. Secondly, the matrix $\mathbf{S}$ as estimated above may not be positive-semi-definite and its $k$ largest eigenvalues may not be positive. Thirdly, the decomposition $\mathbf{GH}^T$ may not be of rank $k$ and although the Eckart–Young theorem may be cited to justify a rank-$k$ least-squares approximation this, coupled with the later appeal to least-squares when estimating $\mathbf{s}$, leads to a solution which has very doubtful status and is certainly not any form of least-squares solution to the original problem. These, and other difficulties, relegate the method described to one which, at best, may provide a useful initial setting to start to an iterative solution.

One such iterative scheme follows from an examination of the normal equations for least-squares squared scaling (see section 17.7) in the unfolding context. Define the residual matrix

$$\mathbf{R} = \mathbf{D} - \mathbf{EU} - \mathbf{UF} + 2\mathbf{XY}^T,$$

then the criterion to be minimized is trace $(\mathbf{RR}^T)$ which leads to the normal equations:

$$\mathbf{RY} = \text{diag} \, (\mathbf{RU}^T)\mathbf{X}$$

$$\mathbf{R}^T\mathbf{X} = \text{diag} \, (\mathbf{R}^T\mathbf{U})\mathbf{Y}.$$

Direct solution of these equations is not possible, but they may be used as the basis of an iterative solution that is disccussed below. First a few results are derived.

$\mathbf{R}$ is invariant to joint rotations and translations of $\mathbf{X}$ and $\mathbf{Y}$, and it is easy to see that the solutions to the normal equations are undetermined to this degree of generality. When there is an exact fit, $\mathbf{R} = 0$ and both sides of the equations vanish. Premultiplying both equations by $\mathbf{1}^T$ (of appropriate length) gives:

$$\mathbf{1}^T\mathbf{R}\mathbf{Y} = \mathbf{1}^T\mathbf{R}^T\mathbf{X}$$

relating the fitted values and the row and column totals of the residual matrix. The residual sum-of-squares is given by:

$$S = \text{Trace } (\mathbf{R}\mathbf{R}^T) = \text{Trace } (\mathbf{D}\mathbf{R}^T - \mathbf{E}\mathbf{U}\mathbf{R}^T - \mathbf{U}\mathbf{F}\mathbf{R}^T + 2\mathbf{X}\mathbf{Y}^T\mathbf{R}^T)$$

which on substituting the results given by the normal equations gives:

$$S = \text{Trace } (\mathbf{R}\mathbf{R}^T) = \text{Trace } (\mathbf{D}\mathbf{R}^T)$$

leading to

$$\text{Trace } (\mathbf{D}\mathbf{D}^T) = \text{Trace } (\mathbf{R}\mathbf{R}^T) + \text{Trace } (\mathbf{D} - \mathbf{R})(\mathbf{D} - \mathbf{R})^T.$$

This shows that the analysis of variance result derived in section 17.7 for least-squares squared scaling of a complete matrix $\mathbf{D}$, also holds for the rectangular matrix $\mathbf{D}$ occurring in unfolding. Note that with unfolding the fitted sum of squares Trace $(\mathbf{D} - \mathbf{R})(\mathbf{D} - \mathbf{R})^T$ may be further broken down into row and column components, in the usual way for two-way tables.

The normal equations suggest a natural alternating least-squares algorithm for solving the unfolding problem iteratively. If we know $\mathbf{Y}$ the first equation gives the least-squares squared scaling estimate of $\mathbf{X}$ and this estimate will reduce (or at least not increase) the residual sum of squares. Similarly knowing $\mathbf{X}$, an estimate of $\mathbf{Y}$ may be determined from the second normal equation. The process may be repeated until satisfactory convergence is found. Unfortunately even when $\mathbf{Y}(\mathbf{X})$ is known it is not easy to solve the first (second) equation for $\mathbf{X}(\mathbf{Y})$.

To simplify the first normal equation it is legitimate to assume that $\mathbf{Y}$ is centred at the origin and referred to principal axes so that $\mathbf{Y}^T\mathbf{Y} = \mathbf{\Lambda}$, the diagonal matrix of eigenvalues of $\mathbf{Y}\mathbf{Y}^T$. The equation then becomes

$$(\mathbf{D} - \mathbf{U}\mathbf{F})\mathbf{Y} + 2\mathbf{X}\mathbf{\Lambda} = \text{diag } \{(\mathbf{D} - \mathbf{E}\mathbf{U} - \mathbf{U}\mathbf{F})\mathbf{U}^T\}\mathbf{X}$$

$$= \text{diag } (\mathbf{D}\mathbf{U}^T - m\mathbf{E} - (\mathbf{1}^T\mathbf{F}\mathbf{1})\mathbf{I})\mathbf{X}.$$

If $\mathbf{E}$ were known, these equations would be linear in the elements of $\mathbf{X}$ and hence easy to solve. In fact we have that

$$x_{ij} = \frac{w_{ij}}{mD_{i.} - (\mathbf{1}^T\mathbf{F}\mathbf{1}) - 2\lambda_j - mE_{ii}}$$

where $\mathbf{W} = (\mathbf{D} - \mathbf{UF})\mathbf{Y}$ and $E_{ii} = \sum_{j=1}^{k} x_{ij}^2$. Note that none of the equations for the $k$ elements in the $i$th row of $\mathbf{X}$ involve the elements in any other row. Greenacre (1978) suggests that these equations be solved by defining new variables $\boldsymbol{\phi}$:

$$\phi_i = mD_{i\bullet} - (\mathbf{1}^T \mathbf{F} \mathbf{1}) - mE_{ii}$$

so we have that

$$x_{ij} = \frac{w_{ij}}{\phi_i - 2\lambda_j}.$$

Substituting for $x_{ij}$ into the definition of $\phi_i$ yields:

$$\phi_i = c_i - m \sum_{j=1}^{k} \frac{w_{ij}}{(\phi_i - 2\lambda_j)^2}$$

where $c_i = mD_i - \mathbf{1}^T \mathbf{F} \mathbf{1}$.

This is a polynomial in $\phi_i$ of degree $2k+1$ so must have at least one real root. The problem is how to find the best solution to this polynomial equation. In problems of this kind it is usual for the smallest root $\phi_0$ to correspond to the smallest residual sum-of-squares (see e.g. oblique Procrustes analysis described in section 17.10). Greenacre reports that he has never found an example in which $\phi_0$ is not the optimal root but no proof that this must be so is known. It is therefore necessary to examine all the real solutions to the polynomial, although some shortcuts are available. For example $\phi_0 \leq 2\lambda_1$; $2\lambda_1 \leq \phi_1$, $\phi_2 \leq 2\lambda_2$; $2\lambda_2 \leq \phi_3$, $\phi_4 \leq 2\lambda_3$ etc. and there is no real root greater than $c_i$. These simple rules, giving bounds on the real roots of the polynomial, allow approximation methods, such as Newton's method for solving polynomials or bisection methods, to be readily used. More subtle rules allow further savings.

For each candidate $\phi_i$ it is necessary to evaluate the residual sum-of-squares. We have that

$$S = \text{Trace} (\mathbf{R}^T \mathbf{R}) = \text{Tr} (\mathbf{D}^T \mathbf{R} - \mathbf{U}^T \mathbf{E} \mathbf{R} - \mathbf{F} \mathbf{U}^T \mathbf{R} + 2 \mathbf{Y} \mathbf{X}^T \mathbf{R})$$

but the first Normal equation implies that $\text{Trace} (\mathbf{X}\mathbf{X}^T\mathbf{R}) = \text{Trace} (\mathbf{U}^T\mathbf{E}\mathbf{R})$ so we have that:

$$S = \text{Trace} (\mathbf{D}^T \mathbf{R} + \mathbf{U}^T \mathbf{E} \mathbf{R} - \mathbf{F} \mathbf{U}^T \mathbf{R})$$

$$= \text{Trace} (\mathbf{D}^T + \mathbf{U}^T \mathbf{E} - \mathbf{F} \mathbf{U}^T)(\mathbf{D} - \mathbf{E} \mathbf{U} - \mathbf{U} \mathbf{F} + \mathbf{X} \mathbf{Y}^T).$$

Isolating the terms in $\mathbf{X}$ gives:

$$S = \text{Trace} (2\mathbf{W}^T \mathbf{X} - \mathbf{U}^T \mathbf{E}^2 \mathbf{U}) + \text{terms not involving } \mathbf{X}$$

$$= 2 \sum_{i,j} w_{ij} x_{ij} - m \sum_{i=1}^{n} E_{ii}^2 + \text{terms not involving } \mathbf{X}$$

and substituting for $x_{ij}$ at the minimum we have:

$$S = 2 \sum_{i,j} \frac{w_{ij}^2}{\phi_i - 2\lambda_j} - m \sum_{i,j} \left(\frac{w_{ij}}{\phi_i - 2\lambda_j}\right)^2 + \text{terms not involving } \mathbf{X}$$

$$= 2 \sum_{i,j} \frac{w_{ij}^2}{\phi_i - 2\lambda_j} + \sum_{i=1}^{n} (\phi_i - c_i) + \text{terms not involving } \mathbf{X}.$$

The contribution from the $i$th row of $\mathbf{D}$ that involves $\phi_i$ is given by:

$$S_i = 2 \sum_{j=1}^{k} \frac{w_{ij}^2}{\phi_i - 2\lambda_j} + \phi_i$$

which gives a simple method for assessing trial values of $\phi_i$ in search of a minimum value of $S_i$.

Using such methods the optimal value of $\phi_i$ for each row of $\mathbf{D}$ may be found and hence a value of $\mathbf{X}$ determined. Centering on this new $\mathbf{X}$ and rotating the coordinate axes to be the principal axes of $\mathbf{X}$, puts $\mathbf{X}$ in a form that can be used in a similar manner to that described above to form a new estimate of $\mathbf{Y}$, and so on until an acceptable degree of convergence is achieved. Much progress has been made but further work is still needed before it can be said that a fully acceptable metric unfolding algorithm is available.

The special problem of unfolding a symmetric matrix $\mathbf{D}$ is one that has not received the attention it deserves. For example when $\mathbf{D}$ is a distance matrix, with zero diagonal elements, its unfolded form is merely given by a conventional ordination in which the points representing rows and columns coincide in pairs. However, if we are prepared to ignore diagonal information, as is trivial to do with non-metric scaling algorithms, there are interesting consequences. For example consider a regular simplex with $n$ vertices and with side $d$, for which the elements of $\mathbf{D}$ are all equal except for the zero diagonal. Then if we do not wish to represent the diagonals, the unfolded form of $\mathbf{D}$ is merely a set of coincident row-points (say point $R$) and a set of coincident column points (say point $C$) where the distance between $R$ and $C$ is $d$. This representation in one dimension is exact. It contrasts dramatically with the $n-1$ dimensions required by conventional ordinations to represent a regular simplex and suggests that symmetric unfolding may offer useful possibilities of dimension-reduction at the expense of representing every point twice—once as a row-point and once as a column-point.

It is instructive to consider the ways that an unfolding can represent a symmetric matrix. With the previous notation we must have $\mathbf{D} = \mathbf{D}^T$, i.e.

$$\mathbf{EU} + \mathbf{UF} - 2\mathbf{XY}^T = \mathbf{UE} + \mathbf{FU} - 2\mathbf{YX}^T.$$

Writing $2\mathbf{G} = \mathbf{E} - \mathbf{F}$ and rearranging gives:

$$\mathbf{GU} - \mathbf{UG} = \mathbf{XY}^T - \mathbf{YX}^T.$$

Writing $\mathbf{g} = \text{diag}(\mathbf{G})\mathbf{1}^T$ this becomes:

$$\mathbf{g1}^T - \mathbf{1g}^T = \mathbf{XY}^T - \mathbf{YX}^T$$

so that the skew-symmetric matrix on the right-hand side must have rank 2. An exhaustive analysis of what is implied by the condition has not been undertaken but some special solutions of importance are known. One obvious solution is $\mathbf{Y} = \mathbf{X}$, so that $\mathbf{g} = 0$. Does a solution $\mathbf{Y} = \mathbf{XH}$ exist where $\mathbf{H}$ is orthogonal? This condition implies $\mathbf{g} = 0$ and $\mathbf{X}(\mathbf{H}^T - \mathbf{H})\mathbf{X}^T = 0$ so that when $\mathbf{X}$ has full column-rank $\mathbf{H} = \mathbf{H}^T$. The only symmetric orthogonal matrices are diagonal matrices with elements $\pm 1$, which correspond to various reflections in the coordinate axes (including reflections through the origin), and Householder transforms $\mathbf{H} = \mathbf{I} - 2\mathbf{u}\mathbf{u}^T$ where $\mathbf{u}$ is a unit vector. This solution represents reflections in the hyperplane with normal $\mathbf{u}$ [see III, § 4.10]. More complicated solutions exist. For example if $\mathbf{Y} = \mathbf{XS}$, where $\mathbf{S}$ is symmetric, then the right-hand side of the equation vanishes and the left-hand side gives $g_i = k$, a constant. Thus diag $\mathbf{X}(\mathbf{I} - \mathbf{S}^2)\mathbf{X}^T = k\mathbf{I}$ and the coordinates $\mathbf{X}$ lie on the general conic with quadratic form $\mathbf{x}^T(\mathbf{I} - \mathbf{S}^2)\mathbf{x} = \mathbf{k}$. The coordinates $\mathbf{Y}$ lie on a related conic with quadratic form $\mathbf{y}^T(\mathbf{S}^{-2} - \mathbf{I})\mathbf{y} = \mathbf{k}$. these two conics have the same (principal) axes. With $\mathbf{D}$ symmetric, it might be thought that the distances between the row-points and the distances between the column-points would be the same, but the last example shows that this need not be so.

As with ordination methods in general, two-dimensional solutions have special interest. Apart from the two-dimensional forms of the above solutions, there are other obvious possibilities. For example the row points might be on the circumference of one circle and the column points on the circumference of a concentric circle with corresponding points on a common radius. Or as a special case of this, the two sets of points might lie on a pair of parallel lines, or we may have three equidistant parallel lines, the central one containing (say) the row points and the column points either side of the other two lines. Although the above solutions have a certain geometrical regularity the disposition of the two sets of points need not be related in any clear way.

The interpretation of two-dimensional symmetric unfoldings is not yet fully understood but some observed phenomena are suggestive:

(i) If the unfolding gives pairs of coincident points, the original distances are probably nearly in two-dimensional form.

(ii) If a reflection-unfolding is obtained, this probably suggests that the original distances are of high-dimensional form.

(iii) Reflections about pairs of axes are sometimes found, suggesting that the sample distances occur in two separate sub-spaces.

The possibilities of this method are still largely unexplored but it seems that it may work best when the dimensionality of the sample is either very low or very high. Intermediate numbers of dimensions may perhaps be handled satisfactorily by doubling the size of $\mathbf{D}$ by repeating its values (ignoring diagonals) so that each sample is represented by two row-names and two column-names. Unfolding this matrix gives four points for every sample. The process may be developed further, encouraging a trade-off between numbers of dimensions needed for a good representation and the number of points in the ordination.

## 17.10. ORTHOGONAL PROCRUSTES ANALYSIS

Previous sections have been concerned with fitting distances $\delta_{ij}$ to observations $d_{ij}$ by optimizing some goodness-of-fit criterion such as that for least-squares scaling. Such criteria may be regarded as comparing the fitted set of coordinates $\mathbf{X}$ to the coordinates $\mathbf{Y}$ (when they exist) that generate the observed distances. The problem of comparing two sets of coordinates $\mathbf{X}$ and $\mathbf{Y}$ whose rows refer to the same samples or populations is a very common one. For example $\mathbf{X}$ and $\mathbf{Y}$ may arise from two different ordinations of the same data or they may arise from the same ordination method used on two sets of data pertaining to the same $n$ objects. How can comparisons of this kind be made? Writing $\delta_{ij}$ and $\varepsilon_{ij}$ for typical distances generated by $\mathbf{X}$ and $\mathbf{Y}$, it is clear that $f_1(\delta, \varepsilon) = \Sigma(\delta_{ij} - \varepsilon_{ij})^2$ and $f_2(\delta, \varepsilon) = \Sigma(\delta_{ij}^2 - \varepsilon_{ij}^2)^2$ are criteria suitable for comparing the two sets. They are invariant to translations and orthogonal transformations of $\mathbf{X}$ and $\mathbf{Y}$, which is clearly a prerequisite for any sensible criterion. Little is available to help judge what values of $f_1$ and $f_2$ indicate good and bad fits, although the relative values of $f_{1,2}(d, \delta)$, $f_{1,2}(d, \varepsilon)$ and $f_{1,2}(\delta, \varepsilon)$ may be of some interest. Plotting $\delta_{ij}$ against $\varepsilon_{ij}$ is worth doing because a linear relationship indicates a good agreement between the configurations, and a curvilinear relationship indicates sections of the distance-scales that one configuration exaggerates or reduces compared to the other configuration.

Another possibility is to refer the coordinates given by $\mathbf{X}$ and $\mathbf{Y}$ to their principal axes. This should be done by all ordination computer programs as it eases the comparison between ordinations in $r$ and $r+1$ dimensions. Without the principal axis representation the $i$th axis of the $r$-dimensional representation need have no relationship to the $i$th axis of the $s$-dimensional relationship. This makes it hard to see what are the effects of adding or dropping a dimension. For example if a dropped dimension has been making a useful contribution to the fit, this may be the effect of only one or two points which would be easy to see in a principal axes representation, but not otherwise.

In an extreme case $\mathbf{X}$ and $\mathbf{Y}$ may be the same configuration but be represented by different coordinates that seem to bear no relationship to each other. It is natural to ask whether the configuration $\mathbf{X}$ can have its axes and origin changed so that it can be seen to be the same as, or similar to, $\mathbf{Y}$. Thus we require an orthogonal matrix $\mathbf{H}$ (representing axes rotation and reflection) and a row-vector $\mathbf{m}$ (representing translation of origin) such that $\mathbf{XH}+\mathbf{1m}$ approximates $\mathbf{Y}$. A suitable criterion to measure the degree of match should itself be invariant to further joint rotations and translations of the two configurations, which suggests some combination of the distances between $\mathbf{Y}$ and $\mathbf{XH}+\mathbf{1m}$. A sum-of-squares criterion is in the spirit of much that has gone before and leads to tractable algebra. Thus we wish to estimate $\mathbf{H}$ and $\mathbf{m}$ to minimize:

$$f_3(\delta, \varepsilon) = \text{Trace } (\mathbf{Y}-\mathbf{XH}-\mathbf{1m})^T(\mathbf{Y}-\mathbf{XH}-\mathbf{1m}).$$

Assuming without loss of generality that both $\mathbf{X}$ and $\mathbf{Y}$ have their centroid at the origin we have that:

$$f_3 = \text{Trace } (\mathbf{Y}-\mathbf{XH})^T(\mathbf{Y}-\mathbf{XH}) + n\Sigma m_i^2$$

so that the effects of translation and rotation can be separated, and $f_3$ is minimized for translation by choosing $m_i = 0$ in all dimensions, i.e. translation is taken care of by ensuring that $\mathbf{X}$ and $\mathbf{Y}$ have the same centroid, here chosen to be the origin. Minimizing $f_3$ over $\mathbf{H}$ is equivalent to maximizing Trace $(\mathbf{Y}^T \mathbf{XH})$. Writing the singular value decomposition $\mathbf{Y}^T \mathbf{X} = \mathbf{U} \mathbf{\Gamma} \mathbf{V}^T$ gives:

$$\text{Trace } (\mathbf{Y}^T \mathbf{XH}) = \text{Trace } (\mathbf{U}\mathbf{\Gamma}\mathbf{V}^T\mathbf{H})$$

$$= \text{Trace } (\mathbf{\Gamma}\mathbf{V}^T\mathbf{HU}).$$

Now $\mathbf{V}^T\mathbf{HU} = \mathbf{Q}$ (say) is an orthogonal matrix. Hence we have

$$\text{Trace } (\mathbf{Y}^T\mathbf{XH}) = \Sigma \gamma_i q_{ii}$$

where the non-zero singular values are positive and $-1 \le q_{ii} \le 1$. An upper bound is therefore $\Sigma \gamma_i$, occurring when $q_{ii} = 1$ for all $i$. This upper bound is attainable, for $\mathbf{Q} = \mathbf{I}$ is an orthogonal matrix corresponding to $\mathbf{H} = \mathbf{VU}^T$ which gives the required transformation. The residual sum-of-squares is given by:

$$f_3(\delta, \varepsilon) = \text{Trace } (\mathbf{X}^T\mathbf{X}) + \text{Trace } (\mathbf{Y}^T\mathbf{Y}) - 2 \text{ Trace } \mathbf{\Gamma}.$$

In the above nothing has been said about the dimensions (number of columns) of $\mathbf{X}$ and $\mathbf{Y}$ but implicit in the statement that both $\mathbf{U}$ and $\mathbf{V}$ are orthogonal and that they commute is an assumption that $\mathbf{Y}^T\mathbf{X}$ is square and hence that $\mathbf{X}$ and $\mathbf{Y}$ have the same number of columns. If this is not so, extra zero columns may be added to the smaller matrix so that the rotation $\mathbf{H}$ takes place in the higher-dimensional space. In this case some of the singular values $\gamma_i$ will be zero and the corresponding values of $q_{ii}$ become irrelevant. The solution found is still optimal but it is no longer unique. The other optimal solutions correspond to arbitrary rotations in the space orthogonal to the smaller space. In section 17.11 the problem of matching $\mathbf{X}$ to $\mathbf{Y}$ when $\mathbf{Y}$ has fewer dimensions than $\mathbf{X}$ is discussed.

The minimization of $f_3(\cdot)$ is termed the *orthogonal Procrustes problem.* In Greek mythology Procrustes was an innkeeper who adjusted the size of his customers, either by stretching them or by lopping off their limbs, so that they fitted his bed. Here we have adjusted $\mathbf{X}$ to fit $\mathbf{Y}$. The advantage of the approach is that the rotated and translated configuration $\mathbf{XH}$ may be plotted together with $\mathbf{Y}$. When both configurations are the same, the two sets of points will coincide, otherwise the relative positions of pairs of like points indicate agreements and discrepancies between the two ordinations or multidimensional scalings.

It should be borne in mind that one of the two sets $\mathbf{X}$ and $\mathbf{Y}$ may be inherently bigger than the other. This possibility is most simply overcome by normalizing $\mathbf{X}$ and $\mathbf{Y}$ so that Trace $(\mathbf{X}^T\mathbf{X}) = $ Trace $(\mathbf{Y}^T\mathbf{Y})$. Alternatively a scale factor $\rho$ can be estimated by minimizing

$$\text{Trace } (\mathbf{Y} - \rho\mathbf{XH})^T(\mathbf{Y} - \rho\mathbf{XH}).$$

The estimate of $\mathbf{H}$ is exactly as above and simple differentiation gives an estimate $\hat{\rho}$ of $\rho$ as:

$$\hat{\rho} = \frac{\text{Trace } (\mathbf{XHY}^T)}{\text{Trace } (\mathbf{X}^T\mathbf{X})} = \frac{\text{Trace } (\boldsymbol{\Gamma})}{\text{Trace } (\mathbf{X}^T\mathbf{X})}.$$

The sums of squares are now related by:

$$\text{Trace } (\mathbf{Y}^T\mathbf{Y}) = \hat{\rho}^2 \, \text{Trace } (\mathbf{X}^T\mathbf{X}) + \text{Trace } (\mathbf{Y} - \hat{\rho}\mathbf{XH})^T(\mathbf{Y} - \hat{\rho}\mathbf{XH})$$

showing that the corrected sums-of-squares amongst the coordinates $\mathbf{Y}$ is equal to the sum of the corrected sum-of-squares amongst the fitted values $\hat{\rho}\mathbf{XH}$ and the residual sum-of-squares. This may be used as the basis of an orthogonal decomposition in an analysis of variance.

Because $f_3(\cdot)$ is invariant to joint rotations of $\mathbf{X}$ and $\mathbf{Y}$ it follows that after rotating $\mathbf{X}$ through $\mathbf{H}$ to fit $\mathbf{Y}$ we can perform the inverse transformation $\mathbf{H}^T$ to both configurations. This restores $\mathbf{X}$ to where it started and transforms $\mathbf{Y}$ to $\mathbf{YH}^T$. Thus if $\mathbf{H}$ optimizes the fit of $\mathbf{X}$ to $\mathbf{Y}$, then $\mathbf{H}^T$ optimizes the fit of $\mathbf{Y}$ to $\mathbf{X}$, both with some value $m_{xy}^2$ (say) of $f_3(\cdot)$. Thus $m_{XY}$ is symmetric:

$$m_{XY}^2 = f_3(\delta, \varepsilon) = f_3(\varepsilon, \delta) = m_{YX}^2.$$

Clearly $m_{XX}^2 = 0$ (a perfect fit) and it is easy to show that $m_{XY}$ satisfies the metric inequality, i.e. for any three configurations $\mathbf{X}$, $\mathbf{Y}$, $\mathbf{Z}$ we have that $m_{XY} \le m_{XZ} + m_{YZ}$. Although $m_{XY}$ is a metric, it can be shown that in general it is not a Euclidean metric, except with particular sets of data.

When $\rho$ is estimated, it is no longer true that $m_{XY}^2 = m_{YX}^2$ and therefore $m_{XY}$ is not then a metric. This asymmetry is untidy and one way of overcoming it occurs as a special case of generalized Procrustes analysis discussed in section 17.12.

The usual interest in Procrustes analysis is to examine the relationships between the two configurations. Occasionally the interest focusses more on the criterion value $f_3(\delta, \varepsilon)$. One such use is described in section 17.12. Davies (1978) has established the asymptotic mean and variance of the criterion when $\delta_{ij}$, and $\varepsilon_{ij}$ are two sets of Mahalanobis distances based on $n$ parent multinormal distributions. Sibson (1979) has investigated the robustness of classical scaling where given squared-distances $d_{ij}^2$ arising from a configuration $\mathbf{Y}$ in $r$ dimensions, are perturbed to become $d_{ij}^2 + \varepsilon f_{ij}$ with classical scaling solution $\mathbf{X}$ in $r$ dimensions. The two configurations can be compared by a Procrustes analysis to give:

$$m_{xy}^2 = \tfrac{1}{4}\varepsilon^2 \Sigma \frac{(\mathbf{e}'\mathbf{Fe}_k)^2}{\mu_j - \mu_k} + \text{terms in } \varepsilon^3 \text{ and higher powers,}$$

where $\mathbf{e}_j$ is the $j$th eigenvector and $\mu_j$ the $j$th eigenvalue of the matrix $\mathbf{M}_0^*$ defined in section 17.6. Summation is over all different pairs $j, k \le r$. That no term linear in $\varepsilon$ occurs suggests that classical scaling is a fairly robust method.

## 17.11.  THE GENERAL COMPARISON OF SCALINGS

Section 17.10 introduced this topic and developed the theme of orthogonal Procrustes analysis. The criterion $f_3$ had the form $\|\mathbf{Y} - \mathbf{XH}\|$ where $\mathbf{H}$ is orthogonal and $\|\mathbf{Z}\|$ is a notation for Trace $(\mathbf{Z}^T\mathbf{Z})$. In this section several other criteria of the same general form will be discussed, but where $\mathbf{H}$ is replaced by matrices with different kinds of restriction. Having abandoned orthogonal matrices, distance properties are not retained after transformation, so that interest lies in the coordinate matrices themselves. To emphasize this the notation $f(\mathbf{X}, \mathbf{Y})$ will be used. Usually $\mathbf{Z}$ can be interpreted as a matrix of residuals. If translation of $\mathbf{X}$ is also required then $\mathbf{X}$ and $\mathbf{Y}$ should both be transformed to have the same centroid, conveniently taken to be the origin, just as with orthogonal Procrustes analysis.

The first such criterion was investigated by Hurley and Cattell (1962) who required $\mathbf{C}$ to minimize

$$f_3(\mathbf{Y}, \mathbf{X}) = \|\mathbf{Y} - \mathbf{XC}\|.$$

These authors coined the term Procrustes analysis. The solution is given by $\mathbf{C} = (\mathbf{X}^T\mathbf{X})^{-1}\mathbf{X}^T\mathbf{Y}$, a result very close indeed to that of classical multiple regression.

Interest in criteria of this kind began in factor analysis where there is an extensive literature concerning the possibility of transforming factor-loadings (which may be expressed as a set of coordinates) into a simpler structure. Cattell and Khanna (1977) give a recent review and ten Berge (1977) gives a very comprehensive account establishing a taxonomy of the procedures that classifies them into 36 types—fortunately many seem unlikely to have applications.

Criterion $f_4(\cdot)$ imposes no restriction on $\mathbf{C}$, except of course that its dimensions match those of $\mathbf{Y}(n \times r)$ and $\mathbf{X}(n \times s)$. Thus $\mathbf{C}$ must have dimensions $(s \times r)$.

Browne (1967) discussed the oblique Procrustes problem which is to find $\mathbf{C}$ that minimizes

$$f_5(\mathbf{Y}, \mathbf{X}) = \|\mathbf{Y} - \mathbf{XC}\|$$

where $\mathbf{C}$ is now restricted to satisfy diag $(\mathbf{C}^T\mathbf{C}) = \mathbf{I}$. This restriction implies that the $i$th column of $\mathbf{C}$ can be regarded as giving the direction cosines of the $i$th of $r$ axes relative to $s$ orthogonal axes. The rows of $\mathbf{X}$ give the coordinates of $n$ points referred to the orthogonal axes and $\mathbf{XC}$ are the projections of these points onto the $r$ axes which will, in general, be obliquely inclined to one another. Criterion $f_5(\cdot)$ expresses that the directions of the oblique axes are to be chosen so that projections match values in a given target matrix $\mathbf{Y}$ as closely as possible, in a least-squares sense. The geometry is illustrated in Figure 17.11.1 which shows a point in $s = 2$ dimension being projected onto oblique axes in $r = 3$ dimensions. It will be made clear below that criterion $f_5(\cdot)$ is *not* a sum of suqares of distances between given and

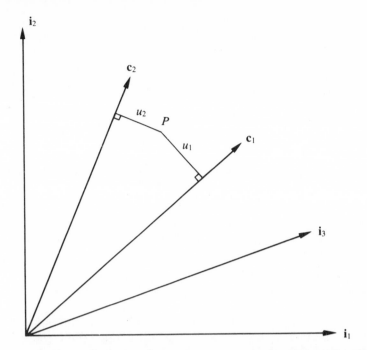

Figure 17.11.1: The point $P$ has coordinates $(x_1, x_2, x_3)$ relative to the orthogonal axes $(\mathbf{i}_1, \mathbf{i}_2, \mathbf{i}_3)$ and coordinates $(u_1, u_2)$ relative to the oblique axes with direction cosines $\mathbf{c}_1$ and $\mathbf{c}_2$ (with respect to the orthogonal axes). The values $u_1$, $u_2$ are measured as the distances of $P$ from the oblique axes. The two sets of coordinates are related by $\mathbf{u} = \mathbf{xC}$ where (in this case) $\mathbf{C}$ is the $3 \times 2$ matrix whose columns are the direction cosines $\mathbf{c}_1$ and $\mathbf{c}_2$ and hence diag $(\mathbf{C}^T\mathbf{C}) = \mathbf{I}$.

transformed points in the oblique space—the criterion $f_8(\,\cdot\,)$ discussed below handles that situation.

To minimize $f_5(\,\cdot\,)$ we first note that we can minimize with respect to the $i$th column $\mathbf{c}_i$ of $\mathbf{C}$, independently of the other columns. That is we fit $\mathbf{Y}_i$, the $i$th column of $\mathbf{Y}$, to $\mathbf{Xc}_i$, where $\mathbf{c}_i^T\mathbf{c}_i = 1$ and can ignore the other columns of $\mathbf{C}$. We shall write $\mathbf{y}$ for $\mathbf{Y}_i$, and $\mathbf{c}$ for $\mathbf{c}_i$ so have to minimize $\|\mathbf{y} - \mathbf{Xc}\|$ which is the same as maximizing $2\mathbf{y}^T\mathbf{Xc} - \mathbf{c}^T\mathbf{X}^T\mathbf{Xc}$ subject to $\mathbf{c}^T\mathbf{c} = 1$. Differentiating with respect to $\mathbf{c}$ with the Lagrangian term $\lambda(\mathbf{c}^T\mathbf{c} - 1)$ gives

$$(\mathbf{X}^T\mathbf{X} - \lambda\mathbf{I})\mathbf{c} = \mathbf{X}^T\mathbf{y}$$

so that $\mathbf{c}^T\mathbf{c} = 1$ gives the polynomial in $\lambda$

$$\mathbf{y}^T\mathbf{X}(\mathbf{X}^T\mathbf{X} - \lambda\mathbf{I})^{-2}\mathbf{X}^T\mathbf{y} = 1.$$

To simplify this polynomial, express $\mathbf{X}^T\mathbf{X}$ in its eigenvector decomposition $\mathbf{X}^T\mathbf{X} = \mathbf{Q}^T\boldsymbol{\mu}\mathbf{Q}$, ($\mathbf{Q}$ orthogonal, $\boldsymbol{\mu}$ diagonal) which on substitution into the previous equation gives:

$$\mathbf{y}^T\mathbf{X}\mathbf{Q}^T(\boldsymbol{\mu} - \lambda\mathbf{I})^{-2}\mathbf{Q}\mathbf{X}^T\mathbf{y} = 1.$$

Writing the known vector $\mathbf{QX}^T\mathbf{y}$ as $\mathbf{z}$, this polynomial becomes:

$$\sum_{i=1}^{r} \frac{z_i^2}{(\mu_i - \lambda)^2} = 1$$

which has degree $2r$ and hence $2r$ roots [see I: § 14.5]. For any root $\lambda_j$ the residual sum-of-squares is

$$e_j = y^T y + \lambda_j - \sum_{i=1}^{r} \frac{z_i^2}{(\mu_i - \lambda_j)}.$$

Hence if $e_j$ and $e_k$ are the residuals sum-of-squares for two roots we have

$$e_j - e_k = (\lambda_j - \lambda_k)\left[ 1 - \sum_{i=1}^{r} \frac{z_i^2}{(\mu_i - \lambda_j)(\mu_i - \lambda_k)} \right].$$

Writing $z_i/(\mu_i - \lambda_j) = u_i$, and $z_i/(\mu_i - \lambda_k) = v_i$ we have that $\Sigma u_i^2 = \Sigma v_i^2 = 1$ so that from Cauchy's inequality,

$$\sum \frac{z_i}{(\mu_i - \lambda_j)(\mu_i - \lambda_k)} = \sum u_i v_i \le 1 \text{ [see IV, § 21.2.4].}$$

Hence if $\lambda_j < \lambda_k$ it follows that $e_j < e_k$, showing that to minimize $f_5(\cdot)$ we require the *smallest* root of the polynomial. Browne (1967) shows that $\lambda_0$ the smallest root satisfies $\lambda_0 < \mu_1$, where $\mu_1$ is the smallest eigenvalue of $\mathbf{X}^T\mathbf{X}$ and recommends a Newton–Raphson iterative technique [see III, § 5.4.1] to solve the polynomial. Cramer (1974) suggests an alternative approach and discusses the problem that arises when the minimum root is not unique and ten Berge and Nevels (1977) give a definitive solution for that case. If a lower bound can be found for $\lambda_0$ then a bisection technique can be used to home-in on $\lambda_0$. I have found this a perfectly satisfactory method, using the following lower bound. Let $z = \max(z_i)$ and let $\lambda = \mu_i - z/r$ then we have

$$\sum_{i=1}^{r} \frac{z_i^2}{(\mu_i - \lambda)^2} - 1 = \sum \frac{z_i^2}{(\mu_i - \mu_1 + z/r)^2} - 1$$

$$\le \sum \frac{z_i^2}{rz^2} - 1 \le 0.$$

Hence $\lambda_0 > \mu_1 - z/r$ gives the required lower bound.

A special form of Browne's oblique Procrustes problem is to minimize $f_6(\mathbf{Y}, \mathbf{X}) = \|\mathbf{Y} - \mathbf{XC}\|$ subject to $\mathbf{C}$ being orthonormal. That is $\mathbf{C}'\mathbf{C} = \mathbf{I}$ so that what were oblique axes must now be orthogonal. When $\mathbf{Y}$ and $\mathbf{X}$ both occupy $r$ dimensions, $\mathbf{C}$ is an orthogonal matrix and we have the orthogonal Procrustes problem discussed and solved in section 17.10. Hence it is only when $\mathbf{Y}$ has dimensions $(n \times r)$ and $\mathbf{X}$ dimensions $(n \times s)$ that there is a problem. $\mathbf{C}$ then has dimensions $(s \times r)$ and we assume $s > r$. Even when $s > r$ the orthonormal estimation of criterion $f_6(\cdot)$ may be better replaced by rotating $\mathbf{Y}$ in the higher-dimensional space of $\mathbf{X}$, which is trivially done by adding $s - r$ zero columns to $\mathbf{Y}$ and using the methods of section 17.10. Green and Gower (1981) discuss situations when $f_6(\,\cdot\,)$ is definitely required. One geometrical interpretation of $f_6(\,\cdot\,)$ is that $\mathbf{X}$ is first orthogonally projected onto the smaller $r$ dimensional space and then these projections are rotated, still in the smaller space, to fit $\mathbf{Y}$. Alternatively we may regard $\mathbf{Y}$ as first rotated in the higher-dimensional space, followed by a projection onto the lower-dimensional space. When $r = 1$ this problem is identical to the oblique Procrustes problem so it is no surprise that a similar equation has to be solved:

$$\mathbf{X}^T\mathbf{X}\mathbf{C} - \mathbf{C}\boldsymbol{\Lambda} = \mathbf{X}^T\mathbf{Y}$$

where now $\boldsymbol{\Lambda}$ is a whole $(r \times r)$ symmetric matrix of Lagrange multipliers. Previously $\boldsymbol{\Lambda}$ was a scalar, and then we have $\mathbf{C}\boldsymbol{\Lambda} = \boldsymbol{\Lambda}\mathbf{C}$ allowing the equation to be solved explicitly for $\mathbf{C}$ in terms of $\lambda$ and hence ultimately leading to a practical solution. This is no longer possible, but the equation is still linear in the elements of $\mathbf{C}$ and so in principle can be solved for $\mathbf{C}$, whence the orthonormality condition $\mathbf{C}'\mathbf{C} = \mathbf{I}$ gives $\frac{1}{2}r(r+1)$ polynomial equations for the $\frac{1}{2}r(r+1)$ elements of $\boldsymbol{\Lambda}$. To solve a set of simultaneous polynomial equations is a difficult numerical problem which is further complicated by not knowing which set of roots corresponds to a minimum sum of squares of $f_6(\,\cdot\,)$. Thus this approach is impracticable. Green and Gower (1981) suggest the following iterative algorithm which seems to work well in practice, although they were unable to show that it must converge to the global minimum.

1. Add $s - r$ zero, or arbitrary, columns to $\mathbf{Y}$.
2. Use the orthogonal Procrustes procedure to fit $\mathbf{X}$ to the new $\mathbf{Y}$.
3. Replace the last $s - r$ columns of $\mathbf{Y}$ by the corresponding columns of the rotated $\mathbf{X}$, and repeat step 2.

This process improves $f_6(\,\cdot\,)$ at each step and must converge. The last $s - r$ columns of $\mathbf{Y}$ and $\mathbf{X}$ will eventually agree and contribute nothing to $f_6(\,\cdot\,)$ and hence only the first $r$ columns of the full orthogonal matrix are required and these estimate the orthonormal matrix.

An interesting link between the orthonormal and oblique Procrustes problems follows by writing $\boldsymbol{\Lambda} = \mathbf{Q}\boldsymbol{\Delta}\mathbf{Q}^T$ the eigenvector decomposition form. The basic equation then may be written:

$$\mathbf{X}^T\mathbf{X}(\mathbf{C}\mathbf{Q}) - (\mathbf{C}\mathbf{Q})\boldsymbol{\Delta} = \mathbf{X}^T\mathbf{Y}\mathbf{Q}$$

showing that if $\mathbf{Q}$ were known, the $i$th column of the orthonormal matrix $\mathbf{C}\mathbf{Q}$, together with $\delta_i$, the $i$th diagonal element of $\boldsymbol{\Delta}$, would comprise an oblique Procrustes solution to fitting $\mathbf{X}$ to $\mathbf{Y}\mathbf{q}_i$. Thus there exists an orthogonal rotation matrix $\mathbf{Q}$ which rotates $\mathbf{Y}$ in the smaller space (i.e. it makes no real change to the problem) to a position in which the oblique Procrustes problem gives an orthonormal matrix that can also minimize $f_6(\,\cdot\,)$. Attempts to base an algorithm to minimize $f_6(\,\cdot\,)$ on iterations to determine $\mathbf{Q}$ have not been successful.

Figure 17.11.1 has illustrated how the coordinates of a point relative to oblique axes can be expressed in terms of the shortest distances of that point from the axes. This is just one way of handling oblique axes and it is not the usual way, which is illustrated in Figure 17.11.2, in which coordinate distances

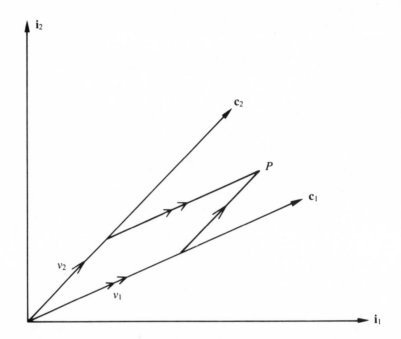

Figure 17.11.2: The point $P$ has coordinates $(x_1, x_2)$ relative to the orthogonal axes $(\mathbf{i}_1, \mathbf{i}_2)$ and the coordinates $(v_1, v_2)$ relative to the oblique axes with direction cosines $\mathbf{c}_1$ and $\mathbf{c}_2$. The two sets of coordinates are related by $\mathbf{x} = \mathbf{v}\mathbf{C}^T$.

are not measured orthogonally to the oblique axes but parallel to them. When the oblique axes are themselves orthogonal, both methods are equivalent to the usual cartesian coordinates. For the projection method the oblique-axes coordinates $\mathbf{u}$ are related to the orthogonal coordinates by

$$\mathbf{u} = \mathbf{x}\mathbf{C}$$

and for the 'parallel' method the relationship is

$$\mathbf{x} = \mathbf{v}\mathbf{C}^T$$

and hence

$$\mathbf{u} = \mathbf{v}\mathbf{C}^T\mathbf{C}$$

relates the two kinds of oblique axes representation. The distance $d_{ij}$ between two points with coordinates $\mathbf{x}_i, \mathbf{x}_j$ is given by

$$d_{ij}^2 = (\mathbf{x}_i - \mathbf{x}_j)(\mathbf{x}_i - \mathbf{x}_j)^T$$

and hence

$$d_{ij}^2 = (\mathbf{v}_i - \mathbf{v}_j)\mathbf{C}^T\mathbf{C}(\mathbf{v}_i - \mathbf{v}_j)^T = (\mathbf{u}_i - \mathbf{u}_j)(\mathbf{C}^T\mathbf{C})^{-1}(\mathbf{u}_i - \mathbf{u}_j)^T$$

where it has been assumed that $\mathbf{C}^T\mathbf{C}$ is non-singular, which implies in particular that $s \geq r$ as was assumed above for oblique Procrustes analysis. The symmetric matrix $\mathbf{C}^T\mathbf{C}$ gives the cosines of the angles between all pairs of oblique axes, and therefore has unit diagonal.

From the above it is clear that to minimize the sums-of-squares of residual distances between a target matrix $\mathbf{Y}$, whose rows are referred to oblique axes by the 'projection' coordinate system, and a matrix $\mathbf{X}$ transformed to that coordinate system, the criterion to minimize is

$$f_7(\mathbf{Y}, \mathbf{X}) = \text{Trace } (\mathbf{Y} - \mathbf{XC})(\mathbf{C}^T\mathbf{C})^{-1}(\mathbf{Y} - \mathbf{XC})^T.$$

This variant of Browne's problem does not seem to have been studied. When $\mathbf{C}$ is orthonormal $f_7(\cdot)$ reduces to $f_6(\cdot)$. When $\mathbf{C}$ is square and non-singular it simplifies to $\|\mathbf{YC}^{-1} - \mathbf{X}\|$, a problem very similar to one discussed by Gruvaeus (1970) except that he imposes the constraint diag $(\mathbf{CC}^T) = \mathbf{I}$ rather than diag $(\mathbf{C}^T\mathbf{C}) = \mathbf{I}$. The corresponding problem with the 'parallel' coordinate system is to minimize

$$f_8(\mathbf{Y}, \mathbf{X}) = \text{Trace } (\mathbf{Y} - \mathbf{XC}^{T-1})\mathbf{C}^T\mathbf{C}(\mathbf{Y} - \mathbf{XC}^{T-1})^T.$$

This is discussed by Browne and Kristof (1969). Criteria of these kinds are usually discussed in a factor-rotation context and it is clear that at most they have only a tenuous relevance to comparing multidimensional scalings. Criterion $f_8(\cdot)$, and probably also $f_7(\cdot)$, requires iterative methods for minimization.

The weighting matrices $(\mathbf{C}^T\mathbf{C})^{-1}$ and $(\mathbf{C}^T\mathbf{C})$ may be replaced by other forms of weighting, often related to models postulating unequal variances amongst residuals. Lissitz, Schönemann and Lingoes (1976) have discussed the weighted form of the orthogonal Procrustes criterion Tr $(\mathbf{Y} - \mathbf{XH})\mathbf{D}^2(\mathbf{Y} - \mathbf{XH})^T$ and Tr $(\mathbf{Y} - \mathbf{XH})^T\mathbf{D}^2(\mathbf{Y} - \mathbf{XH})$ to be minimized for orthogonal $\mathbf{H}$ and given $\mathbf{D}$. Note that the terms in the product no longer commute as in the unweighted case. The second problem is trivial because $\mathbf{D}$ is known and hence $\mathbf{Y}$ and $\mathbf{X}$ may be replaced by $\mathbf{DY}$ and $\mathbf{DX}$ and we may proceed as before. The first problem is much more difficult and leads to equations very like those described by Green and Gower (1981), see above. Lissitz *et al.* note that by replacing the condition $\mathbf{H}$ be orthogonal by requiring $\mathbf{HD}$ to be orthogonal, the difficulties vanish, but this hardly solves the original problem.

A problem which conceivably occurs in a multidimensional-scaling context is when $\mathbf{Y}$ and $\mathbf{X}$ may refer to the same (or similar) objects but in different orders. Thus we may wish to find a permutation matrix $\mathbf{P}$ that permutes the rows of $\mathbf{X}$ to match the rows of $\mathbf{Y}$. Thus we wish to minimize

$$f_9(\mathbf{Y}, \mathbf{X}) = \|\mathbf{Y} - \mathbf{PX}\|.$$

A permutation matrix is a special orthogonal matrix all of whose elements are either zero or unity. There is one and only one unit in every row and column. This $f_9(\,\cdot\,)$ is equivalent to maximizing Trace $(\mathbf{PXY}^T)$ which is a linear function of the elements of $\mathbf{P}$. Now permutation matrices are also a special case of doubly-stochastic matrices, that is matrices of non-negative elements all of whose rows and columns sum to unity. Thus we may consider the following problem:

Maximize Trace $(\mathbf{PXY}^T)$ subject to

$$\left.\begin{array}{c} P_{i.} = 1 \\ P_{.i} = 1 \\ P_{ij} \geq 0 \end{array}\right\} \qquad i, j = 1, 2, \ldots, n.$$

Thus we have to maximize a linear function subject to linear constraints and this is a linear programming problem. The maximum must occur at a vertex of the feasible region bounded by the constraints. But such vertices must correspond to permutation matrices, so giving the required solution.

In practice one is likely to require an orthogonal rotation as well as a permutation matrix. That is we may wish to minimize:

$$f_{10}(\mathbf{Y}, \mathbf{X}) = \|\mathbf{Y} - \mathbf{PXH}\|.$$

This is equivalent to maximizing Trace $(\mathbf{PXHy}^T)$. The solution to this is not known but perhaps an alternating iteration would work, first fixing $\mathbf{P}$ and determining $\mathbf{H}$ as in section 17.10, then fixing $\mathbf{H}$ and determining $\mathbf{P}$ by linear programming as above, and repeating the process until it converges. Each step of the iteration will reduce the criterion $f_{10}(\,\cdot\,)$ so convergence, not necessarily to the global optimum, is assured.

Two-sided Procrustes problems also get some discussion in the literature but here again the relevance to multidimensional scaling is obscure. The main results are listed for their interest and because there is a link with the permutation problem discussed above. The most simple problem of this kind is to minimize

$$f_{11}(\mathbf{Y}, \mathbf{X}) = \|\mathbf{Y} - \mathbf{T}^T\mathbf{XT}\|$$

where $\mathbf{Y}$ and $\mathbf{X}$ are both symmetric and $\mathbf{T}$ is orthogonal. Schönemann (1968) showed that if $\mathbf{Y}$ has eigenvector decomposition $\mathbf{U}^T\mathbf{\Lambda}^Y\mathbf{U}$ and $\mathbf{X}$ has decomposition $\mathbf{V}^T\mathbf{\Lambda}^X\mathbf{V}$, then the estimate of $\mathbf{T}$ is $\mathbf{V}^T\mathbf{U}$. That is $\mathbf{Y} = \mathbf{U}^T\mathbf{\Lambda}^Y\mathbf{U}$ is fitted by $\mathbf{U}^T\mathbf{\Lambda}^X\mathbf{U}$. Effectively the axes of the two matrices are lined up, while the axes lengths are unchanged. Presumably fitting a scaling factor $\rho$ would entail a scaling of the axes of $\mathbf{X}$ so that they were comparable to those of $\mathbf{Y}$. The general two-sided Procrustes problem, also discussed by Schonemann (1968), is more difficult. We have to minimize

$$f_{12}(\mathbf{Y}, \mathbf{X}) = \|\mathbf{Y} - \mathbf{T}^T\mathbf{XS}\|$$

where $\mathbf{Y}$ and $\mathbf{X}$ are square and $\mathbf{T}$ and $\mathbf{S}$ are orthogonal. Starting with the singular-value decompositions

$$\mathbf{Y} = \mathbf{P}\mathbf{\Gamma}^Y \mathbf{Q}^T \quad \text{and} \quad \mathbf{X} = \mathbf{U}\mathbf{\Gamma}^X \mathbf{V}^T$$

leads to the least-squares fit to $\mathbf{Y}$ of $\mathbf{P}\mathbf{\Gamma}^X \mathbf{Q}^T$, i.e. $\mathbf{S} = \mathbf{V}\mathbf{Q}^T$ and $\mathbf{T} = \mathbf{U}\mathbf{P}^T$. The complication comes in that the elements of $\mathbf{\Gamma}^Y$ and $\mathbf{\Gamma}^X$ can be permuted arbitrarily and have arbitrary signs attached. Provided parallel changes are made in the orthogonal matrices, the decompositions are invariant. It turns out that there are many possible orthogonal matrices $\mathbf{T}, \mathbf{S}$ that minimize $f_{12}(\cdot)$ but they all give the same fit to $\mathbf{Y}$.

The link between the permutation problem and two-sided Procrustes problems is clear by examining the forms for $f_{10}(\cdot)$ and $f_{12}(\cdot)$. The only difference is that $\mathbf{T}$ is a general orthogonal matrix while $\mathbf{P}$ is a permutation matrix, a special case of an orthogonal matrix. This suggests an approximate non-iterative solution to minimizing $f_{10}(\cdot)$:

   (i) Find $\mathbf{T}$ by the Schönemann (1968) method.
   (ii) Estimate $\mathbf{P}$ as the nearest permutation matrix to $\mathbf{T}^T$.

Step (ii) is equivalent to minimizing $\|\mathbf{T}^T - \mathbf{P} . \mathbf{I}\|$ which may be done by setting $\mathbf{Y} = \mathbf{T}^T$ and $\mathbf{X} = \mathbf{I}$ in $f_9(\cdot)$ and using its linear programming solution.

Of the criteria $f_p(\mathbf{Y}, \mathbf{X})$ discussed in this section, those with $p = 4, 5, 9, 11$ and 12 have analytical solutions, while the remainder rely on iterative algorithms.

## 17.12.   THREE-WAY SCALING

Suppose data in the form of $m$ different distance matrices $\mathbf{D}_1, \mathbf{D}_2, \ldots, \mathbf{D}_m$ each of order $n$ are available and suppose we write $d_{ijk}$ to denote the distance between the $i$th and $j$th objects in the $k$th matrix. Thus a third suffix now enters into consideration and multidimensional scaling methods for such data are generally referred to as three-way or three-mode scaling methods. The rows and columns of all matrices are assumed to refer to the same $n$ things and the object of the methods discussed in this section is to assess to what extent the different matrices agree with one another. Typically the suffix $k$ will refer to different individuals, or occasions, or possibly different methods of analysis.

As with previous sections there is some ambivalence as to whether to use the matrices $\mathbf{D}_k$ or coordinates $\mathbf{X}_k$ arising from ordination of $\mathbf{D}_k$. Both approaches are used in the following. Which is dominant will be clear from context.

One obvious approach is to use one of the methods discussed in section 17.11 to compare all $\frac{1}{2}m(m-1)$ pairs of matrices forming a matrix $\mathbf{M}$ of order $m$ whose entries are the residual sums-of-squares $f(\mathbf{X}_i, \mathbf{X}_j)$ of the chosen criterion. With many of the criteria $\mathbf{M}$ will not be symmetric but it will have zero-diagonals. With the orthogonal Procrustes statistic $\mathbf{M}$ will be symmetric and, as already mentioned, its values are metrics. Having formed $\mathbf{M}$ it may be analysed by the methods of earlier sections to give an ordination in which near points will refer to pairs of matrices with similar values.

Analyses of the kind just discussed have been used but methods which produce some kind of average configuration, together with an indication of how the matrices $\mathbf{D}_k$(or $\mathbf{X}_k$) diverge from the average, are more popular.

One possibility is to generalize the orthogonal Procrustes approach and seek a set of orthogonal rotations such that $\mathbf{X}_1\mathbf{H}_1, \mathbf{X}_2\mathbf{H}_2, \ldots, \mathbf{X}_m\mathbf{H}_m$ are in some kind of optimal agreement. After these rotations let the $k$th points in the $m$ sets be denoted by $(P_{1k}, P_{2k}, \ldots, P_{mk})$ and let the centroid of these points be $G_k$. Then the optimal fit may be defined as that which minimizes the sums of the squares of the $m$ distances of the points in the $i$th set from their own centroid, summed over all $n$ settings of $k$. The geometry is illustrated in Figure 17.12.1 for three sets of four points. Denoting by $\mathbf{Y}$ the set of coordinates for the $n$ centroids then

$$\mathbf{Y} = \frac{1}{m} \sum_{k=1}^{m} \mathbf{X}_k\mathbf{H}_k.$$

The criterion is to choose the $\mathbf{H}_i$ so that

$$f(\mathbf{X}_1, \mathbf{X}_2, \ldots, \mathbf{X}_m) = \sum \|\mathbf{Y} - \mathbf{X}_k\mathbf{H}_k\|$$

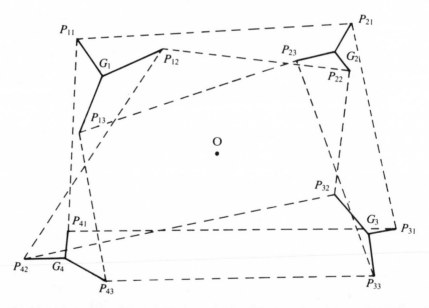

Figure 17.12.1: Generalized procrustes analysis. Three sets of four points have coordinates $P_{i1}, P_{i2}, P_{i3}$ ($i = 1, 2, 3, 4$) with a common centroid O. The centroid of $P_{1k}, P_{2k}$ and $P_{3k}$ is $G_k$. Optimal fit (not shown here) is defined to be that given by the orientations which minimize the sum of squares of residuals $\sum_{i=1}^{4} \sum_{j=1}^{3} (P_{ij}G_i^2)$ marked in the diagram.

is minimum. This is equivalent to minimizing:

$$\Sigma\|\mathbf{X}_k\mathbf{H}_k - \mathbf{X}_l\mathbf{H}_l\|$$

summed over all pairs $k, l$. If $\mathbf{Y}$ were known, we would merely have to rotate each $\mathbf{X}_k$ to optimal fit with $\mathbf{Y}$, exactly as in section 17.10. But $\mathbf{Y}$ is itself dependent on knowing the $\mathbf{H}_k$ so we have to proceed iteratively starting with an initial centroid configuration $\mathbf{Y}$, rotating each $\mathbf{X}_k$ to fit $\mathbf{Y}$, determining orthogonal matrices $\mathbf{H}_k$ which, in turn, generate a new centroid. Details are given in Gower (1975) who termed this kind of analysis *generalized Procrustes analysis* and gave an example. Translation of each set is easily accommodated by placing the centroid of each $\mathbf{X}_k$, and hence also of $\mathbf{Y}$, at the origin. Also each set may be scaled by a factor $\rho_i$. Thus we have to minimize $\Sigma\|\mathbf{Y} - \rho_i\mathbf{X}_k\mathbf{H}_k\|$ where $\mathbf{Y} = (1/m)\Sigma\rho_k\mathbf{X}_k\mathbf{H}_k$. The trivial solution $\rho_i = 0$ may be avoided by making the constraint $\Sigma\rho_k^2\|Xk\| = \Sigma\|\mathbf{X}_k\|$ which preserves the total size of the configurations after scaling. The estimation of the $\mathbf{H}_k$ proceeds as before except that $\mathbf{X}_k$ is everywhere replaced by $\rho_k\mathbf{X}_k$. Scaling is given by:

$$\rho_k^2 = \mathrm{Tr}\,(\rho_k\mathbf{X}_k\mathbf{H}_k\mathbf{Y})\Sigma\|\mathbf{X}_k\|/m\|\mathbf{X}_k\|\,\|\mathbf{Y}_k\|$$

which can be used to update iterative estimates. Alternatively ten Berge (1977) shows that at each stage of iteration the vector $\boldsymbol{\rho}$ of all scaling-factors can be expressed as an eigenvector of a certain matrix, which method gives theoretically better iterative improvement, at the expense of more calculation.

In the above discussion of generalized Procrustes analysis, the position of the whole configuration (i.e. $\mathbf{Y}$, $\mathbf{X}_k\mathbf{H}_k$, $k = 1, 2, \ldots, m$) may itself be rotated without affecting distances. A unique representation is conveniently obtained by referring $\mathbf{Y}$ to its principal axes and adjusting all the rotated $\mathbf{X}_k$ accordingly.

To compare the two configurations set $m = 2$. By fitting $\mathbf{X}_1$ and $\mathbf{X}_2$ to their joint centroid-configuration there is no possibility of the asymmetry $m_{12}^2 \neq m_{21}^2$ found in section 17.10 for scaled orthogonal Procrustes analysis; the problem is overcome by avoiding it. Gower (1975) showed that if $\mathbf{X}_1$ and $\mathbf{X}_2$ are first standardized to have equal (say unit) sum of squares then (i) fitting $\mathbf{X}_1$ and $\mathbf{X}_2$ to their joint centroid, (ii) fitting $\mathbf{X}_1$ to $\mathbf{X}_2$ with scaling and (iii) fitting $\mathbf{X}_2$ and $\mathbf{X}_1$ with scaling, all give the same value of $m_{12}^2$. With this standardization we find $\hat{\rho}_1 = \hat{\rho}_2 = 1$. All this adds up to a strong recommendation to standardize, which is essential when $\mathbf{X}_1$ and $\mathbf{X}_2$ are measured on incommensurable scales with no conceivable interest in any scaling factor $\rho$.

In principle, the generalized Procrustes ideas could be formulated in terms of the other criteria discussed in section 17.10. We could retain the orthogonal rotations $\mathbf{H}_k$ and centroid configuration $\mathbf{Y}$ but, instead of working in terms of squared distances between like points, we could use any criterion of likeness and proceed as follows. The average value of $d_{ijk}$ over the $m$ configurations is given by $D_{ij} = (1/m)\sum_{k=1}^{m} d_{ijk}$ and let the distances in $\mathbf{Y}$ be $\delta_{ij}$. Then we can minimize $f_p(\delta_{ij}, D_{ij})$ where $p$ takes on the suffix of the chosen criterion. An algorithm would have to iterate on the $\mathbf{H}_k$ to find the centroid configuration $\mathbf{Y}$ that minimized $f_p(\cdot)$. What might be the difficulties of this approach seem not to have been investigated. The generalized Procrustes procedure is implemented in the program PINDIS (see e.g. Borg (1977)). However, this program has many other features including the possibility of fitting models of the individual scaling type discussed below.

The most used class of three-way scaling models is that originally described by Carroll and Chang (1970), which is termed Individual Differences Scaling or INDSCAL, the name of the related computer program. This model expresses the fitted values in the form

$$\delta_{ijk}^2 = (\mathbf{x}_i - \mathbf{x}_j)\mathbf{W}_k(\mathbf{x}_i - \mathbf{x}_j)^T$$

where $\mathbf{x}_i$ is the $i$th row of an $n \times r$ matrix $\mathbf{X}$ and $\mathbf{W}_k$ is a diagonal matrix of positive weights. The matrix $\mathbf{X}$ is taken as an average configuration for all $m$ matrices $\mathbf{D}_k$ and its values may be plotted in the usual way for an ordination. $\mathbf{X}$ is often termed the *group average* configuration. The quantities $w_{ik}$ (i.e. the $i$th diagonal value of $\mathbf{W}_k$) are interpreted as how the $k$th individual weights the $i$th coordinate axis of $\mathbf{X}$. The coordinates $(w_{1k}, w_{2k}, \ldots, w_{rk})$ may be plotted for each of the $k = 1, 2, \ldots, m$ individuals. When $r = 2$, the usual case, the points lying near the 45-degree line represent individuals who give equal weight to both axes of the group average. Points lying to one side or other of this line give greater emphasis to one of the axes.

How to fit the individual scaling model has given rise to almost as many criteria as for ordinary multidimensional scaling, and both metric and non-metric approaches have been proposed. Carroll and Chang's (1970) original method was based on observing that in classical metric scaling the doubly-centred matrix $\mathbf{M}_0^*$ (see § 17.6) is decomposed into the scalar product form $\mathbf{X}\mathbf{X}^T$. Similarly the doubly-centred form of $\mathbf{D}_k$ would give $\mathbf{D}_k^* = \mathbf{X}\mathbf{W}\mathbf{X}^T$. We are now assuming that the elements of $\mathbf{D}_k$ have, or have been transformed into, elements $-\frac{1}{2}d_{ijk}^2$. Thus Carroll and Chang (1971) looked for a least-squares solution to minimizing

$$\sum_{k=1}^m \|\mathbf{D}_k^* - \mathbf{W}\mathbf{X}\mathbf{W}^T\|,$$

a criterion termed STRAIN. Note that this use of centred matrices $\mathbf{D}_k^*$ precludes any treatment of missing data. This problem is essentially the same as that considered by Harshman (1972) in his PARAFAC model. Carroll and Chang's solution was to consider the more general problem of minimizing

$$\sum_{ijk} \sum_{s=1}^r (y_{ijk} - a_{is}b_{js}c_{ks})^2$$

over the parameters $a_{is}, b_{js}, c_{ks}$. This may be viewed as a generalization to three dimensions of the problem, solved by the Eckart–Young Theorem, of finding the best rank $r$ fit to a matrix. Carroll and Chang developed an algorithmic solution (termed CANDECOMP) which proceeds iteratively in a simple manner. Initial estimates of $b_{js}$ and $c_{ks}$ are assumed and the values $a_{is}$

are determined using the multiple regression formulae in the usual way. Then $a_{is}$, $c_{ks}$ are held fixed and the $b_{js}$ estimated. Finally $a_{is}$, $b_{js}$ are held fixed and the $c_{ks}$ are estimated. The whole process is repeated as often as necessary, each stage reducing the residual sum of squares. The CANDECOMP procedure can be used to minimize the Individual Scaling criterion by setting

$$a_{is} = x_{is}, \quad b_{js} = x_{js} \quad \text{and} \quad c_{ks} = w_{ks}.$$

This introduces obvious constraints on the general model but these lead to no difficulties. There is no guarantee that the global optimum is reached, although there are grounds in believing that it usually is. An important property of this form of individual scaling is that the group average configuration $\mathbf{X}$ is unique. If its axes are rotated, distance is unchanged but the weights are no longer correct. The uniqueness property is often stressed as an advantage of this method.

The ALSCAL system (see Takane, Young and de Leeuw (1977)) fits the individual scaling model using a different criterion of goodness-of-fit and hence a different computational algorithm from that used by INDSCAL. This criterion (SSTRESS) is to minimize

$$\Sigma (d_{ijk}^2 - \delta_{ij}^2)^2$$

clearly of the least-squares-squared scaling family (section 7). The ALSCAL algorithm is said to be very efficient and also generalizes the situation covered by INDSCAL. It will, for example, allow missing values, replicated information on the same individual(s), admits a wider variety of data and has a non-metric facility.

Another similar, but simpler, model is used in SMACOF-I (Heiser and de Leeuw (1979)) who minimize

$$\Sigma w_{ijk}(d_{ijk} - \delta_{ij})^2$$

where $\delta_{ij}$ are distances in the fitted average configuration and $w_{ijk}$ are *given* weights, similar to those described in section 17.7, that should not be confused with the weights $w_{ks}$ used by INDSCAL and ALSCAL which have to be estimated. The usual breakdown of a sum of squares shows that

$$\sum_{i,j,k} w_{ijk}(d_{ijk} - \delta_{ij})^2 = \sum_{i,j,k} w_{ijk}(d_{ijk} - \bar{d}_{ij})^2 + \sum_{i,j} \bar{w}_{ij}(\bar{d}_{ij} - \delta_{ij})^2$$

where

$$\bar{w}_{ij} = \frac{1}{m} \sum_{k=1}^{m} w_{ijk} \quad \text{and} \quad \bar{d}_{ij} = \frac{\sum_{k=1}^{m} w_{ijk} d_{ijk}}{m \bar{w}_{ij}}.$$

Thus the left-hand side is minimized when $\mathbf{X}$ gives rise to distances $\delta_{ij}$ that minimize $\Sigma \bar{w}_{ij}(\bar{d}_{ij} - \delta_{ij})^2$. This is exactly the least-squares scaling problem of section 7 and the methodology discussed there is directly applicable. With this model the group average configuration $\mathbf{X}$ is readily obtained, but information on the nature of the individual differences is limited to the components of the sum-of-squares $\sum_{i,j,k} w_{ijk}(d_{ijk} - \bar{d}_{ij})^2$. Note that in this derivation there is no symmetry assumption that $d_{ijk} = d_{jik}$ so this last sum-of-squares may be partitioned into components measuring symmetry and departures from symmetry.

Carroll and Chang (1972) generalized their Individual Scaling model so that the matrices $\mathbf{W}_k$ can be any positive-definite symmetric matrix. Thus each individual has his own metric space. The method is therefore termed *idiosyncratic scaling* and its associated program IDIOSCAL. Again the group average matrix $\mathbf{X}$ is estimated but now the individual differences are expressed in the symmetric matrices $\mathbf{W}_k$. The simple plot of weights of individual scaling is not available, but comparison of the $m$ matrices $\mathbf{W}_k$ is not difficult in the usual two-dimensional case. When $r > 2$ we begin to meet difficulties whose solution perhaps requires an individual scaling of the matrices $\mathbf{W}_k$!

The three-factor extensions of two-way multiplicative models, briefly mentioned in section 3, provide yet another way of analysing three-way data, now with emphasis on models for observed data $\mathbf{X}_k$ ($k = 1, 2, \ldots, m$).

This section has been able to give only a brief and incomplete survey of a rapidly-expanding set of techniques. It is particularly difficult to distinguish the various models that have been proposed from the criteria (metric and non-metric) used to fit them and the computer programs written to implement the criteria. The role of data in the form of $m$ matrices $\mathbf{X}_k$ or $m$ distance-type matrices $\mathbf{D}_k$ has been mentioned. Even when $\mathbf{D}_k$ is square we have seen that, for SMACOF-1 at least, there is no constraint that it should be symmetric. The precise nature of data becomes blurred when programs accept rectangular parts of a symmetric distance matrix as is appropriate for multidimensional unfolding (§ 17.9). Three-way forms of unfolding have been studied by De Sarbo (1978).

## 17.13.   THE ANALYSIS OF ASYMMETRY

Many of the methods discussed earlier may be used for analysing rectangular matrices. Thus section 17.3 discusses multiplicative models, section 17.5 correspondence analysis and section 17.9 the distance models of multidimensional unfolding. In this section we shall be concerned with the problems of square asymmetric tables. The methods for general rectangular matrices are still applicable but new models become relevant for those square tables which have the special structure that rows and columns are similarly classified. In such cases the rows and columns usually refer to the same entities but in different modes. For example rows might refer to places of immigration and columns to the same places viewed as centres of emigration, or we might have social classes of fathers for one classification and social class of sons for the other classification. With such data one process often governs symmetric aspects and departures from symmetry are governed by another process. The methods so far discussed make no distinction between these aspects and so tend not to be notably successful for analysing data of this type. Thus in this section we discuss models that include symmetric and asymmetric components. The square $n \times n$ data-matrix will be written as $\mathbf{D}$.

A model due to Baker, Young and Takane (1977) expresses the elements of $\mathbf{D}$ as

$$d_{ij}^2 = \sum_{r=1}^{k} w_{ir}(x_{ir} - x_{jr})^2.$$

The coordinates considered as the rows of the $n \times k$ matrix $\mathbf{X}$ can give the usual type of Euclidean plot. The quantities $w_{ir}$, termed weights, determine the asymmetry because $d_{ij}$ has a different weight from that of $d_{ji}$. In a two-dimensional plot ($k = 2$) the weights $w_{i1}$ and $w_{i2}$ may be indicated at the $i$th point as arrowed lines of appropriate lengths in the East/West and North/South directions respectively. This model is in the same family as the individual scaling model of section 17.12.

The above model is basically of distance type. Harshman (1972) has suggested the asymmetric cross-products model

$$\mathbf{D} = \mathbf{X}\mathbf{R}\mathbf{X}^T$$

where $\mathbf{R}$ has dimensions $s \times s$ and is not symmetric, unless $\mathbf{D}$ is. Clearly there is great simplification if $s$ is small but this need not be so. When a good fit with matrices $\mathbf{X}$ and $\mathbf{R}$ is found, equivalent solutions exist for an arbitrary orthogonal matrix $\mathbf{H}$ with $\mathbf{X}$ replaced by $\mathbf{X}\mathbf{H}$ and $\mathbf{R}$ replaced by $\mathbf{H}^T\mathbf{R}\mathbf{H}$. This opens the door to factor–rotation techniques to derive simpler and perhaps more interpretable solutions. The matrix $\mathbf{X}$, or its rotation, may be plotted as a set of $n$ points to give a fit to what may be regarded as the symmetric aspects of $\mathbf{D}$. Because of the cross-product relation, interpretation is in terms of angles between radius vectors, rather than in terms of distances. A variant of Harshman's model has been studied by Chino (1978) who replaces $\mathbf{R}$ by matrices with very special structure. Thus when $k = 2$, $\mathbf{R}$ is replaced by $\alpha\mathbf{I} + \beta\mathbf{S}_2$ where $\mathbf{S}_2$ is the $2 \times 2$ skew symmetric matrix with $\mathbf{S}_{12} = 1$. Writing the two columns of $\mathbf{X}$ as $\mathbf{x}_1$ and $\mathbf{x}_2$ it is clear that $\mathbf{X}\mathbf{S}_2\mathbf{X}^T = \mathbf{x}_1\mathbf{x}_2^T - \mathbf{x}_2\mathbf{x}_1^T$ a form equivalent to the first term of the least-squares method outlined below. In general $\mathbf{S}_k$ is the skew-symmetric matrix with $\mathbf{S}_{ij} = (-1)^{i+j-1}$ when $j > i$. With $k = 3$, any skew-symmetric matrix of order three has rank 2. In fact $\mathbf{S}_3 = \frac{1}{2}(ab^T - ba^T)$ where $a^T = (1, -2, 1)$ and $b^T = (1, 0, -1)$. Partitioning $\mathbf{X}$ into its three columns $(\mathbf{x}_1, \mathbf{x}_2, \mathbf{x}_3)$ and evaluating $\mathbf{X}\mathbf{S}_3\mathbf{X}^T$ shows that the very specialized model

$$(\mathbf{x}_2\mathbf{x}_3^T - \mathbf{x}_3\mathbf{x}_2^T) + (\mathbf{x}_3\mathbf{x}_1^T - \mathbf{x}_1\mathbf{x}_3^T) + (\mathbf{x}_1\mathbf{x}_2^T - \mathbf{x}_2\mathbf{x}_1^T)$$

is being fitted as the skew-symmetric part of the model. Furthermore the same vectors $\mathbf{x}_1, \mathbf{x}_2, \mathbf{x}_3$ are constrained to fit the symmetric part of the model. Similar remarks apply to the models for higher values of $k$. A related model is described by Escoufier and Grorud (1980).

It is often better to deal quite separately with the symmetric and skew-symmetric components, and this leads to consideration of the breakdown

$$\mathbf{D} = \mathbf{M} + \mathbf{N},$$

where **M** is symmetric and **N** skew-symmetric. It is easy to see that

$$\sum_{i,j} d_{ij}^2 = \sum_{i,j} m_{ij}^2 + \sum_{i,j} n_{ij}^2$$

showing that the total sum-of-squares breaks down into independent components for symmetry and skew-symmetry. This suggests separate least-squares analyses of **M** and **N**. The matrix **M** is open to analysis by any of the methods discussed in earlier sections, including non-metric methods. Usually we shall be interested in low-rank approximations to **N** and these are given by the Eckart–Young Theorem, so the singular value decomposition of **N** is needed. This has the special form for skew-symmetric matrices given by

$$\mathbf{N} = \mathbf{U}\mathbf{\Sigma}\mathbf{J}\mathbf{U}^T$$

where **U** is orthogonal, $\mathbf{\Sigma} = \mathrm{diag}\,(\sigma_1, \sigma_1, \sigma_2, \sigma_2, \ldots)$, showing that the singular values occur in pairs, and

$$\mathbf{J} = \mathrm{diag}\left\{ \begin{pmatrix} 0 & 1 \\ -1 & 0 \end{pmatrix}, \begin{pmatrix} 0 & 1 \\ -1 & 0 \end{pmatrix}, \ldots \right\}.$$

When $n$ is odd, the last elements of $\mathbf{\Sigma}$ is zero and the last elements of **J** is unity.

The singular value decomposition may be written down in summation form as:

$$\mathbf{N} = \sum_{i=1}^{[n/2]} \sigma_1(\mathbf{u}_{2i-1}\mathbf{u}_{2i}^T - \mathbf{u}_{2i}\mathbf{u}_{2i-1}^T),$$

where $[n/2]$ is the smallest integer not greater than $n/2$, and $\mathbf{u}_{2i}$ is the $2i$th column of **U**. The equality of the singular-value pairs implies that fits of odd rank are not unique and only even-order-rank solutions are admissible. Suppose we concentrate on the first term of the expansion of **N** and, to avoid subscripts, write **u** for $\mathbf{u}_1$ and **v** for $\mathbf{u}_2$ then we have:

$$\mathbf{N} = \sigma_1(\mathbf{u}\mathbf{v}^T - \mathbf{v}\mathbf{u}^T),$$

which corresponds with the Chino skew-symmetric component when $k = 2$, found above. This term may be parameterized in many ways by replacing **u** and **v** by *any* two vectors **a** and **b** (not necessarily orthogonal) in the same plane. All such parameterizations are equivalent giving the same fitted sum of squares $2\sigma_1^2$. It is sometimes convenient to choose a parameterization in which $\mathbf{1}^T\mathbf{a} = \mathbf{0}$ and **b** is orthogonal to **a**.

The values $(\mathbf{u}_i, \mathbf{v}_i)$ $i = 1, 2, \ldots, n$ may be taken as coordinates of $n$ points and plotted in two dimensions. The approximation $n_{ij} = \sigma_1(u_i v_j - v_i u_j)$ shows that the area of the triangle subtended at the origin by the $i$th and $j$th points is approximately proportional to $n_{ij}$. The skew symmetry is obtained because the sign of the area depends on whether the vertices of the triangle are traversed in a clockwise or anticlockwise sense. The different parameterizations correspond to different scalings of the two axes, different rotations and different relative scalings, all of which affect areas by, at most, a constant multiplier.

This representation in terms of areas differs markedly from the usual distance interpretation of ordinations. For example two points collinear with the origin give a zero area, even when they are very distant from one another. With distance the locus of points equidistant from a given point $P$ is a circle centre $P$; with areas the locus of points with equal area of a triangle with base $OP$ is a straight line parallel to $OP$. The area of a triangle with vectors $P_i$, $P_j$ and $P_k$ is proportional to $n_{ij} + n_{jk} + n_{ki}$. These considerations have to be kept in mind when interpreting area diagrams.

The area diagram may also be used as a diagnostic tool in a similar manner to the diagnostic biplot (§ 17.4). Consider the most simple model of asymmetry

$$n_{ij} = a_i - a_j$$

where we can assume that $\Sigma a_i = 0$ because the values of $a_i$ are clearly arbitrary up to an additive constant. In matrix form this becomes

$$\mathbf{N} = \mathbf{a1}^T - \mathbf{1a}^T$$

in which $\mathbf{1}^T \mathbf{a} = 0$.

When plotted, one of the coordinates is a constant, so the plot is linear. Thus a linear plot implies the simple skew-symmetric form $a_i - a_j$. When this simple, but important, form of asymmetry does occur, it is worth plotting the values $a_i$ onto any ordination of the symmetric part $\mathbf{M}$, perhaps joining similar values by contours.

The separation of the analysis of the symmetric and skew-symmetric components of $\mathbf{D}$ has been emphasized. This does not imply that attempts should not be made to recombine the two parts and to identify relationships between the parameters of the two parts of the model. This is a subtle problem on which much work is needed. Current progress has been confined to studying explicit models for $\mathbf{D}$ and obtaining, by analytical means, the relationships together with their geometric implications. For further details and references see Gower (1977b) and Constantine and Gower (1982).

J.C.G.

## 17.14.   FURTHER READING AND REFERENCES

Baker, R. F., Young, F. W. and Takane, Y. (1977). An Asymmetric Euclidean Model (available from F. W. Young), Psychometric Laboratory, Dave Hall 013a, University of North Carolina, Chapel Hill, NC 27514.

ten Berge, J. M. F. and Nevels, K. (1977). A General Solution to Mosier's Oblique Procrustes Problem, *Psychometrika* **42**, 593–600.

Barlow, R. E., Bartholomew, D. J., Bremmemr, J. M. and Brunk, H. D. (1972). *Statistical Inference Under Order Restrictions: the Theory and Application of Isotonic Regression*, J. Wiley, Chichester, New York, Brisbane.

Borg, I. (1977). Some Basic Concepts of Facet Theory. In J. C. Lingoes (Ed.) *Geometric Representations of Relational Data*, Ann Arbor, Mathesis Press.

Blumenthal, L. M. (1970). *Theory and Applications of Distance Geometry*, 2nd edition, Chelsea, New York.

Bradu, D. and Gabriel, K. R. (1978). The Biplot as a Diagnostic Tool for Models of Two-way Tables, *Technometrics* **20**, 47–68.

Browne, M. W. (1967). On Oblique Procrustes Rotation, *Psychometrika* **32**, 125–132.

Browne, M. W. and Kristof, W. (1969). On the Oblique Rotation of a Factor matrix to a Specified Pattern, *Psychometrika* **34**, 237–248.

Cailliez, F. (1983). The Analytical Solution to the Additive Constant Problem, *Pyschometrika* **48**, 305–308.

Carroll, J. D. and Chang, J. J. (1970). Analysis of Individual Differences in Multi-dimensional Scaling via an $n$-way Generalization of 'Eckart–Young' Decomposition, *Psychometrika* **35**, 283–319.

Carroll, R. B. and Chang, J. J. (1972). IDIOSCAL (Individual Differences in Orientation Scaling), Paper presented at the Spring meeting of the Psychometric Society, Princeton, New Jersey, April 1972.

Cattell, R. B. and Khanna, D. K. (1977). Principles and Procedures for Unique Rotation in Factor Analysis, Chapter 9 of *Statistical Methods for Digital Computers* (Vol. III of *Mathematical Methods for Digital Computers*) Ed., K. Einstein, A. Ralston and H. S. Wilf. New York, Wiley-Interscience.

Cramer, E. M. (1974). On Browne's Solution for Oblique Procrustes Rotation, *Psychometrika* **39**, 139–163.

Constantine, A. G. and Gower, J. C. (1978). Graphical Representation of Asymmetric Matrices Journal of the Royal Statistical Society, C., *Applied Statistics* **27**, 297–304.

Constantine, A. G. and Gower, J. C. (1981). Models for the Analysis of Interregional Migration, *Environment and Planning A*, **14**, 477–497.

Davies, A. W. (1978). On the Asymptotic Distribution of Gower's $m^2$ Goodness-of-fit Criterion in a Particular Case, *Ann. Inst. Statist. Math.* **30**, 71–79.

Digby, P. G. N. and Gower, J. C. (1981). Ordination Between- and Within-groups Applied to Soil Classification, *Down to Earth Statistics: Solutions Looking for Geological Problems*, Syracuse University Geology Contributions (ed. D. F. Merriam), 63–75.

Eckart, C. and Young, G. (1936). The Approximation of One Matrix by Another of Lower Rank, *Pyschometrika* **1**, 211–218.

Escoufier, Y. and Grorud, A. (1980). Analyse Factorielle des Matrice Carrees non Symmetriques. In: *Data Analysis and Informatics*, 17–19 October 1979 (Eds. E. Diday, L. Lebart, J. P. Pages, R. Tommassone, North-Holland, Amsterdam), pp. 2633–2276.

Gower, J. C. (1975). Generalised Procrustes Analysis, *Psychometrika* **40**, 33–51.

Gower, J. C. (1977). The Analysis of Three-way Grids. In: *Dimensions of Intra-Personal Space*, Vol. 2: The Measurement of Intra-Personal Space by Grid Technique. (Ed. P. Slater), J. Wiley & Son, 163–173.

Gower, J. C. (1977b). The Analysis of Asymmetry and Orthogonality. In *Recent Developments in Statistics* (Eds. F. Brodeau, G. Romier, B. van Cutsem), North-Holland, Amsterdam, pp. 109–123.

Gower, J. C. (1982). Euclidean Distance Geometry, *Mathematical Scientist* **7**, 1–14.

Guttman, L. A. (1968). A General Non-metric Technique for Finding the Smaller Coordinate Space for a Configuration of Points, *Psychometrika* **33**, 469–506.

Green, B. F. and Gower, J. C. (1981). *A Problem with Congruence* (Available on request).

Greenacre, M. J. (1978). Some Objective Methods of Graphical Display of a Data Matrix. Translation of doctoral thesis (Universite de Paris, VI). Published as special report by University of South Africa, Pretoria.

Gruvaeus, G. T. (1970). A General Approach to Procrustes Pattern Rotation, *Psychometrika* **35**, 493–505.

Harshman, R. A. (1972). PARAFAC2: Mathematical and Technical Notes. In: Working Papers in Phonetics 22, University of California at Los Angeles.

Heiser, W. and de Leeuw, J. (1979). How to Use SMACOF-I, *A Program for Metric Multidimensional Scaling*, Department of Datatheory, Faculty of Social Sciences, University of Leiden, Wassenaarseweg 80, Leiden, The Netherlands, 1–63.

Hurley, J. R. and Cattell, R. B. (1962). The Procrustes Program: Producing Direct Rotation to Test an Hypothesized Factor Structure, *Behavioural Science* **7**, 258–262.

Lingoes, J. C. (1971). Some Boundary Conditions for a Monotone Analysis of Symmetric Matrices, *Psychometrika* **36**, 195–203.

Lissitz, R. W., Schonemann, P. H. and Lingoes, J. C. (1976). A Solution to the Weighted Procrustes Problem in which the Transformation is in Agreement with the Loss Function, *Psychometrika* **41**, 547–550.

Mandel, J. (1961). Non-additivity in Two-way Analysis of Variance, *J. Amer. Statist. Assn* **56**, 878–888.

Murray, W. (Ed.) (1972). *Numerical Methods for Unconstrained Optimisation*, Academic Press, London and New York.

Nishisato, S. (1980). *Analysis of Categorical Data: Dual Scaling and its Applications*, University of Toronto Press, Toronto, Buffalo, London.

Pearson, K. (1901). On Lines and Planes of Closest Fit to a System of Points, *Phil. Mag. ser.* 6 **2**, 559–572.

Ramsay, J. O. (1977). Maximum Likelihood Estimation in Multidimensional Scaling, *Psychometrika* **42**, 241–266.

Sibson, R. (1978). Studies in the Robustness of Multidimensional Scaling: Procrustes Statistics, *J. Roy. Statist. Soc.* **40**, 234–238.

Sibson, R. (1979). Studies in the Robustness of Multidimensional Scaling: Perturbational Analysis of Classical Scaling, *J. Roy. Statist. Soc. B* **41** 217–229.

Schönemann, P. H. (1968). On Two-sided Orthogonal Procrustes Problems, *Psychometrika* **33**, 19–33.

Takane, Y., Young, F. and de Leeuw, J. (1976). Non-metric Individual Differences Multidimensional Scaling: an Alternating Least Squares Method with Optimal Scaling Features, *Psychometrika* **42**, 7–67.

Torgerson, W. S. (1958). *Theory and Methods of Scaling*, New York, Wiley.

Tukey, J. W. (1949). One Degree of Freedom for Non-additivity, *Biometrics* **5**, 232–242.

# CHAPTER 18

# *Time Series*

## 18.1.  INTRODUCTION

A time series is simply the set of measurements of a variable—let us call it $X$—recorded as time progresses. In theory, the measurements may be continuously recorded, but more usually they are obtained at equally spaced points in time and indexed to provide a sample (of size $n$)

$$\mathbf{x} = (x_1, x_2, \ldots x_n)'.$$

Interest in a time series focuses on describing or modelling its structure. If this is done without reference to any other observed variable it is called a univariate analysis, and we devote most of this chapter to such methods. The aim of the analyst generally extends beyond modelling, although the model may supply him directly with the information he requires, such as the magnitude of a cyclical component. Commonly, a model is used to extrapolate or forecast a time series, and a useful criterion for selecting between several models for a time series is their predictive ability. Other applications such as seasonal adjustment, signal extraction and smoothing, generally require good time series models. Finally, the models may be used to simulate [see II, § 5.6] longer stretches of observations, for use in the study of a large system to which the time series variable acts as an input. We limit ourselves to forecasting applications in this chapter.

A statistical approach is necessary in time series analysis. Measurement error is common, and stochastic fluctuations are likely to be inherent in the system which is being observed, be it environmental, economic, engineering or biological. An empirical approach is almost inevitable in a univariate analysis. A complete mathematical model of the system being observed is likely to be of limited value if only one variable is measured. Relatively simple empirical models, with sufficient flexibility to fit the data, and a good reputation established by several years of successful application, will be presented in this chapter.

We have implicitly assumed that a time series has structure, i.e. the observations are not a totally unrelated set of numbers. This structure may be evident to the eye, in such features as trends or cycles—although the eye is good at seeing structure where there is pure randomness! We assume that this structure may be described by a model involving few parameters in comparison with the number of observations, and furthermore that it extends in time beyond the range of the observations—which is particularly important in forecasting applications. The time series is therefore a subset of observations of a stochastic process [see II, Chapter 18]:

$$\{\ldots X_{-1}, X_0, X_1, X_2, \ldots\}.$$

It is common for this process to be called the time series, as well as the data. From this point on we shall also, for notational convenience use $x_t$ to represent both the observations and the stochastic process, relying on the context to indicate the meaning.

   Figure 18.1.1 shows graphs of five time series, chosen to illustrate different features. The logarithms of airline passenger totals (hereafter the 'airline' series) is a monthly series, showing a strong trend and annual pattern. The national pig population census figures (hereafter the 'gilt' series—a predictor of the breeding herd size) are obtained quarterly, and seem to swing from a peak to a trough and back about every two to three years—an irregular cycle. The variations in the earth's rotation rate (hereafter the 'daylength' series) is an annual series which shows a slight overall trend, with some large, long period, variations about the trend. The 'random' series is a random sample of Normal deviates [see II, § 11.4], corrected for their mean. The 'random walk' [see II, § 18.3] series is simply the cumulative total of the 'random' series, and is introduced to show how such series gives the impression of cyclical behaviour.

## 18.2.   CLASSICAL REGRESSION MODELS FOR THE TIME SERIES

### 18.2.1.   Seasonal Effects Model

   It is natural to suppose that strong features of trend and seasonality might be modelled by components which are deterministic functions of time. For example, the airline series might be modelled as

$$x_t = a + bt + \alpha_1 M_{1,t} + \ldots + \alpha_{12} M_{12,t} + e_t \qquad (18.2.1)$$

where $(a + bt)$ is a linear trend, and $M_{1,t}, \ldots, M_{12,t}$ are indicator variables, one for each month of the year. Thus $M_{1,t} = 0$ except that every January $M_{1,t} = 1$, and so on. Then $\alpha_1$ measures the deviation of the January values from the trend, and so on, giving 12 separate 'monthly effects'. The final term $e_t$ represents the error, which we expect to be small in comparison with the main trend and seasonal effects. In order to be uniquely parameterized the model in fact requires a constraint, such as the seasonal effects summing to zero. If the model is fitted by standard least squares regression [see §§ 6.5 and

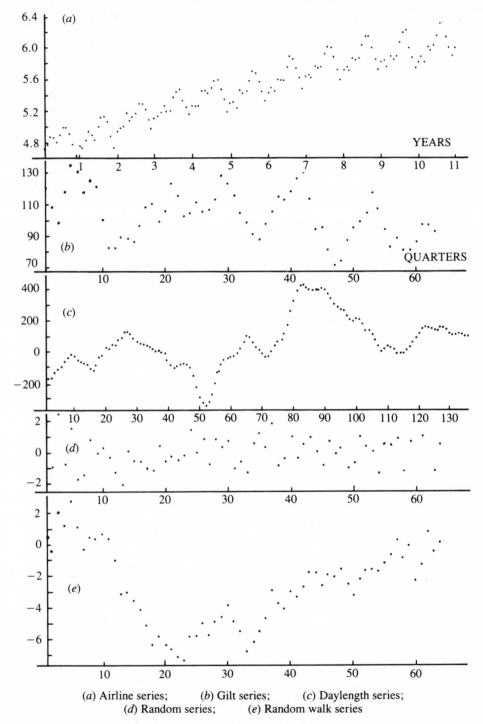

(a) Airline series;　　　(b) Gilt series;　　　(c) Daylength series;
(d) Random series;　　　(e) Random walk series

Figure 18.1.1: Examples of time series. (a) Airline series; (b) Gilt series; (c) Daylength series; (d) Random series: .(e) Random walk series.

8.2], the fit looks good as illustrated in Figure 18.2.1(a). However, close examination of the residual errors [see § 82.4] is particularly called for in time ·series analysis. The observations are not obtained from independent experiments, and successive errors must be suspected as possibly being related. For this example the residuals are shown in (b) in this figure. They are evidently not random, with long stretches of constant sign, some gradual and some sudden changes in level, and the occasional outlier. A model which allowed for possible changes in trend and seasonal pattern might be an improvement. The fitted model does in fact extrapolate quite well for one year, as shown in Figure 18.4.1, but our concern over the error structure precludes us from making any precise confidence statement about the forecasts.

### 18.2.2.   Cyclical Component Model

The seasonal pattern of the airline series is complex because of the effects of winter, spring and summer holidays. For some seasonal series, a good fit may be obtained with fewer parameters if the seasonal variation is smooth, by using sinusoidal [see IV, § 2.12] waves of the appropriate period as regression components, as explained in section 18.3.2. In other examples, the period or wavelength of the cycle may not be precisely known. Figure 18.2.2(a) shows measurements of the magnitude of the variable star T Ursa Majoris, obtained by averaging observations over successive ten-day intervals. There is a strong cycle evident which we might represent by a regression model

$$x_t = \mu + R \cos(wt + \phi) + e_t \qquad (18.2.2)$$

where $\mu$ is the mean, $\omega$ the angular frequency (in radians/unit time), $R$ is the amplitude (not a random variable) and $\phi$ the phase of the wave. Again $e_t$ is an error term. As it stands with $\mu$, $\omega$, $R$ and $\phi$ as unknown parameters, this model is linear in $\mu$ and $R$, but not in $\omega$ and $\phi$. It may however be written

$$x_t = \mu + A \cos \omega t + B \sin \omega t + e_t \qquad (18.2.3)$$

where $A = R \cos \phi$ and $B = -R \sin \phi$, in which form it is linear in $\mu$, $A$ and $B$ which may be estimated by regression if $\omega$ can first be determined with a fair degree of precision. Figure 18.2.2(a) shows five cycles in 130 time units so we use as an initial value for the period, 26, and for the angular frequency $2\pi/26 = \omega_0$. We then set up two regression vectors, or variates, containing the values

$$\cos \omega_0 t, \sin \omega_0 t, \qquad t = 1 \ldots 131, \qquad (18.2.4)$$

and using least squares regression [see § 8.2] obtain the estimates (± their standard errors [see Definition 3.1.1]):

$$\hat{A} = 0{\cdot}247 \pm 0{\cdot}078, \qquad \hat{B} = -2{\cdot}277 \pm 0{\cdot}078.$$

From these, $\hat{R} = 2{\cdot}290$ and $\hat{\phi} = 83.8°$ or $1{\cdot}46$ radians.

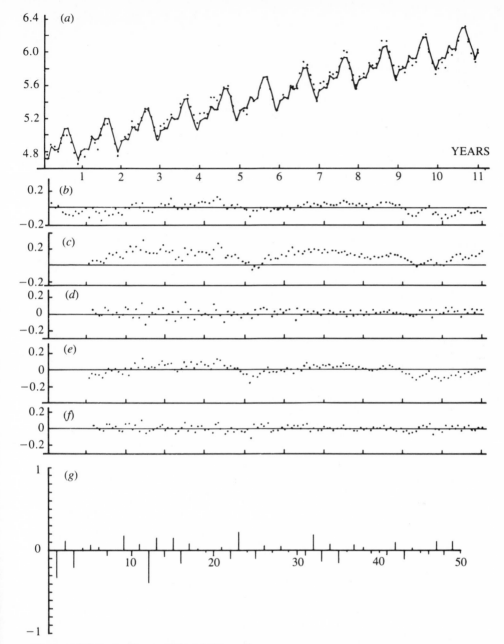

Figure 18.2.1: Analyses of the airline series.
(a) Fit of trend and monthly effects model (18.2.1) to airline series.
(b) Residual errors after fitting model (18.2.1).
(c) Seasonal differences of the airline series.
(d) The first differences of the seasonal differences of the airline series.
(e) One step ahead errors from the seasonal EWMA prediction of the airline series.
(f) One step ahead errors from prediction of the airline series by the Box–Jenkins. Model (18.7.43).
(g) Sample autocorrelations of the doubly differenced series in (d).

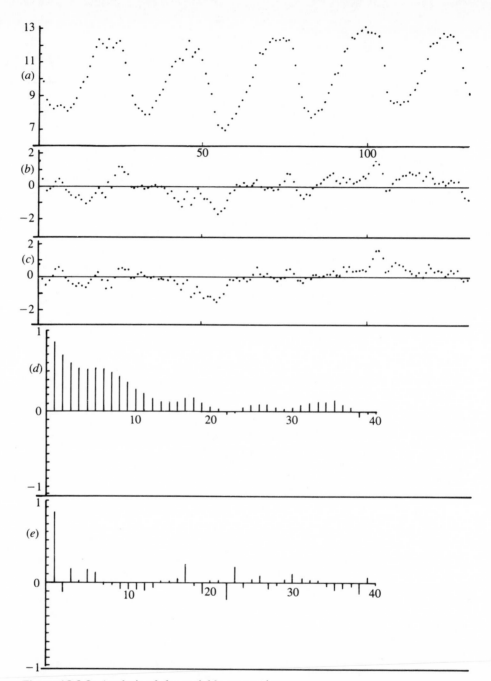

Figure 18.2.2: Analysis of the variable star series.
  (a) Measurements related to the magnitude of the variable star $T$. Ursa Majoris.
  (b) Residual errors after fitting a single cyclical component to the measurements.
  (c) Residual errors after fitting a further harmonic component and refining the frequency.
  (d) Sample autocorrelations of the residuals in (c).
  (e) Sample partial autocorrelations of the residuals in (c).

Again we examine the residuals shown in Figure 18.2.2(b). The model accounts for 86·8% of the variance of the data, so there is a considerable residual variance. The residuals appear to be far from random and we consider two possible explanations for the structure which remains. Firstly, the frequency $\omega_0$ may be in error by an amount large enough to produce a poor 'match' with the observations. Possibly more important, the lack of symmetry of the observed waveform suggests that a 'harmonic' of the fundamental wave, i.e. of twice the frequency, may improve the fit. We take up these points in order.

The frequency value $\omega_0 = 2\pi/26$ which has been used, can be refined quite simply using the regression approach. Let the true frequency $\omega$ be represented as $\omega_0 + \delta$ where we assume that $\delta$ is small. By elementary calculus [see IV, § 3.6] we approximate as follows:

$$R \cos (\omega t + \phi) \approx R \cos (\omega_0 t + \phi) - \{Rt \sin (\omega_0 t + \phi)\}\delta$$

$$= A \cos \omega_0 t + B \sin \omega_0 t + \delta\{t(-A \sin \omega_0 t + B \cos \omega_0 t)\}.$$

We therefore extend the regression equation (18.2.3) by introducing the new vector

$$t(-A \sin \omega_0 t + B \cos \omega_0 t), \qquad t = 1, \ldots, 131 \qquad (18.2.5)$$

with components calculated using the values of $A$ and $B$ already estimated. Carrying out the regression yielded $\hat{\delta} = 0 \cdot 00337 \pm 0 \cdot 00084$, and a small but significant improvement in fit. The improved frequency estimate is $\hat{\omega} = \omega_0 + \hat{\delta} = 2\pi/26 + 0 \cdot 00337 = 0 \cdot 02450$, corresponding to a period of $25 \cdot 64$ time units (256·4 days).

The harmonic of the fundamental wave is estimated by introducing further regression vectors

$$\cos 2\omega_0 t, \quad \sin 2\omega_0 t \quad t = 1 \ldots 131 \qquad (18.2.6)$$

at twice the fundamental frequency, in addition to those in (18.2.4). In fact this was done with $\omega_0$ replaced by the revised frequency $\hat{\omega}$. The improvement in fit was again small but significant, the amplitude of the harmonic being estimated as 0·34, the residual variance being 10% of that of the original data. The frequency was again revised, but with no appreciable change. The residuals shown in Figure 18.2.2(c) are noticeably reduced, but still cannot be considered to be anything like random. They are analyzed further in section 18.11.

In this particular example deterministic sinusoidal components are unquestionably present. Later examples illustrate the fact that deterministic functions should be introduced with caution, and if a good fit can only be achieved using a relatively large number of terms, the results should be treated with suspicion.

## 18.3.   THE PERIODOGRAM OF A TIME SERIES

### 18.3.1.   Harmonic Components of Periodic Time series

A periodic time series is one which repeats its values exactly after an integer period $p$, i.e. $x_{t+p} = x_t$ for all $t$. In section 18.3.2 we show that such a series may be represented by a mean level plus a linear combination of $(p-1)$ sinusoidal waves—the harmonic components. These waves have fundamental period $p$ and harmonics with periods $p/2, p/3, \ldots$, corresponding to $1, 2, 3, \ldots$ whole waves within the period $p$. Thus with $p = 12$ as may be appropriate for monthly data, besides the mean level or constant term a pair of sine and cosine terms is associated with each of the periods 12, 6, 4, 3 and 12/5. Finally, at the period 2, the sine term is omitted since it is identically zero; $\sin(\pi t) = 0$ for integer $t$. This leaves the cosine term to make up the 12, and we note that $\cos(\pi t) = (-1)^t$, so this term simply alternates between $+1$ and $-1$. Collecting the first 12 values, $t = 1, 2, \ldots, 12$ of each of these component terms into column vectors $\mathbf{c}_0, \ldots, \mathbf{c}_6, \mathbf{s}_1, \ldots, \mathbf{s}_5$ defined by

$$c_{j,t} = \cos(2\pi j t/12), \qquad s_{j,t} = \sin(2\pi j t/12)$$

we may represent them in matrix form (recall $\cos \pi/6 = \sqrt3/2$, $\sin \pi/6 = \tfrac{1}{2}$) as

| $t$ | $\mathbf{c}_0$ | $\mathbf{c}_1$ | $\mathbf{c}_2$ | $\mathbf{c}_3$ | $\mathbf{c}_4$ | $\mathbf{c}_5$ | $\mathbf{c}_6$ | $\mathbf{s}_1$ | $\mathbf{s}_2$ | $\mathbf{s}_3$ | $\mathbf{s}_4$ | $\mathbf{s}_5$ |
|---|---|---|---|---|---|---|---|---|---|---|---|---|
| 1 | 1 | $\frac{\sqrt3}{2}$ | $\frac{1}{2}$ | 0 | $-\frac{1}{2}$ | $-\frac{\sqrt3}{2}$ | $-1$ | $\frac{1}{2}$ | $\frac{\sqrt3}{2}$ | 1 | $\frac{\sqrt3}{2}$ | $\frac{1}{2}$ |
| 2 | 1 | $\frac{1}{2}$ | $-\frac{1}{2}$ | $-1$ | $-\frac{1}{2}$ | $\frac{1}{2}$ | 1 | $\frac{\sqrt3}{2}$ | $\frac{\sqrt3}{2}$ | 0 | $-\frac{\sqrt3}{2}$ | $-\frac{\sqrt3}{2}$ |
| 3 | 1 | 0 | $-1$ | 0 | 1 | 0 | $-1$ | 1 | 0 | $-1$ | 0 | 1 |
| 4 | 1 | $-\frac{1}{2}$ | $-\frac{1}{2}$ | 1 | $-\frac{1}{2}$ | $-\frac{1}{2}$ | 1 | $\frac{\sqrt3}{2}$ | $-\frac{\sqrt3}{2}$ | 0 | $\frac{\sqrt3}{2}$ | $-\frac{\sqrt3}{2}$ |
| 5 | 1 | $-\frac{\sqrt3}{2}$ | $\frac{1}{2}$ | 0 | $-\frac{1}{2}$ | $\frac{\sqrt3}{2}$ | $-1$ | $\frac{1}{2}$ | $-\frac{\sqrt3}{2}$ | 1 | $-\frac{\sqrt3}{2}$ | $\frac{1}{2}$ |
| 6 | 1 | $-1$ | 1 | $-1$ | 1 | $-1$ | 1 | 0 | 0 | 0 | 0 | 0 |
| 7 | 1 | $-\frac{\sqrt3}{2}$ | $\frac{1}{2}$ | 0 | $-\frac{1}{2}$ | $\frac{\sqrt3}{2}$ | $-1$ | $-\frac{1}{2}$ | $\frac{\sqrt3}{2}$ | $-1$ | $\frac{\sqrt3}{2}$ | $-\frac{1}{2}$ |
| 8 | 1 | $-\frac{1}{2}$ | $-\frac{1}{2}$ | 1 | $-\frac{1}{2}$ | $-\frac{1}{2}$ | 1 | $-\frac{\sqrt3}{2}$ | $\frac{\sqrt3}{2}$ | 0 | $-\frac{\sqrt3}{2}$ | $\frac{\sqrt3}{2}$ |
| 9 | 1 | 0 | $-1$ | 0 | 1 | 0 | $-1$ | $-1$ | 0 | 1 | 0 | $-1$ |
| 10 | 1 | $\frac{1}{2}$ | $-\frac{1}{2}$ | $-1$ | $-\frac{1}{2}$ | $\frac{1}{2}$ | 1 | $-\frac{\sqrt3}{2}$ | $-\frac{\sqrt3}{2}$ | 0 | $\frac{\sqrt3}{2}$ | $\frac{\sqrt3}{2}$ |
| 11 | 1 | $\frac{\sqrt3}{2}$ | $\frac{1}{2}$ | 0 | $-\frac{1}{2}$ | $-\frac{\sqrt3}{2}$ | $-1$ | $-\frac{1}{2}$ | $-\frac{\sqrt3}{2}$ | $-1$ | $-\frac{\sqrt3}{2}$ | $-\frac{1}{2}$ |
| 12 | 1 | 1 | 1 | 1 | 1 | 1 | 1 | 0 | 0 | 0 | 0 | 0 |

The corresponding waves are graphed in Figure 18.3.1. The columns $\mathbf{c}_j$ for $j = 2, \ldots 5$ are simply derived from $\mathbf{c}_1$, by stepping through its values $j$ at a time, returning to the beginning as necessary. It is now clear why no wavelength of less than two units (or frequency greater than $\pi$) need concern us, since if we step through column $\mathbf{c}_1$ seven values at a time we obtain $\mathbf{c}_5$ again. Similarly $\mathbf{s}_7$ would be $-\mathbf{s}_5$. In terms of the absolute frequency $f$, defined as the number of waves per sampling interval (the reciprocal of the period, or $1/2\pi$ times the angular frequency), a sinusoidal wave of frequency $f$ cannot be distinguished from one of frequency $1-f, 1+f, 2-f, 2+f, \ldots$ by sampling at integer

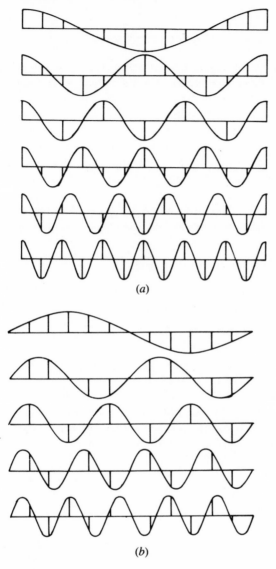

(a)

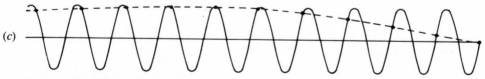

(b)

(c)

Figure 18.3.1: Sinusoidal waves.
   (a)  Cosine waves ⎫    Showing harmonics of fundamental period 12,
   (b)  Sine waves   ⎬       sampled at unit time intervals.
   (c)  A simplified representation of the height of a tide (solid line) sampled at 6 a.m.
        and 6 p.m. each day, with the long period wave (dashed line) which might
        conventionally be drawn through the sample points.

values of $t$, i.e.

$$\cos(2\pi[k \pm f]t) = \cos 2\pi ft, \qquad \sin(2\pi[k \pm f]t) = \pm\sin 2\pi ft.$$

It is therefore conventional to assume that $f$ is in the range $0 \le f \le 0.5$, since any frequency greater than this gives a sampled wave identical with one in this range. The highest frequency $f = .5$ corresponds to a period of two sampling intervals and is called the Nyquist frequency. Such a convention may lead to an incorrect assumption, but is designed to give the correct answer provided the sampling interval is sufficiently small—half the period of the shortest wavelength suspected. Figure 18.3.1(c) shows the points which may be recorded if a person measured the height of the tide at 6 a.m. and 6 p.m. each day. The above convention would lead to the erroneous assumption of a frequency $f = \frac{1}{28}$ (one tide in 14 days) whereas $f = 1 - \frac{1}{28}$ (27 tides in 14 days). Measuring the height every six hours would give the correct frequency.

### 18.3.2. Harmonic Regression Analysis

Consider the harmonic vectors defined for a general period $p$ by

$$\left. \begin{array}{ll} \mathbf{c}_j = \{\cos 2\pi jt/p\}_{t=1}^{p}; & j = 0, \ldots, \frac{1}{2}p \\ \mathbf{s}_j = \{\sin 2\pi jt/p\}_{t=1}^{p}; & j = 1, \ldots, \frac{1}{2}p - 1 \end{array} \right\} \qquad (18.3.1)$$

provided $p$ is even. If $p$ is odd, the upper limit on $j$ is $\frac{1}{2}(p-1)$ in both cases, the alternating vector being absent.

It may be shown that these vectors form an orthogonal set [see I, § 10.2], and the sum of squares of elements for each vector is $p/2$, except for $\mathbf{c}_0$ and (only in the case of $p$ even) $\mathbf{c}_{p/2}$, for which the value is $p$. Any vector of $p$ elements, say $\mathbf{x} = (x_1, x_2, \ldots, x_p)$ may therefore be expressed as a unique linear combination of the harmonic vectors, i.e.

$$x_t = \mu + \sum_{j=1}^{q} \{A_j \sin 2\pi jt/p + B_j \cos 2\pi jt/p\} + \nu(-1)^t, \qquad t = 1, \ldots, p \qquad (18.3.2)$$

where $q = [\frac{1}{2}(p-1)]$, the integer part of $\frac{1}{2}(p-1)$, and the last term is omitted if $p$ is odd.

Now note that if the RHS of (18.3.2) is extrapolated for $t > p$ all the harmonic terms take the same values over the ranges $t = p+1, \ldots 2p$ and $t = 2p+1, \ldots, 3p$ etc., as they do for $t = 1, \ldots, p$. The RHS of (18.3.2) is therefore periodic with period $p$ by definition of the harmonic components. If $x_1, \ldots, x_p$ are the first $p$ values of a periodic time series with period $p$, the representation in (18.3.2) will hold for all $t$.

The application of this result is to the modelling of a time series with an apparently strong fixed periodic pattern with integer period $p$. Thus for a monthly time series of length $n$, taking $p = 12$, the harmonic regression vectors $\mathbf{c}_0, \ldots, \mathbf{c}_6, \mathbf{s}_1, \ldots, \mathbf{s}_5$, extended in time up to the full length $n$ of the series by repeating the values for each successive year, are an attractive alternative to the indicator variates $M_{j,t}$ considered in 18.2.1. If the full set is used, the model is equivalent. The attraction is that fewer components might suffice, the high frequency components being negligible for 'smooth' series. The orthogonality property is retained if $n$ is a multiple of $p$, but this condition is not necessary to the application.

### 18.3.3.  The Periodogram

The properties of harmonic regression may be usefully exploited in a more general context, where a time series of moderate to large length $n$ is suspected of having one or more sinusoidal components of unknown frequencies (as opposed to a periodic component of known and relatively small period $p$). The length $n$ is substituted for the period $p$ in (18.3.2) giving a fairly fine division of frequencies $\omega_j = 2\pi j/n$, $j \le m = [\frac{1}{2}(n-1)]$. The time series may then be represented (taking $\nu = 0$ if $n$ is odd) as

$$x_t = \mu + \sum_{j=1}^{m} \{A_j \sin \omega_j t + B_j \cos \omega_j t\} + \nu(-1)^t, \qquad t = 1, \ldots, n. \quad (18.3.3)$$

The orthogonality properties [see IV, § 20.4] allow the coefficients to be directly expressed as $\mu = \bar{x} = (1/n)\sum_1^n x_t$, $\nu = (1/n)\sum(-1)^t x_t$ and

$$A_j = (2/n)\sum x_t \sin \omega_j t, \qquad B_j = (2/n)\sum x_t \cos \omega_j t; \qquad j = 1 \ldots m. \quad (18.3.4)$$

The amplitude $R_j$ of the component at frequency $\omega_j$ is given by

$$R_j^2 = A_j^2 + B_j^2,$$

and a sum of squares decomposition (loosely termed Analysis of Variance [see § 8.3]) also follows:

$$\sum_1^n (x_t - \bar{x})^2 = (n/2) \sum_1^m R_j^2 + n\nu^2. \quad (18.3.5)$$

The representation (18.3.3) should not be taken as a model for the series, since it would imply that $x_t$ were periodic, with period $n$. Rather, it should be viewed as a means of transforming the data with the hope that the amplitudes $R_j$ may reveal hitherto non-obvious features which may be interpreted in terms of frequency components. In order that we can do this intelligently we shall investigate the properties of $R_j^2$, $j = 1 \ldots m$ for various possible models which may give rise to the time series $x_t$. We first give a definition which is motivated by the foregoing discussion.

DEFINITION 18.3.1.   The Periodogram of $x_1, \ldots, x_n$ over the range $0 \le \omega \le \pi$ is

$$I_n(\omega) = (2/n)\left\{ \left( \sum_1^n x_t \sin \omega t \right)^2 + \left( \sum_1^n x_t \cos \omega t \right)^2 \right\}$$

$$= (2/n)\left| \sum_1^n x_t \exp(i\omega t) \right|^2.$$

At a harmonic frequency $\omega_j$, $I_n(\omega_j) = (n/2)R_j^2$, the component in the 'analysis of variance' (18.3.5) associated with that frequency. The definition merely extends this finite set of quantities over a continuous range of frequencies. The alternative expression involving complex exponentials [see IV, (9.5.8)] is convenient for some algebraic manipulation.

### 18.3.4.   Effect of Mean Correction on the Periodogram

The value of the periodogram at the origin is $I_n(0) = 2n\bar{x}^2$. Since the data mean $\bar{x}$ usually incorporates a somewhat arbitrary reference level for the observations, it is common practice (as in analysis of variance) to 'mean-correct' the data, i.e. to replace $x_t$ by $(x_t - \bar{x})$ in the Definition 18.3.1. This makes $I_n(0) = 0$. The orthogonality properties ensure that adding or subtracting any constant to or from the data has no effect on the periodogram at the harmonic frequencies $\omega_j$. Intermediate frequencies are affected, however. Failure to mean-correct the data leads to values of $I_n(\omega)$ of the same order as $I_n(0)$, i.e. of the order $n\bar{x}^2$, throughout the interval $0 < \omega < \omega_1$. Within succeeding intervals $\omega_j < \omega < \omega_{j+1}$, $j = 1, 2, \ldots$ $I_n(\omega)$ rises in sidelobes to values of the order $n\bar{x}^2/(\pi^2 j^2)$, and it may be calculated that for any fixed frequency $\omega$ well away from the origin the effect is (at worst) of the order $\bar{x}/\sqrt{n}$. The general effect is seen in Figure 18.3.2(a), which is the periodogram of a constant, $x_t = 1$, $t = 1, \ldots, 64$.

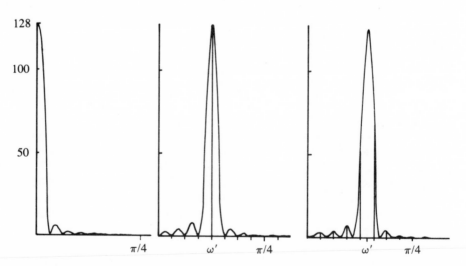

Figure 18.3.2: Periodograms of simple series with harmonic frequencies emphasized.
   (a) Periodogram of $x_t = 1, t = 1 \ldots 64$.
   (b) Periodogram of $x_t = R \cos (\omega' t + \phi)$, $t = 1 \ldots 64$ with $\omega' = 2\pi/16$, $\phi = \pi/3$, $R = 2$.
   (c) Periodogram of $x_t = R \cos (\omega' t + \phi), t = 1 \ldots 64$ with $\omega' = 2\pi(4 \cdot 5)/64$, $\phi = \pi/3$, $R = 2$.

### 18.3.5.   Periodograms for Simple Models

These will be calculated at a somewhat finer frequency division than the harmonic frequencies, which will however be emphasized in the Figures.

$$\text{Model (i)}\quad x_t = R\cos(\omega' t + \phi), \qquad t = 1, \ldots, n.$$

$I_n(\omega)$ is shown for $R = 2$, $n = 64$, $\omega' = 2\pi/16$ and $\phi = \pi/3$ in Figure 18.3.2(b). Because $\omega'$ is a harmonic frequency, the series is a simple example of the representation (18.3.3) with only one harmonic component corresponding to $j = 4$. Consequently $I_n(\omega_j) = 0$ except for $I_n(\omega_4) = (n/2)R^2 = 128$. Note again however, the sidelobes at intermediate frequencies, similar to those in Figure 18.3.2(a).

Model (ii).  This is the same as for model (i) but with $\omega'$ changed to $2\pi(4 \cdot 5)/64$. The periodogram in (c) of Figure 18.3.2(c) looks similar to that in (b), but the maximum is not precisely at $\omega'$, and does not have the precise value 128. This is because the vectors

$$\{\sin \omega' t\}_1^n, \qquad \{\cos \omega' t\}_1^n$$

are not exactly orthogonal, and do not have sums of squares exactly equal to $(n/2)$ when $\omega'$ is not a harmonic frequency. Nevertheless, the periodogram peak gives an adequate indication of the frequency and squared amplitude, or power. If only the harmonic frequencies are inspected the peak is for the most part divided between frequencies $\omega_4$ and $\omega_5$. Such a periodogram would naturally lead to the interpretation that the series $x_t$ contained a single component wave of frequency between $\omega_4$ and $\omega_5$ but it could coincidently consist of a mixture of components at the harmonic frequencies. In general a cluster of sinusoidal components, with frequencies within a band whose width is of the order $2\pi/n$, will be difficult to distinguish using the periodogram.

Model (iii).  $x_1, \ldots, x_n$ is a random sample from a Normal distribution with expectation $\mu$ and variance $\sigma^2$. The 'analysis of variance' (18.3.5) may then be interpreted as a decomposition into independent components so that $(n/2)R_j^2/\sigma^2$ has a chi-squared [see § 2.5.4(a)] distribution on 2 d.f. (degrees of freedom), there being no true effects in the model. In other words, $I_n(\omega_j)$ for $j = 1, \ldots, m$ form a random sample from the Exponential distribution [see II, § 11.2] with mean $2\sigma^2$. This is illustrated for $n = 64$ and $\sigma^2 = 1$ in Figure 18.3.3(a) which is the periodogram of the series in Figure 18.1.1(d). It is only to be expected that in a random sample of size 31 there should be some extreme values, which an incautious observer may interpret as significant sinusoidal components where none exist. The danger is increased by plotting the periodogram over a continuous frequency range. The largest (and quite meaningless) peak in Figure 18.3.3(a) would be completely avoided if the harmonic frequencies alone were considered.

Model (iv).  The series used is $y_1, \ldots, y_n$ where $y_1 = x_1 - \bar{x}$ and

$$y_t = y_{t-1} + (x_t - \bar{x}), \qquad t = 2, \ldots, n. \tag{18.3.6}$$

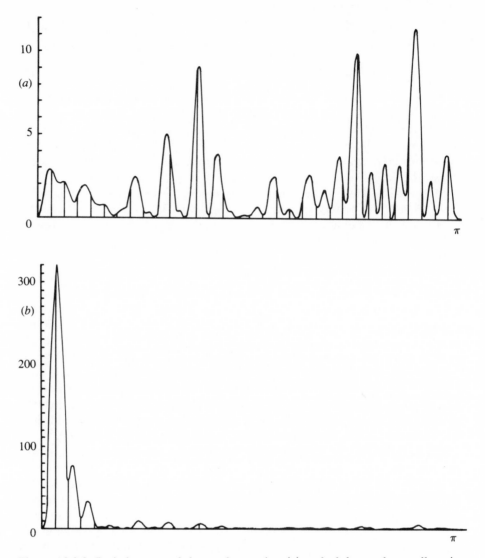

Figure 18.3.3: Periodograms of the random series, (a) and of the random walk series (b) with harmonic frequencies emphasized.

Thus $y_t$ is the cumulative sum of the series in model (iii) above, after first correcting for the mean. This produces a random walk [see II, § 18.3] constrained to return to the origin, since $y_n = 0$. An alternative to mean correcting $x_t$ which would achieve the same constraint, is to subtract the trend line joining the origin to the last point in an unconstrained random walk. Since some sort of trend correction is often recommended before calculating periodograms, this action is not atypical, and in fact allows us to demonstrate a close relationship between the periodograms of $y_t$ and $x_t$ at the harmonic frequencies. If the series $(x_t - \bar{x})$, $t = 1, \ldots, n$ were (artificially) extended in a periodic manner, then the fact that $y_n = 0$ would cause $y_t$ as defined by (18.3.6) for $t > n$, to be similarly periodic, so that

$$y_t - y_{t-1} = x_t - \bar{x} \quad \text{for all } t.$$

Taking a component $R_j \cos (\omega_j t + \phi_j)$ in the harmonic representation of $y_t$ then leads to a component in $y_t - y_{t-1}$ given by [see IV, § 2.12]

$$R_j \cos (\omega_j t + \phi_j) - R_j \cos (\omega_j [t-1] + \phi_j)$$
$$= 2R_j \sin (\tfrac{1}{2}\omega_j) \cos (\omega_j t + \phi_j - \tfrac{1}{2}\omega_j + \pi/2).$$

Thus the frequency is unchanged, but the amplitude is reduced by the frequency dependent (but not data dependent) factor of $2 \sin (\tfrac{1}{2}\omega_j)$, to give the corresponding harmonic component of $(x_t - \bar{x})$. Consequently

$$I_n(\omega_j, y) = I_n(\omega_j, x)/4 \sin^2 (\tfrac{1}{2}\omega_j), \quad j = 1, \ldots, m$$

which from the results for model (iii) constitute a set of independent exponentially distributed random variables with means $2\sigma^2/4 \sin^2 (\tfrac{1}{2}\omega_j)$. For small values of $j$ these may be particularly large. Figure 18.3.3(b) shows such a periodogram for the series in Figure 18.1.1(e). Without knowledge of the purely stochastic structure of the series it might be tempting to suspect that the large peak at low frequency indicates a deterministic sinusoidal component.

### 18.3.6.   The Interpretation of Periodograms

It is evident that difficulties may arise in interpreting a periodogram of data arising as the sum of series such as we have described above. Their frequency components may reinforce or cancel each other, but on average their periodograms sum (as do variances). A deterministic sinusoidal component may be recognized by a spike in the periodogram with height proportional to $n$ and width to $1/n$, at the appropriate frequency. It should therefore stand out above the general level of the periodogram due to stochastic sources, which level tends not to change as $n$ increases. However, stochastic models can give rise to high peaks also, so that the interpretation cannot be clear unless further data becomes available, or other criteria concerning the nature of the series are used. As an example, Figure 18.3.4 shows part of the periodogram of the daylength series in Figure 18.1.1(c), after trend correction. It is tempting to explain some of the peaks as cycles of astronomical origin, but we shall show that a simple stochastic model may account equally well for these features. The similarity with Figure 18.3.3(b) is quite evident.

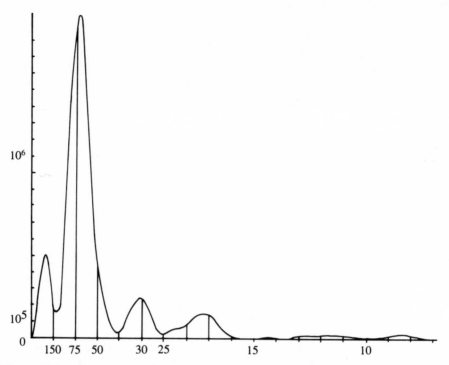

Figure 18.3.4: Part of the periodogram of the full daylength series (150 points) with harmonic frequencies emphasized and some periods indicated.

## 18.4.   DIFFERENCING OPERATIONS

In a time series context it is natural to represent deterministic functions of time by simple recurrence equations [see I, § 14.13] which show how successive values are related. For example the trend

$$x_t = a + bt$$

may be represented in two simple ways:

$$x_t = x_{t-1} + b$$

$$x_t = x_{t-1} + (x_{t-1} - x_{t-2}) = 2x_{t-1} - x_{t-2},$$

the first of these retaining one fixed constant, the second eliminating both. Using the backward difference operator defined by

$$\nabla x_t = x_t - x_{t-1}, \qquad \nabla^2 x_t = \nabla(\nabla x_t) = (x_t - x_{t-1}) - (x_{t-1} - x_{t-2})$$

we may conveniently write the above as

$$\nabla x_t = b, \qquad \nabla^2 x_t = 0.$$

If a time series $y_t$ appears to have a trend component, it may therefore be simpler, and more natural, to analyze the series $\nabla y_t$ of successive increases, or even $\nabla^2 y_t$. Furthermore, if $y_t$ appears to have a random walk component $w_t$ such as is illustrated in Figure 18.1.1(e), this will also be simplified, since by definition $\nabla w_t$ is a random series as in Figure 18.1.1(d).

A purely periodic series such as may form a component of a seasonal time series with period $s$, has the property

$$x_t = x_{t-s}.$$

Introducing the seasonal backward difference operator defined by

$$\nabla_s x_t = x_t - x_{t-s}$$

such a seasonal component satisfies $\nabla_s x_t = 0$. Given a seasonal time series $y_t$ it may therefore be simpler to analyze $\nabla_s y_t$, in which any deterministic seasonal component will be removed. For the airline series, the seasonal differences are shown in Figure 18.2.1(c). It is of interest to compare these with the residual series in Figure 18.2.1(b) obtained from the model (18.2.1). Both are the consequence of 'deseasonalizing operations'. In Figure 18.2.1(b) a fixed seasonal pattern (the same for each year) has been fitted over the whole series, then subtracted out (together with the trend). In Figure 18.2.1(c), for each year a different seasonal pattern has been subtracted out—simply the previous years values. Because the seasonal pattern is so marked and stable, the results are very similar, though Figure 18.2.1(c) still has a positive mean level (the mean annual increase) deriving from the trend. It is worth remarking however that apart from the mean level, Figure 18.2.1(c) is also the seasonal difference of Figure 18.2.1(b). The effect of subtracting out any fixed seasonal pattern is lost after seasonal differencing.

If a series contains both trend and seasonal components, they may both be removed by applying the operators $\nabla$ and $\nabla_s$ in succession (in either order), i.e.

$$\nabla \nabla_s y_t = (y_t - y_{t-s}) - (y_{t-1} - y_{t-s-1}).$$

The results of this for the airline series are shown in Figure 18.2.1(d), which is simply the first differences of Figure 18.2.1(c). This is a series with mean close to zero and fairly random appearance.

The emphasis on the use of these operators to simplify series by removing structure, needs to be balanced by a constructive interpretation. A series consisting purely of a fixed trend and seasonal pattern certainly has all its structure removed by these operations since $\nabla \nabla_s x_t = 0$. However, rewriting this equation as

$$x_t = x_{t-s} + (x_{t-1} - x_{t-s-1}) \tag{18.4.1}$$

shows how such a series may be indefinitely extended, starting with at least $s+1$ successive values. This rule may be applied to forecast, for example, the last year of the monthly airline data. In words, the rule takes the latest available annual increase—from December 1958 to December 1959, and applies this to the January, February, ... values for 1959, to produce forecasts for January,

February, ... of 1960. The results are shown in Figure 18.4.1, and are comparable with those obtained by extrapolating the fitted model (18.2.1), being better for the first four months. These two procedures are extremes, the model (18.2.1) placing similar weight on the whole of the available data, the rule (18.4.1) placing weight only on the most recent (13 months) data. We shall see in 18.7.6 how a compromise between these extremes is possible.

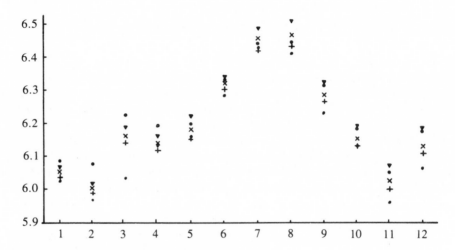

Figure 18.4.1: Forecasts of the airline series.
   ● Actual series values.
   ○ Forecasts by extrapolation of trend + monthly effects model (18.2.1)
   ▼ Forecasts derived from extrapolation equation (18.4.1) associated with differencing $\nabla\nabla_{12}$.
   × Forecasts derived from the seasonal EWMA model (18.7.42).
   + Forecasts derived from the Box–Jenkins model (18.7.43).

## 18.5.   STATIONARY TIME SERIES

The previous section leads on to the question of how we might determine the optimum weighting to be applied to observed values in order to construct forecasts. This optimum weighting must depend upon the statistical correlation between present and future values, and our way forward is to investigate the correlation structure of the observed series, after any necessary correction for fixed trends, cycles and seasonality using either regression or differencing methods.

### 18.5.1.   The Autocorrelation Function (acf)

A common assumption is that a time series is stationary, i.e. generated by some process which is not changing with time, and which has reached statistical equilibrium. In practice, the statistics of main interest are the first and second moments [see § 2.1.2], so for a time series $x_t$ the assumptions are:

(i) the expectation and variance [see II, chapters 8, 9] of the series are constant in time:

$$E(x_t) = \mu_x, \qquad \text{var}(x_t) = \sigma_x^2, \quad \text{for all } t; \qquad (18.5.1)$$

(ii) the covariance [see II, § 9.6.1] between any two terms of the series depends only upon the absolute time difference between them, or the separation lag:

$$\text{cov}(x_t, x_{t+k}) = \gamma_x(k) \quad \text{for all } t. \qquad (18.5.2)$$

By convention $\gamma_x(0) = \sigma_x^2$. The autocorrelation function (acf) is defined by

$$\rho_x(k) = \gamma_x(k)/\sigma_x^2, \quad k > 0; \qquad \rho_x(0) = 1. \qquad (18.5.3)$$

[c.f. II, § 22.2]. It is also useful to consider the acf as a doubly infinite sequence, taking $\rho_x(-k) = \rho_x(k)$.

Given observations $x_1, \ldots, x_n$ it is common to use the sample mean and the sample variance

$$\bar{x} = (1/n) \sum_1^n x_t, \qquad s^2 = (1/n) \sum_1^n (x_t - \bar{x})^2 \qquad (18.5.4)$$

and sample autocovariance and autocorrelation coefficients defined by

$$C_x(k) = (1/n) \sum_1^{n-k} (x_t - \bar{x})(x_{t+k} - \bar{x}); \qquad r_x(k) = C_x(k)/s_x^2; \quad k = 1, \ldots, n-1, \qquad (18.5.5)$$

taking $r_x(0) = 1$ [see § 2.1.2(b), (c)]. These should be looked upon as descriptive sample statistics. They are averages over time, not over independent realizations of the process. Thus $r_x(1)$ would not necessarily be the best estimate of $\rho_x(1)$ if we were to assume some parametric model for $x_t$. Nevertheless, with some appreciation of their limitations, these sample statistics are most useful. To determine their properties, further distributional assumptions need to be made about $x_t$. The following results hold for a Gaussian series—i.e. all marginal and joint distributions [see II, § 13.1] of the time series elements are assumed to be Normal. It is also assumed that $\sum |\rho_x(k)|$ is finite. This may not appear to be an unduly restrictive assumption since for most practical purposes $\rho_x(k)$ will be negligible for values of $k$ beyond some (suitably large) lag. If, however, there has been a failure to remove some deterministic component, such as a sinusoidal wave, from the series, the sample autocorrelations $r_x(k)$ will reflect this by failing to die out at high lags.

The following approximations are only true for fixed $k$ and large $n$, the errors being of the order $1/n$.

$$E(\bar{x}) = \mu_x, \qquad \text{var}(\bar{x}) \approx (1/n)\sigma_x^2 \left\{ 1 + 2 \sum_1^\infty \rho_x(j) \right\} \qquad (18.5.6)$$

$$E(s_x^2) \approx \sigma_x^2, \qquad \text{var}(s_x^2) \approx (1/n)2\sigma_x^4 \left\{1 + 2\sum_1^\infty \rho_x(j)^2\right\} \qquad (18.5.7)$$

$$E(r_x(k)) \approx \rho_x(k),$$

$$\text{var}(r_x(k)) \approx (1/n)\left\{1 + 2\sum_1^\infty \rho_x(j)^2\right\} + (1/n)\left\{\sum_{-\infty}^\infty \rho_x(j+k)\rho_x(j-k)^2\right\}.$$

$$(18.5.8)$$

The sample quantities are therefore consistent estimators [see § 3.3.1(c)], although for small $n$, $r_x(k)$ has appreciable bias [see Definition 3.3.2]. These formulae are of limited use in such tasks as constructing confidence intervals for $\rho_x(k)$ [see § 4.2] or testing for significant values [see § 5.1], since they involve the very quantities which are being estimated. A cautious 'bootstrap' approach is necessary. A starting point is the hypothesis that $\rho_x(k) = 0$ for all $k > 0$, from which S.E.$(r_x(k)) \approx 1/\sqrt{n}$. Values outside the range $\pm 2/\sqrt{n}$ are then considered. Take for example the sample acf shown in Figure 18.5.1(b) of the first differences of the daylength series, shown in Figure 18.5.1(a). Differencing has been applied because of the suspicion that the daylength series has a random walk component. The first 140 points have been used, the remainder being kept back as a check upon forecasts which we shall make.

Our first remark is that $|r_x(k)| > 2/\sqrt{n}$ (=0·17) for $k = 1, 2, 3, 12, 13, 14, 27 \ldots 33$. The very fact that so many lags appear, means that we are underestimating the S.E. (standard error: see § 2.1.2(c)). Proceeding cautiously let us revise our hypothesis by assuming merely that $\rho_x(1)$ and $\rho_x(2)$ are non-zero, corresponding to the two outstanding values of $r_x(1) = 0·76$, $r_x(2) = 0·52$.

Now using the formula (18.5.8) taking $\rho_x(k) = 0$ for $k > 2$ and estimating $\rho_x(1)$ and $\rho_x(2)$ by $r_x(1)$ and $r_x(2)$ gives

$$2\text{S.E.}[r_x(k)] \approx 0·17\{1 + 2(0·76^2 + 0·52^2)\} = 0·28, \quad \text{for } k > 2$$

Note that in general for tests of the form $\rho_x(j) = 0$ for $j > K$, the second term in the formula (18.5.8) for var $(r_x(k))$ will disappear for $k > K$. Using the new limits there is a slight indication that there may be significant values at lags 29, 30 with autocorrelations of 0·29, but when inspecting over 40 coefficients we would expect two extreme values by chance, and we remain sceptical about their significance. Our conclusion that only $\rho_x(1)$ and $\rho_x(2)$ are non-zero therefore appears consistent with the data. We shall see later that it is rather conservative, and needs to be extended. We have nevertheless a useful starting point from which to proceed to the fitting of parametric models which may be tested in a more precise manner, as considered in section 18.7.3. We finally comment on the temptation to attach meaning to the cycle which appears as a strong pattern to the eye in the sample acf. To understand this, consider the formula (independent of $n$),

$$\text{correlation}\{r_x(k), r_x(k+i)\} \approx \left\{\sum_{-\infty}^\infty \rho_x(j)\rho_x(j+i)\right\}\bigg/\left\{\sum_{-\infty}^\infty \rho_x(j)^2\right\} \quad (18.5.9)$$

which is appropriate for $k > K$ under the hypothesis that $\rho_x(j) = 0$ for $j > K$ (in our case $K = 2$). Thus the sequence $r_x(k)$ for $k > 2$ itself appears to be a stationary autocorrelated time series, but with an autocorrelation stronger than that of the original time series $x_t$ and extending up to a lag (between the coefficients) of $i = 4$. The values of $r_x(k)$ are only independent if separated by five lags or more; and follow a smooth pattern at intermediate lags, giving the appearance of waviness or cycles.

### 18.5.2.  The Relationship between the Periodogram and acf

The periodogram of a time series is very simply related to the sample acf—it is another way of looking at the same information. Assuming the series $x_t$ to be already mean corrected

$$\tfrac{1}{2} I_n(\omega) = s_x^2 \left\{ 1 + 2 \sum_1^{n-1} r_x(k) \cos \omega k \right\}, \qquad 0 \le \omega \le \pi \qquad (18.5.10)$$

—a cosine series whose coefficients depend upon the sample acf.

This result is derived from Definition 18.3.1 by first expressing, for example

$$\left( \sum_1^n x_t \sin \omega t \right)^2 = \left( \sum_{t=1}^n x_t \sin \omega t \right) \left( \sum_{s=1}^n x_s \sin \omega s \right)$$

then expanding and collecting terms in a double sum:

$$\tfrac{1}{2} I_n(\omega) = (1/n) \sum_{t=1}^n \sum_{s=1}^n x_t x_s (\sin \omega t \sin \omega s + \cos \omega t \cos \omega s)$$

$$= (1/n) \sum_{t=1}^n \sum_{s=1}^n x_t x_s \cos (\omega[s - t]).$$

Now collecting terms $x_t x_s$ for which $|s - t| = k$, for $k = 0, \ldots, (n-1)$ gives

$$(1/n) \left\{ \sum_1^n x_t^2 \right\} + 2 \sum_{k=1}^{n-1} \left\{ (1/n) \sum_{t=1}^{n-k} x_t x_{t+k} \right\} \cos \omega k$$

from which (18.5.10) follows directly.

The acf can be simply derived from the periodogram:

$$s_x^2 r_x(k) = (1/n) \left\{ \tfrac{1}{2} I_n(0) + \sum_{v=1}^{n-1} I_n(\omega_v) \cos k\omega_v + \tfrac{1}{2} I_n(\pi) \right\} \qquad (18.5.11)$$

where $\omega_v = \pi v / n$, $v = 1 \ldots n-1$ are the harmonic and mid-harmonic frequencies. This inversion formula can be derived from (18.3.4), and can provide an indirect but efficient means of computing the acf of large data sets, via the periodogram. Algorithms for fast Fourier transforms are used in this task.

### 18.5.3.   The Spectrum and Sample Spectrum

The representation (18.5.10) of the periodogram in terms of the sample acf directly motivates a definition involving the theoretical acf.

DEFINITION 18.5.1.   *Spectrum.*   The spectrum of a time series with acf $\rho_x(k)$ is

$$f_x(\omega) = (1/\pi)\sigma_x^2 \left\{ 1 + 2 \sum_1^\infty \rho_x(k) \cos \omega k \right\}, \qquad 0 \le \omega \le \pi.$$

DEFINITION 18.5.2.   *Sample spectrum.*   The sample spectrum of observations $x_1, \ldots, x_n$ is

$$f_x^*(\omega) = (1/\pi)s_x^2 \left\{ 1 + 2 \sum_1^{n-1} r_x(k) \cos \omega k \right\} = (1/2\pi)I_n(\omega), \qquad 0 \le \omega \le \pi.$$

The factor $(1/2\pi)$ which is all that distinguishes the sample spectrum from the periodogram, ensures that

$$\int_0^\pi f_x(\omega) \, d\omega = \sigma_x^2. \qquad (18.5.12)$$

This describes how the total variance $\sigma_x^2$ may be considered to be distributed over the range $[0, \pi]$ with density $f_x(\omega)$, by analogy with the analysis of variance (18.3.5) which may be expressed (for $n$ even)

$$\sum_1^m I_n(\omega_j) = \sum_1^n (x_i - \bar{x})^2, \qquad (18.5.13)$$

—a distribution of the sample variance among the harmonic components. In different contexts the spectrum may be normalized by different factors, e.g. $(1/\pi)$ is replaced by 2 if the absolute frequency $f = \omega/2\pi$ is used over the range $[0, \cdot 5]$. The fact that $f_x(\omega) \ge 0$ will shortly be demonstrated.

The equivalence of the spectrum and acf is demonstrated by an inversion formula analogous to (18.5.11). Because the spectrum is an infinite cosine series we need to use the integral orthogonality properties of the cosine function [see II, § 20.4] to extract its coefficients:

$$\sigma_x^2 \rho_x(k) = \int_0^\pi f_x(\omega) \cos \omega k \, d\omega, \qquad k = 0, 1, 2, \ldots. \qquad (18.5.14)$$

The definition of the spectrum is basically motivated by the large sample property

$$E\{f_x^*(\omega)\} \to f_x(\omega) \quad \text{as } n \to \infty \qquad (18.5.15)$$

which not surprisingly follows directly from the fact that $E\{r_x(k)\} \to \rho_x(k)$.

The more surprising and significant result is that although it is also true that $r_x(k) \to \rho_x(k)$ in a probabalistic sense for each fixed $k$ as $n \to \infty$, so that each term of the sum in Definition 18.5.2 converges to the corresponding term in

Definition 18.5.1, in no sense does $f_x^*(\omega) \to f_x(\omega)$. In fact the sample spectrum always has a fixed distribution about the theoretical spectrum:

**THEOREM 18.5.1**   (*Distribution of sample spectrum*).   *For large n, $f_x^*(\omega_j)$, $j = 1, \ldots, m$ are independent exponential variates with expectations $f_x(\omega_j)$, where $\omega_j$ are the harmonic frequencies.*

We do not prove this theorem, but point out that in the special case when $\rho_x(k) = 0$ for $k > 0$, i.e. $x_1, \ldots, x_n$ from a random sample and the spectrum $f_x(\omega) = \sigma_x^2 / \pi$ is constant, we have already demonstrated its truth when investigating the periodogram properties for model (iii) in 18.3.5. The simplest assumptions necessary for the theorem are those used in 18.5.1, but these may be broadened considerably, as outlined in 18.6.4.

The fact that the sample spectrum is not a consistent estimator [see § 3.3.1(c)] is of great practical importance, and not immediately obvious from Definition 18.5.2. The point is that as $n$ increases further terms enter into the sum in that definition, and the construction of consistent estimators of the spectrum requires that the terms in the sum be in some way restricted. Such smoothing methods are described in 18.10.2; however it is not difficult for the eye to see the underlying spectral shape when inspecting the sample spectrum or periodogram.

### 18.5.4.   Properties of the Spectrum and acf

A direct consequence of (18.5.15) is that $f_x(\omega) \geq 0$, since the sample spectrum or periodogram is by definition non-negative. Otherwise the spectrum may be any sufficinetly smooth continuous function. Continuity of the spectrum is a consequence of our condition that $\sum |\rho_x(k)| < \infty$. Conversely this condition holds if $f_x(\omega)$ is sufficiently smooth. The spectrum is therefore a natural way of characterizing the structure of stationary time series. Broad peaks in the spectrum reflect irregular cyclic behaviour in the series, and sharp peaks the more regular cycles, which will be difficult to distinguish from deterministic cycles given small samples from the series. Indeed, such deterministic components if not previously extracted from the series are referred to as the discrete part of the spectrum by analogy with discrete probability distributions, and in contrast to the continuous spectral density $f_x(\omega)$. A series with discrete and continuous components has a mixed spectrum—the variable star data is an example which is further analysed in section 18.11.

In contrast the acf is subject to the usual constraints upon correlations. These are summarized by considering the correlation matrix $R_n$ [see Definition 16.1.1] of $n$ successive terms $x_1, \ldots x_n$ say,

$$R_n = \begin{pmatrix} 1 & \rho_x(1) & \rho_x(2) & \cdots & \rho_x(n-1) \\ \rho_x(1) & 1 & \rho_x(1) & \cdots & \rho_x(n-2) \\ \vdots & & & \ddots & \rho_x(1) \\ \rho_x(n-1) & & \cdots & \rho_x(1) & 1 \end{pmatrix}. \qquad (18.5.16)$$

The requirement that this highly structured matrix be positive definite [see Definition 16.1.3] for all $n$ imposes necessary and sufficient conditions upon the acf which are exactly equivalent to the requirement that the spectrum be positive. Take for example an acf satisfying $\rho_x(k) = 0$ for $k > 1$. Then the range of possible values for $\rho_x(1)$ is $[-0·5, 0·5]$. This is seen by considering the spectrum $f_x(\omega) = \sigma_x^2\{1 + 2\rho_x(1) \cos \omega\}$, which will be negative at one end or the other of the range $[0, \pi]$ unless $|2\rho_x(1)| \leq 1$. If $\rho_x(k) = 0$ for $k > 2$ then $\rho_x(1)$ may not exceed $1/\sqrt{2}$, and to attain this value requires $\rho_x(2) = \frac{1}{4}$. The values of $r_x(1)$ and $r_x(2)$ observed for the first difference of the daylength data in 18.5.1 are $0·76$ and $0·52$, well outside the possible range even allowing for sampling fluctuations, and make us aware of the need to relax the assumption $\rho_x(k) = 0$ for $k > 2$.

### 18.5.5.  The Partial Autocorrelation Function (pacf)

Yet another method of usefully representing the information in the acf arises from the viewpoint of prediction. Given the acf $\rho_x(k)$ of a stationary time series we have all the information required to construct the linear least squares (one step ahead) predictor of $x_t$ in terms of a finite number of previous values, say $x_{t-1}, \ldots x_{t-k}$. If the time series were Gaussian this could be interpreted as the conditional expectation [see II, § 8.9] of $x_t$ given $x_{t-1}, \ldots, x_{t-k}$, which would be in this case a linear regression upon these values. Asking for the linear least squares predictor allows us to relax the Gaussian assumption, whilst still giving the same answer in the Gaussian case, the coefficients in the predictor being determined by the correlations between the variables, i.e. $\rho_x(1), \ldots \rho_x(k)$. We write the prediction or regression equation

$$x_t = \phi_{k,1}x_{t-1} + \phi_{k,2}x_{t-2} + \ldots + \phi_{k,k}x_{t-k} + e_{k,t} \qquad (18.5.17)$$

where the linear combination of $x_{t-1}, \ldots, x_{t-k}$ is the predictor, and $e_{k,t}$ is the predictor error. The first subscript $k$ of the coefficients $\phi_{k,j}, j = 1 \ldots k$, and of $e_{k,t}$ is to emphasize that as further terms, e.g. $x_{t-(k+1)}$ are introduced into the predictor, the existing coefficients are liable to change. We have omitted any constant term from the predictor by assuming that $\mu_x = 0$, otherwise $x_t$ should simply be replaced by $x_t - \mu_x$. We denote the predictor error variance var $(e_{k,t})$ by $\sigma_k^2$, and the ratio $\sigma_k^2/\sigma_x^2$ by $v_k$.

Ostensibly a set of $k$ linear least squares equations need to be solved to determine the coefficients, but the autocorrelation structure of time series allows a rapid recursive method of determining the solution. This has been, and still is, of great practical importance, particularly in subjects such as geophysics where large amounts of data provide good acf estimates. The order $k$ of the predictor may then be several hundreds.

The recursive cycle of equations for calculating the coefficients is now stated. It is initialized by $v_0 = 1$. The first cycle is

$$\phi_{1,1} = \rho_x(1); \qquad v_1 = v_0\{1 - \phi_{1,1}^2\} \qquad (18.5.18)$$

and in general for $k > 1$

$$\phi_{k+1,k+1} = \{\rho_x(k+1) - \phi_{k,1}\rho_x(k) - \ldots - \phi_{k,k}\rho_x(1)\}/v_k \quad (18.5.19)$$

$$\phi_{k+1,j} = \phi_{k,j} - \phi_{k+1,k+1}\phi_{k,k+1-j}; \quad j = 1, \ldots, k \quad (18.5.20)$$

$$v_{k+1} = v_k\{1 - \phi_{k+1,k+1}^2\} \quad (18.5.21)$$

We outline a derivation of these equations based on standard least squares theory [see Chapter 8] which we present in a more general context.

Suppose the least squares regression [see § 6.5.1] of a dependent variable upon a set of variables $x_1, \ldots, x_k$ has been determined, with error $e_k$. If a new variable $x_{k+1}$ is introduced which is uncorrelated with $x_1, \ldots, x_k$, then the existing coefficients are unchanged, and the coefficient $\alpha_{k+1}$ of $x_{k+1}$ may be expressed in the forms

$$\alpha_{k+1} = \text{cov}\,(y, x_{k+1})/\text{var}\,(x_{k+1}) = \text{cov}\,(e_k, x_{k+1})/\text{var}\,(x_{k+1}) \quad (18.5.22)$$

$$= \rho\{\text{var}\,(e_k)/\text{var}\,(x_{k+1})\}^{1/2}$$

where

$$\rho = \text{correlation}\,(e_k, x_{k+1}). \quad (18.5.23)$$

Moreover the error variance is reduced by that of the new term to give

$$\text{var}\,(e_{k+1}) = \text{var}\,(e_k) - \alpha_{k+1}^2\,\text{var}\,(x_{k+1}) = \text{var}\,(e_k)(1 - \rho^2). \quad (18.5.24)$$

If the new variable $x_{k+1}$ is not uncorrelated with $x_1, \ldots, x_k$ it may be replaced by $x'_{k+1}$ obtained by orthogonalization, i.e. correcting $x_{k+1}$ by subtracting its linear predictor in terms of the existing variables $x_1, \ldots, x_k$. The correlation $\rho$ in (18.5.23) is then called the partial correlation [see II, § 13.4.6(iii)] between $x_{k+1}$ and $y$, and in a Gaussian context is indeed the correlation conditional upon $x_1, \ldots, x_k$.

We apply these results to time series, starting from equation (18.5.17). Introducing $x_{t-(k+1)}$ as the new variable, we cannot assume this to be uncorrelated with $x_{t-1}, \ldots, x_{t-k}$, so construct the corrected variable $x'_{t-(k+1)}$. But this is simply

$$x'_{t-(k+1)} = x_{t-(k+1)} - \phi_{k,1}x_{t-k} - \phi_{k,2}x_{t-k+1} - \ldots - \phi_{k,k}x_{t-1} \quad (18.5.25)$$

i.e. the same coefficients are applied to the same variables, but in reverse order, to construct the backwards predictor to correct $x_{t-(k+1)}$.

It is this last step which exploits the special structure of stationary time series to computational advantage. It follows directly from the fact that correlation between the variables depends only on the absolute time lag between them. Furthermore, since $x'_{t-(k+1)}$ is a (backward) prediction error

$$\text{var}\,(x'_{t-(k+1)}) = \text{var}\,(e_{k,t}) = \sigma_x^2 v_k \quad (18.5.26)$$

The new coefficient of $x_{t-(k+1)}$ is then given, using (5.22), by

$$\phi_{k+1,k+1} = \text{cov}\,(x_t, x'_{t-(k+1)})/\text{var}\,(x'_{t-(k+1)}). \quad (18.5.27)$$

Using (18.5.25) and (18.5.26) together with (18.5.27) directly gives the first equation of the cycle, (18.5.19). Adding the new term $\phi_{k+1,k+1}x'_{t-(k+1)}$ onto the previous predictor amends the existing coefficients as in the second equation of the cycle, (18.5.20). Finally, using (18.5.26), the new coefficient as given by (18.5.23) is equated with the correlation: $\phi_{k+1,k+1} = \rho$. Then (18.5.24) furnishes the last step of the cycle, (18.5.21).

The partial autocorrelation between $x_t$ and $x_{t-(k+1)}$ given $x_{t-1}, \ldots x_{t-k}$ is simply the coefficient $\phi_{k+1,k+1}$, which therefore measures the additional information supplied by $x_{t-(k+1)}$ in the linear prediction of $x_t$. The pacf is the sequence of these coefficients. The only constraints they satisfy are $|\phi_{k,k}| < 1$, otherwise they may take any set of values—and if such a set of values were prescribed, a corresponding valid acf could be constructed by reversing the recursive cycle. However, the almost invariable existence of a positive lower limit for the predictor error variance $\sigma_x^2 v_k$ implies that $\sum \phi_{k,k}^2$ is convergent, and in particular $\phi_{k,k} \to 0$.

### 18.5.6. The Sample pacf

If sample values $r_x(k)$ are used in place of $\rho_x(k)$, the sample pacf $\hat{\phi}_{k,k}$ may be constructed. Their statistical properties are required in order to decide when the order $k$ is sufficiently large to provide an adequate description of the structure of the time series, possibly with the intention of using the coefficients $\hat{\phi}_{k,j}$ for predicting future values.

Under the null hypothesis that the true values satisfy

$$\phi_{k,k} = 0 \quad \text{for } k > K, \tag{18.5.28}$$

then, given estimates based on a sample $x_1, \ldots, x_n$,

$$E(\hat{\phi}_{k,k}) \approx 0, \quad \text{var}(\hat{\phi}_{k,k}) \approx 1/n,$$
$$E(\hat{\sigma}_k^2) = (1 - k/n)\sigma_k^2 \quad \text{for } k > K. \tag{18.5.29}$$

One approach is to set limits at $\pm 2/\sqrt{n}$ about 0 on the graph of $\hat{\phi}_{k,k}$ and look for the lag $K$ beyond which the graph effectively lies within these limits. Since the approximations in (18.5.29) require that $k \ll n$, the selected value of $K$ must be relatively low if any direct application to prediction is to be effective. Another approach is to graph the bias corrected estimate of $\sigma_k^2$ given by $(1 - k/n)^{-1}\hat{\sigma}_k^2$, and to find the lag $K$ where this levels out, indicating that increasing the order is not improving the fit. More recently the Final Prediction Error (FPE) criterion $(1 + k/n)(1 - k/n)^{-1}\hat{\sigma}_k^2$ has been used. The factor $(1 + k/n)$ allows for the increase in prediction error variance arising from use of the estimated coefficients $\hat{\phi}_{k,j}$ to construct the predictor. It introduces a penalty on an excessive order $k$. The selected order $K$ is that for which the FPE is a minimum.

It is wise to use these procedures as means of exploring the data for possible structure, rather than making final decisions about the prediction order. As

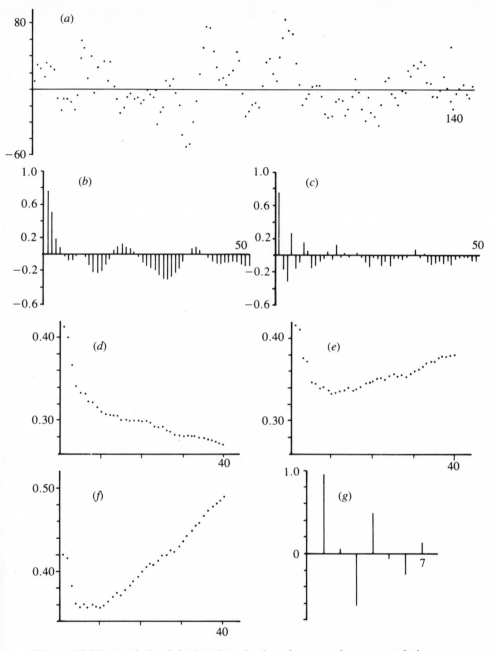

Figure 18.5.1: Analysis of daylength series based on sample autocorrelations.
    (a) First differences of the daylength series.
    (b) Sample autocorrelations of series (a).
    (c) Sample partial autocorrelations of series (a).
    (d) Prediction error variance estimates $\hat{\sigma}_k^2$ for linear predictor of order $k$.
    (e) Bias corrected estimates $(1 - k/n)^{-1}\sigma_k^2$.
    (f) Final prediction error estimates $(1 + k/n)(1 - k/n)^{-1}\sigma_k^2$.
    (g) Coefficients $\phi_{K,j}$, $j = 1 \ldots 7$ for linear predictor of order $K = 7$.

an example, Figures 18.5.1(c), (d), (e), (f) show graphs of $\hat{\phi}_{k,k}$, $\hat{\sigma}_k^2$, $(1-k/n)^{-1}\hat{\sigma}_k^2$ and $(1+k/n)(1-k/n)^{-1}\hat{\sigma}_k^2$ for the first differences of the daylength series. The selected order $K$ in this case would be 7, using the last criterion. The corresponding predictor coefficients are shown in Figure 18.5.1(g). It is evident that shortly after levelling out, the bias corrected variance sequence starts to increase steadily, the bias correction factor being overgenerous for larger values of $k$. Forecasts of the last 10 points of the daylength series constructed using the predictor coefficients are shown in Figure 18.5.2. The

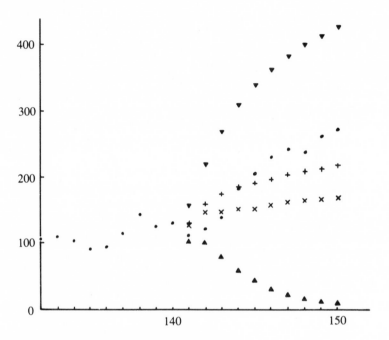

Figure 18.5.2: Forecasts of the daylength series.
  ● Actual series values.
  × Forecasts derived from the linear least squares predictor of order 7 for $\nabla x_t$.
  + Forecasts derived from the trend + ARMA (2, 2) model (18.11.2).
  ▼▲ Upper and lower 90% probability limits for forecasts +.

predictions for more than one step ahead were made by treating each successive forecast as it is were an actual observation, and repeatedly applying the one-step ahead predictor. The forecasts of the original series were then recovered by successively adding on those of the differenced series. The drastic loss of forecast accuracy as the lead time increases is typical of a series such as this, which has something of the nature of a random walk. Forecasts based on a more rigorous model building procedure, described in section 18.11 are shown for comparison on the same figure, together with the actual series values.

## 18.6. THE GENERAL LINEAR MODEL (GLM)

### 18.6.1. Definition and Properties

The simplest stationary time series is a sequence $\{a_t\}$ of independent and identically distributed random variables. Forming a new time series $\{x_t\}$ as a linear combination of present and past values of $a_t$ induces autocorrelation. The general linear model (GLM) supposes that a given time series $\{x_t\}$ has arisen in this manner, i.e. that

$$x_t = \psi_0 a_t + \psi_1 a_{t-1} + \psi_2 a_{t-2} + \ldots = \sum_0^\infty \psi_i a_{t-i}. \qquad (18.6.1)$$

The series $a_t$ is not directly observed. It is often termed 'white noise' because the noise which is often heard in loud speakers (in the absence of any true signal) is characterized by an electrical signal with similar properties. Thus when analysed into its frequency components these are uniformly represented across the spectrum, just as a white light may be analysed into an approximately uniform spectrum. As we shall see, the autocorrelated series $a_t$ no longer has a uniform spectrum, so is sometimes called 'coloured noise', and the operation (18.6.1) by which it is derived, a 'filter'. Although noise often refers to unwanted contamination of data by observation error, 'noise models' such as the GLM, are finding an increasing role in the modeling of many phenomena.

This model is of great use in theoretical investigations of the distributional properties of sample autocorrelations and spectra. The finite parameter models to be introduced in the following sections, are of great practical importance, and are special cases of the GLM. Its correlation properties are soon established. Assuming that var $(a_t) = \sigma_a^2$ is finite,

$$\text{var}(x_t) = \left( \sum_0^\infty \psi_i^2 \right) \sigma_a^2, \qquad (18.6.2)$$

so for the model to be meaningful $\sum \psi_i^2$ must converge. It is usual to assume the stronger condition that $\sum |\psi_i|$ converges. Then

$$\gamma(k) = \text{cov}(x_t, x_{t+k}) = \text{cov}(\sum \psi_i a_{t-i}, \sum \psi_j a_{t+k-j}) = \left( \sum_0^\infty \psi_i \psi_{i+k} \right) \sigma_a^2, \quad (18.6.3)$$

which follows because the term $\psi_i a_{t-i}$ in $x_t$ is correlated only with the term $\psi_j a_{t+k-j}$ in $x_{t+k}$ for which $j = i + k$.

Other simple properties derived directly from (6.1) are cov $(a_t, x_{t-k}) = 0$, cov $(a_t, x_t) = \psi_0 \sigma_a^2$, cov $(x_t, a_{t-k}) = \psi_k \sigma_a^2$ where $k \geq 1$.

### 18.6.2. Linear Operations on Time Series

The GLM is an example of the application of a linear operation (or linear filter) to one series, to produce another. Such operations are common in time series analysis. Consider the equation

$$v_t = \sum_j \alpha_j u_{t-j} \tag{18.6.4}$$

where we ask only that the time series $\{u_t\}$ satisfy $E(|u_t|) < K$ say, for all $t$, and the coefficients $\alpha_i$ satisfy $\sum |\alpha_i| \le M$. Then $\{v_t\}$ is well defined, satisfying $E(|v_t|) \le KM$. The range of the index $i$ may be from $-\infty$ to $\infty$, though in most contexts 0 to $\infty$ is more usual. A useful notation for such operations involves the backward shift operator $B$ which acts on a time series $\{u_t\}$ to produce $\{u_{t-1}\}$. We write

$$Bu_t = u_{t-1}, \qquad B^j u_t = u_{t-j}. \tag{18.6.5}$$

The operation (18.6.4) is therefore represented

$$v_t = \sum_j \alpha_j B^j u_t = \alpha(B) u_t \tag{18.6.6}$$

where the operator

$$\alpha(B) = \sum_j \alpha_j B^j \tag{18.6.7}$$

is, formally, a power series [see IV, § 1.10] in $B$. The notation is algebraically convenient since if further:

$$w_t = \sum_j \beta_j v_{t-j} = \beta(B) v_t \tag{18.6.8}$$

then

$$w_t = \beta(B)\alpha(B) u_t. \tag{18.6.9}$$

By this we mean that $w_t$ may be expressed in terms of $u_t$:

$$w_t = \sum_j \delta_j u_{t-j} = \delta(B) u_t \tag{18.6.10}$$

where

$$\delta(B) = \beta(B)\alpha(B). \tag{18.6.11}$$

Thus the results of successive linear operations corresponds to formal products of the operators.

One characteristic of linear operations is that when applied to a sinusoidal series they leave it unchanged apart from its amplitude and phase. Thus, without loss of generality, taking

$$u_t = \cos \omega t \tag{18.6.12}$$

the series defined by (18.6.4) is

$$v_t = \sum_j \alpha_j \{\cos \omega t \cos \omega j + \sin \omega t \sin \omega j\}$$

$$= C(\omega) \cos \omega t + S(\omega) \sin \omega t = R(\omega) \cos \{\omega t + \phi(\omega)\}, \tag{18.6.13}$$

where

$$C(\omega) = \sum \alpha_j \cos \omega_j, \; S(\omega) = \sum \alpha_j \sin \omega_j \qquad (18.6.14)$$

are the real and imaginary parts [see IV: § 9.5.8] of

$$\sum \alpha_j e^{i\omega j} = \alpha(e^{i\omega}), \qquad (18.6.15)$$

the result of substituting $e^{i\omega}$ for the formal operator $B$ in (18.6.7).

The *gain $R(\omega)$*, defined by

$$R(\omega)^2 = C(\omega)^2 + S(\omega)^2 = |\alpha(e^{i\omega})|^2 \qquad (18.6.16)$$

is of particular interest, considered as a function of frequency over the range $0 \le \omega \le \pi$. Much of classical time series analysis in applications such as seasonal adjustment and smoothing, is concerned with the design of filters of particular shapes, so as to remove or to extract certain frequency components of a series. Furthermore, the relationship (18.6.6) may be inverted provided only that $R(\omega) \ne 0$ for all $\omega$, i.e. given this condition, it is possible to write

$$u_t = v(B) v_t, \qquad \sum |v_i| < \infty \qquad (18.6.17)$$

where $v(B)\alpha(B) \equiv 1$.

### 18.6.3.  Linear Operations and the Spectrum

The effect of linear operations upon the correlation structure of stationary time series is simple. Assuming that $\{u_t\}$ in (6.4) is stationary with $\text{cov}\,(u_t, u_{t+k}) = \gamma_u(k)$, then it follows directly that

$$\text{cov}\,(v_t, v_{t+k}) = \gamma_v(k) = \sum_i \sum_j \alpha_i \alpha_j \gamma_u(k - i + j) \qquad (18.6.18)$$

which may be conveniently expressed in a form similar to (18.6.6):

$$\gamma_v(k) = \alpha(B)\alpha(B^{-1})\gamma_u(k) \qquad (18.6.19)$$

where $B$ is now interpreted as acting upon the index $k$. Expressing the product

$$\alpha(B)\alpha(B^{-1}) = A(B) = \sum_l A_l B^l \qquad (18.6.20)$$

where

$$A_l = A_{-l} = \sum_i \alpha_i \alpha_{i+l} \qquad (18.6.21)$$

we obtain a symmetric linear filter relating the covariance sequences:

$$\gamma_v(k) = \sum_l A_l \gamma_u(k - l) = A(B)\gamma_u(k). \qquad (18.6.22)$$

For example, if

$$v_t = \tfrac{1}{3}\{u_t + u_{t-1} + u_{t-2}\} \tag{18.6.23}$$

—a one sided, equal weight, three point filter, so that

$$\alpha_0 = \alpha_1 = \alpha_2 = \tfrac{1}{3},$$

then

$$A_0 = \alpha_0^2 + \alpha_1^2 + \alpha_2^2 = \tfrac{3}{9}, \qquad A_1 = \alpha_0\alpha_1 + \alpha_1\alpha_2 = \tfrac{2}{9}, \qquad A_2 = \alpha_0\alpha_2 = \tfrac{1}{9}$$

and

$$\gamma_v(k) = \tfrac{1}{9}\gamma_u(k+2) + \tfrac{2}{9}\gamma_u(k+1) + \tfrac{3}{9}\gamma_u(k) + \tfrac{2}{9}\gamma_u(k-1)\tfrac{1}{9}\gamma_u(k-2) \tag{18.6.24}$$

—a filter spanning five lags.

The effect of the linear operation on the spectrum is simply expressed as

$$f_v(\omega) = R(\omega)^2 f_u(\omega), \tag{18.6.25}$$

the intuitive interpretation being that the linear operation multiplies the variance of each frequency component of the series, by the squared gain at that frequency. We derive (18.6.25) by introducing the covariance generating function

$$\Gamma_u(B) = \sum_{-\infty}^{\infty} \gamma_u(k) B^k \tag{18.6.26}$$

(formally resembling a probability generating function, for which see II, § 12.1) whence (18.6.22) is equivalent to

$$\Gamma_v(B) = A(B)\Gamma_u(B) = \alpha(B)\alpha(B^{-1})\Gamma_u(B). \tag{18.6.27}$$

We then substitute $B = e^{i\omega}$ and use the fact that

$$\Gamma_u(e^{i\omega}) = \sum_{-\infty}^{\infty} \gamma_u(k)\, e^{i\omega k} = \gamma_u(0) + 2\sum_{1}^{\infty} \gamma_u(k) \cos k\omega = 2\pi f_u(\omega)$$

$$\tag{18.6.28}$$

and

$$\alpha(e^{i\omega})\alpha(e^{-i\omega}) = |\alpha(e^{i\omega})|^2 = R(\omega)^2. \tag{18.6.29}$$

On cancelling the factor $2\pi$ this gives (18.6.25). A useful result is that

$$R(\omega)^2 = A(e^{i\omega}) = \sum_{-\infty}^{\infty} A_l\, e^{i\omega l} = A_0 + 2\sum_{1}^{\infty} A_l \cos l\omega, \tag{18.6.30}$$

i.e. a cosine series. For example the filter (18.6.23) has squared gain

$$R(\omega)^2 = \tfrac{1}{9}\{3 + 4\cos\omega + 2\cos 2\omega\}.$$

In the particular case of the GLM, rewriting (18.6.1) as

$$x_t = \psi(B)a_t \tag{18.6.31}$$

the covariance properties (18.6.3) are a special case of (18.6.22). The spectrum of $\{x_t\}$ is proportional to the squared gain $R_\psi(\omega)^2$ of the operator $\psi(B)$, since $f_a(\omega) = \sigma_a^2/\pi$ is constant, and (18.6.25) gives

$$f_x(\omega) = R_\psi(\omega)^2 \sigma_a^2 / \pi. \tag{18.6.32}$$

### 18.6.4.   Linear Operations on a Finite Sample

Given a finite sample $u_1, u_2, \ldots, u_n$, it is not generally possible to calculate $v_1, v_2, \ldots v_n$ using (18.6.4). For example, the filter defined by (18.6.23) will only allow $v_3, \ldots v_n$ to be calculated exactly. Alternatively, approximations to $v_1, v_2$ may be computed either by assuming that the unknown values $u_{-1}$, $u_0$ (which are necessary for their computation) are zero, or extrapolating values for them in some way.

In general the magnitude of transient errors, or end-effects, so introduced will depend on how rapidly the sequence of filter weights decay. Given samples $u_1, \ldots u_n$ and $v_1, \ldots v_n$ of series which are related by (18.6.4) for all $t$, then their sample autocovariances are related in a manner analogous to (18.6.22), i.e.

$$C_v(k) \simeq \sum_l A_l C_u(k-l) \tag{18.6.33}$$

and their sample spectra in a manner analogous to (18.6.25), i.e.

$$f_v^*(\omega) \simeq R(\omega)^2 f_u^*(\omega). \tag{18.6.34}$$

The approximations, being due again to the end effects, are of order $(1/n)$ if the filter is finite or decays sufficiently rapidly (e.g. geometrically).

This last result applied to the GLM gives

$$f_x^*(\omega) \simeq R_\psi(\omega)^2 f_a^*(\omega)$$

or

$$f_x^*(\omega)/f_x(\omega) \simeq f_a^*(\omega)/f_a(\omega), \text{ provided } f_x(\omega) \neq 0. \tag{18.6.35}$$

The GLM therefore allows the distributional properties of the sample spectrum of $\{x_t\}$ to be directly related to those of the series $\{a_t\}$. The periodogram properties established in section 18.3.5 demonstrate that the RHS of (18.6.35), and hence the LHS, from a random sample form the exponential distribution when taken at the harmonic frequencies $\omega_j = 2\pi j/n$, providing a proof for Theorem 18.5.1. Furthermore, the distribution of $\{a_t\}$ need not be Normal, it being possible by use of the Central Limit Theorems [see II, § 17.3] to relax this condition, since $\{a_t\}$ is an independent series.

As an illustration, Figure 18.6.1(a) shows the sample spectrum of an independent series $\{a_t\}$, together with its theoretical spectrum $f_a(\omega) = 1/\pi$. Figure 18.6.1(b) shows the sample spectrum of $x_t$ defined by

$$x_t = \tfrac{1}{3}(a_t + a_{t-1} + a_{t-2}) \tag{18.6.36}$$

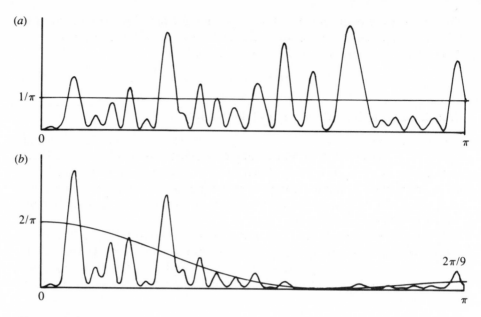

Figure 18.6.1: Illustration of the similar pattern of fluctuation of sample spectra of filtered series.
  (a) Sample spectrum of a random series $a_t$, $t = 1 \ldots 64$ and its theoretical spectrum $f_a(\omega) = 1/\pi$.
  (b) Sample spectrum of the series $x_t$, $t = 1 \ldots 64$ obtained by application of the filter in (18.6.36) to the series $a_t$, together with its theoretical spectrum.

together with its theoretical spectrum

$$f_x(\omega) = \left(\frac{1}{9\pi}\right)\{3 + 4 \cos \omega + 2 \cos 2\omega\}.$$

Both samples are of length $n = 64$. Note the almost identical pattern of fluctuation of the sample spectra relative to the theoretical spectra. Note also that $f_x(\omega) = 0$ at $\omega = \frac{2}{3}\pi$, i.e. the filter in (6.36) has zero gain at this frequency.

### 18.6.5.  Constraints on the GLM

The GLM involves a one sided operator, acting on present and past values of $a_t$. This constraint, together with a further condition upon the operator $\psi(B)$ in (18.6.31) allows a useful interpretation of the series $\{a_t\}$. This is the invertibility condition:

$$\psi(B) \neq 0 \quad \text{for } |B| \leq 1 \tag{18.6.37}$$

where we are now allowing real or complex numbers to be substituted for $B$. The result of this condition is that the inverse operator exists and is also one sided, so that $a_t$ is a linear function of present and past values of $x_t$:

$$a_t = \pi_0 x_t - \pi_1 x_{t-1} - \pi_2 x_{t-2} - \ldots$$

or

$$a_t = \pi(B) x_t$$

where $\pi(B) = \{\psi(B)\}^{-1}$ and $\Sigma|\pi_i| < \infty$. It is conventional in the GLM to take $\psi_0 = \pi_0 = 1$ since $\sigma_a^2$ allows for scaling, so that

$$x_t = \{\pi_1 x_{t-1} + \pi_2 x_{t-2} + \ldots\} + a_t. \tag{18.6.38}$$

Viewed like this we have a regression equation, the error $a_t$ being uncorrelated with the regression variables $x_{t-1}, x_{t-2}, \ldots$ because $\text{cov}(a_t, x_{t-k}) = 0$ for $k > 0$, as stated in section 18.6.1. Consequently, $\{\pi_1 x_{t-1} + \pi_2 x_{t-2} + \ldots\}$ is the linear least squares prediction of $x_t$ in terms of all past values, and $a_t$ is the corresponding one step ahead prediction error, or innovation series. As an example consider a very simple case of the GLM

$$x_t = a_t - \theta a_{t-1} = (1 - \theta B) a_t.$$

The only zero of $(1 - \theta B)$ is at $B = 1/\theta$, which lies outside $|B| \leq 1$ provided $|\theta| < 1$. In that case the required inverse is

$$(1 - \theta B)^{-1} = 1 + \theta B + \theta^2 B^2 + \ldots$$

giving

$$x_t = -\{\theta x_{t-1} + \theta^2 x_{t-2} + \ldots\} + a_t. \tag{18.6.39}$$

In the case that $|\theta| > 1$ it is still possible to obtain an inverse

$$(1 - \theta B)^{-1} = -\theta^{-1} B^{-1} (1 - \theta^{-1} B^{-1})^{-1} = -\theta^{-1} B^{-1} - \theta^{-2} B^{-2} - \ldots$$

i.e.

$$a_t = -\theta^{-1} x_{t+1} - \theta^{-2} x_{t+2} - \ldots$$

but this has no predictive interpretation.

It is reasonable to ask whether an invertible GLM is always appropriate for a stationary time series. The simplified answer in practice is yes, provided all the deterministic sinusoidal components have been removed, and if the independence property required of $\{a_t\}$ is relaxed to absence of correlation: $\rho_a(k) = 0$ for $k \geq 1$. The GLM equations (18.6.1) and (18.6.38) therefore provide a plausible starting point for finite parameter time series models. The innovation series $\{a_t\}$ for any given stationary series may be constructed in theory as the limit of the prediction errors $\{e_{k,t}\}$ in (18.5.17) as the order $k$ of the finite predictor is increased, and the coefficients $\psi_j$ then extracted as $\text{cov}(x_t, a_{t-k})/\sigma_a^2$, to give the representation (18.6.1).

### 18.6.6.   Prediction Using the GLM

The form of the GLM is particularly suitable for presenting the principles of forecasting several steps ahead. Given observations of $x_t$ for $t \leq n$ and an invertible GLM (18.6.1), we can also derive (in theory) the values of $a_t$ for $t \leq n$ using (18.6.38). Any future value $x_{n+k}$ for $k > 0$ may then be written:

$$x_{n+k} = a_{n+k} + \psi_1 a_{n+k-1} + \ldots + \psi_{k-1} a_{n+1}$$

$$+ \psi_k a_n + \psi_{k+1} a_{n-1} + \ldots \qquad (18.6.40)$$

where on the RHS we separate on the two lines the components of $x_{n+k}$ depending respectively on the future (unknown) terms $a_t$ for $t > n$, and on the past and present (known) terms $a_t$ for $t \leq n$. To obtain the least squares forecast $\hat{x}_n(k)$ of $x_{n+k}$, we merely note that the future terms $a_{n+1}, \ldots a_{n+k}$ are uncorrelated with the observations $x_n, x_{n-1}, \ldots$ and hence their least squares forecasts are all zero. This leaves the second line of (18.6.40) as the forecast of $x_{n+k}$

$$\hat{x}_n(k) = \psi_k a_n + \psi_{k+1} a_{n-1} + \ldots \qquad (18.6.41)$$

and the forecast error is the first line:

$$a_n(k) = x_{n+k} - \hat{x}_n(k) = a_{n+k} + \psi_1 a_{n+k-1} + \ldots + \psi_{k-1} a_{n+1}. \qquad (18.6.42)$$

In principle therefore we know both the forecast value, and its error variance

$$\mathrm{var}\{a_n(k)\} = (1 + \psi_1^2 + \ldots + \psi_{k-1}^2)\sigma_a^2, \qquad (18.6.43)$$

from which probability limits for $x_{n+k}$ may be constructed under Gaussian assumptions. The practical implications will become evident in the following sections on parametric modelling.

## 18.7.   THE MOVING AVERAGE (MA) MODEL

### 18.7.1.   Model Definition

A class of simple finite parameter time series models is obtained by assuming that the GLM (18.6.1) involves only a finite number of terms, i.e. that $\psi_k = 0$ for $k > q$ say. We rename the remaining coefficients to emphasize this assumption. The result is called the *moving average* model of order $q$, or MA($q$) model:

$$x_t = a_t - \theta_1 a_{t-1} - \ldots - \theta_q a_{t-q}$$

$$= (1 - \theta_1 B - \ldots - \theta_q B^q)a_t = \theta(B)a_t. \qquad (18.7.1)$$

We recall the assumption that $a_t$ is a sequence of I.I.D. [see § 1.4.2(i)] random variables with $E(a_t) = 0$, $\mathrm{var}(a_t) = \sigma_a^2$. Replacing $x_t$ by $x_t - \mu$ in (18.7.1) allows for a non-zero mean $\mu$ of $x_t$.

The invertibility condition (18.6.37) is now expressed as $\theta(B) \neq 0$ for $|B| < 1$, i.e. if we factorize the polynomial:

$$\theta(B) = (1 - s_1 B)(1 - s_2 B) \ldots (1 - s_q B)$$

then it is equivalent to $|s_i| < 1$, $i = 1 \ldots q$. This factorization can assist in the invertion of the operator by use of partial fractions [see I, § 14.10], e.g. if $q = 2$ and $\theta(B) = (1 - rB)(1 - sB)$,

$$\theta(B)^{-1} = (r - s)^{-1}\{r(1 - rB)^{-1} - s(1 - sB)^{-1}\}$$

$$= (r - s)^{-1}\left\{\sum_0^\infty r^{k+1}B^k - \sum_0^\infty s^{k+1}B^k\right\}$$

giving $\pi_k = -(r - s)^{-1}(r^{k+1} - s^{k+1})$ in the representation (18.6.38). We have assumed for convenience that $r$, $s$ are real and distinct. If they are real and equal,    then    $\pi_k = -(k+1)r^k$,    and    if    they    are    complex    $\pi_k = -r^k \sin\{(k+1)\lambda\}/\sin\lambda$, where $r$ and $\lambda$ are given by $r^2 = -\theta_2$, $2r \cos\lambda = \theta_1$.

### 18.7.2.   Characteristic Property of the MA(q) Model

The acf of $x_t$ as derived from (18.7.1) is a special case of (18.6.3),

$$\gamma_x(k) = \begin{cases} \left(-\theta_k + \sum_1^{q-k} \theta_i\theta_{i+k}\right)\sigma_a^2 & \text{for } k \le q \\ 0 & \text{for } k > q \\ \sigma_x^2 = (1 + \theta_1^2 + \ldots + \theta_q^2)\sigma_a^2. \end{cases} \tag{18.7.2}$$

The finite extent of the acf is a characteristic property, and a stationary time series for which $\gamma_x(k) = 0$ when $k > q$ can always be represented by an invertible MA($q$) model, provided only that the spectrum of the series is strictly positive, and with the qualification that if $x_t$ is not Gaussian then the $a_t$ are not necessarily independent—though they are uncorrelated.

For $q = 1$, $\rho_x(1) = -\theta_1/(1 + \theta_1^2)$ and $\rho_x(k) = 0$ for $k > 1$. As $\theta_1$ passes over its allowable range between $-1$ and $+1$, it is easily seen that $\rho_x(1)$ covers the range between $-\frac{1}{2}$ and $\frac{1}{2}$, as a monotonic decreasing function of $\theta_1$. This is the complete range of values for $\rho_x(1)$ as described in section 18.5.3 given the constraint that $\rho_x(k) = 0$ for $k > 1$, and the condition that the spectrum be strictly positive.

It is simple therefore to spot whether a MA($q$) model with a moderately small order $q$ might be appropriate for an observed time series, by inspecting its sample acf for a cut off point. For example the discussion in section 18.5.1 suggests that a MA(2) model may be appropriate for the day-length series. However, we discovered in § 18.5.3 that assuming $\rho_x(k) = 0$ for $k > 2$ and taking $\rho_x(1) = 0.76$, $\rho_x(2) = 0.52$, which are the corresponding sample of values, gives an inconsistent set in the sense that the corresponding spectrum would have some negative values, e.g. at $\omega = 2\pi/3$. Equivalently, there is no set of MA(2) model parameters which would yield these acf values exactly. Even if we allow the possibility of an MA(3) model and take $\rho_x(3) = r_x(3) = 0.21$ with $\rho_x(k) = 0$ for $k > 3$ we still do not obtain a consistent set, but examination of the corresponding spectrum in this case would show only

marginally negative values over a short frequency range. Moreover we might
spot that taking $\theta_1 = \theta_2 = \theta_3 = -1$ in an MA(3) model

$$x_t = a_t + a_{t-1} + a_{t-2} + a_{t-3} \tag{18.7.3}$$

would yield acf values, using (7.2), of

$$\rho_x(1) = 0 \cdot 75, \qquad \rho_x(2) = 0 \cdot 5, \qquad \rho_x(3) = 0 \cdot 25$$

which are well within sampling limits of the values

$$r_x(1) = 0 \cdot 76, \qquad r_x(2) = 0 \cdot 52, \qquad r_x(3) = 0 \cdot 21,$$

which we have been trying to match. The only misgiving is that (18.7.3) is
not strictly invertible without infinitesimal adjustment to its parameters such
as taking $\theta_i = -0 \cdot 999$, the corresponding spectrum having zeros at $\omega = \pi/2$,
$\pi$ but being otherwise positive. It is also suggestive of the possibility that much
of the correlation may have been induced by a uniform smoothing filter
previously applied to the data—compare (18.7.3) with (18.6.23). This discussion also suggests that a more reliable and efficient method of estimating
moving average parameters is required, in comparison with the procedure of
finding parameters which match the sample autocorrelations.

### 18.7.3.   Efficient Estimation for MA($q$) Models

The general method is well illustrated by considering the MA(1) case only.
We suppose that we have observations $x_1, x_2, \ldots, x_n$ and use the maximum
likelihood (ML) method [see Chapter 6]. We therefore need to evaluate the
probability density function (pdf) [see II, § 10.1.1] for the model, which is
most easily done by transforming from the sequence $x_t$ to the sequence $a_t$
which we now assume to be independent Normal $(0, \sigma_a)$ [see § 1.4.2(ii)].
Because our data is finite we need to introduce $a_0$ in order that we may
regenerate $a_1, a_2, \ldots, a_n$ by rewriting the model equation

$$x_t = a_t - \theta a_{t-1} \tag{18.7.4}$$

in the form

$$a_t = x_t + \theta a_{t-1}, \qquad t = 1 \ldots n \tag{18.7.5}$$

—where for convenience we write $\theta$ rather than $\theta_1$.
  The recursive equations (7.5) may be explicilty written as

$$a_1 = x_1 + \theta a_0$$

$$a_2 = x_2 + \theta x_1 + \theta^2 a_0$$

$$\vdots$$

$$a_t = x_t + \theta x_{t-1} + \ldots + \theta^{t-1} x_1 + \theta^t a_0 \tag{18.7.6}$$

$$\vdots$$

$$a_n = x_n + \theta x_{n-1} + \ldots + \theta^{n-1} x_1 + \theta^n a_0$$

where we have merely used successive substitution, e.g.

$$a_2 = x_2 + \theta a_1 = x_2 + \theta(x_1 + \theta a_0) = x_2 + \theta x_1 + \theta^2 a_0.$$

The equations (18.7.6) show that the relationship is a linear transformation from $a_0, x_1, \ldots x_n$ to $a_0, a_1, \ldots a_n$ with Jacobian 1 [see IV, (5.12.2)], though it is highly non-linear in $\theta$. Thus the pdf is

$$f(a_0, x_1, \ldots x_n) = f(a_0, a_1, \ldots, a_n)$$

$$= (2\pi\sigma_a^2)^{-(n+1)/2} \exp\left\{-(1/2\sigma_a^2) \sum_{t=0}^{n} a_t^2\right\}. \qquad (18.7.7)$$

This may be calculated directly using (18.7.5), from any data set $x_1, \ldots x_n$ and for any proposed value of $\theta$, providing some value for $a_0$ can be supplied. The ML method is then to search for the value of $\theta$ which would make (18.7.7) a maximum, or equivalently would minimize the sum of squares

$$S(\theta|a_0) = \sum_{t=0}^{n} a_t^2. \qquad (18.7.8)$$

The minimum may be found in this simple case by graphing $S$ over the range $-1 < \theta < 1$. For the general MA($q$) case many standard optimization [see III, Chapter 11] routines are available which have proved very successful in this application.

We must consider further the problem of providing a value for $a_0$. A simple solution is to set $a_0 = 0$ which is its expected value had we no observations at all of the series. Writing $\tilde{a}_t$ for the approximation to $a_t$ obtained by replacing $a_0$ with $\tilde{a}_0 = 0$, we see from (18.7.6) that

$$a_t - \tilde{a}_t = \theta^t a_0 \qquad (18.7.9)$$

which tends to 0 as $t$ increases. For values of $\theta$ not too close to 1 and $n$ fairly large, the error introduced by the solution will be relatively small whatever the true value of $a_0$ might have been. We call this the conditional least squares solution since it leads to the minimization of

$$S(\theta|a_0 = 0) = \sum_{t=1}^{n} \tilde{a}_t^2. \qquad (18.7.10)$$

A second approach is to treat $a_0$ as a nuisance parameter in the sum of squares (18.7.8), and to include it along with $\theta$ as a free parameter in the minimization. This is extremely convenient and readily generalized to MA($q$) models, at the expense of doubling the number of parameters. It is the method which has been used to provide the estimates in the examples which follow. However, since $a_0$ enters linearly into the terms $a_t$ it is possible to relieve the minimization routine of some of its work by explicitly calculating and substituting for the value $\hat{a}_0$ of $a_0$ which minimizes $S$ for any given value of $\theta$. The objective function is then

$$S(\theta) = S(\theta|\hat{a}_0) = \min_{a_0} S(\theta|a_0). \qquad (18.7.11)$$

Illustrating this for the MA(1) case, we make the dependence upon $a_0$ explicit by writing, following (18.7.6) and (18.7.9),

$$a_t = \tilde{a}_t + \theta^t a_0$$

since $\tilde{a}_1, \ldots \tilde{a}_n$ depend upon $x_1, \ldots, x_n$ only. Then

$$\sum_0^n a_t^2 = \sum_1^n \tilde{a}_t^2 + 2\left(\sum_1^n \theta^t \tilde{a}_t\right) a_0 + \left(\sum_0^n \theta^{2t}\right) a_0^2. \qquad (18.7.12)$$

Now any positive quadratic

$$Q(z) = Az^2 + 2Bz + C \qquad (18.7.13)$$

can be written as

$$A(z - \hat{z})^2 + Q(\hat{z}) \qquad (18.7.14)$$

where $\hat{z} = -B/A$ is the value of $z$ which minimizes $Q(z)$. The minimum in (18.7.12) w.r.t. $a_0$ is given by

$$\hat{a}_0 = -K^{-1}\left(\sum_1^n \theta^t \tilde{a}_t\right) \qquad (18.7.15)$$

where corresponding to $B$ in (18.7.13) we have

$$K = \sum_0^n \theta^{2t} = (1 - \theta^{2n+2})/(1 - \theta^2). \qquad (18.7.16)$$

Corresponding to (18.7.14) we obtain in place of (18.7.12)

$$\sum_0^n a_t^2 = K(a_0 - \hat{a}_0)^2 + \sum_0^n \hat{a}_t^2 \qquad (18.7.17)$$

or equivalently

$$S(\theta|a_0) = K(a_0 - \hat{a}_0)^2 + S(\theta). \qquad (18.7.18)$$

The extra computation required beyond the calculation of $\tilde{a}_1 \ldots \tilde{a}_n$ required for the conditional least squares solution, is the calculation of $K$ and $\hat{a}_0$ as in (18.7.15) and (18.7.16), then $\hat{a}_1, \ldots \hat{a}_n$ according to

$$\hat{a}_t = x_t + \theta\hat{a}_{t-1}, \qquad t = 1 \ldots n$$

starting with $\hat{a}_0$. Finally,

$$S(\theta) = \sum_0^n \hat{a}_t^2. \qquad (18.7.19)$$

This second approach provides, by minimization of $S(\theta)$, what we call the exact least squares estimate of $\theta$. A little further consideration provides the exact ML estimates. We use the identity (18.7.17) to expand the pdf (18.7.7)

$$f(a_0, x_1, \ldots, x_n) = (K/2\pi\sigma_a^2)^{1/2} \exp\{-(K/2\sigma_a^2)(a_0 - \hat{a}_0)^2\}$$
$$\times K^{-1/2}(2\pi\sigma_a^2)^{-n/2} \exp\left\{-(1/2\sigma_a^2)\sum_0^n \hat{a}_t^2\right\}. \tag{18.7.20}$$

On the RHS, only the first line contains $a_0$, and the factor $K^{1/2}$ has been introduced to arrange that the whole expression on this line yields the value 1 when integrated w.r.t. $a_0$. This leaves the second line, which contains the compensating factor $K^{-1/2}$, as the marginal pdf of the data:

$$f(x_1, x_2, \ldots x_n) = K^{-1/2}(2\pi\sigma_a^2)^{-n/2} \exp\{-S(\theta)/2\sigma_a^2\}. \tag{18.7.21}$$

In fact, the first line of (18.7.20) is interpreted as the conditional distribution of the unknown quantity $a_0$, given the observations, i.e. a Normal pdf with mean $\hat{a}_0$ and variance $K^{-1}\sigma_a^2$.

Returning to (18.7.21), the log likelihood [see § 6.2.1] may be obtained and maximized w.r.t. $\sigma_a^2$, giving a monotonic function of

$$K^{1/n}S(\theta) \tag{18.7.22}$$

whose minimum is the exact ML estimate. Note that $K$ is a function of $\theta$ only, not of the data, and as $n$ increases $K^{1/n}$ tends to 1, so although this factor is important for small data sets, and is not difficult to calculate in this case, it is often dispensed with when $n$ is moderate to large with little effect upon the estimate, and a useful saving in computation for a general MA($q$) model. This implies the use of the exact least squares estimate.

Asymptotically, i.e. as $n \to \infty$, the estimates described are essentially the same, and their properties have been shown to parallel those of maximum likelihood estimates obtained in the context of I.I.D. observations. Thus they are asymptotically unbiased and Normally distributed, with variances well estimated by the inverse Hessian of the negative log likelihood evaluated at the MLE [see § 6.2.5(c)]. This information is generally supplied by optimization routines. Also, tests based on the residual sum of squares $S(\hat{\theta})$ may be carried over from standard linear regression if applied cautiously. The fact that the residuals $\hat{a}_t$ may be interpreted as the one step ahead prediction errors which would be obtained by using the fitted model to predict successive values of the observations, reassures one that $S(\hat{\theta})$ is a meaningful measure of how good, or bad, a model is. It is used to estimate

$$\hat{\sigma}_a^2 = (n - d)^{-1}S(\hat{\theta})$$

where $d$ is the number of degrees of freedom associated with the estimated parameter, e.g. $d = 1$ for the MA(1) model, plus one if the mean $\mu$ is also estimated.

As an example, two models were fitted to the differenced daylength series. The MA(2) model was

$$x_t = \underset{(\pm 3\cdot 6)}{3\cdot 1} + a_t + \underset{(\pm 0\cdot 05)}{0\cdot 91\,a_{t-1}} + \underset{(\pm 0\cdot 05)}{0\cdot 79\,a_{t-2}}$$

with parameter S.E.'s (standard errors [see Definition 3.1.1]) indicated. The residual sum of squares RSS = 33870, and $\hat{\sigma}_a^2 = 249\cdot 0$.

The MA(3) model was

$$x_t = \underset{(\pm 4\cdot 1)}{3\cdot 2} + a_t + \underset{(\pm 0\cdot 09)}{0\cdot 99\,a_{t-1}} + \underset{(\pm 0\cdot 09)}{0\cdot 93\,a_{t-2}} + \underset{(\pm 0\cdot 09)}{0\cdot 18\,a_{t-3}}$$

with RSS = 32813 and $\hat{\sigma}_a^2 = 243\cdot 1$.

The ratio of the reduction in RSS to $\hat{\sigma}_a^2$ is $1057/243 = 4\cdot 34$, not to be regarded as significant since $4\cdot 34$ is only just in excess of the upper 5% point on the chi-squared distribution with one d.f. Similarly, the new parameter $\theta_s$ is equal to just twice its own standard error, so we might conclude that the MA(3) model is at best a marginal improvement on the MA(2) model. Note that the model yields a value of $\hat{\sigma}_a^2/\hat{\sigma}_x^2 \approx 0\cdot 34$, so the one-step ahead forecast achieves a 66% reduction in variance, but for more than three steps ahead the MA(3) model implies no reduction in forecast variance. Furthermore, the estimate of the series mean is not significantly, casting doubt on the proposition that the original daylength series contains an overall trend.

### 18.7.4.   Forecasting Using MA Models

Because the MA($q$) model is a finite form of the GLM (18.6.1), the forecast $\hat{x}_n(k)$ of $x_{n+k}$ given $x_t$ for $t < n$ is simply expressed in terms of $a_t$ for $t \leq n$, by adaptation of (6.41):

$$\hat{x}_n(k) = \begin{cases} -\theta_k a_n - \ldots - \theta_q a_{n+k-q} & \text{for } k \leq q \\ 0 & \text{for } k > q. \end{cases} \qquad (18.7.23)$$

So for example if $q = 1$,

$$\hat{x}_n(1) = -\theta_1 a_n, \qquad \hat{x}_n(k) = 0 \quad \text{if } k > 1. \qquad (18.7.24)$$

The inverted model from (6.38) has been presented in this case as (6.39), giving also

$$\hat{x}_n(1) = -(\theta x_n + \theta^2 x_{n-1} + \theta^3 x_{n-2} + \ldots) \qquad (18.7.25)$$

which explicitly expresses the dependence upon previous observations.

In practice the observations are finite in extent: $x_1 \ldots x_n$, but presuming that the model parameters have been estimated, the residuals $\hat{a}_t$ resulting from the estimation as in (18.7.19) provide the values to use in place of $a_t$ in (18.7.23). If $n$ is large and $\theta$ not close to 1, i.e. $\theta^n$ is small, then $\tilde{a}_t$ may suffice.

An alternative way of coping with the finite extent of the data is a neat and efficient method known as *back-forecasting*, which we illustrate, again for the

MA(1) model. For convenience we resort to writing simply $x_t$ for forecasts outside the range $1 \leq t \leq n$, and $a_t$ for the corresponding residuals or forecast errors. We know that

$$x_t = 0 \quad \text{for } t > n+1; \qquad a_t = 0 \quad \text{for } t > n. \qquad (18.7.26)$$

Now as we previously exploited in section 18.5.4, the structure of the time series, viewed in the reverse direction as $t$ decreases, is exactly parallel to that in the forward direction, i.e. we may alternatively represent

$$x_t = b_t - \theta b_{t+1} \qquad (18.7.27)$$

where $b_t$ is the error in predicting $x_t$ using future values $x_{t+1}, x_{t+2}, \ldots$. In parallel with (18.7.26) we get

$$x_t = 0 \quad \text{for } t < 0; \qquad b_t = 0 \quad \text{for } t < 1 \qquad (18.7.28)$$

and since $a_t$ is a linear combination of $x_t, x_{t-1}, \ldots$ according to the inverted model forms we may get the further results

$$a_t = 0 \quad \text{for } t < 0, \qquad b_t = 0 \quad \text{for } t > n+1$$

so that outside the range $0, \ldots, n+1$ all quantities are zero knowing only $x_1, \ldots x_n$. The argument then runs:

If $x_0$ were known, we could regenerate $a_0, \ldots a_n$ using

$$a_t = x_t + \theta a_{t-1}, \quad t = 0, \ldots n \qquad (18.7.29)$$

since we know that $a_0 = 0$. Then we could generate (using $a_{n+1} = 0$)

$$x_{n+1} = -\theta a_n. \qquad (18.7.30)$$

Similarly, if $x_{n+1}$ were known, we could regenerate $b_{n+1}, \ldots b_1$ using

$$b_t = x_t + \theta b_{t+1}, \qquad t = n+1, \ldots 1 \qquad (18.7.31)$$

since we know that $b_{n+1} = 0$. Then we could generate (using $b_0 = 0$)

$$x_0 = -\theta b_1. \qquad (18.7.32)$$

This argument is not circular. Starting with a guess for $x_0$, say 0, these last four equations can be used in a cycle which rapidly converges to the correct values of $x_0$ and $x_{n+1}$, the error decreasing by the factor $\theta^{2n+2}$ in every cycle. The procedure is therefore widely used, not only in forecasting, but as a means of calculating $a_0$—more correctly $\hat{a}_0$—in the estimation context of section 18.7.3. It readily generalizes to large values of $q$, avoiding the solution of large sets of simultaneous equations. The explicit form of the finite predictor may be readily derived for the MA(1) model. Replacing $x_t$ by 0 for $t < 0$ in (18.7.25) and writing $x_{n+1}$ for $\hat{x}_n(1)$ gives

$$x_{n+1} = -\theta x_n - \theta^2 x_{n-1} - \ldots - \theta^n x_1 - \theta^{n+1} x_0. \qquad (18.7.33)$$

A parallel equation in reversed time gives

$$x_0 = -\theta x_1 - \theta^2 x_2 - \ldots - \theta^n x_n - \theta^{n+1} x_{n+1}. \tag{18.7.34}$$

Substitution of $x_0$ from (18.7.34) into (18.7.33) gives

$$x_{n+1} = -\{(\theta - \theta^{2n+1}) x_n + \ldots + (\theta^n - \theta^{n+2}) x_1\}/(1 - \theta^{2n+2}).$$

We comment that the coefficient of $x_1$ gives the pacf for the MA(1) model:

$$\phi_{n,n} = -\theta^n (1 - \theta^2)/(1 - \theta^{2n+2}).$$

An approximately geometrical decay pattern such as this, if apparent in a sample pacf, serves to confirm the appropriates of the MA(1) model.

### 18.7.5.  The EWMA Predictor

A forecasting procedure of great popularity and convenience of use, is simply to form an average of all past values, but with a geometric or exponential weighting so as to place greatest weight on the most recent values. Thus given $w_t$ for $t \le n$, the forecast of the next value is

$$\bar{w}_{n+1} = (1 - \theta)(w_n + \theta w_{n-1} + \theta w_{n-2} + \ldots) \tag{18.7.35}$$

where the factor $(1 - \theta)$ ensures that this is a true average, i.e. the weights sum to 1. This is termed the exponentially weighted moving average (EWMA) though the 'MA' usage does not accord with that of our MA($q$) models—in this case it is an infinite moving average applied to past series values.

In practice the EWMA is calculated by a simple updating equation

$$\bar{w}_{n+1} = (1 - \theta) w_n + \theta \bar{w}_n = w_n - \theta(w_n - \bar{w}_n). \tag{18.7.36}$$

Thus as each new observation $w_n$ comes to hand, the forecast $\bar{w}_{n+1}$ of the next one is calculated. The parameter $\theta$ may be chosen so as to minimize the forecast errors on an initial data set.

If we assume that the prediction errors $a_t = w_t - \bar{w}_t$ are I.I.D., corresponding to an optimal situation in which it would be impossible to construct any better forecasting scheme, then from (18.7.36)

$$w_{n+1} = \bar{w}_{n+1} + a_{n+1} = w_n - \theta a_n + a_{n+1} \tag{18.7.37}$$

Writing $t$ for $n+1$ and letting $w_t - w_{t-1} = x_t$, we get

$$\nabla w_t = x_t = a_t - \theta a_{t-1}. \tag{18.7.38}$$

Thus the EWMA predictor is appropriate to a situation in which the first difference of the series follows a MA(1) model. A simple structural model conforms with this idea. Typically $w_t$, which might be the weekly sales of some product, will have an underlying level which drifts up and down slowly, in an unpredictable fashion, i.e. may be represented by a random walk

$$u_t = u_{t-1} + \alpha_t \tag{18.7.39}$$

where the $\alpha_t$ are I.I.D. Actual sales will be subject to random fluctuations superimposed upon this underlying level, due to the weather for example,

$$w_t = u_t + \beta_t \qquad (18.7.40)$$

where $\beta_t$ are also I.I.D. and independent of $\alpha_t$. Then

$$x_t = w_t - w_{t-1} = \alpha_t + \beta_t - \beta_{t-1}$$

so that

$$\text{var}\,(x_t) = \sigma_\alpha^2 + 2\sigma_\beta^2,$$

$$\text{cov}\,(x_t, x_{t+1}) = -\sigma_\beta^2, \text{cov}\,(x_t, x_{t+k}) = 0 \quad \text{for } k > 1.$$

Thus $x_t$ is stationary with

$$\rho_x(1) = -1/(2+r), \quad \rho_x(k) = 0 \quad \text{for } k > 1 \qquad (18.7.41)$$

where $r$ is the variance ratio $\sigma_\alpha^2/\sigma_\beta^2$. Consequently $x_t$ may be represented by a MA(1) model, the parameter $\theta$ depending upon $r$.

The series $w_t$ is said to follow an IMA or integrated moving average model, since it is obtained by cumulative summation of $x_t$:

$$w_t = w_0 + x_1 + x_2 + \ldots + x_t.$$

A simple extension is to allow $x_t$ to have a non-zero mean $\mu$ so that $E(w_t) = E(w_0) + t\mu$, incorporating a linear trend.

### 18.7.6. Box–Jenkins Seasonal Model for the Airline Series

We have already noted how differencing of the form

$$w_t = \nabla \nabla_s x_t$$

can simplify a time series by removing a strong trend and seasonal pattern. An appealing extension of the EWMA predictor to seasonal time series automatically leads to a model which incorporates this differencing. For the airline series consider predicting the January values by an EWMA of previous January figures only, similarly for February, March and the other months. Hopefully the same smoothing parameter $\Theta$ and trend constant $\mu$, will suffice for all the months. The result of fitting such a model is illustrated in Figure 18.2.1(e) by the one step ahead forecasts errors $\alpha_t$ which arise. They are similar to the errors from fitting the classical model, shown in Figure 18.2.1(b). They are also closely related to the differenced series $\nabla_{12} x_t$ shown in Figure 18.2.1(c), since

$$\nabla_{12} x_t = \mu + \alpha_t - \Theta\alpha_{t-12}. \qquad (18.7.42)$$

which is the seasonal version of the IMA model (18.7.38), the parameter values being $\mu = 0\cdot123$, $\Theta = 0\cdot546$.

The forecast produced by this model for the final year of the observations are also shown in Figure 18.4.1—they are on the whole better than those produced by the classical model. It may in fact be shown that in the limit as $\Theta \to 1$, the model (18.7.42) is equivalent to the classical model (18.2.1), so it is only to be expected that introducing the free parameter $\Theta$ should allow a better fit.

In fact the trend in the series is slowly varying, so besides a more or less random component to the errors $\alpha_t$ in Figure 18.2.1(e) there is a slowly varying bias. Now the seasonal predictor so far considered makes no use of the most recent 11 months observations, so it is a natural step to improve the forecasts by predicting this bias in the forecast error, using the most recently observed errors—say by an ordinary EWMA with parameter $\Theta$ applied on a monthly basis. This predicted bias is then used to correct the seasonal forecasts to produce final forecasts. The errors $a_t$ from such a scheme are shown in Figure 18.2.1(f) to be appreciably smaller than any other set.

The model implied by this two-step process may be derived as follows. The error series $\alpha_t$ from the seasonal EWMA is related to $x_t$ by

$$\nabla_{12} x_t = \mu + \alpha_t - \Theta\alpha_{t-12} = \mu + (1 - \Theta B^{12})\alpha_t,$$

and the error series $a_t$ from the final EWMA is related to $\alpha_t$ by

$$\nabla \alpha_t = a_t - \theta a_{t-1} = (1 - \theta B)a_t.$$

Taken together these give

$$w_t = \nabla\nabla_{12} x_t = (1 - \theta B)(1 - \Theta B^{12})a_t \qquad (18.7.43)$$

where we assume now that $a_t$ is an I.I.D. sequence. This is the celebrated airline model proposed by Box and Jenkins (1976), incorporating the double differencing operation $\nabla\nabla_{12}$ which we introduced in section 18.4. The values of both $\theta$ and $\Theta$ were adjusted to give the best fit using the exact least squares estimates $\theta = 0.34$, $\Theta = 0.63$. Note that the weights attached to previous data by the seasonal EWMA part of the model, effectively die out over a span of 5 years, whereas those associated with the nonseasonal part die out over a span of only two or three months. This reflects the fact that the seasonal pattern is fairly stable, but changes in level and trend must be swiftly adapted to.

Confirmation that this model is appropriate for the series is found by inspection of the sample acf of the differenced series $w_t$, which follows a seasonal MA model with the acf properties

$$\rho_w(1) = -\theta/(1 + \theta^2), \qquad \rho_w(12) = -\Theta/(1 + \Theta^2)$$

$$\rho_w(11) = \rho_w(13) = \rho_w(1)\rho_w(12)$$

and otherwise $\rho_w(k) = 0$.

In practice $r_w(1)$ and $r_w(12)$ supply the most pronounced evidence in favour of this acf pattern, as shown in Figure 18.2.1(g).

The forecast function for the airline model also has a simple structure. Because $w_t$ may be viewed as following a MA(13) model in (18.7.43), it follows that $\hat{w}_n(k) = 0$ for $k > 13$, leading to

$$\nabla\nabla_{12}\hat{x}_n(k) = 0 \quad \text{for } k > 13,$$

where the operators act on $k$. Recalling (18.4.1), this implies that $\hat{x}_n(k)$ consists of a regular seasonal pattern plus linear trend, established by the initial set for $k = 1, \ldots, 13$. The particular pattern followed will depend most heavily upon recent data, and will be adapted as new data comes to hand and $n$ is increased.

## 18.8.   THE AUTOREGRESSIVE (AR) MODEL

### 18.8.1.   Model Definition

A further class of simple finite parameter models is obtained by assuming that the inverted form (18.6.38) of the GLM is finite in extent, i.e. the series is best predicted by a finite number of past values. Thus $\pi_k = 0$ for $k > p$, say, and we again rename the remaining coefficients. The result is called the autoregressive model of order $p$, or AR($p$) model:

$$x_t = \phi_1 x_{t-1} + \phi_2 x_{t-2} + \ldots + \phi_p x_{t-p} + a_t \qquad (18.8.1)$$

or

$$(1 - \phi_1 B - \phi_2 B^2 - \ldots - \phi_p B^p) x_t = \phi(B) x_t = a_t, \qquad (18.8.2)$$

where $a_t$ is again a sequence of I.I.D. random variables with $E(a_t) = 0$, var $(a_t) = \sigma_a^2$, and $x_t$ may be replaced by $x_t - \mu$ to allow for a non-zero mean $\mu$ of $x_t$.

Because $\phi(B)$ is the inverse of $\psi(B)$ in the GLM (18.6.31), it must itself satisfy the condition $\phi(B) \neq 0$ for $|B| < 1$, i.e. if we factorize

$$\phi(B) = (1 - r_1 B)(1 - r_2 B) \ldots (1 - r_p B) \qquad (18.8.3)$$

then $|r_i| < 1$ for $i = 1 \ldots p$.

Viewing (18.8.2) as defining a stochastic process which starts at some time origin in the remote past, the above condition is equivalent to assuming that $x_t$ eventually reaches statistical equilibrium or stationarity, and is therefore called the stationarity condition.

Autoregressive models are particularly appropriate for (and were originally proposed for) stochastic systems exhibiting momentum and restoring forces, to use a mechanical analogy. In particular, second order models, with $p = 2$, have been very successful in representing behaviour of an approximately cyclical nature, such as might be typified by a pendulum subjected to small random impulses. The swinging motion would be evident, though with constantly fluctuating amplitude and phase.

### 18.8.2.   First and Second Order Examples

Consider the AR(1) model, using $\phi$ rather than $\phi_1$ for convenience,

$$x_t = \phi x_{t-1} + a_t, \qquad |\phi| < 1. \tag{18.8.4}$$

Writing this as

$$(1 - \phi B) x_t = a_t$$

we obtain

$$x_t = (1 - \phi B)^{-1} a_t = (1 + \phi B + \phi^2 B^2 + \ldots) a_t$$

$$= a_t + \phi a_{t-1} + \phi^2 a_{t-2} + \ldots \tag{18.8.5}$$

showing how the dependence of $x_t$ on past values $a_{t-k}$ decays geometrically. It is possible to derive the correlation properties from (18.8.5), but it is more direct to use the model (18.8.4), recalling from section 18.6.1 the property of the GLM that cov $(a_t, x_{t-1}) = 0$,

$$\text{var}(x_t) = \phi^2 \, \text{var}(x_{t-1}) + \sigma_a^2$$

and assuming stationarity, $(1 - \phi^2)\sigma_x^2 = \sigma_a^2$, or

$$\sigma_x^2 = \sigma_a^2 / (1 - \phi^2). \tag{18.8.6}$$

Further, multiplying both sides of (18.8.4) by $x_{t-k}$ for $k > 0$ and taking expectations gives

$$\gamma_x(k) = \phi \gamma_x(k-1) \tag{18.8.7}$$

leading to

$$\gamma_x(k) = \phi^k \gamma_x(0), \qquad \rho_x(k) = \phi^k. \tag{18.8.8}$$

An autocorrelation pattern with this simple geometric decay has found particular popularity in econometric modelling, where an AR(1) model is often assumed for the structure of the error process in a linear regression relating economic variables.

The properties of the AR(2) model

$$x_t = \phi_1 x_{t-1} + \phi_2 x_{t-2} + a_t \tag{18.8.9}$$

depend qualitatively upon the roots of

$$\phi(B) = (1 - \phi_1 B - \phi_2 B^2) = (1 - rB)(1 - sB).$$

In the GLM form (18.6.1) this AR(2) model has coefficients

$$\psi_k = (r - s)^{-1}(r^{k+1} - s^{k+1})$$

which are derived in a manner exactly parallel to the calculation of the coefficients $\pi_k$ for the MA(2) model in section 18.7.1, assuming $r$ and $s$ to be real. The acf of $x_t$ may be shown to follow a similar pattern:

$$\rho_x(k) = Ar^{k+1} + Bs^{k+1}$$

where

$$A = (s^2 - 1)/\{(s - r)(1 + rs)\}, \qquad B = (r^2 - 1)/\{(r - s)(1 + rs)\}.$$

This mixture of geometric decay patterns is replaced by a damped sinusoidal pattern in the case that $\phi(B)$ has complex roots. Setting

$$\phi_1 = 2r \cos \lambda, \qquad -\phi_2 = r^2 \qquad (18.8.10)$$

the acf may be written in terms of $\lambda$ and $r$ as

$$\rho_x(k) = Ar^k \cos(k\lambda - \nu)$$

where

$$\tan \nu = \{(1 - r^2)/(1 + r^2)\} \cot \lambda, \qquad A = 1/\cos \nu.$$

This sinusoidal wave, which may be most pronounced if $r$ is close to 1, reflects the general appearance of the series $x_t$ as being approximately cyclical. The gilt series of Figure 18.1.1 has something of this appearance. The corresponding feature of the spectrum of the model in this case, is a peak in the spectrum at a frequency $\omega'$ which will be cloase to $\lambda$, again provided $r$ is fairly close to 1. In fact

$$\cos \omega' = \{(1 + r^2)/2r\} \cos \lambda = -\phi_1(1 - \phi_2)/4\phi_2. \qquad (18.8.11)$$

As the decay factor $r \to 1$ and $\sigma_a^2 \to 0$, the model (18.8.9) in fact becomes indistinguishable from a pure cosine wave

$$x_t = R \cos(\lambda t + \eta)$$

which may be represented by the second order recursion

$$x_t = (2 \cos \lambda) x_{t-1} - x_{t-2}.$$

Thus in practice it may be difficult in a finite data set to distinguish a pure cosine wave from an autoregressive model—particularly if it forms just one component of a series which may also incorporate appreciable observational error.

### 18.8.3. Characteristic Properties of the AR($p$) Model

The autocorrelation function in the general case is a mixture of damped exponential and sinusoidal patterns, corresponding to the real and complex factors of $\phi(B)$. These properties are derived from the model (18.8.1) by multiplying by $x_{t-k}$ for $k > 0$, taking expectations and dividing by $\sigma_x^2$. Since $E(a_t x_{t-k}) = 0$, the result is

$$\rho_x(k) = \phi_1 \rho_x(k-1) + \ldots + \phi_p \rho_x(k-p), \qquad k > 0. \qquad (18.8.12)$$

The important Yule–Walker equations are those corresponding to $k = 1, \ldots, p$. They are linear in both the parameters $\phi_1, \ldots, \phi_p$ and in $\rho_x(1) \ldots \rho_x(p)$, and may be used to determine one set in terms of the other, e.g. if $p = 2$, the equations

$$\rho_x(1) = \phi_1 + \phi_2 \rho_x(1), \qquad \rho_x(2) = \phi_1 \rho_x(1) + \phi_2$$

may be solved for

$$\rho_x(1) = \phi_1/(1 - \phi_2), \qquad \rho_x(2) = \phi_2 + \phi_1^2/(1 - \phi_2).$$

Also useful is the variance ratio

$$\sigma_a^2/\sigma_x^2 = 1 - \phi_1 \rho_x(1) - \ldots - \phi_p \rho_x(p) \tag{18.8.13}$$

which is obtained in the same manner as (18.8.12) but with $k = 0$, and using $E(a_t x_t) = \sigma_a^2$, from (18.6.1).

For $k > p$, (18.8.12) shows how successive terms of the acf may be calculated, the general theory of difference equations [see I, § 14.13] leading to the stated result concerning the damped exponential/sinusoidal pattern.

The main characteristic property of the AR($p$) model is, almost by definition, the finiteness of its pacf:

$$\phi_{k,k} = 0 \quad \text{for } k > p, \tag{18.8.14}$$

since the model (18.8.1) was derived from (18.6.38) by assuming the coefficient of $x_{t-k}$ to be 0 for $k > p$.

Thus if the cut off lag of the sample pacf, as described in section 18.5.6 is low, say up to 3 or 4, there is strong evidence that an AR model of that order is appropriate. Indeed, the recursive procedure described in section 18.5.5 for calculating finite predictor coefficients, if stopped at $k = p$, provides the solution of the Yule–Walker equations for the parameters $\phi_1, \ldots, \phi_p$ ($\equiv \phi_{p,1}, \ldots, \phi_{p,p}$) and variance ratio $\sigma_a^2/\sigma_x^2$ ($\equiv v_p$) given $\rho_x(1), \ldots, \rho_x(p)$. Using sample values $r_x(1), \ldots, r_x(p)$ instead, fairly good estimates of $\hat{\phi}_1, \ldots, \hat{\phi}_p$ are obtained, e.g. $\hat{\phi}_1 = r_x(1)$ in the case $p = 1$.

It may be proved that given any set of autocorrelation values $\rho_x(1), \ldots, \rho_x(p)$, the solution of the Yule–Walker equations supplies a valid set of autoregressive parameters $\phi_1, \ldots, \phi_p$, in the sense that the stationarity condition is automatically satisfied, provided only that the matrix $R_n$ in (5.16) is positive definite for $n = p + 1$. This condition is readily verified if the recursive construction procedure is used by checking that $|\phi_{k,k}| < 1$ for $k = 1, \ldots, p$. This result is in contrast to the situation noted in 17.7.2 for MA($q$) models, where it is possible that no MA parameters exist which correspond to a given set $\rho_x(1), \ldots, \rho_x(q)$. This does not of course mean that AR($p$) models are always to be preferred, as an example will show, but it does give them a certain appeal. Since the sequence $r_x(k)$ is always positive definite, the AR parameter estimates derived from them are always valid.

### 18.8.4.   Efficient Estimation for the AR($p$) Model

We consider the AR(1) model only, which well illustrates the general case. In order to use the Maximum Likelihood method we evaluate the pdf for observations $x_1, \ldots, x_n$ as

$$f(x_1, x_2, \ldots, x_n) = f(x_1)f(x_2|x_1)f(x_3|x_2) \ldots f(x_n|x_{n-1}). \quad (18.8.15)$$

This decomposition is possible because the AR(1) process is a lag$-1$ Markov Chain [see II, § 19.2].

Now conditional upon $x_{t-1}$ assume $x_t$ to be Normal with mean $\phi x_{t-1}$ and variance $\sigma_a^2$, according to the model (18.8.4), so

$$f(x_1, x_2, \ldots, x_n) = f(x_1) \prod_2^n (2\pi\sigma_a^2)^{-1/2} \exp\{-(1/2\sigma_a^2)(x_t - \phi x_{t-1})^2\}$$

$$(18.8.16)$$

$$= f(x_1)(2\pi\sigma_a^2)^{-(n-1)/2} \exp\left\{-(1/2\sigma^2) \sum_2^n (x_t - \phi x_{t-1})^2\right\}.$$

$$(18.8.17)$$

If we were to treat $x_1$ as a fixed quantity, and thereby lose one degree of freedom, the likelihood of $\phi$ would be equivalent to the sum of squares

$$S(\phi|x_1) = \sum_2^n (x_t - \phi x_{t-1})^2 = \sum_2^n a_t^2. \quad (18.8.18)$$

The minimum of this w.r.t. $\phi$ is simply

$$\hat{\phi} = \left(\sum_1^{n-1} x_t x_{t+1}\right) \bigg/ \left(\sum_1^{n-1} x_t^2\right). \quad (18.8.19)$$

This is very close to the sample autocorrelation $r_x(1)$, for which the upper limit in the second sum is $n$, neglecting mean correction. One possibly important difference is that (18.8.19) may give rise to a value of $\hat{\phi} > 1$ in borderline situations, unlike $r_x(1)$. This is avoided by including the information from $x_1$, using the fact that its marginal distribution has mean 0 and variance $\sigma_x^2 = \sigma_a^2/(1 - \phi^2)$, i.e.

$$f(x_1) = (1 - \phi^2)^{1/2}(2\pi\sigma_a^2)^{-1/2} \exp\{-(1/2\sigma_a^2)(1 - \phi^2)x_1^2\}. \quad (18.8.20)$$

Incorporating this in (18.8.17) and maximizing w.r.t. $\sigma_a^2$, gives a function of

$$(1 - \phi^2)^{-1/n} S(\phi) \quad (18.8.21)$$

where

$$S(\phi) = \left\{(1 - \phi^2)x_1^2 + \sum_2^n (x_t - \phi x_{t-1})^2\right\}. \quad (18.8.22)$$

The factor $(1 - \phi^2)^{-1/n}$ ensures that a minimum exists within the allowable range $-1 < \phi < 1$, although iterative methods are now required to find it, the estimate $r_x(1)$ providing a good starting point. Although from (18.8.22), $x_1$ appears to have a special role, it is easily checked that the expression for $S(\phi)$

remains unchanged when the data are reversed in order; in fact

$$S(\phi) = x_1^2 + x_n^2 + (1+\phi^2) \sum_2^{n-1} x_t^2 - 2\phi \sum_1^{n-1} x_t x_{t+1}. \qquad (18.8.23)$$

This symmetry may be exploited in various closed form expressions for the likelihood of general AR($p$) models.

A further point concerns the commonly required estimation of a mean $\mu$ for the series, in which case $x_t$ is replaced by $x_t - \mu$ in $S(\phi)$. For a given value of $\phi$, the estimate of $\mu$ given by minimizing (18.8.23) is

$$\hat{\mu} = \left\{ x_1 + x_n + (1-\phi) \sum_2^{n-1} x_t \right\} \Big/ (\{2 + (n-2)(1-\phi)\}. \qquad (18.8.24)$$

If $\phi$ is close to 1, this can differ appreciably from $\bar{x}$ in small samples, so ML estimation of $\mu$ is to be recommended, rather than simply correcting the data by $\bar{x}$.

Finally, the factor $(1-\phi^2)^{-1/n}$ in (18.8.21) is often dispensed with for the same reasons that $K^{1/n}$ was dispensed with in (18.7.22), giving what we again call the least squares, rather than the ML estimate.

As an example, an AR(3) model was fitted to the differenced daylength series, to provide a comparison with the MA(3) model fitted in section 18.7.3. Writing $w_t$ for $x_t - \mu$, the result was

$$w_t = \underset{(\pm 0.08)}{0.85} \ w_{t-1} + \underset{(\pm 0.12)}{0.12} \ w_{t-2} - \underset{(\pm 0.08)}{0.32} \ w_{t-3} + a_t$$

with $\hat{\mu} = 2.7 \pm 4.0$, $RSS = 38043$ and $\hat{\sigma}_a^2 = 281.8$.

Though not as good a fit as the MA(3) model, it may be considered a good approximation.

A further example, of an AR(1) model, is given in section 18.11 for representing the structure of the errors, shown in Figure 18.2.2(c), after fitting the sinusoidal components of the variable star data. An AR(2) model is also a good fit to the pig series, provided a quarterly seasonal component is also taken into account as described in section 18.11.

### 18.8.5.  Forecasting Using AR($p$) Models

Autoregressive models are particularly suitable for constructing forecasts. The finite extent of the data $x_1, \ldots, x_n$ is of no concern given the very mild restriction that $p \le n$. Then we proceed according to the principle of replacing future values of $a_t$ by 0, as presented for the GLM in section 18.6.5, to get

$$\hat{x}_n(1) = \phi_1 x_n + \ldots + \phi_p x_{n+1-p}$$
$$\hat{x}_n(2) = \phi_1 \hat{x}_{n+1} + \phi_2 x_n + \ldots + \phi_p x_{n+2-p}$$

etc. Forecasts already produced are used in the autoregressive equation to produce the next forecast, as the lead time increases. Thus for the AR(1) model

$$\hat{x}_n(k) = \phi^k x_n.$$

For the AR(2) model with complex factors, the forecast function $\hat{x}_n(k)$, $k = 1, 2, \ldots$ is a damped sinusoidal wave, fitted through the last two observations—imagine the pendulum, referred to in section 18.8.1, slowly swinging to a halt in the absence of the impulses. An example of this forecasting procedure is shown in Figure 18.5.2, as described in section 18.5.5 for the finite predictor of order 7.

Calculation of the forecast error limits requires derivation of the coefficients $\psi_k$ for use in (18.6.43), by inverting and expanding

$$\{\phi(B)\}^{-1} = \psi(B) = 1 + \psi_1 B + \psi_2 B^2 + \ldots$$

examples of which may be found in section 18.8.2.

## 18.9.  AUTOREGRESSIVE MOVING AVERAGE (ARMA) MODELS

### 18.9.1.  Model Definition and Properties

In a certain sense, MA($q$) and AR($p$) models may approximate with arbitrary precision any stationary time series for which the GLM is appropriate—by choosing $q$ or $p$ sufficiently large. The aim must be to construct models which approximate well using few parameters, and the acheivement of this is greatly assisted by an autoregressive moving average model, or ARMA($p, q$) model

$$x_t = \phi_1 x_{t-1} + \phi_2 x_{t-2} + \ldots + \phi_p x_{t-p}$$
$$+ a_t - \theta_1 a_{t-1} - \theta_2 a_{t-2} - \ldots - \theta_q a_{t-q} \qquad (18.9.1)$$

or

$$\phi(B) x_t = \theta(B) a_t \qquad (18.9.2)$$

where $\phi(B)$ and $\theta(B)$ are the operators previously defined for the AR($p$) and MA($q$) models and satisfy the stationarity and invertibility conditions respectively, and $a_t$ is also used as before. This model might be appropriate, for example, when the time series is the sum of two or more independent component series, which separately follow either AR or MA models, but which are not themselves directly observed. To take a simple case, if $u_t$ follows an AR(1) model and $v_t$ a MA(1) model, then their sum $x_t = u_t + v_t$ will typically follow an ARMA(1, 2) model.

An ARMA model should be considered whenever the characteristic properties of MA($q$) or AR($p$) models, i.e. finite order acf or pacf, are not suggested by inspection of the sample quantities. Perhaps 'finite' should be interpreted as less than five! A geometric or sinusoidal decay pattern of a sample acf may still be used to suggest the presence of AR terms in the model with corresponding order $p$, and the pacf may in theory be used in an analogous way to suggest a value for $q$, though in practice the pacf tends to damp down rather rapidly. The patterns of theoretical acf and pacf for the ARMA(1, 1) model

$$x_t = \phi x_{t-1} + a_t - \theta a_{t-1}$$

are shown in Figure 18.9.1. Note that $\phi_{1,1} = \rho_x(1)$ and has the sign of $(\phi - \theta)$. For lags greater than one, the decay pattern of the acf depends upon the sign of $\phi$, that of the pacf depends upon $\theta$.

In estimation and forecasting the ideas already introduced for AR($p$) and MA($q$) models are combined, with essentially no new features.

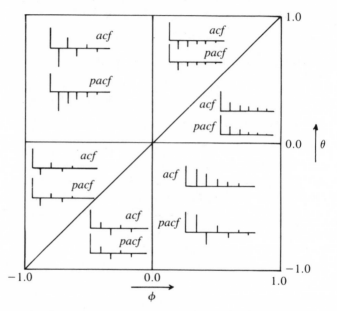

Figure 18.9.1: Qualitative features of the autocorrelations (acf) and partial autocorrelations (pacf) of the ARMA (1, 1) model $x_t = \phi x_{t-1} + a_t - \theta a_{t-1}$.

### 18.9.2.   An Example

An ARMA(1, 2) model was fitted to the differenced daylength series. This was considered as an extension of the MA(2) model, and an alternative to the MA(3) model. The AR term is also suggested by the approximate geometric decay of the acf values at lags $k = 2, 3, 4$ shown in Figure 18.5.1(b), although sampling fluctuations are relatively large over this range. With two MA terms this decay pattern would be expected to start at lag 2. The fitted model was

$$w_t = 0{\cdot}33\ w_{t-1} + a_t + 0{\cdot}70\ a_{t-1} + 0{\cdot}72\ a_{t-2}$$
$$\phantom{w_t = }{\scriptstyle(\pm 0{\cdot}10)} \phantom{w_{t-1} + a_t + } {\scriptstyle(\pm 0{\cdot}08)} \phantom{a_{t-1} + } {\scriptstyle(\pm 0{\cdot}07)}$$

where again $w_t = x_t - \mu$, and $\hat{\mu} = 3{\cdot}3 \pm 4{\cdot}7$, $RSS = 3221$ and $\hat{\sigma}_a^2 = 238{\cdot}7$, being the best so far acheived. The natural question to ask is whether the model may be still further improved by extending the orders. Certainly, given sufficient computing facilities, it is sensible to test the model in this way. One should proceed in a cautious manner since badly conditioned optimization situations can occur if both $p$ and $q$ are simultaneously and unnecessarily extended, reflecting degeneracy in the model. Analysis of the residuals and their sample acf for any remaining structure is also to be recommended, and this is done for a slightly refined version of the above model in 18.11.

A comparison of the ARMA(1, 2) model above may be made with the MA(3) model by noting the values of its first three coefficients $\psi_1 = 1 \cdot 03$, $\psi_2 = 1 \cdot 06$, $\psi_3 = 0 \cdot 35$, which are similar to $-\theta_1$, $-\theta_2$, $-\theta_3$. Also note the similarity of the finite predictor coefficients in Figure 18.5.1(g) with the first seven terms of the $\pi_k$ coefficients of the ARMA(1, 2) model: $1 \cdot 03$, $0 \cdot 00$, $-0 \cdot 74$, $0 \cdot 52$, $0 \cdot 17$, $-0 \cdot 49$, $0 \cdot 22$. Such comparisons usefully reveal similarities in apparently dissimilar models.

## 18.10. SPECTRUM ESTIMATION

### 18.10.1. The Difficulties

The sample spectrum $f_x^*(\omega)$ of an observed stationary time series $x_1, \ldots, x_n$, as defined in section 18.5.3, supplies $[(n-1)/2]$ independent estimates of the true spectrum $f_x(w)$, one at each of the frequency points $\omega_j = 2\pi j/n$, $j = 1, \ldots, [(n-1)/2]$. Each of these has just two degrees of freedom and therefore, no matter how large $n$ may be, they are not consistent estimates of $f_x(\omega)$, i.e.

$$f_x^*(\omega) \not\to f_x(\omega) \text{ as } n \to \infty.$$

Whatever means is used to derive a consistent estimator must essentially seek to represent the spectrum by a smaller number of parameters, i.e. much smaller than $(n/2)$.

### 18.10.2. Direct Methods of Estimation

One method still in use is motivated by the fact that if many independent realizations of the series were available, their sample spectra could be averaged to produce a consistent estimate. Given the single realization $x_1, \ldots, x_n$, the method disects this into $K$ subseries of length $m$ say, where $Km = n$. The sample spectra $f_k^*(\omega)$ of the subseries $k = 1, \ldots, m$ are then averaged. Assuming that the autocorrelation of $x_t$ dies away sufficiently rapidly for the series to be treated as independent, the resulting estimates at frequencies $\nu_k = 2\pi k/m$, $0 < \nu_k < \pi$ will be called $\bar{f}_x(\nu_k)$, and are independent with expectation $f_x(\nu_k)$ and variance $K^{-1} f_x(\nu_k)^2$ provided $m$ is also large—a bias of order $1/m$ is in fact present. Supposing $n$ is fixed, a balance is to be sought between choosing $m$ larger and $K$ smaller which gives more frequency point and lower bias with the disadvantage of increased variance, and the reverse situation

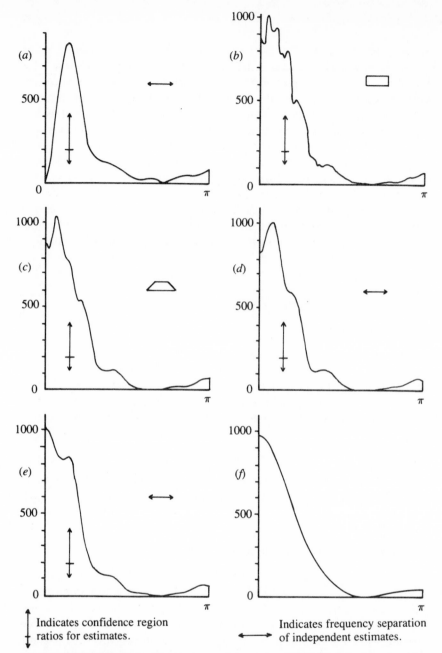

Indicates confidence region
ratios for estimates.

Indicates frequency separation
of independent estimates.

Figure 18.10.1: Smoothed spectrum estimates for differenced daylength series.

 (a) Average sample spectrum of 10 subseries of length 14.

 (b) Sample spectrum of full series averaged over frequency range with uniform
   weighting shown.

 (c) As for (b) but with trapezium weighting pattern shown.

 (d) Using Parzen lag window with truncation lag 26.

 (e) Derived from linear predictor coefficients of order 7.

 (f) Derived from ARMA (1, 2) model.

with $m$ smaller and $K$ larger. This procedure was applied to the first 140 points of the differenced daylength series using 10 subseries of length 14, the results being shown in Figure 18.10.1(a). The main defect in this is the dip to 0 at $\omega = 0$ caused by separate mean correction of all the subseries, which could have been avoided by mean correcting the full length series only.

A second method, presently very popular, is to start with the sample spectrum $f_x^*(\omega_j)$ based on $x_1, \ldots, x_n$ and to construct estimates at the coarser frequency division $\nu_k = 2\pi k/m$ by averaging the values which are closest to each of these points, e.g.

$$\tilde{f}_x(\nu_k) = K^{-1} \sum_{l=-L}^{L} f_x^*(\nu_k + 2\pi L/n) \qquad (18.10.1)$$

where $K = 2L + 1 = n/m$. The statistical properties are the same as for $\bar{f}_x(\nu_k)$. It is clear that $f_x(\omega)$ should only be slowly varying over the frequency range of each average, otherwise appreciable bias will be introduced, particularly at a peak or trough of the spectrum. A weighted average with weights tapering off at the extremes over a slightly extended frequency range, generally gives a more pleasing result. In practice the sample spectrum $f_x^*(\omega)$ is computed at a frequency division which is much finer than $2\pi/n$, and the average (18.10.1) modified to take in all these values so that it becomes effectively an integral average over the frequency range of length $2\pi/m$ centred upon $\nu_k$. The result for the differenced daylength series using $m = 14$ again, is shown in Figure 18.10.1(b), the computation having also been carried out for a finer division of frequencies $\nu$, though it must be appreciated that the estimates at intermediate frequencies are highly correlated with those at the neighbouring frequencies $\nu_k$. Figure 18.10.1(c) shows the less ragged results using tapered weights.

The classical method of computing spectral estimates is closely related to the preceding two, but is designed to reduce computation—important in the days before cheap automatic computing, and still worth considering. In order to introduce it, note first that the subsample spectrum estimates of the first method has an expected value which is more precisely given, allowing for small sample bias, by

$$E\{\bar{f}_x(\nu)\} = (1/\pi)\sigma_x^2 \left\{ 1 + 2 \sum_{1}^{m-1} (1 - k/m)\rho_x(k) \cos \nu k \right\}.$$

The higher order terms are thus deflated by the factor $(1 - k/m)$ and eliminated for $k \geq m$. A similar result holds for the averaging or smoothing procedure (18.10.1) in the second method. Thus a general class of spectral estimates may be produced using a set of weights $w_k$ called the lag window, as

$$\hat{f}_x(\omega) = (1/\pi)s_x^2 \left\{ 1 + 2 \sum_{1}^{M} w_k r_x(k) \cos \omega k \right\}, \qquad 0 \leq \omega \leq \pi. \quad (18.10.2)$$

It is convenient for direct computation from this formula to use weights which reduce to 0 beyond the truncation point $M$. The estimate may also be expressed as a convolution of the sample spectrum with a weight function called the frequency window:

$$W(\omega) = 1 + 2 \sum_1^M w_k \cos \omega_k \tag{18.10.3}$$

so that

$$\hat{f}_x(\omega') = \int W(\omega) f_x^*(\omega' + \omega) \, d\omega, \tag{18.10.4}$$

a form which is useful for investigating the statistical properties of the estimate. A practical consideration is to achieve a well concentrated unimodal positive window $W(\omega)$ for any given value of $M$. Various window shapes have been proposed, for which the lag window is of the form

$$w_k = w(k/M), \qquad k = 0, 1, \ldots, M.$$

and $w(\alpha)$ is defined for $0 \le \alpha \le 1$. Varying $M$ then controls the amount of smoothing, the number of frequency points for which the spectral estimates are approximately independent being proportional to $M$, and the variance of each estimate to $M/n$. For example the Parzen window defined by

$$w(\alpha) = \begin{cases} 1 - 6\alpha^2 + 6\alpha^3, & 0 \le \alpha \le \frac{1}{2} \\ 2(1-\alpha)^3, & \frac{1}{2} < \alpha \le 1 \end{cases}$$

has the property

$$\operatorname{var}\{\hat{f}_x(\omega)\} \simeq 0 \cdot 54 (M/n) f_x(\omega)^2.$$

Choosing $M = 26$ for this window results in spectral estimates for the differenced daylength series which have a variance very close to that obtained using the previous method. They are shown in Figure 18.10.1(d), and are very similar to those in Figure 18.10.1(c).

### 18.10.3.  Indirect Methods of Estimation

These are a consequence of fitting parametric models to the data. The most popular approach is to use the sample acf to construct the finite predictor coefficients $\hat{\phi}_{K,j}$ by the recursive method in section 18.5.4. The order $K$ may be determined by the FPE criterion of section 18.5.5 or chosen beforehand. The result is treated as an AR($K$) model with residual variance $\sigma_a^2 = \hat{\sigma}_K^2/(1 - K/n)$ and autoregressive operator

$$\phi(B) = 1 - \hat{\phi}_{K,1} B - \ldots - \hat{\phi}_{K,K} B^K.$$

The corresponding spectrum derives from the fact that $\psi(B) = \phi(B)^{-1}$, so using (18.6.32) and (18.6.16) gives the estimate

$$\hat{f}_{AR}(\omega) = (\sigma_a^2/\pi)/|\phi(e^{i\omega})|^2. \tag{18.10.5}$$

This may be computed using sine and cosine formula as in (18.6.14) and (18.6.16), or by expressing the denominator as a cosine series of order $K$ following (18.6.21) and (18.6.30). If $K$ is fixed beforehand, and provided both $K$ and $n/K$ are large, then

$$E\{\hat{f}_{AR}(\omega)\} \simeq f_x(\omega), \qquad \mathrm{var}\{\hat{f}_{AR}(\omega)\} \simeq (2K/n)f_x(\omega)^2$$

In fact, asymptotically the estimate is similar to that which could be obtained using spectral smoothing methods with rectangular lag window $W_k = 1$, $k = 1, \ldots, K$. This window is rarely used in practice because the corresponding frequency window is not strictly positive, and negative spectrum estimates can arise, but the autoregressive spectral estimate (18.10.5) is always positive. Figure 18.10.1(e) shows the results derived from the differenced daylength series using the predictor of order $K = 7$, for which the variance is close to those of the estimates in Figures 18.10.1(a), (b), (c).

If an ARMA model has been fitted to the data, by efficient procedures, then the corresponding spectrum may be computed. If spectral estimation is the main aim, this is a rather lengthy operation for it to be generally recommended. On the other hand, the spectrum can give insight into the model. The spectrum for the ARMA model (18.9.2) is, using $\psi(B) = \theta(B)\{\phi(B)\}^{-1}$,

$$f_x(\omega) = (\sigma_a^2/\pi)|\theta(e^{i\omega})|^2/|\phi(e^{i\omega})|^2. \qquad (18.10.6)$$

The numerator $|\theta(e^{i\omega})|^2$ and denominator $|\phi(e^{i\omega})|^2$ may again be evaluated using (18.6.14) and (18.6.16). Of interest is the flexibility of (18.10.6) in approximating the variety of shapes which spectra may have. If $\theta(B)$ has a zero close to the boundary of the unit circle $B = e^{i\omega}$, i.e. a zero of modulus close to 1, then the spectrum will have a near zero value at this point. Similarly, if $\phi(B)$ has a zero with modulus close to 1, then the spectrum will have a peak at this point. The presence of both numerator and denominator allows particularly sharp dips or peaks in the spectrum, giving the ARMA model a capacity for approximating spectral shapes which is much superior to that of AR or MA models alone.

The numerator and denominator may also be represented, following (18.6.21) and (18.6.30), as cosine series and therefore polynomials in $c = \cos \omega$, of degrees $q, p$ respectively. Thus (18.10.6) is a rational function of $\cos \omega$, such functions being well known for their flexibility in approximation. Figure 18.10.1(f) shows the spectrum of the ARMA(1, 2) model fitted to the differenced daylength series. Its statistical properties may be derived, with some difficulty, from those of the ARMA parameter estimates.

### 18.10.4.  The Use of Spectral Analysis

The technique has been most useful for the evidence given by spectral peaks, both large and small, broad and narrow, of hidden cycles in the data which were masked by observational error or other strong features. We have warned against the risk of being misled by random peaks in the periodogram, but the technique of smoothing and awareness of the statistical properties guard against this. The long data records available in many engineering and environmental situations are particularly appropriate for spectral analysis. Smoothed spectral estimates are also useful in suggesting the orders $p, q$ for an ARMA model to be fitted to the data.

## 18.11.  TIME SERIES REGRESSION MODELS

The problems of establishing valid relationships between two time series are manifold. It is quite possible for each series to depend upon one or more past values of the other. If they are part of a larger system of interdependent series, the other series being unobserved, then it may not be possible to establish causal relationships. We therefore limit ourselves in this section to consideration of a single time series with explanatory variables which model deterministic components such as cycles, trends and seasonality. The models will be of the form

$$x_t = c + mv_t + e_t \qquad (18.11.1)$$

where $c$ and $m$ are constants, $v_t$ is the explanatory variable, and $e_t$ follows an ARMA model. We shall see from examples that sometimes the major component is the deterministic part, sometimes it is the ARMA part, but it is always essential to have both parts correctly modelled.

Our first example is the variable star data of Figure 18.2.2(a). The periodogram of the data is shown on a logarithmic scale in Figure 18.11.1(a). Besides the peaks at frequencies 0·245 and 0·491 corresponding to the main cycle in the data and the first harmonic, there is a more general rise of the level towards lower frequencies. Now consider the residuals in Figure 18.2.2(c) obtained after fitting the cyclical components of the series. Their acf and pacf are shown in Figures 18.2.2(d), (e), and are typical of those of an AR(1) model. The cyclical components model described in section 18.2.2 was therefore refitted with an AR(1) model simultaneously fitted to the error component. The residual variance was reduced so that 97·5% of the total variance was accounted for. The parameter estimates changed slightly, but not significantly, though their standard errors were increased. In particular the standard error of the estimated frequency of the cycle almost doubled. This could be important if further observations were made at a later time in order to test for a change in frequency. The estimated coefficients of the cosine and sine parts of the fundamental cycle were

$$\hat{A} = 0·82 \pm 0·21, \qquad \hat{B} = -2·22 \pm 0·16$$

and for the first harmonic were

$$\hat{A}_1 = 0·31 \pm 0·08, \qquad \hat{B}_1 = -0·19 \pm 0·09.$$

The frequency estimate was $\hat{\omega} = 0·2452 \pm 0·0012$, corresponding to a period of 256.26 days. The autoregressive parameter was $\hat{\phi} = 0·86 \pm 0·05$.

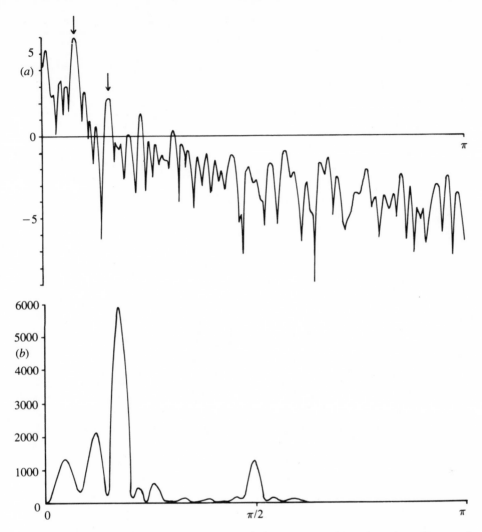

Figure 18.11.1: (a) Log periodogram of variable star series with main cycle and harmonic indicated. (b) Periodogram of Gilt series with seasonal component at $\pi/2$ clearly evident.

The new residuals were examined and no evidence found for any remaining autocorrelation, so the above analysis was assumed to be adequate. This is an example of a series with a mixed spectrum as described in section 18.5.4.

The second example is the daylength series. The ARMA(1, 2) model fitted to the differenced series may be written in terms of the original series $x_t$ as

$$(1 - \phi B)[(1 - B)\{x_t - \mu t\}] = \{1 - (1 + \phi)B + \phi B^2\}\{x_t - \mu t\} = (1 - \theta_1 B - \theta_2 B^2)a_t$$

where we have expressed $\nabla = 1 - B$ and incorporated the mean $\mu$ of the differenced series as a trend $\mu t$ in the original series. The purpose of this is to allow a possible relaxation of the differencing operation by fitting an ARMA(2, 2) model to the original series, together with a trend term,

$$x_t = c + \mu t + e_t \tag{18.11.2}$$

where $e_t$ is ARMA(2, 2).

On fitting this model the residual variance was slightly reduced to 233·1 from 238·7 for the ARMA(2, 1) model. One degree of freedom was lost and the corresponding chi-squared statistic was 4·22, significant at the 5% level. The estimates were

$$\hat{c} = -173 \pm 137, \qquad \hat{\mu} = 2\cdot 80 \pm 1\cdot 48, \qquad \hat{\phi}_1 = 1\cdot 31 \pm 0\cdot 13,$$

$$\hat{\phi}_2 = -0\cdot 35 \pm 0\cdot 13 \qquad \hat{\theta}_1 = -0\cdot 68 \pm 0\cdot 09, \qquad \hat{\theta}_2 = 0\cdot 72 \pm 0\cdot 07.$$

The factorization of

$$\phi(B) = (1 - 1\cdot 31 B + 0\cdot 35 B^2) = (1 - 0\cdot 38 B)(1 - 0\cdot 93 B)$$

shows that the random walk element of the previous model as typified by the operator $\nabla = 1 - B$, has now been replaced by a strongly autoregressive element represented by the factor $(1 - 0\cdot 93 B)$. The residuals from this latest model appear to be random, though containing a few rather extreme values, and their acf showed no evidence of any remaining structure. Forecasts of the final 10 points of the series (not used for fitting the model) together with error limits are shown in Figure 18.5.2), and are reasonably close to the actual values.

Our third example is the gilt series. The periodogram of the data, in Figure 18.11.1(b) indicates a broad peak in the spectrum at lower frequencies which might correspond to an AR(2) model, with a sharp peak at $\omega = \pi/2$ suggesting a seasonal component of period 4. This is to be expected in a quarterly series but tends to be masked by the large amplitude swings of the series shown in Figure 18.1.1(b). The acf and pacf in Figure 18.11.2 suggest autoregressive behaviour, but not pure autoregressive, since the pacf persists up to lag 7, presumably because of the seasonality present. A model of the form

$$x_t = \alpha_1 Q_{1,t} + \alpha_2 Q_{2,t} + \alpha_3 Q_{3,t} + \alpha_4 Q_{4,t} + e_t \tag{18.11.3}$$

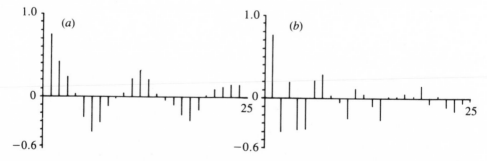

Figure 18.11.2: Sample acf (a), and sample pacf (b) for Gilt series.

was then fitted, with $Q_{j,t}$ being an indicator variable taking the value one in quarter $j$, but otherwise zero. The error term $e_t$ was represented by an ARMA(2, 1) model, the moving average term being included for comparison with an earlier model which omitted the quarterly components, and which needed this term. The necessity for the quarterly components was shown by a striking reduction in the residual variance from 68·1 for this earlier model, to the value 39·3 for model (11.3). The ARMA model parameter estimates were

$$\hat{\phi}_1 = 1\cdot55 \pm 0\cdot16, \qquad \hat{\phi}_2 = -0\cdot74 \pm 0\cdot14, \qquad \hat{\theta}_1 = 0\cdot28 \pm 0\cdot22$$

the MA parameter $\theta_1$ evidently being no longer necessary. The AR parameters imply a peak in the spectrum at a frequency corresponding to the period 14·8 quarters, using (18.8.11).

The residual series and its acf showed no evidence of remaining structure, but a further model was fitted which represents the seasonal structure in a manner similar to that in the airline series model (7.43), motivated by the seasonal EWMA predictor. This allows for a changing rather than fixed seasonal component, the non-seasonal ARMA(2, 1) structure being retained to produce the model

$$(1 - \phi_1 B - \phi_2 B^2)\{\nabla_4 x_t - \mu\} = (1 - \theta B)(1 - \Theta B^4). \qquad (18.11.4)$$

The residual variance came down slightly to 36·9. Two degrees of freedom were lost, associated with the new parameters $\Theta$, and $\mu$ which allows for a mean annual increase (or decrease). The corresponding chi-squared test statistic based on reduction in residual sum of squares was 5·6, and not significant. The parameter estimates were

$$\hat{\phi}_1 = 1\cdot62 \pm 0\cdot09, \qquad \hat{\phi}_2 = -0\cdot86 \pm 0\cdot09, \qquad \hat{\mu} = -1\cdot4 \pm 0\cdot8,$$
$$\hat{\theta} = 0\cdot41 \pm 0\cdot17, \qquad \hat{\Theta} = 0\cdot68 \pm 0\cdot10.$$

Although there is no strong evidence favouring (18.11.4), its flexibility for adapting to changes in seasonal pattern might favour its use in the longer term.

G.T.W.

## 18.12.  FURTHER READING AND REFERENCES

The applications of time series methods are very diverse, and each field tends to have its specialist texts, often with different terminology. Two books which have found wide acceptance are *Spectral Analysis and Its Applications* by Jenkins and Watts, and *Time Series Analysis, Forecasting and Control* by Box and Jenkins. The former gives a thorough account of spectral methods, including multivariate applications, and the latter covers the methodology of ARMA modelling, with extensions to models for the dependency of a series on one or more other series.

A more recent text is *Fourier Analysis of Time Series: an Introduction* by Bloomfield, which gives a good account of transformation methods. The book *Forecasting Economic Time Series* by Granger and Newbold is excellent in its field. *Physical Applications of Stationary Time Series* by Robinson, deals mainly with signal extraction applications in Geophysics. An up to date coverage of the subject is given in *Spectral Analysis and Time Series* by Priestley.

The practice of time series analysis almost inevitably requires good computer programs which are now available from many sources. The computations carried out for the examples in this Chapter used the software library and GENSTAT package distributed by the Numerical Algorithms Group, Banbury Road, Oxford, U.K.

Detailed references for the books mentioned above are as follows.

Bloomfield, P. (1976). Fourier Analysis of Time Series, Wiley.

Box, G. E. P. and Jenkins, G. M. (1976). *Time Series Analysis, Forecasting and Control*, Second edition, Holden-Day, San Francisco.

Granger, C. W. J. and Newbold, P. (1977). *Forecasting Economic Time Series*, Academic Press.

Jenkins, G. M. and Watts, D. G. (1968). *Spectral Analysis and Its Applications*, Holden-Day, San Francisco.

Priestley, M. B. (1981). *Spectral Analysis and Time Series*, Academic Press.

Robinson, E. A. (1980). *Physical Application of Stationary Time Series*, Griffin.

# CHAPTER 19

# *Decision Theory*

## 19.1. BASIC IDEAS

### 19.1.1. Mathematical Framework

The basic elements of a decision problem can be formalized mathematically in the following way:

we specify a set $\mathcal{A}$ [see I, § 1.1], called the *action space*, which consists of all the possible actions, $a \in \mathcal{A}$, available to the decision-maker;

we specify a set $\Theta$, called the *parameter space*, which consists of all possible 'states of nature', $\theta \in \Theta$, one and only one of which will occur (this 'true' state of nature being unknown to the decision-maker at the time he has to choose an action);

we specify a function $L$, called the *loss function* which has domain $\Theta \times \mathcal{A}$ (the set of all ordered pairs $(\theta, a)$, $\theta \in \Theta$, $a \in \mathcal{A}$ [see I, § 1.2.6]) and codomain $\mathbb{R}$ (the real line) [see I, Definition 1.4.1]; the pairs $(\theta, a)$ are called the *consequences* (of taking action $a$ when $\theta$ is the 'true' state of nature);

we *observe a random variable*, $X$, whose possible realizations are $x \in \mathcal{X}$ (*the sample space* [see II, § 3.1]) and whose distribution has a probability density function [see II, § 10.1.1] belonging to the family $\{f(x|\theta); \theta \in \Theta\}$;

we define the set $D$, called the *decision space*, to consist of all mappings $d$ from $\mathcal{X}$ to $\mathcal{A}$ [see I, Definition 1.4.1].

The interpretation is as follows. At the time he chooses his action, the decision-maker is unaware of the 'true' state of nature and thus is unaware of the actual consequence of his action (if he chooses $a \in \mathcal{A}$, the actual consequence $(\theta, a)$ is unknown since $\theta \in \Theta$ is unknown). The decision-maker does, however, know the 'loss' that would result from each of the possible consequences $(\theta, a)$ contingent on his choosing $a \in \mathcal{A}$ and the world turning out to be $\theta \in \Theta$ (of course, the 'loss' might actually be a 'gain', in which case the numerical value attached to $L(\theta, a)$ would be negative; alternatively, we could work, instead with a 'gain' or 'utility' function). To help modify or reduce his uncertainty about $\theta$, the decision-maker acquires information in the form of an observation of a random variable, $X$, whose probability distribution involves $\theta$. Knowing that $X = x$, and knowing the form of $f(x|\theta)$, the decision-maker can extract 'information' about $\theta$ which can help guide him in the choice of a general 'strategy', which defines, for each $X = x$, the choice of $a$. Formally, the decision-maker chooses an $a \in \mathcal{A}$ on the basis of having observed an $x \in \mathcal{X}$. The choice of a general 'strategy', which defines, for each $X = x$, the choice of $a \in \mathcal{A}$, is equivalent to choosing a *decision function, $d \in D$*. Once chosen, $d$ specifies the action to be taken for all possible $X = x$.

Decision theory can therefore be viewed as the study of how to select a $d$ from $D$. This involves two different considerations: first, the 'philosophical' problem of what criterion to use for comparing elements of $D$; secondly, the 'technical' problem of how to identify the optimal $d$, having decided on a criterion.

Sometimes, a decision problem is viewed as a 'game against Nature'. It is as though Nature chooses an element $\theta \in \Theta$ and then the decision-maker, without knowing which $\theta$ Nature has chosen, chooses an element $a \in \mathcal{A}$. The result of these two choices is that the decision-maker loses an amount $L(\theta, a)$ (measured in some appropriate units, not necessarily money). The possibility of observing a random variable $X$ having pdf $f(x|\theta)$ provides the decision-maker with some limited information about Nature's choice. The choice of a decision function can be seen as a 'strategy' for 'playing the game'.

Although we shall return to this in more detail later (§ 19.2), let us note immediately that the two main topics of statistical inference—namely, estimation and hypothesis testing—are both special cases of the general decision problem structure given above.

In the case of *estimation*, we have $\mathcal{A} = \Theta$ (often, the real line, or a subset of it), since the required action in this case is to choose a parameter value (i.e. an estimate). The form of loss function will depend on the practical exigencies of the problem being modelled, but typical choices might be $L(\theta, a) = |\theta - a|$ or $L(\theta, a) = (\theta - a)^2$. The decision functions $d: \mathcal{X} \to \mathbb{R}$ are usually called *estimators* and then $d(x)$ becomes the *estimate* of $\theta$ given data $X = x$ [c.f. Definition 3.1.1].

In the case of *hypothesis testing* [see § 5.12], the parameter space $\Theta$ is thought of as broken up into $\Theta = \Theta_0 \cup \Theta_1$, where $\Theta_0$ and $\Theta_1$ are disjoint subsets. The characteristic feature of $\mathcal{A}$ in this case is that it consists of only two elements, $\mathcal{A} = \{a_0, a_1\}$, defined such that action $a_0$ is the rejection of the *hypothesis* that $\theta \in \Theta_1$ and action $a_1$ is the rejection of the hypothesis that $\theta \in \Theta_0$. The loss function depends on the nature of the sets $\Theta_0$ and $\Theta_1$. In the case of a simple null versus a simple alternative hypothesis [see § 5.2.1(c), (d)], $\Theta = \{\theta_0, \theta_1\}$, we might, for example, take $L(\theta_0, a_0) = L(\theta_1, a_1) = 0$ (since in these cases the correct action has been taken), choosing the remaining values $L(\theta_0, a_1)$, $L(\theta_1, a_0)$ to reflect the relative seriousness of the two types of error.

The structure of this chapter is as follows. In the remainder of section 19.1 we shall discuss alternative criteria for assessing decision rules and then give some basic results which indicate which criteria lead to 'good' decisions. In section 19.2 some of these ideas will be illustrated for *statistical* decision problems (estimation and hypothesis testing).

It is clear that the 'loss' function (or, equivalently, the 'utility' or 'negative-loss' function) plays a fundamental role in this decision theory framework. Working in terms of the 'utility' function, we shall, in section 19.3, examine the way in which the choice of such a function is related to risk attitudes.

In section 19.4, we shall give a brief introduction to an important class of problems where a sequence of decisions follow each other in time, later decisions depending on the consequences of earlier ones. The useful pictorial device of a 'decision tree' will be explained and illustrated.

Finally, in section 19.5 we shall return briefly to the question of 'how' to justify criteria of decision-making, starting from a small number of 'self-evident' postulates, or axioms. Suggestions for further reading are given in section 19.6.

### 19.1.2.   Minimax and Bayes Decision Rules

It might be thought that the choice of an optimal decision function would be straightforward, since we simply wish to choose a $d \in D$ such that the loss is minimized, no matter what state of nature occurs.

However, a moment's thought reveals that this is never possible unless we *know* the true state of nature—in which case, we cannot really be said to have a decision problem. To illustrate this, suppose we are facing the problem of estimating an unknown real parameter $\theta$ [see § 3.1] with a loss function of the form $L(\theta, a) = (\theta - a)^2$. Suppose that we have observed $X = x$ and that $a = d(x)$ is the estimate of $\theta$ specified by $d$. If the true parameter value is $\theta$, we thus suffer a loss of $(\theta - d(x))^2$. If, in fact, $\theta = \theta_0$, we should need to take $d(x) = \theta_0$ in order to minimize the loss; whereas if $\theta = \theta_1$, we should need to take $d(x) = \theta_1$. But we do *not* know the value of $\theta$! Hence we cannot select $d(x)$ in order to minimize the loss—it simply is not a properly defined mathematical problem.

One possible way of assessing the 'goodness', or otherwise, of a decision function $d$, in terms of a quantity which *can* be calculated, is to look for a measure of how well the strategy performs 'on average'. The following definition will prove useful.

The function $R$ with domain $\Theta \times D$, codomain $\mathbb{R}$ and defined by

$$R(\theta, d) = \int_{\mathscr{X}} L(\theta, d(x)) f(x|\theta)\, dx \qquad (X \text{ continuous})$$

[see IV, Definition 4.1.3] and

$$R(\theta, d) = \sum_{x \in \mathscr{X}} L(\theta, d(x)) f(x|\theta) \qquad (X \text{ discrete})$$

[see IV, (1.7.1)] is called the *Risk function* of $d$, evaluated at $\theta$.

The risk $R(\theta, d)$ therefore measures the *expected loss* of using decision function $d$ if the true state of nature is $\theta$ (the expectation being with respect to the distribution specified by $f(x|\theta)$), so that, denoting the expectation operator by $E_{X|\theta}$ [see II, § 10.4.1], we could write

$$R(\theta, d) = E_{X|\theta}[L(\theta, d(X))].$$

To avoid the tiresome chore of having to distinguish continuous or discrete cases throughout, we shall either use this general expectation form, or, usually, just quote the integral (continuous) form.

The expectation operator $E_{X|\theta}$ can be applied to any function $h(X, \theta)$ whose expectation with respect to $X$ exists, so that

$$E_{X|\theta}\{h(X, \theta)\} = \int_{\mathscr{X}} h(x, \theta) f(x|\theta) \, dx.$$

We shall also use the variance operator $V_{X|\theta}$, defined as

$$V_{X|\theta}\{h(X, \theta)\} = E_{X|\theta}\{h(X, \theta)\}^2 - [E_{X|\theta}\{h(X, \theta)\}]^2.$$

If $\theta$ is a one-dimensional parameter, the graphs of $R(\theta, d)$ against $\theta$, for various $d$, provide insight into the relative performances of the decision rules, as measured by these risk functions.

EXAMPLE 19.1.1.   Consider Figure 19.1.1. This shows the risk functions of the following decision rules for the problem of estimating $\theta$ using $L(\theta, a) = (\theta - a)^2$, given $X = (X_1, \ldots, X_n)$ a random sample from $N(\theta, 1)$, with $n$ large:

$$d_1(X_1, \ldots, X_n) = \frac{1}{n}(X_1 + \ldots + X_n) = \bar{X}$$

$$d_2(X_1, \ldots, X_n) = \text{median}\,\{X_1, \ldots, X_n\}$$

$$d_3(X_1, \ldots, X_n) \equiv 0$$

Thus, $d_1$ is the sample mean, $d_2$ the sample median, $d_3$ a decision rule which says 'ignore the data and always estimate $\theta$ to be zero'. To calculate the risk functions, we note, for example, that

$$R(\theta, d_1) = E_{X|\theta}(\bar{X} - \theta)^2 = V_{X|\theta}(\bar{X}) = \frac{1}{n}$$

(since $\bar{X}$ is Normal $(\theta, 1/\sqrt{n})$ in this case), and that

$$R(\theta, d_3) = E_{X|\theta}(0 - \theta)^2 = E_{X|\theta}(\theta^2) = \theta^2.$$

The result for the median follows from the asymptotic (large $n$) result—which we shall simply quote—that median $\{X_1, \ldots, X_n\}$ is Normal $(\theta, \sqrt{(\pi/2n)})$.

Intuitively, the graphs suggest the following. Given our assumptions (Normal distribution, quadratic loss), we should *never* consider using the median as an estimator since the risk function for the mean is lower for *all* values of $\theta$.

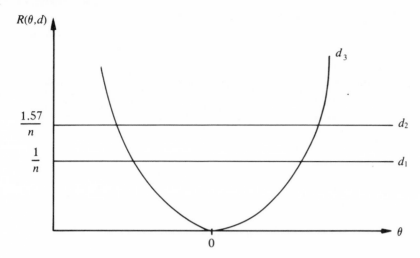

Figure 19.1.1: The risk functions $d_1, d_2, d_3$ of Example 19.1.1.

However, the mean is not necessarily better than the 'bigoted' estimator $d_3$, since if, in fact, $\theta$ is sufficiently near to 0 the latter would have smaller risk. The risk functions on their own do not tell us how to choose between $d_1$ and $d_3$, although they do indicate that $d_2$ should never be chosen (at least one of $d_1, d_3$ is always better).

As a further illustration of the difficulty of choosing between decision rules, consider

EXAMPLE 19.1.2.  Suppose $X = (X_1, \ldots, X_n)$ is a random sample from $N(\theta, \sqrt{\theta})$, where $\theta > 0$; i.e. a Normal distribution having both mean and variance equal to $\theta$. We wish to estimate $\theta$ using $L(\theta, a) = (\theta - a)^2$.

Consider, for purposes of illustration, the two decision functions

$$d_1(X) = \frac{1}{n}(X_1 + \ldots + X_n) = \bar{X}$$

and

$$d_2(X) = \frac{1}{n-1} \sum_{i=1}^{n} (X_i - \bar{X})^2.$$

Routine expectation calculations give

$$R(\theta, d_1) = E_{X|\theta}(\theta - \bar{X})^2 = V_{X|\theta}(\bar{X}) = \theta/n,$$

[see II, § 9.2.4] where $V_{X|\theta}$ is the variance operator defined above, and

$$R(\theta, d_2) = E_{X|\theta}\{\theta - d_2(X)\}^2 = V_{X|\theta}\{d_2(X)\},$$

since $E_{X|\theta}[d_2(X)] = \theta$, and so [see (2.5.22)]

$$R(\theta, d_2) = \frac{2\theta^2}{(n-1)}.$$

It follows that $R(\theta, d_1) < R(\theta, d_2)$ if $\theta > (n-1)/(2n)$ and $R(\theta, d_1) > R(\theta, d_2)$ if $\theta < (n-1)/(2n)$ with equality if $\theta = (n-1)/(2n)$. The graphs of the risk functions are sketched in Figure 19.1.2 for the case $n = 2$.

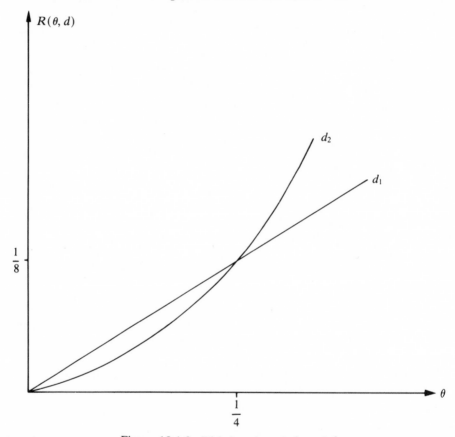

Figure 19.1.2: Risk function of $d_1$ and $d_2$.

Again, we see the dilemma. If we know that $\theta < \frac{1}{4}$ it would clearly be optimal to use $d_2$ rather than $d_1$. If we knew that $\theta > \frac{1}{4}$ the converse would be true. However, we *do not know* the value of $\theta$ and thus we must bring in some further *criterion* to help us choose.

In general, a specific decision problem leads to a vast number of possible decision functions (the set $D$ is very large) and so we cannot proceed as in the above examples by comparing a small number of decision rules graphically. Rather, we must seek general criteria which enable us to select an 'optimal' rule from the whole class $D$.

In this section, we shall study just two such criteria—the so-called Minimax and Bayes approaches. To motivate our definitions, let us first consider Figure 19.1.3 which displays two (hypothetical) risk functions corresponding to rules $d_1$ and $d_2$ for some (unspecified) decision problem with one-dimensional $\Theta$.

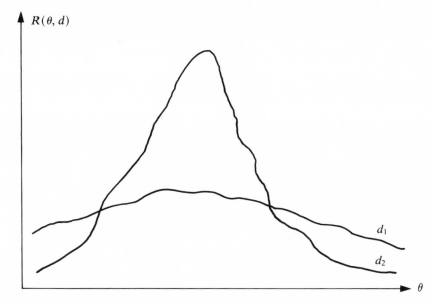

Figure 19.1.3: Two hypothetical decision rules.

For most values of $\theta$, $d_2$ has smaller risk than $d_1$; but for some values of $\theta$ $d_2$ has considerably higher risk. What should one do in such a situation?

Two possible approaches are the following:

(a) Guard yourself against the worst! It is preferable to be cautious and choose $d_1$ since this will protect you against the worst that might happen; it minimizes the maximum potential risk.

(b) Take into account other information you may possess and use your judgement about which values of $\theta$ might reasonably be expected to occur. If you strongly believe that $\theta$ will be in the range where $d_2$ does very badly, then choose $d_1$. If, on the other hand, you regard values of $\theta$ in that range as unlikely to occur, choose $d_2$. In any case, introduce your *personal beliefs* about $\theta$ into the analysis.

These two alternative intuitive responses may be formalized as follows. For (a), we choose $d^*$ such that

$$\sup_{\theta \in \Theta} R(\theta, d^*) = \inf_{d \in D} \sup_{\theta \in \Theta} R(\theta, d),$$

[see I, § 2.6.3]. In other words, we choose the decision function whose maximum risk is the smallest of all the possible maximum risks (as we range over $D$). For obvious reasons, $d^*$ is called the *Minimax decision function*. For (b), we shall assume that beliefs about $\theta$ can be represented in the form of a probability density function $p(\theta)$ with domain $\Theta$ (for further discussion of this topic, see Chapter 15 on Bayesian Statistics and section 19.5.2 of this chapter). For example, these beliefs might have shapes like $p_1$ and $p_2$ of Figure 19.1.4, where the range of $\theta$ shown is the same as that of Figure 19.1.3.

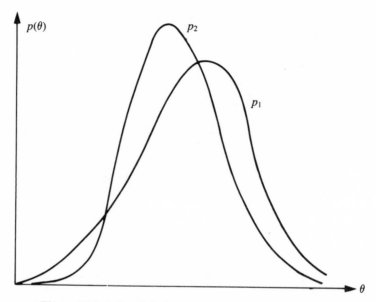

Figure 19.1.4: Possible forms of belief distribution for $\theta$.

Intuitively, someone with beliefs of the form $p_1$ might well feel that $d_2$ was a dangerous choice and that $d_1$ was to be preferred. Conversely, someone with beliefs $p_1$ might feel that $d_2$ was preferable. We incorporate such judgements formally as follows.

The risk function $R(\theta, d)$ expresses the expected loss of using $d$, conditional on $\theta$ being the true state of nature. Considering this as a function of $\theta$ for fixed $d$, we can calculate *its expected value, with respect to our distribution of belief, $p(\theta)$*. We then define the *Bayes risk $r(d)$* of a decision function $d$ to be

$$r(d) = \int_{\Theta} R(\theta, d)p(\theta)\, d\theta \qquad (\theta \text{ continuous})$$

or

$$r(d) = \sum_{\theta \in \Theta} R(\theta, d)p(\theta) \qquad (\theta \text{ discrete}).$$

It is now natural to seek a decision function which minimizes this overall expected loss $(r(d) = E_{\theta}E_{X|\theta}(L(\theta, d(X))))$. We call $d^*$ a *Bayes Decision*

*Function* if

$$r(d^*) = \inf_{d \in D} r(d).$$

Note that for a given decision problem, $d^*$ is not unique since, in particular, it depends on the choice of $p(\theta)$. It is better therefore to say that $d^*$ is a Bayes decision function *with respect to* $p(\theta)$. In the estimation context, $d(X)$ is called a *Bayes estimator* and $d(x)$ the *Bayes estimate* corresponding to $X = x$.

As an illustration of the Minimax and Bayes approaches, consider

EXAMPLE 19.1.3.  For the situation considered in Example 19.1.1, we have

$$\sup_{\theta} R(\theta, d_1) = 1/n$$

$$\sup_{\theta} R(\theta, d_2) = 1 \cdot 57/n$$

$$\sup_{\theta} R(\theta, d_3) = \infty$$

and so $d_1$ is the minimax decision rule.

Now suppose that we are sure that $\theta$ lies in the interval between $\pm \frac{1}{10}$ and that within this range we do not regard any particular value of $\theta$ as being more plausible than any other. If we agree to represent this belief by a uniform distribution over the interval $(-\frac{1}{10}, +\frac{1}{10})$, we have $p(\theta) = 5$ over the range, and

$$r(d_1) = \int_{-1/10}^{1/10} \left(\frac{1}{n}\right) \cdot 5 \, d\theta = \frac{1}{n}$$

$$r(d_2) = \int_{-1/10}^{1/10} \frac{1 \cdot 57}{n} \cdot 5 \, d\theta = \frac{1 \cdot 57}{n}$$

$$r(d_3) = \int_{-1/10}^{1/10} \theta^2 \cdot 5 \, d\theta = \frac{1}{300}.$$

If $n > 300$, the Bayes decision function is $d_1$; if $n = 300$, $d_1$ and $d_3$ have equal Bayes risks, and if $n < 300$, $d_3$ is preferred (all with respect to this particular choice of $p(\theta)$).

However, if our prior beliefs (§ 15.3.3) had been different, for example uniform over the interval $(-1, +1)$, the three Bayes risks would have been

$$r(d_1) = \frac{1}{n}, \qquad r(d_2) = \frac{1 \cdot 57}{n}, \qquad r(d_3) = \frac{10}{3}$$

so that $d_3$ would be the *least favoured* from the Bayes point of view.

We have introduced two possible approaches to choosing decision rules, the minimax approach and the Bayes approach, and have shown that they typically lead to different answers (although for some choices of prior distribution both approaches *can* lead to the same result). We shall now discuss a more 'neutral' condition which serves to separate decision functions into the 'sheep' and the 'goats'.

### 19.1.3.  Admissible Decisions

The general picture that we have seen emerge is summarized in Figure 19.1.5. It is clear that $d_1$ and $d_2$ (as portrayed in the figure) should not be considered further. Both $d_3$ and $d_4$ have uniformly (in $\theta$) smaller risks and it only remains to choose between *them*. In colloquial terms, $d_1$ and $d_2$ are 'worthless', but $d_3$ and $d_4$ are 'worth considering'. We can express this idea more formally as follows.

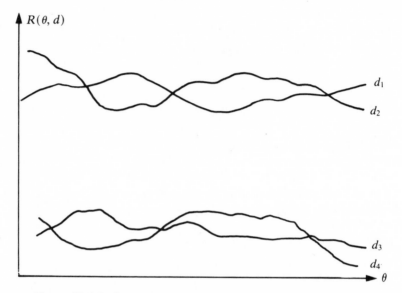

Figure 19.1.5: General picture of risk functions for several $d$'s.

Given a decision function $d' \in D$, if there exists another decision function $d'' \in D$ satisfying

$$R(\theta, d'') \le R(\theta, d')   \text{ for all } \theta \in \Theta$$

and

$$R(\theta, d'') < R(\theta, d')   \text{ for some } \theta \in \Theta,$$

then $d'$ is said *to be dominated* by $d''$. Equivalently, $d''$ *dominates* $d'$.

A decision function which is dominated by another is said to be *inadmissible*. Otherwise, it is said to be *admissible*.

In Figure 19.1.5, $d_1$ and $d_2$ are dominated by both $d_3$ and $d_4$ and so are inadmissible. Within the class $D = \{d_1, d_2, d_3, d_4\}$, $d_3$ and $d_4$ are admissible since neither dominates the other. If $D$ is a larger set, however, of which $\{d_1, d_2, d_3, d_4\}$ is just a subset, $d_3$ and $d_4$ may themselves be dominated by some other decision functions—and hence be inadmissible. Admissibility is always relative to the specification of $D$.

In many ways, it would be more natural to begin a study of decision theory with this concept of admissibility. An obvious clarification could then be achieved by eliminating the inadmissible rules and concentrating on the choice among the remaining admissible ones. Much of the fundamental work in decision theory has therefore been directed towards the problem of identifying the nature of the class of admissible decision functions. In the next but one section, we shall give a brief sketch of some of the kinds of general results that can be obtained.

Before doing that, however, it is interesting to consider the special case where $\Theta = \{\theta_1, \ldots, \theta_n\}$, a finite-dimensional parameter space (and more particularly still the case $k = 2$). This will enable us to give a geometric representation, and hence interpretation, of the concepts we have discussed so far.

### 19.1.4.   Geometric Interpretation

Before proceeding to the geometric interpretation, we shall need to introduce one further idea. It can be motivated by asking the question 'should decisions ever be made by tossing a coin?' So fas as 'rational' decision-making is concerned, intuition says no. But consider the following table which defines a loss function for a problem where $\Theta = \{\theta_1, \theta_2\}$, $\mathscr{A} = \{a_1, a_2, a_3\}$.

*Table of losses*

|            | $a_1$ | $a_2$ | $a_3$ |
|------------|-------|-------|-------|
| $\theta_1$ | 4     | 1     | 3     |
| $\theta_2$ | 1     | 4     | 3     |

Should we ever take action $a_3$? At first sight, it does not seem possible to rule it out of consideration, since if $\theta_1$ were true $a_3$ would be preferred to $a_1$, whereas if $\theta_2$ were true $a_3$ would be preferred to $a_2$.

However, suppose we consider the *randomized action* corresponding to tossing a (fair) coin and then choosing $a_1$ if Heads occur, $a_2$ if Tails occur. Then, for this randomized action, *if $\theta_1$ is true*, the expected loss is given by

$$\tfrac{1}{2}L(\theta_1, a_1) + \tfrac{1}{2}L(\theta_1, \theta_2) = \tfrac{1}{2} \times 4 + \tfrac{1}{2} \times 1 = \tfrac{5}{2}$$

and *if $\theta_2$ is true*, by

$$\tfrac{1}{2}L(\theta_2, a_1) + \tfrac{1}{2}L(\theta_2, a_2) = \tfrac{1}{2} \times 1 + \tfrac{1}{2} \times 4 = \tfrac{5}{2}.$$

Since $\tfrac{5}{2} < 3$, we see that in both cases the randomized action is preferable to $a_3$.

This suggests that randomization may be a valuable device. In fact, it is more convenient to think of randomizing the *decision rules* $d \in D$ rather than the *actions* $a \in \mathscr{A}$, and we shall write $\delta_\alpha = \alpha d_1 + (1 - \alpha) d_2$ $(0 \le \alpha \le 1)$ to indicate a randomized decision which chooses $d_1$ with probability $\alpha$ and $d_2$ with probability $1 - \alpha$. We shall define the risk of such a rule to be

$$R(\theta, \delta_\alpha) = \alpha R(\theta, d_1) + (1 - \alpha) R(\theta, d_2).$$

More generally, if $\alpha = (\alpha_1, \ldots, \alpha_m)$ with $\alpha_1 + \ldots + \alpha_m = 1$, $\alpha_i \geq 0$, we could define $\delta_\alpha = \alpha_1 d_1 + \ldots + \alpha_m d_m$, a randomization over elements of $D$ (or we could even, by taking $\alpha$ as the 'label' of a probability density, think of randomizing over a continuous 'region' of $D$ in certain cases). In the case of a randomization over $m$ elements we define

$$R(\theta, \delta_\alpha) = \alpha_1 R(\theta, d_1) + \ldots + \alpha_m R(\theta, d_m).$$

Having admitted the possibility of considering all randomizations of the elements of $D$, let us call the set of all randomized decision rules $D^*$; clearly $D \subseteq D^*$, and we shall denote a general element of $D^*$ by $\delta$.

An important aid in providing a geometric interpretation for the case of $\Theta = \{\theta_1, \ldots, \theta_n\}$ is the *risk set* $\mathscr{S}$ which is defined as follows:

$$\mathscr{S} = \{(y_1, \ldots, y_k) \in \mathbb{R}^k \text{ such that } y_j = R(\theta_j, \delta), j = 1, \ldots, k, \text{ for some } \delta \text{ in } D^*\}.$$

Here $\mathbb{R}^k$ ('$k$-dimensional space') denotes the set of all ordered $k$-ples $(y_1, y_2, \ldots, y_k)$ of real numbers; so that, for example, $\mathbb{R}^3 = \mathbb{R} \times \mathbb{R} \times \mathbb{R}$ [see I, § 1.2.6]. Thus $\mathscr{S}$ is the subset of $\mathbb{R}^k$ consisting of points whose component along the $j$th axis is the $\theta_j$ component of risk, $R(\theta_j, \delta)$ for *some* (randomized) decision rule $\delta$.

The important thing to note is that $\mathscr{S}$ is a *convex* set—that is, the line joining any two points in $\mathscr{S}$ never passes 'outside' the set. This is illustrated for $k = 2$ in Figure 19.1.6. It is easy to see how to prove convexity. All points on the line joining the risk points for $d_1$, $d_2$ (say) correspond to the risk point for some randomization of $d_1$ and $d_2$. All such points lie in $\mathscr{S}$ (by definition of the latter) and so $\mathscr{S}$ is convex.

We can now use this fact to give a geometric interpretation of the minimax and Bayes approaches.

*The Minimax approach.* For a given $\delta \in D^*$, the quantity $\sup_{\theta \in \Theta} R(\theta, \delta)$ becomes, for $\Theta = \{\theta_1, \ldots, \theta_k\}$, simply $\max_j y_j$ where $y = (y_1, \ldots, y_k)$ is the risk point corresponding to $\delta$. The minimax approach compares decision rules in terms of $\max_j y_j$, and so all decision rules leading to the same value of this quantity are equally regarded under the minimax criterion. In two dimensions, however, the locus of points $(y_1, y_2)$ with $\max (y_1, y_2)$ equal to some specified constant has the form of a 'wedge'. Since the minimax approach seeks to minimize the value of $\max_j y_j$, the minimax decision rule must be the point (or points) where the 90° wedge touches the *lower* boundary (i.e. the south-west boundary) of the risk set $\mathscr{S}$. This is illustrated in Figures 19.1.7(a) and 19.1.7(b), the latter showing why such a minimax rule might not be unique (it depends on the shape of $\mathscr{S}$, which, in turn, depends on the decision problem).

The same interpretation carries over straightforwardly to $k$-dimensions by simply generalizing the notion of a 90° wedge.

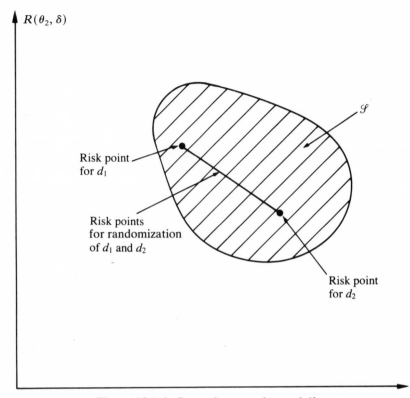

Figure 19.1.6: General convex form of $\mathscr{S}$.

*The Bayes approach.*  With $\Theta = \{\theta_1, \ldots, \theta_k\}$, a prior distribution of belief is specified by any $p = (p_1, \ldots, p_k)$ such that $p_i \geq 0$, $i = 1, \ldots, k$, and $p_1 + \ldots + p_k = 1$. The Bayes risk of a (randomized) decision rule $\delta$ is then given by

$$r(\delta) = \sum_{j=1}^{k} p_j R(\theta_j, \delta) = \sum_{j=1}^{k} p_j y_j.$$

This defines a hyperplane in $k$-dimensions. For simplicity, if $k = 2$, all points $y = (y_1, y_2)$ giving the same value of the Bayes risk lie on a straight line of the form $p_1 y_1 + p_2 y_2 = \text{constant}$. Since $0 \leq p_1, p_2 \leq 1$ and $p_1 + p_2 = 1$, any such straight line must be in a NW–SE direction. But the Bayes approach seeks to minimize $p_1 y_1 + p_2 y_2$: hence it follows that the Bayes decision rule has risk point where the line $p_1 y_1 + p_2 y_2 = \text{constant}$ forms the tangent to the lower boundary of $\mathscr{S}$. This is illustrated in Figures 19.1.8(a) and 19.1.8(b), the latter showing why Bayes decision rules may not be unique.

The same interpretation carries over straightforwardly to $k$-dimensions if we think, instead, of tangent hyperplanes.

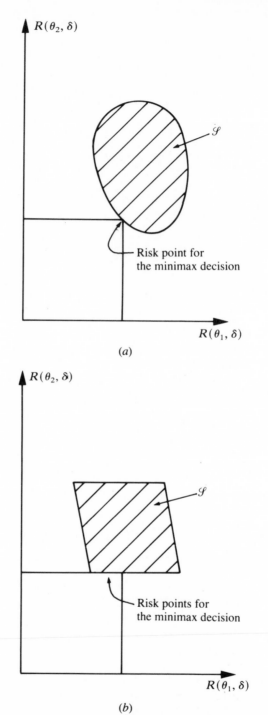

(a)

(b)

Figure 19.1.7: Geometric interpretation of the minimax approach.

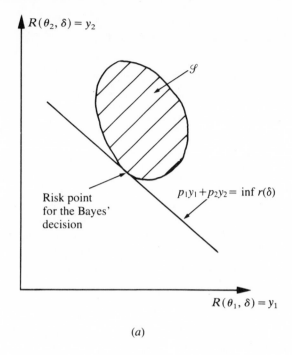

(a)

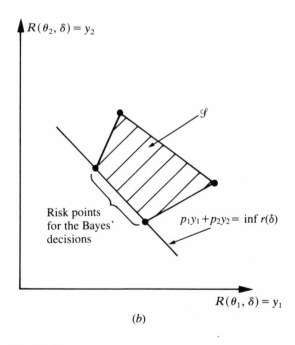

(b)

Figure 19.1.8: Geometric interpretation of the Bayes' approach.

EXAMPLE 19.1.4.  Suppose that for a decision problem with $\Theta = \{\theta_1, \theta_2\}$, $D = \{d_1, d_2, d_3, d_4, d_5\}$, the risk function $R(\theta_i, d_j)$ is defined by the following table.

|            | $d_1$ | $d_2$ | $d_3$ | $d_4$ | $d_5$ |
|------------|-------|-------|-------|-------|-------|
| $\theta_1$ |   0   |   4   |   2   |   1   |   5   |
| $\theta_2$ |   4   |   5   |   0   |   1   |   4   |

The five risk points corresponding to the elements of $D$ are shown in Figure 19.1.9, together with $\mathscr{S}$. It is clear that $\mathscr{S}$ consists of all (and only) the risk points that can be reached by randomizing among the five initial risk points.

The *admissible rules* correspond to points in $\mathscr{S}$ which have no points in $\mathscr{S}$ to the SW of them (including $S$ and $W$; so that we cannot find a $\delta$ which reduces at least one component of risk without increasing the other). In this case, the set of admissible rules corresponds to points lying between '$d_1$' and '$d_4$', and '$d_4$' and '$d_3$'; in other words, it consists of all randomizations of $d_1$, $d_4$ or $d_4$, $d_3$.

The *minimax rule* is $d_4$, since the 90° wedge hits the lower boundary of $\mathscr{S}$ (uniquely) at '$d_4$'.

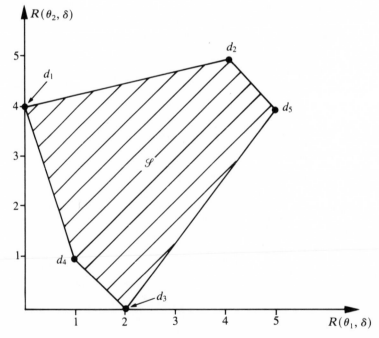

Figure 19.1.9: $\mathscr{S}$ based on the $R(\theta, \delta)$ table.

*The Bayes rule* depends on the choice of $(p_1, p_2)$. By thinking of a vertical line ($p_1 = 1$) moving around to a horizontal line ($p_1 = 0$), and considering the points of contact with the lower boundary, we easily obtain the following summary.

| $p_1$ | Bayes decision |
|---|---|
| $> \frac{2}{3}$ | $d_1$ (unique) |
| $= \frac{2}{3}$ | any randomization of $d_1, d_4$ |
| $< \frac{2}{3}, > \frac{1}{2}$ | $d_4$ (unique) |
| $= \frac{1}{2}$ | any randomization of $d_4, d_3$ |
| $< \frac{1}{2}$ | $d_3$ (unique) |

Although in some cases we *could* take a randomized Bayes decision, it is not *necessary* to do so since, for $p_1 = \frac{2}{3}$, either of $d_1, d_4$ could be chosen and, for $p = \frac{1}{2}$, either of $d_4, d_3$ could be chosen.

### 19.1.5. Some Basic Theorems

The following results give an indication of how we set about trying to establish general relations between the concepts of admissibility and Bayes and Minimax rules, and what those relations are.

Before we consider the first theorem, however, let us just note that it is possible to have two decision functions $\delta_1 \in D^*$, $\delta_2 \in D^*$ such that $\delta_1 \neq \delta_2$, but $R(\theta, \delta_1) = R(\theta, \delta_2)$ for all $\theta \in \Theta$. This could happen if the definitions of $\delta_1, \delta_2$ only varied at a finite number of points, for a continuous random variable $X$. Since $R(\theta, \delta_i)$, $i = 1, 2$, are defined as integrals, their values will be equal. In such a case, we shall say that $\delta_1, \delta_2$ are *equal up to equivalence* (i.e. they cannot be distinguished on the basis of $R(\theta, \delta_i)$, $i = 1, 2$). Using this idea, we have

THEOREM 19.1.1. *For arbitrary $\Theta$, $A$, $L$, and with $\mathcal{X}$ as the sample space of a continuous random variable $X$, if the Bayes decision function, $\delta^*$ with respect to a prior distribution $p(\theta)$, is unique up to equivalence, then $\delta^*$ is admissible.*

*Proof* (assuming $p(\theta)$ is a density; the case of discrete $\Theta$ follows similarly). Assume that $\delta^*$ is inadmissible. Then there exists $\delta \in D^*$ such that

$$R(\theta, \delta) \leq R(\theta, \delta^*) \quad \text{for all } \theta \in \Theta$$

and

$$R(\theta_0, \delta) < R(\theta_0, \delta^*) \quad \text{for some } \theta_0 \in \Theta.$$

Hence,

$$r(\delta) = \int_\Theta R(\theta, \delta) p(\theta) \, d\theta \leq \int_\Theta R(\theta, \delta^*) p(\theta) \, d\theta = r(\delta^*).$$

But $\le$ cannot be $<$ (since this would contradict the fact that $\delta^*$ is Bayes), and $\le$ cannot be $=$ (since this would contradict the fact that $\delta^*$ is *unique Bayes up to equivalence*). We thus have a contradiction and so $\delta^*$ is admissible.

THEOREM 19.1.2.   *If* $\Theta = \{\theta_1, \ldots, \theta_k\}$ *and* $\delta^* \in D^*$ *is the Bayes' decision with respect to* $(p_1, \ldots, p_k)$, *where* $p_i > 0$, $i = 1, \ldots, k$, *then* $\delta^*$ *is admissible.*

[*Interpretation.*   If, for finite $\Theta$, a Bayes' decision is based on a non-bigoted prior distribution (no $p_i = 0$), then it is admissible.]

*Proof.*   Start by assuming that it is *not* admissible—this will be seen to lead to a contradiction and thus to be a false assumption. If $\delta^*$ is not admissible, there exists $\delta \in D^*$ such that

$$\left.\begin{array}{l} R(\theta_j, \delta) \le R(\theta_j, \delta^*) \quad \text{for all } j = 1, \ldots k \\ R(\theta_i, \delta) < R(\theta_i, \delta^*) \quad \text{for some } i \end{array}\right\} \tag{$*$}$$

(since there exists a $\delta$ dominating $\delta^*$). But,

$$r(\delta) = \sum_{j=1}^{k} R(\theta_j, \delta)p_j < \sum_{j=1}^{k} R(\theta_j, \delta^*)p_j = r(\delta^*),$$

the central inequality following from ($*$) together with the fact that $p_j > 0$ for all $j$.

But $r(\delta^*) = \inf_\delta r(\delta)$, so that we have a contradiction.

In order to see what can go wrong if the condition $p_j > 0$, for all $j$, is dropped, consider the risk set $\mathscr{S}$ as shown in Figure 19.1.10. If $p_1 = 1$ (so that $p_2 = 0$), all the points on the left-hand boundary of $\mathscr{S}$ correspond to Bayes decisions. But only the lower SW-vertex of the rectangle corresponds to an admissible decision.

THEOREM 19.1.3.   *If* $\Theta = \{\theta_1, \ldots, \theta_k\}$ *is finite and* $\delta^*$ *is admissible, then there exists* $p = (p_1, \ldots, p_k)$, $p_i \ge 0$ *for all* $i$, $p_1 + \cdots + p_k = 1$, *such that* $\delta^*$ *is a Bayes decision with respect to* $p$.

[*Interpretation.*   For problems with finite $\Theta$, the class of admissible decisions is a subset of the class of Bayes decisions.]

*Outline proof* $(k = 2)$.   Let $x$ denote the risk point corresponding to $\delta^*$ and let $Q_x$ denote the set of all points in the plane to the SW (including S and W) of $x$. Let $Q'_x$ be the set $Q_x$ with the point set $\{x\}$ removed.

Since $\delta^*$ is admissible, there are no points in $\mathscr{S}$ strictly to the SW of $x$. Hence $\mathscr{S}$ and $Q'_x$ are two *disjoint* sets [see I, § 1.2.1]; they are also both *convex* (§ 19.1.4).

A famous theorem (the separating hyperplane theorem) tells us that (in our case; $k = 2$) there exists a line $p_1 y_1 + p_2 y_2 = $ constant as shown in Figure 19.1.11.

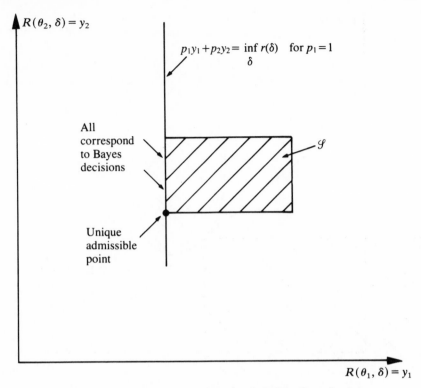

Figure 19.1.10: Example to show inadmissible Bayes' rules.

Such a line has $p_1$, $p_2 \geq 0$ and (if necessary by dividing the equation through by $p_1 + p_2$) we can claim without loss of generality that $p_1 + p_2 = 1$. But this line forms the tangent to $\mathscr{S}$ at $x$ and so $\delta^*$ is a Bayes decision with respect to $(p_1, p_2)$.

These first three theorems give an idea of the kinds of connections that exist between admissible decisions and Bayes decisions. The next two theorems explore similar connections between admissible decisions and minimax decisions. A more detailed account is given in the book *Mathematic Statistics: A Decision Theoretic Approach* by T. S. Ferguson (Academic Press, 1967).

THEOREM 19.1.4. *If, for a given decision problem, $\delta^*$ is the unique minimax decision, then $\delta^*$ is admissible.*

*Proof.* Suppose not. Then there exists $\delta \in D^*$ such that

$$R(\theta, \delta) \leq R(\theta, \delta^*) \quad \text{for all } \theta \in \Theta$$

and

$$R(\theta, \delta) < R(\theta, \delta^*) \quad \text{for some } \theta \in \Theta.$$

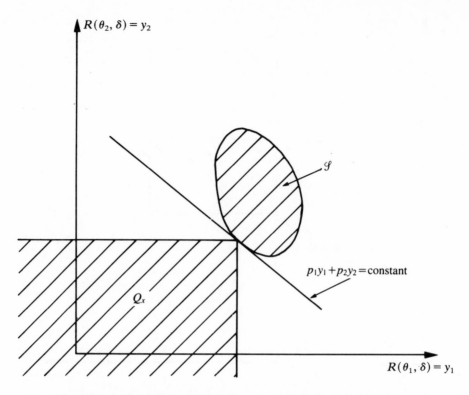

Figure 19.1.11: Sketch 'proof' of why an admissible rule is Bayes.

But this implies that

$$\sup_{\theta \in \Theta} R(\theta, \delta) \leq \sup_{\theta \in \Theta} R(\theta, \delta^*).$$

Strict inequality would contradict the fact that $\delta^*$ is minimax; equality would contradict the fact that $\delta^*$ is unique minimax. Hence our initial hypothesis of inadmissibility was false.

**THEOREM 19.1.5.**   *If $\delta^*$ is admissible and $R(\theta, \delta^*) = constant$, for all $\theta \in \Theta$, then $\delta^*$ is minimax.* [*N.B.* This suggests a strategy for finding minimax decision functions: first identify an admissible rule and then check if it has constant risk.]

*Proof.*   Suppose not. Then there exists $\delta \in D^*$ such that

$$\sup_{\theta \in \Theta} R(\theta, \delta) < \sup_{\theta \in \Theta} R(\theta, \delta^*).$$

But if $R(\theta, \delta^*) = $ constant, this implies that

$$R(\theta, \delta) < R(\theta, \delta^*)] \quad \text{for all } \theta \in \Theta,$$

and would contradict the fact that $\delta^*$ is admissible.

The main classification theorem of decision theory states (more or less—there are, of course, a number of mathematical regularity conditions) that, in general,

*in order to be admissible, a decision rule has to be Bayes.*

This motivates the finding of a simpler method of calculating Bayes decision rules than directly searching for $\inf_{\delta \in D^*} E_\theta E_{X|\theta}\{L(\theta, d(X))\}$. The next theorem is important in identifying such a simple method.

THEOREM 19.1.6.          (*Non-rigorous statement*).  *The Bayes' decision function $\delta^*$, with respect to prior beliefs $p(\theta)$, is determined by $\delta^*(x) = a^*$, where $a^*$ minimizes*

$$\int_\Theta L(\theta, a)p(\theta|x) \, d\theta,$$

*and $p(\theta|x) = f(x|\theta)p(\theta)/\int f(x|\theta)p(\theta) \, d\theta$.*

[*Interpretation.*  For any $X = x$, choose the action $a^*$ which minimizes $E_{\theta|x}[L(\theta, a)]$, the *posterior expected loss.*]

*Proof (non-rigorous)*

$$r(\delta) = \int_\Theta R(\theta, \delta)p(\theta) \, d\theta$$

$$= \int_\Theta \left\{ \int_{\mathscr{X}} L(\theta, \delta(x))f(x|\theta) \, dx \right\} p(\theta) \, d\theta.$$

Assuming that the functions involved permit the interchange of the order of integration (and this is where greater rigour would be required to give a watertight argument), we have

$$r(\delta) = \int_{\mathscr{X}} \left\{ \int_\Theta L(\theta, \delta(x))p(\theta|x) \, d\theta \right\} f(x) \, dx, \qquad (*)$$

since, by the definition of conditional p.d.f.'s,

$$f(x|\theta)p(\theta) = p(\theta|x)f(x).$$

Finding a $\delta^*$ to minimize (*) is equivalent to minimizing the inner integral for each $X = x$ as stated in the theorem.

The calculation of Bayes decisions can therefore be carried out in two steps. First, find $p(\theta|x)$ using Bayes' Theorem (see Chapter 15); then minimize *the posterior expected loss.*

## 19.2.   STATISTICS AND DECISION THEORY

### 19.2.1.   Estimation

As we noted in section 19.1.1, the problem of estimation of an unknown parameter can be regarded as a special case of the general statistical decision problem when $\Theta = \mathscr{A}$ (the action required is to choose an element of the parameter space). Moreover, so far as Bayes decision functions are concerned, Theorem 19.1.6 of the previous section established that such a decision function, $\delta$, is defined by choosing, for each $X = x$, the action $a = \delta(x)$ which minimizes the posterior expected loss

$$\int_\Theta L(\theta, \delta(x))p(\theta|x)\, d\theta.$$

The following theorems establish the general forms of *Bayes estimates* for various standard choices of loss function $L(\theta, a)$.

THEOREM 19.2.1.   *If* $L(\theta, a) = (\theta - a)^2$, *so-called quadratic loss, the Bayes estimate is given by the mean of the posterior distribution.*

*Proof.*   We want to choose $a = \delta(x)$, to minimize

$$\int_\Theta (\theta - a)^2 p(\theta|x)\, d\theta.$$

Differentiating with respect to $a$ and equating to zero, we obtain

$$a = \frac{\int_\Theta \theta p(\theta|x)\, d\theta}{\int_\Theta p(\theta|x)\, d\theta} = \int_\Theta \theta p(\theta|x)\, d\theta,$$

since $\int_\Theta p(\theta|x)\, d\theta = 1$.

THEOREM 19.2.2.   *If* $L(\theta, a) = |\theta - a|$, *so-called absolute-value loss, the Bayes estimate is given by a median of the posterior distribution.*

*Proof.*   (Consider the case $\Theta = \mathbb{R}$).   We want to choose $a$ to minimize the posterior expected loss,

$$\int_{-\infty}^{\infty} |\theta - a| p(\theta|x)\, d\theta = \int_{-\infty}^{a} (a - \theta)p(\theta|x)\, d\theta + \int_{a}^{\infty} (\theta - a)p(\theta|x)\, d\theta.$$

Differentiating with respect to $a$ and equating to zero, and recalling that

$$\frac{d}{da} \int_{a}^{\infty} g(x)\, dx = -g(a), \qquad \frac{d}{da} \int_{-\infty}^{a} g(x)\, dx = g(a),$$

we obtain

$$\int_{-\infty}^{a} p(\theta|x)\, d\theta = \int_{a}^{\infty} p(\theta|x)\, d\theta = \tfrac{1}{2}$$

(since the two integrals must add to 1). Hence, $a$ is a median [see II, § 10.3.3] of the posterior distribution.

THEOREM 19.2.3.   *If $L(\theta, a) = 1$ for $\theta \neq a$, 0 for $\theta = a$, so-called zero-one loss, the Bayes estimate is given by the mode of the posterior distribution.*

*Proof.*   (*Case of discrete $\Theta$ only*).   The loss function is of the form

$$L(\theta, a) = \begin{cases} 1, & \text{if } a \neq \theta \\ 0, & \text{if } a = \theta \end{cases}$$

and we wish to minimize

$$\sum_{\theta \in \Theta} L(\theta, a) p(\theta | x).$$

If we were to choose $a = \theta^*$, the posterior expected loss would be

$$\sum_{\theta \in \Theta \setminus \{\theta^*\}} p(\theta | x) = 1 - p(\theta^* | x).$$

This is clearly minimized when $a = \theta^*$ is chosen to maximize $p(\theta | x)$. But this is precisely the *mode* (most likely value) of the posterior distribution.

A number of particular examples of Bayes estimates using quadratic loss were given in Chapter 15.

In general, the derivation of Minimax estimates is not so straightforward; operations like sup and inf do not lead to obvious manipulations like the integration (or summation) and differentiation involved in finding Bayes solutions. However, Theorem 19.1.5 can often be used as a basis for finding Minimax solutions.

EXAMPLE 19.2.1.   Let us suppose that $X = x$ is the observed number of successes in $n$ independent trials each having success probability $\theta$. We shall find the Minimax estimate of $\theta$ with respect to quadratic loss.

We first note that Theorem 19.1.5 tell us to find an admissible estimator with constant risk. We also recall that Theorem 19.1.1 says that a unique Bayes decision function is admissible. What we attempt to do, therefore, is to find the form of a unique Bayes estimate and then examine whether there are circumstances under which such an estimate has constant risk. If so, we shall have found a Minimax estimator.

To carry out the first step, we recall from section 15.5.1 that if $\theta$ is taken to have a Beta prior distribution with parameters $\alpha$ and $\beta$, so that

$$p(\theta) \propto \theta^{\alpha - 1} (1 - \theta)^{\beta - 1},$$

the posterior distribution for $\theta$ has the same form, but with $\alpha$ replaced by $\alpha + x$ and $\beta$ replaced by $\beta + n - x$. Since the mean of a Beta $(\alpha + x, \beta + n - x)$ distribution is $(\alpha + x)/(\alpha + \beta + n)$, and since this is the unique Bayes estimate with respect to quadratic loss, the unique Bayes estimate corresponding to the prior parameters $\alpha$ and $\beta$ is $(\alpha+x)/(\alpha+\beta+n)=\delta(x)$, say.

We now identify the form of $R(\theta, \delta)$ and examine whether there is a choice of $\alpha$ and $\beta$ which makes the risk constant (for all $\theta$). By definition,

$$R(\theta, \delta) = E_{X|\theta}[(\alpha+X)/(\alpha+\beta+n)-\theta]^2$$

$$= (\alpha+\beta+n)^{-2}E_{X|\theta}[(X-n\theta)+\alpha(1-\theta)-\beta\theta]^2$$

$$= (\alpha+\beta+n)^{-2}\{E_{X|\theta}(X-n\theta)^2+[\alpha(1-\theta)-\beta\theta)]^2\}$$

since

$$E(X) = n\theta,$$

causing the cross-product term to disappear,

$$= (\alpha+\beta+n)^{-2}\{n\theta(1-\theta)+\alpha^2(1-\theta)^2-2\alpha\beta\theta(1-\theta)+\beta^2\theta^2\}$$

since

$$E(X-n\theta)^2 = V(X) = n\theta(1-\theta),$$

$$= (\alpha+\beta+n)^{-2}\{\theta^2((\alpha+\beta)^2-n)+\theta[n-2\alpha(\alpha+\beta)]+\alpha^2\}$$

after some algebra rearrangement. This risk is independent of $\theta$, i.e. the coefficient of $\theta^2$ and $\theta$ are zero, if $\alpha=\beta=\frac{1}{2}\sqrt{n}$. The Minimax estimator of $\theta$ is therefore

$$\frac{x+\sqrt{n}/2}{n+\sqrt{n}}.$$

### 19.2.2.  Tests of Simple Hypotheses versus Simple Hypotheses

We mentioned in section 19.1.1 the way in which the statistical problem of hypothesis testing can be viewed as a special case of a decision problem. We shall now illustrate this in detail by considering the special case of a simple null hypothesis versus a simple alternative hypothesis. If $H_0: \theta=\theta_0$ denotes the null and $H_1: \theta=\theta_1$ the alternative [see § 5.12.2] the parameter space $\Theta$ consists of just the two elements $\{\theta_0, \theta_1\}$. As we mentioned before, for a hypothesis testing problem the action space also consists of just two elements, $\mathscr{A}=\{a_0, a_1\}$, having the interpretation: $a_0$ denotes rejection of $H_1$; $a_1$ denotes rejection of $H_0$.

Let us further suppose that the loss function is defined by the following table.

|       | $\theta_0$ | $\theta_1$ |
|-------|------------|------------|
| $a_0$ | $0$        | $L_{01}$   |
| $a_1$ | $L_{10}$   | $0$        |

and that we observe a random variable $X$, having distribution with density $f(x|\theta)$.

We shall first study the form of the Bayes' decision given prior probabilities $p(\theta_0) = \pi_0$, $p(\theta_1) = \pi_1$, $\pi_0 + \pi_1 = 1$.

Recalling from Theorem 19.1.6 that the Bayes decision for given $X = x$ can be found by choosing the action which minimizes the posterior expected loss, we calculate the following.

*Posterior expected loss if we take $\delta(x) = a_0$:*

$$0 \times p(\theta_0|x) + L_{01} \times p(\theta_1|x) = L_{01}\pi_1 f(x|\theta_1)/[\pi_0 f(x|\theta_0) + \pi_1 f(x|\theta_1)];$$

*if we take $\delta(x) = a_1$:*

$$L_{10} \times p(\theta_0|x) + 0 \times p(\theta_1|x) = L_{10}\pi_0 f(x|\theta_0)/[\pi_0 f(x|\theta_0) + \pi_1 f(x|\theta_1)];$$

where Bayes' Theorem has been used to rewrite $p(\theta_0|x)$, $p(\theta_1|x)$.

It follows that $\delta(x) = a_1$ should be chosen if

$$L_{10}\pi_0 f(x|\theta_0) < L_{10}\pi_1 f(x|\theta_1),$$

in other words, if,

$$\frac{f(x|\theta_0)}{f(x|\theta_1)} < \frac{L_{10}\pi_1}{L_{01}\pi_0}.$$

This has the form: 'reject $H_0$ if the likelihood ratio of $\theta_0$ to $\theta_1$ is less than a certain quantity'. This form may already be familiar as the form of the Neymon–Pearson Lemma [see § 5.12.2]. Here, however, we note that the 'cut-off' quantity on the right-hand side of the inequality is defined in terms of relative losses $(L_{01}/L_{01})$ and the prior odds on $H_1$ to $H_0(\pi_1/\pi_0)$.

Further insight into the relationship (and a method for finding a Minimax test) can be obtained by specializing to the case $L_{01} = L_{10} = 1$ and working with risks in a geometric framework.

### 19.2.3. The Neyman–Pearson Lemma

With $L_{01} = L_{10} = 1$, we first calculate the risk $R(\theta, \delta)$, where $\delta$ is defined such that $\mathcal{X} = \{\mathcal{X}_0, \mathcal{X}_1\}$, and

$$\delta(x) = a_i \quad \text{for } x \in \mathcal{X}_i.$$

Then,

$$R(\theta_0, \delta) = \int_{\mathcal{X}} L(\theta_0, \delta(x)) f(x|\theta_0)\, dx$$

$$= \int_{\mathcal{X}_1} f(x|\theta_0)\, dx = P_r(\tilde{X} \in \mathcal{X}_1|\theta_0)$$

$$= \textit{Probability of rejecting } H_0 \textit{ when } H_0 \textit{ is true. } (\alpha)$$

Recall (§ 5.12.2) that $\alpha$ is the *probability of a Type I error*. We see that

$$R(\theta_1, \delta) = \int_{\mathscr{X}} L(\theta_1, \delta(x)) f(x|\theta_1) \, dx$$

$$= \int_{\mathscr{X}_0} f(x|\theta_1) \, dx = P_r(\tilde{X} \in \mathscr{X}_0|\theta_1)$$

$$= \text{\textit{Probability of accepting } } H_0 \text{ \textit{when} } H_0 \text{ \textit{is false.}} \quad (\beta)$$

Recall (§ 5.12.2) that $\beta$ is the *probability of a Type II error*.

   To draw the risk set $\mathscr{S}$ (§ 19.1.4), we first note that the risk points $(0, 1)$, $(1, 0)$ are attained by tests which *always* accept, reject $H_0$, respectively (irrespective of $x$); these points therefore belong to $\mathscr{S}$. We also note that $\mathscr{S}$ has a basic symmetry (about the line through $(0, 1)$, $(1, 0)$) since any test corresponding to $\mathscr{X} = \{\mathscr{X}_0, \mathscr{X}_1\}$ has a 'symmetric image' corresponding to $\mathscr{X} = \{\mathscr{X}_1, \mathscr{X}_0\}$; in other words, interchanging $a_0$ and $a_1$ in its definition. Recalling the convexity property, we see that for this testing problem the geometric representation has the typical form shown in Figure 19.2.1.

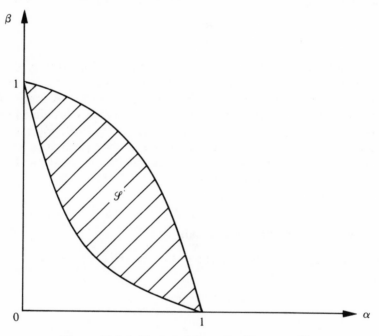

Figure 19.2.1: Typical risk set for $H_0$ versus $H_1$.

   As we have already seen in section 19.1.4, the risk points corresponding to *Bayes tests* are the tangent points of lines running NW–SE with slopes determined by $-\pi_0/\pi_1$. For such a risk set, the *Minimax test* corresponds to the point where a 45° line meets the lower (SW) boundary of $\mathscr{S}$. In particular, we see that a Minimax test defines equal Type I and Type II error probabilities ($\alpha = \beta$). The *admissible tests* are those with risk points lying on the lower (SW) boundary of $\mathscr{S}$. It is of interest to note that this boundary consists of points which, *for fixed $\alpha$, minimize $\beta$*. But these are points characterized by the *Neyman–Pearson approach* to finding tests (§ 5.12): in other words *maximize power* $(1 - \beta)$ for fixed *size* $(\alpha)$. Moreover, the *Neyman–Pearson lemma* can now be given a decision-theoretic statement:

*Neyman–Pearson Lemma.* With zero-one loss function, the *admissible tests* for simple $H_0 = \theta = \theta_0$ versus simple $H_1 : \theta = \theta_1$ are defined by

$$\{\delta \in D^*; \delta(x) = a_1 \text{ if } f(x|\theta_0)/f(x|\theta_1) < k, \text{ for some } k \in \mathbb{R}\}.$$

## 19.3. RISK ATTITUDES AND UTILITY THEORY

### 19.3.1. Risk Aversion

In this section we shall consider some of the basic ideas which enable us to investigate mathematically the problem of 'risk-taking'. By this, we mean the problem of deciding whether to remain with the status quo or to enter an uncertain situation which may result in some form of 'gain', but may also result in some form of 'loss'. Obvious practical examples include offering insurance (premium income may provide a 'gain', but having to pay out following a 'disaster' may result in an overall 'loss'), farming (assets are exchanged for seed or stock, whose yield may provide 'gain', but which could lead to a 'loss' if drought or disease occurs) or various forms of monetary gambling (an entry fee or 'stake' is paid and a 'win' or 'loss' occurs as a result of some agreed contingency occurring or not: for example, particular horse winning a race, card being dealt from a pack, number coming up on a roulette wheel, etc.).

For simplicity, we shall work throughout this section in terms of money units: it is important to remember, however, that the ideas (possibly with minor variations) extend to any situation involving risk.

In Figure 19.3.1, below, we present, schematically, a simple decision problem which captures the essence of all 'risk-taking' problems. One action involves no uncertainty and preserves the status quo; the other leads to a situation of uncertainty which 'may go one way or the other'—leading to an improved situation if there is a favourable outcome, leading to a worse situation if there is an unfavourable outcome.

Using our earlier notation (Section 19.1.1), we have:

$$\mathscr{A} = \{\text{Do not gamble, gamble}\} = \{a_1, a_2\}$$

$$\Theta = \{\text{No change, win, lose}\} = \{\theta_1, \theta_2, \theta_3\}$$

and the consequences are

$$(\theta_1, a_1) \equiv \text{current assets unchanged}$$

$$(\theta_2, a_2) \equiv \text{current assets increased by prize money.}$$

$$(\theta_3, a_2) \equiv \text{current assets decreased by stake money.}$$

Throughout this section, we shall work in terms of the Bayesian approach to decision making. This requires the specification of the following quantities for the example shown in Figure 19.3.1: first, the specification of probabilities for $\{\theta_1, \theta_2, \theta_3\}$, given $a_1$ or given $a_2$; secondly, the assignment of values to the consequences.

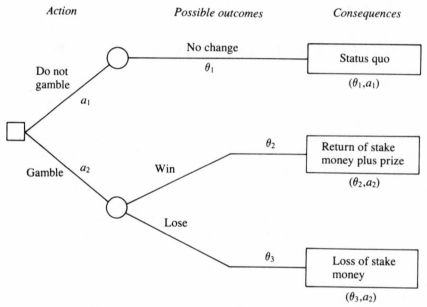

Figure 19.3.1: A simple 'risk-taking' decision problem.

In the introduction (§ 19.1.1) we talked in terms of a *loss function, $L: \Theta \times \mathscr{A} \to \mathbb{R}$* [see I, § 1.4.1], attaching a value to consequences, so that $L(\theta_i, a_j)$ measured the 'loss' experienced if action $a_j$ were chosen and the outcome turned out to be $\theta_i$. In this section, we shall work, instead, with a *utility function, $U: \Theta \times \mathscr{A} \to \mathbb{R}$*, having the interpretation that $U(\theta_i, a_j)$ is the 'positive value' or 'utility' experienced if action $a_j$ and outcome $\theta_i$ are the consequence in question. If we like, we can think of 'utility' as 'negative loss'—or vice-versa. Which form we choose to work with is simply a matter of convenience: in 'statistical' problems, like estimation, we are almost certainly going to get the answer 'wrong' and so a 'loss' scale is appropriate; in, for example, 'investment' problems we aspire to make decisions resulting in an overall gain and so 'utility'

is a more convenient scale. In the former case, the Bayes solution is to seek
to minimize expected loss; in the latter, it is to maximize expected utility.

To provide a concrete example of the situation described in Figure 19.3.1,
suppose that your current monetary assets are £$C$, that the stake money is £$S$
and that the net prize is equal to £$P$. Let us further suppose that, given $a_1$,
$p(\theta_1) = 1$ and, given $a_2$, $p(\theta_2) = p(\theta_3) = \frac{1}{2}$.

In monetary terms, the consequences are

$$(\theta_1, a_1) = C, \qquad (\theta_2, a_2) = C + P, \qquad (\theta_3, a_2) = C - S,$$

so that the required utilities are defined, for *any* C, P and S, by defining a
utility function $U : X \to \mathbb{R}$, where $X$ is the range of possible asset positions that
might arise from such a gamble.

It then follows that the expected utilities of the two actions are given by

| Action | Expected Utility |
|--------|------------------|
| $a_1$ | $U(C)$ |
| $a_2$ | $\frac{1}{2}U(C+P) + \frac{1}{2}U(C-S).$ |

If we proceed by maximizing expected utility, the optimal action is defined by
the following:

Gamble ($a_2$)
Indifferent                 if $U(C) \begin{cases} < \\ = \\ > \end{cases} \frac{1}{2}U(C+P) + \frac{1}{2}U(C-S)$
Do not gamble ($a_1$)

This situation is illustrated in Figure 19.3.2(a), (b), where we have assumed
that $S > 0$, $P > S$.

Now let us consider for a moment the special case $P = S$, so that, *in monetary
terms*, we have a *fair* gamble.

$$C = \tfrac{1}{2}(C+S) + \tfrac{1}{2}(C-S).$$

From Figure 19.3.2, and the inequality condition, we can ask under what
circumstances the decision-maker would feel able, or unable, to enter the
gamble (i.e. to choose $a_2$). We see that $a_2$ will be chosen if

$$U(C+S) - U(C) > U(C) - U(C-S),$$

and that $a_1$ will be chosen if

$$U(C+S) - U(C) < U(C) - U(C-S),$$

with indifference if we have equality.

Now suppose we assume that, for the range of $x$ values under study, the
decision-maker's utility function is such that for *any* choice of C and S he
will either always choose $a_1$ or always choose $a_2$, or always be indifferent. It

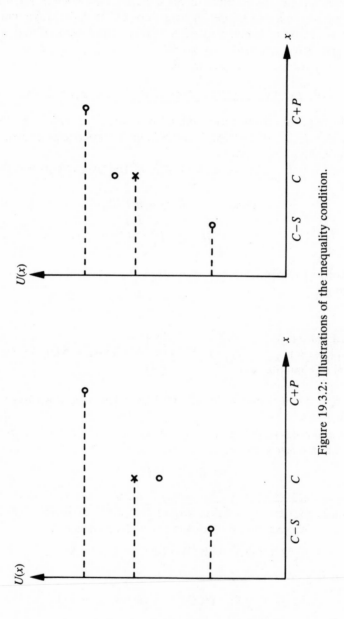

Figure 19.3.2: Illustrations of the inequality condition.

is then easy to see that the shape of his utility function $U(x)$ (assumed here to be continuous) must be as shown in Figure 19.3.3(a), (b), (c).

The form in Figure 19.3.3(a) represents *decreasing marginal utility* for money: additional increments of a fixed amount ($S$, say) produce less and less additional utility when added to greater assets (for example, the move from $C-S$ to $C$ produces greater additional utility than the move from $C$ to $C+S$). Such a form of utility function leads the decision-maker always to prefer $C$ for sure, rather than a (fair) monetary bet on $C+S$, $C-S$. A decision-maker having such a utility function is thus said to be *risk-averse*: the status quo is preferred to an uncertain situation whose expected monetary outcome is equal to the status quo (a fair bet).

In order for a gamble on $C+S$, $C-S$ to be preferred to $C$ for sure, the risk-averse decision-maker requires the probability on the $C+S$ outcome to be increased. If this probability is denoted by $\pi$, the decision-maker will prefer to gamble if

$$U(C) < \pi U(C+S) + (1-\pi)U(C-S),$$

in other words, if

$$\frac{\pi}{1-\pi} > \frac{U(C)-U(C-S)}{U(C+S)-U(C)} > 1.$$

This latter inequality simply reflects the fact that a risk-averse decision-maker requires an odds ratio in his favour and exceeding the (adverse) ratio of utility increments.

The form in Figure 19.3.3(b) represents *increasing marginal utility* for money: additional increments of a fixed amount ($S$, say) produce more and more additional utility when added to greater assets (for example, the move from $C$ to $C+S$ produces more additional utility than the move from $C-S$ to $C$). Such a form of utility function leads the decision-maker to prefer always to enter a fair monetary bet (with expected monetary outcome $C$) rather than to remain with $C$ for sure. Such a decision-maker is said to be a *risk-taker*.

It is easy to see that a risk-taker will sometimes be prepared to enter a gamble on $C+S$, $C-S$ even when the odds on $C+S$, $\pi/(1-\pi)$, are *unfavourable*. Since such a decision-maker will gamble if

$$U(C) < \pi U(C+S) + (1-\pi)U(C-S);$$

this only requires that

$$\frac{\pi}{1-\pi} > \frac{U(C)-U(C-S)}{U(C+S)-U(C)}$$

and from Figure 19.3.3(b) we see that the right-hand is strictly less than 1; this implies the existence of values of $\pi < \frac{1}{2}$ for which the risk-taker prefers to gamble rather than choose the status quo.

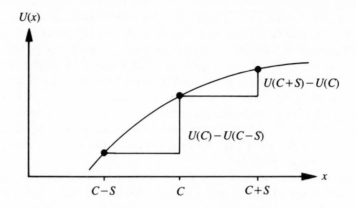

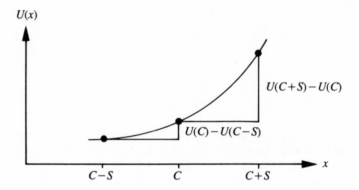

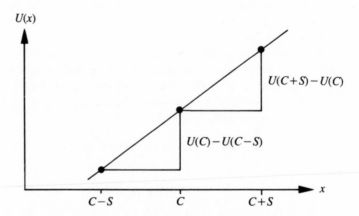

Figure 19.3.3: Forms of utility function which, for $P = C$, result in the optimal decisions: (a) never gamble; (b) always gamble; (c) always indifferent.

The situation described by Figure 19.3.3(c) implies an indifference to entering monetarily fair bets or remaining with the status quo. We note that if $S$ is very small, in relation to $C$, then, locally (i.e. in the region of $C$) both the curves in Figure 19.3.3(a) and (b) will look (approximately) like that in Figure (c) (since in small enough regions the continuous curves are well approximated by straight lines). In particular, this explains why a risk-averse person may not mind entering monetarily fair gambles if the stake (and prize) is *small enough.*

### 19.3.2.  One-dimensional Utility Functions

In this section, we shall indicate how further considerations can lead to a more precise mathematical understanding and representation of the kinds of utility function discussed in the previous section.

For the purpose of illustration, we shall confine attention to the case of *decreasing marginal utility for money.*

If we assume the utility function to be continuous, then in the case of the fair monetary bet between $C+S$ and $C-S$ we have seen that

$$U(C) > \tfrac{1}{2} U(C+S) + \tfrac{1}{2} U(C-S),$$

and so there exists an amount $\pi_C$, $0 < \pi_C < S$, such that

$$\pi(C - \pi_C) = \tfrac{1}{2} U(C+S) + \tfrac{1}{2} U(C-S).$$

This is illustrated in Figure 19.3.4.

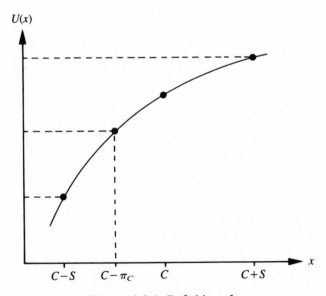

Figure 19.3.4: Definition of $\pi_C$.

This number $\pi_C$ could be thought of as the *insurance premium* that the decision-maker would be willing to pay in order to avoid having to exchange the status quo, $C$, for the (fair but uncertain) gamble between $C+S$ and $C-S$.

The argument clearly generalizes, for the case of decreasing marginal utility, to *any* gamble involving the prospect of a move from $C$ to $C+X$, where $X$ is a random variable with mean zero (in the simple case above, $X = \pm S$ with probabilities equal to $\frac{1}{2}$). The insurance premium $\pi_C$ is defined by the equation

$$U(C - \pi_C) = E[U(C + X)],$$

where expectation, $E(\cdot)$, is taken over the probability distribution of $X$.

An understanding of the nature of $\pi_C$ which itself provides a measure of risk aversion can be obtained by supposing that the variance of $X$ ($\sigma^2$, say) is fairly small, so that transactions take place in a local neighbourhood of $C$ (implying that $\pi_C$, too, is small). If we now expand both sides of the above equation, using Taylor's series [see IV, § 3.6], we obtain

$$U(C - \pi_C) \approx U(C) - \pi_C U'(C)$$

and

$$E[U(C + X)] \approx E\{U(C) + XU'(C) + \tfrac{1}{2}X^2 U''(C)\}$$
$$\approx U(C) + \tfrac{1}{2}\sigma^2 U''(C)$$

In the first expression we have ignored terms in $\pi_C^2$ and above; in the second, we have ignored terms in $E(X^3)$ and above. If these assumptions are reasonable, we have, on equating the two expressions,

$$\pi_C = \tfrac{1}{2}\sigma^2\left[-\frac{U''(C)}{U'(C)}\right] = \tfrac{1}{2}\sigma^2\left[\frac{d}{dC}\{-\log U'(C)\}\right].$$

The quantity $-U''(C)/U'(C)$ is therefore seen to play a fundamental role in defining a measure of (local) risk aversion. The greater this quantity is, the more premium $\pi_C$ the decision-maker is willing to pay (thus indicating his high level of risk aversion).

The form of $\pi_C$ can aid us in choosing a mathematical form for $U$. For example, suppose we know (or assume) that, for a given uncertain outcome $X$, the decision-maker's risk aversion is independent of $C$, so that $\pi_C$ should not depend on $C$. This would imply that

$$-U''(C)/U'(C) = k \quad \text{(say)}$$

where $k$ is a constant.

This differential equation is easily solved, and if we define the upper and lower points of the utility scale to be $U(\infty) = 1$, $U(0) = 0$, we obtain the solution

$$U(x) = 1 - e^{-kx} \quad (0 < x < \infty).$$

This is a strong conclusion: it says that if a decision-maker (over the range of *x*-values under consideration) has decreasing marginal utility for money, and constant aversion to risk, then there is a well-defined mathematical form for the utility function which requires only the specification of a single constant. As this constant increases, the shape of the utility function changes, becoming more and more steep initially, as illustrated in Figure 19.3.5.

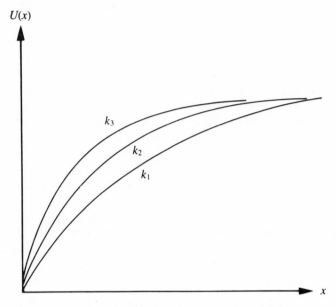

Figure 19.3.5: Forms of utility functions: $k_3 > k_2 > k_1$.

In fact, most decision-makers would appear to be decreasingly risk-averse; in other words, $\pi_C$ decreases with $C$. This reflects a greater willingness to take chances when one's asset position is stronger.

The form derived for $\pi_C$ can be used to investigate the risk-aversion characteristics implied by various suggested forms of utility function. For example, the form $U(x) = \log(x)$ may be considered (at least for some ranges of possible asset position). Straightforward differentiation shows that, in this case,

$$-U''(C)/U'(C) = 1/C$$

and hence that the function represents decreasing risk aversion.

In this section we have just provided a brief introduction to the way in which mathematical ideas can be used to study and represent utilities. For a much more extensive account, the reader should consult *Decision-making with Conflicting Objectives* by H. Raiffa and R. L. Keeney.

### 19.3.3.  Assessment of Utility Functions

Clearly, the form of utility for money (as for much else) will vary from individual to individual and we require an operational device for learning about the form of an individual's utility function.

One such device is the following. Suppose an individual has current assets of £$C$ and we wish to study the form of this utility function for money in the range from $C$ to $C+1000$. Let us now pose the following sequence of questions to the individual.

First, we ask him to suppose that he is in possession of a lottery ticket which entitles him to enter a fifty-fifty gamble with prizes £1000 or £0, respectively. We now ask him to consider selling this lottery ticket and request him to specify (honestly! no 'game-playing', no cheating!) the minimum sum which he would accept as payment for the ticket. Let us call this sum £$S_2$.

Next, we ask him to start again and suppose that, in fact, the lottery consisted of a fifty-fifty gamble on £$S_2$ and £0. What would he now accept as minimum payment for the ticket? Let us suppose that the answer is £$S_1$.

Finally, we ask him to repeat the exercise again, but this time with a fifty-fifty lottery between £$S_2$ and £1000. Let us suppose that the minimum payment required in this case is £$S_3$.

What have we learnt from such an interrogation? In order to fix utilities on a concrete scale, let us first agree to set $U(C)=0$ and $U(C+1000)=100$, say.

The answer to the first question now tells us that (provided we assume that the person is, at least approximately, taking Bayes decisions with respect to an underlying utility function $U(\cdot)$),

$$U(C+S_2)=\tfrac{1}{2}U(C+1000)+\tfrac{1}{2}U(C)$$
$$=\tfrac{1}{2}\times100+\tfrac{1}{2}\times0=50.$$

In other words, $S_2$ is the sum such that $C+S_2$ has utility 50 on the agreed scale.

Similarly, the answers to the other questions tell us that

$$U(C+S_1)=\tfrac{1}{2}U(C+S_2)+\tfrac{1}{2}U(C)$$
$$=\tfrac{1}{2}\times50+\tfrac{1}{2}\times0=25,$$

and

$$U(C+S_3)=\tfrac{1}{2}U(C+S_2)+\tfrac{1}{2}U(C+1000)$$
$$=\tfrac{1}{2}\times50+\tfrac{1}{2}\times100=75.$$

This information may now be represented graphically as shown in Figure 19.3.6. These (hypothetical) responses would indicate a decreasing marginal utility form whose general shape is indicated with a dotted line.

By this means, we can find an empirical form for $U$, or, if we make further assumptions such as constant risk aversion (§ 19.3.2), we could try fitting an implied mathematical form to the empirical data. Of course, the procedure is not limited to these particular three questions. Indeed, it would normally be

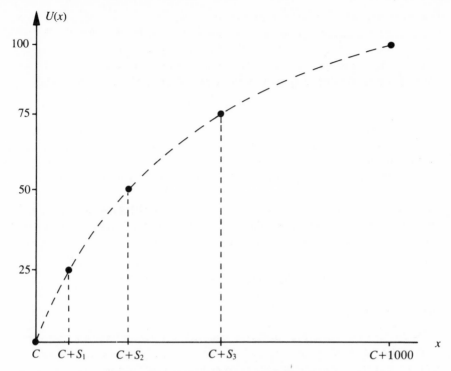

Figure 19.3.6: Assessment of a utility curve.

prudent to ask further questions to provide checks on the reliability of earlier answers. For example, we could pose a question regarding a fifty-fifty lottery on $£S_1$ and $£S_3$ and check if the answer given was close to $£S_2$ (as it should be if the answers conform to the existence of an underlying $U$).

We have discussed just one of several possible techniques for assessing utility functions. For further discussion, the reader is advised to read the article 'Utility and its Measurement' by J. Hull, P. G. Moore and H. Thomas, *J. R. Statist. Soc. A.*, 1973, pp. 226–247.

### 19.3.4.   Higher-dimensional Utility Functions

In previous sections, we have discussed examples where the final consequence could be measured in terms of a *single* quantity (and, in fact, we took this to be money). In general, however, consequences tend to involve several different factors. In industry, for example, consequences may be partly monetary, partly environmental; in health problems, consequences may involve trade-offs between pain, disability and mortality risk. We may think therefore of a consequence as corresponding to a vector of measured effects on different factors, so that

$$(\theta, a) \equiv (x_1, x_2, \ldots, x_m).$$

Now methods such as that outlined in the previous section enable us to assess separate, one-dimensional utility functions $U_1, \ldots, U_m$ for the different factors, so that assigning an overall utility to the consequence $(\theta, a)$ reduces to the problem of specifying a functional form for *combining* $U_1(x_1), \ldots, U_m(x_m)$.

Various possibilities suggest themselves: for example, we could take

$$U(\theta, a) = w_1 U_1(x_1) + \ldots + w_m U_m(x_m)$$

where the 'weights' $w_i$ reflect the importance attached to the $i$th factor. On the other hand, there may be subtle trade-offs in the utilities arising from the practical implications of combining, say, factor $i$ at level $x_i$ with factor $j$ at level $x_j$, and these trade-offs may necessitate a *non-linear* combination of the individual utilities.

This leads to a difficult and subtle area of utility theory and one which we shall not pursue further here. The interested reader is again referred to the book by Raiffa and Keeney (mentioned in § 19.3.2).

## 19.4.   SEQUENTIAL DECISIONS

### 19.4.1.   Basic Ideas

The subject of sequential decision-making is a very extensive one and we shall confine ourselves in this section to indicating the nature of the basic problem, developing in detail just one very valuable technique—that of decision trees.

For concreteness, consider the following problem. An initial action $a^{(1)}$ is to be taken and afterwards the world will be an (uncertain) state $\theta^{(1)}$; a further action $a^{(2)}$ is then to be taken, leading to an uncertain state $\theta^{(2)}$, and resulting in a final consequence $(\theta^{(1)}, \theta^{(2)}, a^{(1)}, a^{(2)})$. How should the decision-maker approach the problem of taking the initial decision?

The situation is represented in Figure 19.4.1, where a square node denotes a decision point (decision-maker chooses an action) and a circular node denotes an uncertainty point (an uncontrollable 'state of nature' occurs). Let us suppose

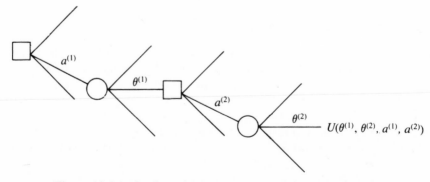

Figure 19.4.1: Outline of two-stage sequential decision problem.

that the utility attached to the final consequence is $U(\theta^{(1)}, \theta^{(2)}, a^{(1)}, a^{(2)})$, so that the 'decision tree' has the following form.

Although we have drawn a finite fan of possible actions emanating from each decision point, and a finite fan of possible outcomes emanating from each uncertainty point, there could, of course, be a continuous range of options or outcomes involved and so Figure 19.4.1 should be interpreted as purely a schematic outline of the problem.

The essence of the sequential problem is that we cannot sensibly decide what to do at the initial stage until we have thought through all the possible consequences $\theta^{(1)}, \theta^{(2)}, a^{(1)}, a^{(2)}$. We therefore begin at the right-hand end of the tree and ask what we would do at the *second* decision node *if* we had initially chosen $a^{(1)}$ and the outcome had been $\theta^{(1)}$. This would give the tree shown in Figure 19.4.2. For a given choice of $a^{(2)}$, the probability distribution

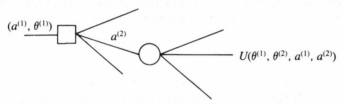

Figure 19.4.2: Modification of Figure 19.4.1 when initial outcome is known to have been $\theta^{(1)}$.

for the uncertain second outcome will have the form $p(\theta^{(2)}|a^{(1)}, \theta^{(1)}, a^{(2)})$, and using the integral form (continuous range of $\theta^{(2)}$) for convenience, the optimal action and resulting expected utility are defined by

$$U^*(\theta^{(1)}, a^{(1)}) = \max_{a^{(2)}} \int p(\theta^{(2)}|a^{(1)}, \theta^{(1)}, a^{(2)}) U(\theta^{(1)}, \theta^{(2)}, a^{(1)}, a^{(2)}) \, d\theta^{(2)}.$$

This is the maximized utility, given $\theta^{(1)}, a^{(1)}$. The situation facing the decision-maker at the first stage can now be represented as in Figure 19.4.3.

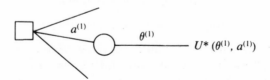

Figure 19.4.3: First-stage decision (assuming optimal second-stage action).

The full solution to the problem is thus seen to be defined by

$$\max_{a^{(1)}} \int p(\theta^{(1)}|a^{(1)}) U^*(\theta^{(1)}, a^{(1)}) \, d\theta^{(1)}.$$

If the problem is extended to *n*-stages, the structure of the solution is easily seen to remain the same. Starting from the right-hand side of the tree, and passing successively through uncertain and decision nodes, we repeat the process of taking expectation and maximizing. The complications in implementing this procedure are of two kinds. First, we note that, in general, there is an implicit use of *all* possible 'histories' of the process (i.e. all combinations of actions that might be taken and outcomes that might arise): secondly, probability distributions for outcomes at a given stage must be conditional on the history of the process 'so far'. The computations involved can become very complicated indeed, unless the structure of the problem allows some simplification of the mathematical recursions required.

Consideration of special cases and the mathematical techniques available for their solution can be found, for example, in M. H. DeGroot *Optimal Statistical Decisions*, and the reader is referred to that work for further details.

In the case of finite sets of outcomes and actions, however, problems can be represented and solved routinely using the simple decision tree formulation. This technique will be discussed in the next section.

### 19.4.2.   Decision Trees

We shall illustrate the use of decision trees by considering a specific problem.

Suppose a company is contemplating a small investment which might prove profitable if demand for a certain manufactured article rises, but which may be embarrassing if demand falls. The company could simply decide straightaway to invest or not to invest—or it could purchase some market research with a view to learning more about the relative likelihood of demand rising or falling. Let us suppose that the market research report will simply give a prediction—demand up, or demand down—and that the final outcome (state of nature) can be simply thought of as demand *actually* up or demand *actually* down.

The structure of the resulting decision tree is shown in Figure 19.4.4.

In our earlier notation (§ 19.1) we have

$$\mathcal{A} = \{a_1, a_2, a_3\} = \{\text{Buy Market Research, Invest, Do not invest}\}$$

$$\Theta = \{\theta_1, \theta_2\} = \{\text{Demand 'up', Demand 'down'}\}$$

$$X = \{x_1, x_2\} = \{\text{Forecast 'up', Forecast 'down'}\}.$$

We must now introduce probabilities for the various uncertainties involved, and we shall suppose the following assessments are made.

The market research company is known to be 80% successful at forecasting rising markets and 70% successful at forecasting falling markets. We have therefore, in the above notation,

$$p(x_1|\theta_1) = 0\cdot8, \qquad p(x_2|\theta_1) = 0\cdot2$$
$$p(x_1|\theta_2) = 0\cdot3, \qquad p(x_2|\theta_2) = 0\cdot7.$$

We shall further suppose that initially (that is, without further market research information) the company assesses there to be a 60% chance of

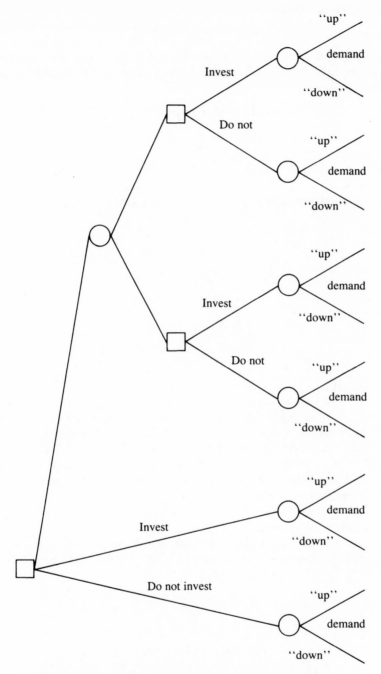

Figure 19.4.4: Decision tree for the investment problem.

demand increasing, so that

$$p(\theta_1) = 0{\cdot}6, \qquad p(\theta_2) = 0{\cdot}4.$$

In fact, the company needs a direct assessment of the uncertain outcome that would result *if* it bought market research and this is provided by simple probability calculus;

$$p(x_1) = p(x_1|\theta_1)p(\theta_1) + p(x_1|\theta_2)p(\theta_2)$$

$$= 0{\cdot}8 \times 0{\cdot}6 + 0{\cdot}3 \times 0{\cdot}4 = 0{\cdot}6$$

$$p(x_2) = 1 - p(x_1) = 0{\cdot}4.$$

So far as revision of beliefs about $\theta_1$, $\theta_2$ *given* the market research information is concerned, we have, by Bayes' Theorem [see § 15.2],

$$p(\theta_1|x_1) = p(x_1|\theta_1)p(\theta_1)/p(x_1)$$

$$= (0{\cdot}8 \times 0{\cdot}6)/0{\cdot}6 = 0{\cdot}8$$

$$p(\theta_2|x_1) = 1 - p(\theta_1|x_1) = 0{\cdot}2$$

and

$$p(\theta_1|x_2) = p(x_2|\theta_1)p(\theta_1)/p(x_2)$$

$$= (0{\cdot}2 \times 0{\cdot}6)/0{\cdot}4 = 0{\cdot}3$$

$$p(\theta_2|x_2) = 1 - p(\theta_1|x_2) = 0{\cdot}7.$$

So far as the utilities of the various outcomes are concerned, we shall assume that, over the range concerned, the company's utility function for money is approximately linear. We shall also assume that (expressed in £'000) if the demand rises the company expects to make a net profit of 1020, whereas if demand falls it expects to make a net profit 980, assuming that it. invests; if investment does not take place, the company expects a net profit of 1000. The cost of market research (expressed in £'000) is $C$.

Figure 19.4.5 shows the decision tree of Figure 19.4.4 with all relevant numerical information displayed.

We now recall that, starting from the right-hand side of the tree, we calculate expected utilities. For the $a_1$ decision branch, the result of doing this is shown in Figure 19.4.6, where, for example,

$$992 - C = (1020 - C) \times 0{\cdot}3 + (980 - C) \times 0{\cdot}7$$

and

$$1012 - C = (1020 - C) \times 0{\cdot}8 + (980 - C) \times 0{\cdot}2.$$

Now applying the maximization of expected utility principle, we see that given $x_1$, the optimal action is $a_2$, whereas, given $x_2$, the optimal action is $a_3$. This results in Figure 19.4.7.

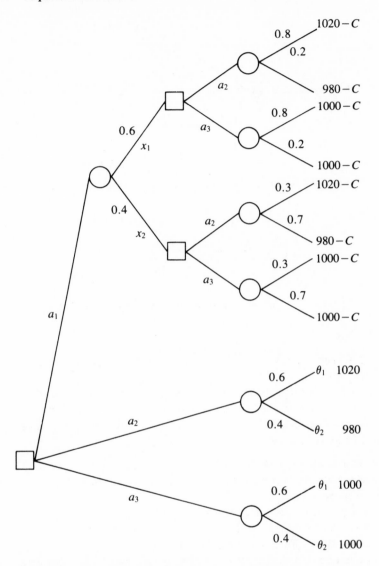

Figure 19.4.5: Investment problem with probabilities and utilities.

We now apply the expectation step again, to obtain an expected utility for $a_1$ of

$$(1012 - C) \times 0 \cdot 6 + (1000 - C) \times 0 \cdot 4 = 1007 \cdot 2 - C.$$

The expected utilities for $a_2$, $a_3$ are given by

$(a_2)$ $\qquad\qquad\qquad 1020 \times 0 \cdot 6 + 980 \times 0 \cdot 4 = 1004$

$(a_3)$ $\qquad\qquad\qquad 1000 \times 0 \cdot 6 + 1000 \times 0 \cdot 4 = 1000.$

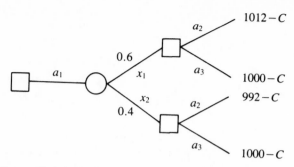

Figure 19.4.6: Initial (expectation) calculation for the $a_1$ branch.

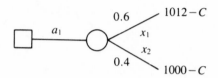

Figure 19.4.7: Second (maximization) step for the $a_1$ branch.

The initial decision is now solved by comparing the expected utilities as shown in Figure 19.4.8.

The initial decision is now clear; $a_3$ should never be chosen, and $a_1$ is preferable to $a_2$ only if the market research fee is not greater than $C = 3 \cdot 2$.

This was just one specific example, but all decision trees are formulated and analysed in the same way.

(1) Write out the logic of the tree in chronological order, describing decision nodes and uncertainty nodes, together with the branches that occur at each node;
(2) enter probabilities on to the uncertain branches, taking care to carry out the appropriate conditioning for each branch;
(3) attach utility values to the final branches;
(4) proceeding from right to left across the tree, take expectations at uncertain nodes and maximize at decision nodes, thus determining the best actions and their expected utilities.

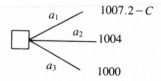

Figure 19.4.8: Expected utilities of the initial actions.

## 19.5.   AXIOMATIC APPROACHES

### 19.5.1.   Coherence Axioms for Decision-making

In section 19.1, we noted that there is no neutral, purely mathematical solution to the problem of choosing an 'optimal' action. We then put forward two suggested procedures, the Bayes and Minimax approaches, for selecting an optimal decision.

However, it could be argued that all this is very arbitrary. Why consider these particular approaches? And, in any case, how do we know that we are being sensible in adopting our original mathematical framework? How do we know that it is sensible to assume the existence of a loss, or utility, function?

An alternative approach to the whole subject is to start from much more primitive concepts (not yet assuming the existence of utility functions, or limiting the choice of approach to choosing an action) and try to *deduce* the form of structure required and optimal approach implied.

To do this, we make a list of axioms (or 'self-evident' postulates) and then attempt to deduce their implications. In this section, we shall not attempt a detailed rigorous approach of this kind, but rather attempt to give the 'flavour' of such an approach and such arguments. In this section, we shall discuss a particular, formal axiom system which lays down postulates for 'rational preferences' among consequences. In the next section, we shall give a more informal discussion of the way in which the notion of 'rational degree of belief' could be analysed.

Suppose that $\mathscr{C}$ is the set of all 'consequences' that can arise in a particular decision problem. In the notation of section 19.1, $\mathscr{C} = \Theta \times \mathscr{A}$. Now let $\mathscr{P}$ denote the set of all probability distributions defined over $\mathscr{C}$ so that $\mathscr{P}$ denotes the set of all possible uncertainties regarding consequences. We now assume that the decision-maker has preferences among the elements of $\mathscr{P}$ and we write $P_1 < P_2$ to denote that $P_2$ is strictly preferred to $P_1$, $P_1 \leqslant P_2$ to denote that $P_1$ is not strictly preferred to $P_2$ and $P_1 \sim P_2$ to denote that neither of $P_1, P_2$ is preferred to the other.

It is assumed that preferences obey the following two axioms:

A1.   If $P_1, P_2$ are elements of $\mathscr{P}$ then either $P_1 < P_2$ or $P_2 < P_1$ or $P_1 \sim P_2$.
A2.   If $P_1, P_2, P_3$ are elements of $\mathscr{P}$ such that $P_1 \leqslant P_2$ and $P_2 \leqslant P_3$ then $P_1 \leqslant P_3$.

In other words, we assume that any two uncertain situations regarding consequences can be compared and that preferences are transitive.

Now let us make the further assumption

A3.   If $P_1, P_2, P_3$ are elements of $\mathscr{P}$ and $\alpha$ is any number such that $0 < \alpha < 1$, then $P_1 < P_2$ if and only if $\alpha P_1 + (1 - \alpha)P_3 < \alpha P_2 + (1 - \alpha)P_3$.

This assumption formalizes the intuitive idea that if two situations of uncertainty about consequences have a common component the comparison of the two should not depend on this component. We further assume that:

A4. If $P_1, P_2, P_3$ are elements of $\mathcal{P}$ such that $P_1 < P_2 < P_3$, then there exist numbers $\alpha, \beta$ $(0 < \alpha < 1, 0 < \beta < 1)$ such that

$$P_2 < \alpha P_3 + (1-\alpha)P_1 \quad \text{and} \quad P_2 > \beta P_3 + (1-\beta)P_1.$$

This assumption formalizes the idea that 'absolute heaven' and 'absolute hell' are not perceived to be among the consequences. For example,

$$P_1 < P_2 < P_3 \quad \text{and} \quad P_2 < \alpha P_3 + (1-\alpha)P_1$$

says that, although $P_2 > P_1$, there exists some $1 - \alpha$ (however tiny) so that the mixture $\alpha P_3 + (1-\alpha)P_1$ is still preferred to $P_2$. If $P_1$ were an 'absolute hell', this would not be so. Similar remarks apply to the other inequality.

These assumptions, and the discussion of them, should serve to indicate how one sets about formalizing intuitive notions about 'rational' or 'coherent' preferences. What we now do is to attempt to deduce from such assumptions (and some others which ensure that mathematical properties are observed— remember that not all sets can necessarily be assigned probabilities!) the form that rational decision-making procedures should take.

The answer that emerges is that rational decision-making requires us to

(a)  assume the existence of a utility function,
(b)  act so as to maximize expected utility,

where 'rational' means conforming to the axioms laid down for preferences.

In short, axiom systems of this kind (and there are many variations in the precise list of axioms chosen) lead to the conclusion that the Bayesian approach to decision-making is required if we are to act in conformity with the axiom system.

A detailed account of a system like the one outlined above can be found in M. H. DeGroot *Optional Statistical Decisions* (Chapter 7).


### 19.5.2.  Degrees of Belief as Probabilities

In Chapter 15, and in many sections in this chapter, we have assumed, in developing Bayesian inference and decision procedures, that 'degrees of belief' could be represented by values, or densities, following the rules of the probability calculus. Some readers may find this entirely reasonable, without the need for further discussion; others may require at least an outline argument in support of this assumption.

One such argument is the following. Suppose that an individual—you, for example—is faced with a situation of uncertainty involving an event $E$ whose outcome ($E$ or not-$E$) is unknown. In any practical situation of interest to you, you will have some feeling—let us call it 'degree of belief'—regarding the relative 'plausibilities' of $E$ or $\sim E$ occurring. It is this feeling which we wish first to give a quantitative form, and then to show that such quantities

must (under a further condition, which we shall make clear) obey the rules
of probability.

First, a possible operational scheme for converting a degree of belief into
a number is the following.

Consider a gamble which offers you £S if E occurs and £0 if ~E occurs.
What sum of money, £C, say, would leave you *indifferent* between possession
of C for certain and entering the gamble (just once)?

Clearly, if C is 'very small' (as a fraction of S) you would prefer to take
the gamble; if, on the other hand, C were large (as a fraction of S), you would
prefer to accept C for sure rather than enter the gamble. We shall assume
that as you think through reactions to a 'small' C, getting larger, or a 'large'
C, getting smaller, you will perceive that for some intermediate value of C
you are indifferent between C for sure and the gamble. Let us call this
intermediate value £C*. (Of course, in practice there will be a 'fuzzy' interval
of indifference, rather than a sharp value, but we are used to 'idealizing' a
little with *all* forms of measurement—for example, we base much science and
technology on the assumptions that bodies have precise 'length' or 'tem-
perature', although in practice these can only be measured up to some 'fuzzy'
interval.)

We now *define* your (revealed) degree of belief, p, in E to be the number
such that

$$C^* = pS.$$

We note first that this corresponds to our intuitive understanding; low degrees
of belief make C* small; high degrees of belief make C* large.

In order to ensure that you are *honest* in your revealed indifference value
C*, we can add the following condition.

Suppose that when you state C* you do not know whether you will be
called upon *to gamble* (in accordance with your stated value you should be
willing to pay C* to enter a gamble with E leading to £S and ~E leading to
£0), or *to act as bookmaker* (offering this gamble for a stake of C*). This
implies that you must state a C* which leaves you indifferent in both the
contexts shown in Figure 19.5.1.

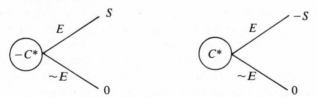

Figure 19.5.1: Honest C* implies indifference between (a) and (b).

Having now indicated a way in which an honest, quantitative degree of
belief for an event E can be arrived at, we wish to examine the ways in which
degrees of beliefs for different events relate to each other.

Consider the following situation. The collection of events $E_1, E_2, \ldots, E_N$ are exclusive and exhaustive and using the above procedure you are to specify degrees of belief $p_1, p_2, \ldots, p_N$ in the individual events. After you have chosen $p_1, p_2, \ldots, p_N$, an 'opponent' will be free to choose amounts $S_1, S_2, \ldots, S_N$ on the assumption that you will be willing to pay entry fees, $p_1 S_1, p_2 S_2, \ldots, p_N S_N$ to a gamble in which you will receive $C_i$ if event $E_i$ occurs.

How must the $p_i$'s 'fit together', or 'cohere', if you are to avoid the situation where the opponent can choose the $S_i$'s in such a way that he will certainly win? Any specification of the $p_i$'s such that you could certainly lose, we shall call *irrational*. Our question then becomes 'what rules must *rational* degrees of belief obey?'

We note first that $0 \le p_i \le 1$, for all $i = 1, \ldots, N$, since, recalling the definition of degree of belief, any choice outside this range would enable the opponent to choose his role (as gambler, or bookmaker) so as to ensure that he won. Specifically, a choice of $p_i > 1$ would imply that you were willing to pay $> S_i$ to enter a gamble with maximum prize of $S_i$; a choice of $p_i < 0$ would imply that you were willing to *pay* your opponent to enter a gamble which would either yield him (from you!) 0 or $S_i$.

Secondly, we note that *your* 'gains', $G_i$ (which may, of course, be negative), satisfy the system of linear equations

$$G_i = S_i - (p_1 S_1 + \ldots + p_N S_N)$$

if $E_i$ occurs, $i = 1, \ldots, N$. (The total entry fee is $\sum p_j S_j$, the 'payoff' is $S_i$ if $E_i$ occurs.)

Written out in full, the system is

$$\begin{bmatrix} G_1 \\ G_2 \\ \vdots \\ G_N \end{bmatrix} = \begin{bmatrix} 1-p_1 & -p_2 & \cdots & -p_N \\ -p_1 & 1-p_2 & & -p_N \\ \vdots & & \ddots & \vdots \\ -p_1 & -p_2 & & 1-p_N \end{bmatrix} \begin{bmatrix} S_1 \\ S_2 \\ \vdots \\ S_N \end{bmatrix}$$

and the problem can be reformulated as follows. The opponent is given the $p_i$'s and wishes to choose the $S_i$'s in order to give $G_i$'s values which he (the opponent) desires. In particular, he would wish to fix negative $G_i$'s (remember they are *your* 'gains'). For a given set of $G_i$'s and $p_i$'s, can he solve the equations and find suitable $S_i$'s? A *rational* set of $p_i$'s is one that prevents him being able to solve these equations. We now recall—from linear algebra—that an explicit solution *can* be found if the matrix (defined by the $p_i$'s) can be inverted. We also recall that this is *not* possible *if the determinant is zero* [see I, § 6.12].

An easy calculation shows that the determinant is equal to $1 - (p_1 + \ldots + p_N)$, so that a collection of rational degrees of belief for the exhaustive and exclusive events $E_1, \ldots, E_N$ must satisfy

$$p_1 + \ldots + p_N = 1.$$

We see therefore that rational degrees of belief satisfy the basic rules of probability:

they have values in the range from 0 to 1; they satisfy the addition axiom.

To complete an outline justification of regarding degrees of belief as probabilities, we must consider the concept of 'conditional' degrees of belief—or the way in which degrees of belief change when we acquire new information.

Let us consider two events $E'$ and $E''$, and let

$$\left.\begin{array}{l} \pi_1 = p(E'|E'') \\ \pi_2 = p(E' \text{ and } E'') \\ \pi_3 = p(E'') \end{array}\right\} \begin{array}{l} \textit{denote degrees of} \\ \textit{belief of} \end{array} \begin{array}{l} E' \text{ given the occurrence of } E'' \\ \text{the occurrence of both } E' \text{ and } E'' \\ \text{the occurrence of } E''. \end{array}$$

You are now to gamble (the outcomes being defined in terms of $E'$, $E''$) on the understanding that if $E''$ does not occur then bets involving $E'$ given $E''$ are 'called-off' (i.e. the entry fee is returned). How should $\pi_1$, $\pi_2$, $\pi_3$ relate to one another if you are to avoid an opponent fixing stakes which lead you to certainly lose?

Suppose that we consider the three contingencies $E'$ given $E''$, $E'$ and $E''$, $E''$, respectively, with amounts $S_1$, $S_2$, $S_3$ as 'pay-offs'. It is easily seen that the following table of 'gains' (for you) arises from the stated outcomes.

| Actual outcome | Gain |
|---|---|
| $E'$ and $E''$ | $G_1 = (1 - \pi_1)S_1 + (1 - \pi_2)S_2 + (1 - \pi_3)S_3$ |
| $\sim E'$ and $E''$ | $G_2 = -\pi_1 S_1 - \pi_2 S_2 + (1 - \pi_3)S_3$ |
| $\sim E''$ | $G_3 = -\pi_2 S_2 - \pi_3 S_3$ |

This may be rewritten in the form

$$\begin{bmatrix} G_1 \\ G_2 \\ G_3 \end{bmatrix} = \begin{bmatrix} 1 - \pi_1 & 1 - \pi_2 & 1 - \pi_3 \\ -\pi_1 & -\pi_2 & 1 - \pi_3 \\ 0 & -\pi_2 & -\pi_3 \end{bmatrix} \begin{bmatrix} S_1 \\ S_2 \\ S_3 \end{bmatrix}$$

and the argument we used previously tells us that the $\pi_i$'s must be chosen so that the determinant is zero if irrationality is to be avoided.

In this case, the determinant is equal to $\pi_2 - \pi_1 \pi_3$ and so the degrees of belief must satisfy

$$p(E'|E'') = \frac{p(E' \text{ and } E'')}{p(E'')}.$$

This establishes that rational, conditional degrees of belief satisfy the usual definition of conditional probability.

It could be objected that if the sums $S_i$ used in the above arguments are large then the procedure for defining degrees of belief would be invalidated by the phenomenon of risk aversion discussed in section 19.3.1. This is true, and so one must understand the above argument as either conducted with 'small' $S_i$, so that utility is (approximately) linear, or with the $S_i$ defined on a preassessed utility scale (assessed, for example, using the method of section 19.3.3).

A.F.M.S.

## 19.6.  FURTHER READING AND REFERENCES

The following two books provide a thorough, non-mathematical treatment of the basic ideas of decision theory.

Lindley, D. V. (1971). *Making Decisions*, Wiley.

Raiffa, H. (1968). *Decision Analysis: Introductory Lectures on Choices under Uncertainty*, Addison-Wesley.

For more advanced mathematical developments, the reader should consult:

DeGroot, M. H. (1970). *Optimal Statistical Decisions*, McGraw-Hill.

Ferguson, T. S. (1967). *Mathematical Statistics: A Decision-Theoretic Approach*, Academic Press.

# CHAPTER 20

# *Kalman Filtering*

## 20.1. HISTORICAL BACKGROUND

Kalman filtering, first developed by Kalman (1960) and Kalman and Bucy (1961), is a method of predicting the behaviour of a signal $s(t)$ given observations $y(t)$ which are subject to error $v(t)$, so that $y(t) = s(t) + v(t)$. The term *filtering* refers to the removal as much as possible of the error term $v(t)$ to give a prediction of the true signal $s(t)$. The procedure is used for models of a particular type in which the behaviour of $s(t)$ follows a differential or difference equation which is itself subject to random disturbances.

While filtering methods had been developed prior to the introduction of the Kalman filter, notably by Kolmogorov (1941) and Wiener (1949), the computational aspects of these methods were rather less elegant than the recursive approach employed by the Kalman filter. Furthermore, applications to vector-valued and non-stationary processes present no essential difficulties in the latter case.

Stemming from the original, seminal papers of Kalman (1960) and Kalman and Bucy (1961), a vast literature on Kalman filtering has accumulated. Several books and papers give a comprehensive treatment of the subject; amongst these, the texts by Kwakernaak and Sivan (1972) and Gelb (1974) offer good treatments of the subject at a relatively elementary mathematical level; a more rigorous mathematical treatment is given by Jazwinski (1970). A comprehensive survey paper with an extensive bibliography and a historical account of the development of the theory has been written by Kailath (1974); a useful exposition of the theory with some interesting insights has also been given by Willens (1978). Harrison and Stevens (1976) have presented Kalman filtering from a Bayesian statistical viewpoint. Details of these references will be found in section 20.9.

Applications of the Kalman filter have been both numerous and diverse; perhaps its most famous role was in the historical Apollo XI mission when it was used to carry out the tracking of the spacecraft and the lunar module. Apart from applications in aerospace engineering and navigation (e.g. Schmidt, 1966; Bucy and Joseph, 1968), usage has been reported in the fields of industrial process control (e.g. Mehra and Wells, 1971; Bohlin, 1976), in communications theory (e.g. Snyder, 1969), in economics and socio-economic studies (e.g. Stevens, 1974, Athans, 1974, Mehra, 1978), in cyclone forecasting (Takeuchi, 1976), and in hydrology and water resources (e.g. Chiu, 1978; Wood and Szollosi Nagy, 1980; O'Connell, 1980, IAHS, 1980, amongst others). It is the last-mentioned field of application that will provide the setting for the exposition on Kalman filtering presented here, which will seek to cover the subject from the standpoint of a prospective user with a modest background in probability, statistics and matrix algebra.

Perhaps it might be useful to the reader at the outset to describe qualitatively how the Kalman filter can be utilized in hydrology. In order to provide a basis for short-term decision-making in the forecasting and control of floods, on-line monitoring schemes are being used increasingly to transmit measurements of rainfall and river flow at various points in a river basin to a centrally located computer. The rainfall measurements are then typically fed into a model relating rainfall to river flow from which forecasts of future river flow may be derived; however, measurements of rainfall and river flow, particularly the former, are both recognized to be subject to error, and so a method of filtering out this noise is required. Furthermore, at each time point, new data on river flow and rainfall become available, and so a method of recursively updating the model forecasts is required. The Kalman filter meets both these requirements, while also furnishing an estimate of the uncertainty in forecasts of future flow; as the Kalman filter is itself a recursive algorithm, it is particularly well suited to real-time applications of this kind.

In section 20.2, state-space models, which provide the necessary framework for applying the Kalman filter, are introduced, and their formulation illustrated through a simple example for the discrete linear case. Section 20.3 covers the development of the discrete linear Kalman filter equations; the derivation presented is geared to providing insight into the operation of the filter. The assessment of filter performance is treated in section 20.4, and some statistical tests for this purpose are described; the problem of filter divergence is also discussed. Section 20.5 deals with the application of the Kalman filter to continuous time and non-linear systems; the latter case gives rise to the extended Kalman filter. For many applications, the structure and parameters of the state-space model will be unknown: section 20.6 describes some procedures for model identification and parameter estimation. Two applications of the Kalman filter, one for the linear and one for the non-linear case, are described in section 20.7: both of these are concerned with the problem of forecasting streamflow from rainfall in real-time. Some concluding remarks are given in section 20.8, and a bibliography in section 20.9.

*Some points of notation*

The reader will see that vectors and matrices are not distinguished typographically from scalar quantities in this chapter. It will always be clear from the context which is which. The transpose of a vector $x$ is denoted by the symbol $x^T$, and that of a matrix $A$ by $A^T$.

The dispersion or variance-covariance matrix of a vector random variable $x$ [see II, Definition 9.6.3], for which various symbols are used in the literature, is here written 'var $(x)$'.

## 20.2. STATE-SPACE MODELS

### 20.2.1. System and Measurement Equations

A prerequisite to the application of the Kalman filter is that the behaviour of the system under study be described through a quantity known as the system 'state', which may be defined in terms of a first order differential or difference equation (according to whether time $t$ is continuous or discrete) known as the system equation. Here, the case of a discrete-time linear model will be considered as this facilitates the derivation of the Kalman filter equations while also corresponding to the most natural case for implementation on a digital computer; the system equation is then

$$x_{t+1} = F_t x_t + w_t, \tag{20.2.1}$$

where $x_t$ is an $(n \times 1)$ vector of state variables (giving rise to the term 'state space'), $F_t$ is an $(n \times n)$ state transition matrix, and $w_t$ is an $(n \times 1)$ vector of system noise (or model error) which can be thought of as driving the process $x_t$. Associated with the system equation is the *measurement equation*

$$y_t = H_t x_t + v_t, \tag{20.2.2}$$

where $y_t$ is an $(m \times 1)$ vector of measurements, $H_t$ is an $(m \times n)$ measurement matrix describing the linear combination of state variables which comprise $y_t$ in the absence of noise, and $v_t$ is an $(m \times 1)$ vector of measurement noise. To complete the model specification, a description of the noise terms $w_t$ and $v_t$ is required; their means and variance-covariance matrices may be written as

$$E\{w_t\} = 0; \quad Q = \text{var}\,(w_t) = E\{w_t w_t^T\},$$
$$E\{v_t\} = 0; \quad R = \text{var}\,(v_t) = E\{v_t v_t^T\}, \tag{20.2.3}$$

where $E\{.\}$ denotes the expectation operator and it is assumed that the system and measurement noises have zero means. It is further assumed that the noise processes are not serially or cross-dependant; in particular,

$$E\{w_t w_k^T\} = 0, \quad E\{v_t v_k^T\} = 0, \quad t \neq k, \tag{20.2.4}$$

and

$$E\{w_t v_k^T\} = 0 \quad \text{for all } t, k.$$

Finally, it is customary to assume that the noises $w_t$ and $v_t$ have Gaussian ('Normal') distributions [see II, § 11.4]; however, this condition is not mandatory in deriving the Kalman filter, but in the Gaussian case the linear estimator employed by the Kalman filter is the optimum estimator (in the mean-square error sense) among all possible estimators. In the non-Gaussian case, non-linear estimators can, in principle, perform better, but they will usually prove very difficult to apply; under such conditions, the Kalman filter will provide the linear unbiased minimum variance estimator [see § 3.3.2].

Given the state-space representation of a discrete linear dynamic system (equations (20.2.1) and (20.2.2)) the problem is then to estimate the state $x_t$ from the noisy observations $y_1, y_2, \ldots, y_t$; in this context, three distinct problems may be distinguished:

(i)  *filtering*: the measurements $y_1, \ldots, y_t$ are used to form an estimate $\hat{x}_t$ of the state at time $t$;

(ii)  *smoothing*: the measurements $y_1, \ldots, y_t$ are used to form an estimate $\hat{x}_t$ of $x_s$ at some past time point for $1 \le s < t$;

(iii)  *prediction*: the measurements $y_1, \ldots, y_t$ are used to form an estimate $\hat{x}_s$ of $x_s$ at some future time point for $s > t$.

In many practical problems, decision-making will proceed from the latest information available on system behaviour, and so filtering and prediction will be of primary concern; both of these problems are handled recursively by the Kalman filter, which makes it particularly well suited for use in real-time decision-making.

### 20.2.2.  A Simple Example

To illustrate the formulation of models within the state space framework a process $Q_t$ (e.g. streamflow at a given point on a river) is assumed to be described by the autoregressive scheme

$$Q_t = a\ Q_{t-1} + b\ Q_{t-2} + c\ Q_{t-3} + w_{t-1}, \qquad (20.2.5)$$

where $w_{t-1}$ is noise due to the imperfect representation of $Q_t$ by $Q_{t-1}$, $Q_{t-2}$ and $Q_{t-3}$, and $a$, $b$, $c$ are assumed known (Weiss, 1980).

Although $Q_t$ alone is of interest, it is not an appropriate choice for a state, since $Q_t$ is a function not only of $Q_{t-1}$ but also of $Q_{t-2}$ and $Q_{t-3}$ as well. However, if at time $(t-1)$ the vector

$$x_{t-1} = \begin{bmatrix} Q_{t-1} \\ Q_{t-2} \\ Q_{t-3} \end{bmatrix}$$

is considered, then $x_{t-1}$ contains all the information in the system about $Q_t$ and

$$Q_t = (abc) \begin{bmatrix} Q_{t-1} \\ Q_{t-2} \\ Q_{t-3} \end{bmatrix} + w_{t-1}. \qquad (20.2.6)$$

A system equation is then obtained by writing

$$
\begin{bmatrix} Q_t \\ Q_{t-1} \\ Q_{t-2} \end{bmatrix} = \begin{bmatrix} a & b & c \\ 1 & 0 & 0 \\ 0 & 1 & 0 \end{bmatrix} \begin{bmatrix} Q_{t-1} \\ Q_{t-2} \\ Q_{t-3} \end{bmatrix} + \begin{bmatrix} w_{t-1} \\ 0 \\ 0 \end{bmatrix}, \tag{20.2.7}
$$

which is of the form

$$
x_t = F_{t-1} x_{t-1} + w_{t-1} \tag{20.2.8}
$$

with

$$
F_{t-1} = \begin{bmatrix} a & b & c \\ 1 & 0 & 0 \\ 0 & 1 & 0 \end{bmatrix}.
$$

Although equation (20.2.8) looks simpler than equation (20.2.5), this simplicity is misleading since the latter is a *vector* equation involving the matrix $F_{t-1}$ and the vectors $x_t$, $x_{t-1}$ and $w_{t-1}$.

The measurement equation corresponding to the system equation (20.2.7) is formulated as follows. At time $t$, a measurement of $Q_t$ (but not of the other components of $x_t$) is observed. This measurement $y_t$ is assumed to be subject to error and accordingly

$$
y_t = Q_t + v_t, \tag{20.2.9}
$$

where the measurement inaccuracy is the noise $v_t$. This can be expressed in terms of the state as

$$
y_t = (1 \quad 0 \quad 0) \begin{bmatrix} Q_t \\ Q_{t-1} \\ Q_{t-2} \end{bmatrix} + v_t, \tag{20.2.10}
$$

which in vector notation is

$$
y_t = H_t x_t + v_t, \tag{20.2.11}
$$

where

$$
H_t = (1 \quad 0 \quad 0).
$$

It will be assumed throughout that the average of $v_t$ is 0: if, however, the mean value of $v_t$ were known to be $\bar{v}$, the measurement $y_t$ could be redefined by subtracting $\bar{v}$.

The system equation can be used for prediction as follows. If $x_{t-1}$ is assumed known, then it is possible to calculate the quantity

$$
a \; Q_{t-1} + b \; Q_{t-2} + c \; Q_{t-3}
$$

which differs from $Q_t$ only by the noise $w_{t-1}$. Since $w_{t-1}$ is independent of $x_{t-1}$ (in other words $x_{t-1}$ contains no information on $w_{t-1}$), $w_{t-1}$ is predicted by its mean value of zero.

The prediction of $Q_t$ is then

$$\hat{Q}_t = a\ Q_{t-1} + b\ Q_{t-2} + c\ Q_{t-3},\qquad(20.2.12)$$

and the prediction of $x_t$ is therefore

$$\hat{x}_t = \begin{bmatrix} \hat{Q}_t \\ Q_{t-1} \\ Q_{t-2} \end{bmatrix} = \begin{bmatrix} a & b & c \\ 1 & 0 & 0 \\ 0 & 1 & 0 \end{bmatrix} \begin{bmatrix} Q_{t-1} \\ Q_{t-2} \\ Q_{t-3} \end{bmatrix}\qquad(20.2.13)$$

which in vector notation is

$$\hat{x}_t = F_{t-1} x_{t-1},\qquad(20.2.14)$$

The error in the prediction (20.2.14) is

$$x_t - \hat{x}_t = \begin{bmatrix} w_{t-1} \\ 0 \\ 0 \end{bmatrix},\qquad(20.2.15)$$

which is the system noise at time $t$.

## 20.3.   DERIVATION OF THE KALMAN FILTER FOR A DISCRETE LINEAR DYNAMIC SYSTEM

### 20.3.1.   Assumptions

In section 20.2.1, it was stated that, if the system and measurement noises are Gaussian, then the Kalman filter provides the best linear unbiased estimate of the state $x_t$ and that this will be the best estimator among all possible estimators. Under these assumptions, the derivation of the filter equations follows naturally from distributional considerations. In the derivation presented here, which helps to provide more insight into the operation of the filter, no distributional assumptions are made, and the Kalman filter is derived to provide the best linear unbiased estimate of the state $x_t$; the result is the same as that derived from Gaussian distributional considerations.

### 20.3.2.   The Kalman Gain

In the example presented in section 20.2.2, the prediction of $x_t$ was made under the assumption of perfect knowledge of $x_{t-1}$. However, because of measurement noise, the true value of $x_{t-1}$ will never be known, and at time $(t-1)$ only the measurements $y_1, \ldots, y_{t-1}$ will be available from which $x_{t-1}$ can be estimated, and the prediction of $x_t$ will be based on that estimate of $x_{t-1}$. Let $\hat{x}_{t|t-1}$ denote such an estimate, where the subscript $t|t-1$ denotes the estimate made at time $t$ using the information available up to time $t-1$, and assume for the moment that this estimate is available. A new measurement $y_t$ now becomes available; this in itself constitutes an estimate of $x_t$ but it is, of course, subject to noise. To obtain the best unbiased linear estimate of $x_t$, a weighted linear sum of the two available estimates is formed to yield

$$\hat{x}_{t|t} = L_t \hat{x}_{t|t-1} + K_t y_t \tag{20.3.1}$$

where $L_t$ and $K_t$ are time-varying weighting matrices to be specified by imposing on the filter the conditions that the estimate of the state at each time point should be unbiased and of minimal variance [cf. § 3.2.2].

We first impose the condition of unbiasedness.

If the errors of estimates of the state are defined as

$$\tilde{x}_{t|t} = \hat{x}_{t|t} - x_t,$$
$$\tilde{x}_{t|t-1} = \hat{x}_{t|t-1} - x_t, \tag{20.3.2}$$

then $\hat{x}_{t|t}$, $\hat{x}_{t|t-1}$ and $y_t$ can be replaced in (20.3.1) to give

$$\tilde{x}_{t|t} = [L_t + K_t H_t - I] x_t + L_t \tilde{x}_{t|t-1} + K_t v_t, \tag{20.3.3}$$

where $I$ is the identity matrix. By definition, $E\{v_t\} = 0$, and if $E\{\tilde{x}_{t|t-1}\} = 0$, the updated estimator will be unbiased, that is $E\{\tilde{x}_{t|t}\} = 0$, if the quantity $[L_t + K_t H_t - I] = 0$. An unbiased estimate of $x_t$ is therefore ensured by having

$$L_t = I - K_t H_t, \tag{20.3.4}$$

which, when substituted into (20.3.1) gives

$$\hat{x}_{t|t} = (I - K_t H_t) \hat{x}_{t|t-1} + K_t y_t. \tag{20.3.5}$$

Alternatively

$$\hat{x}_{t|t} = \hat{x}_{t|t-1} + K_t [y_t - H_t \hat{x}_{t|t-1}] \tag{20.3.6}$$

with corresponding estimation error

$$\tilde{x}_{t|t} = [I - K_t H_t] \tilde{x}_{t|t-1} + K_t v_t. \tag{20.3.7}$$

Equation (20.3.6) gives the relation between the updated estimate $\hat{x}_{t|t}$ and the previous estimate $\hat{x}_{t|t-1}$. It may be written in the form

$$\hat{x}_{t|t} = \hat{x}_{t|t-1} + K_t v_t,$$

whence the weighting matrix $K_t$ is referred to as *the Kalman gain* and the quantity

$$v_t = y_t - H_t \hat{x}_{t|t-1} \tag{20.3.8}$$

is referred to as the *filter innovation*, since it represents the new information incorporated in the measurement $y_t$, which is potentially of use in estimating $x_t$.

The Kalman gain clearly plays a crucial role in the operation of the filter, and so an optimum choice of $K_t$ is clearly desirable to ensure a minimum variance estimate of the state $x_t$.

### 20.3.3.  Optimum Choice of the Kalman Gain

To describe the uncertainty in the state estimates, the variance-covariance matrices of $\tilde{x}_{t|t}$ and $\tilde{x}_{t|t-1}$ are required; these are defined as

$$P_{t|t} = E\{\tilde{x}_{t|t}\tilde{x}_{t|t}^T\},$$
$$P_{t|t-1} = E\{\tilde{x}_{t|t-1}\tilde{x}_{t|t-1}^T\}. \qquad (20.3.9)$$

An expression for $P_{t|t}$ can be obtained by substituting for $\hat{x}_{t|t}$ from (20.3.7) into (20.3.9) to give

$$P_{t|t} = [I - K_t H_t]P_{t|t-1}[I - K_t H_t]^T + K_t R K_t^T, \qquad (20.3.10)$$

where $R$ is the variance-covariance matrix of $y_t$, as defined in (20.2.3). The terms $E\{\tilde{x}_{t|t-1}v_t^T\}$ and $E\{v_t\tilde{x}_{t|t-1}^T\}$ are zero since $v_t$ is uncorrelated with $v_{t-1}$ and thus cannot affect the estimate $\hat{x}_{t|t-1}$. The diagonal entries in the matrix $P_{t|t}$ are the variances of the components of the estimate $\hat{x}_{t|t}$ of the state vector, and it is these that we have to minimize. A typical one of these variances, the $r$th, say, is the $(r, r)$ entry in the matrix

$$(I - KH)A(I - H^T K^T) + KRK^T$$
$$= A - KHA - AH^T K^T + K(HAH^T + R)K^T,$$

where, for the moment, we have omitted the time subscripts and replaced $P_{t|t-1}$ by $A$. This $(r, r)$ entry is

$$J_r = A_{rr} - 2\sum_s K_{rs}(HA)_{sr} + \sum_s \sum_j K_{rs}(HAH^T + R)_{sj}K_{rj}, \qquad (20.3.11)$$

the subscripts here serving to identify matrix elements. To find the minimizing value of $K$ we solve the equations

$$\partial J_r/\partial K_{uv} = 0 \qquad (20.3.12)$$

for each $u$ and $v$, and for each $r$. The resulting scalar equations may be reassembled in matrix form to give the result

$$-AH^T + K(HAH^T + R) = 0,$$

whence

$$K = AH^T(HAH^T + R)^{-1}.$$

Thus, reverting to the standard notation, the optimal value of $K_t$ is

$$K_t = P_{t|t-1}H_t^T[H_t P_{t|t-1}H_t^T + R]^{-1}. \qquad (20.3.13)$$

Substituting for $K_t$ in (20.3.10) results in

$$P_{t|t} = P_{t|t-1} - P_{t|t-1}H_t^T[H_t P_{t|t-1}H_t^T + R]^{-1}H_t P_{t|t-1}$$
$$= [I - K_t H_t]P_{t|t-1}, \qquad (20.3.14)$$

which is the variance-covariance matrix of the state estimates corresponding to the optimum value of $K_t$ given by (20.3.13). This equation also provides a convenient computational means of updating the variance-covariance matrix of the state to take account of the measurement made at time $t$.

### 20.3.4.   State Prediction

So far, attention has focussed on how the filtered estimate of the state and of the variance-covariance matrix of its error of estimate can be derived after the measurement $y_t$ has been observed. In accord with (20.2.1), the forecast at time $t$ of the state at time $t+1$ is taken to be

$$\hat{x}_{t+1|t} = F_t \hat{x}_{t|t}. \tag{20.3.15}$$

By subtracting the system equation (20.2.1) from this, the error of prediction, $\tilde{x}_{t+1|t}$, is obtained as

$$\tilde{x}_{t+1|t} = F_t \tilde{x}_{t|t} + w_t, \tag{20.3.16}$$

from which the variance-covariance matrix $P_{t+1|t}$ of the prediction error can be derived, in a similar way to the derivation of (20.3.10), as

$$P_{t+1|t} = F_t P_{t|t} F_t^T + Q, \tag{20.3.17}$$

where $Q$ is the dispersion matrix of $w_t$.

Again, note the convenient recursive form for obtaining $P_{t+1|t}$ from $P_{t|t}$.

### 20.3.5.   Interpretation of Filter Equations

The expression (20.3.13) giving the optimal value of $K_t$ may be simplified by rewriting it in the form

$$K_t[H_t P_{t|t-1} H_t^T + R] = P_{t|t-1} H_t^T$$

or

$$(K_t H_t - I) P_{t|t-1} H_t^T + K_t R = 0.$$

Thus

$$K_t R = (I - K_t H_t) P_{t|t-1} H_t^T$$

whence, using (20.3.14),

$$K_t R = P_{t|t} H_t^T.$$

The simplified expression for $K_t$ is then

$$K_t = P_{t|t} H_t^T R^{-1}. \tag{20.3.18}$$

An examination of this equation provides some useful insight into the operation of the filter (Gelb, 1974). Suppose that $H_t$ is the identity matrix; in this case, both $P_{t|t}$ and $R$ are $(n \times n)$ matrices. If $R^{-1}$ is a diagonal matrix (no cross correlation between noise terms), $K_t$ results from multiplying each column of $P_{t|t}$ by the inverse of the appropriate variance of the measurement noise, and is thus proportional to the uncertainty in the state estimate, and inversely proportional to the measurement noise. If measurement noise is large and state estimate errors are small, the innovation $v_t$ in (20.3.8) is due mainly to the measurement noise, and the elements of $K_t$ will be relatively small; thus, in equation (20.3.6), only small changes to the state estimate will be made. On the other hand, small measurement noise and large uncertainty in the state estimates suggests that $v_t$ contains considerable information about errors in the state estimates; thus the elements of $K_t$ will be relatively large.

The uncertainty in the state projection, given by equation (20.3.17), is seen to be influenced by the variance-covariance matrix of the system noise. The magnitude of the system noise will reflect the 'goodness' of the model used to represent the system dynamics; if $Q$ is large, reflecting a poor model, then the uncertainty in $\hat{x}_{t+1|t}$ will be large.

If $Q$, $R$, $F_t$, $H_t$ are known constant matrices, and initial estimates $\hat{x}_{0|0}$ and associated error variance-covariance matrix $P_{0|0}$ are available, the filter equations can be applied recursively at each time point; initially, the matrices $P_{t|t}$, $P_{t|t-1}$ and $K_t$ will vary over time but these will, for a time invariant system, eventually converge to steady state values independent of future observations. Thus, these matrices are fixed properties of the filter.

### 20.3.6.   Summary

Before the filter can start its recursive operations, an initial estimate of the state, denoted as $\hat{x}_{0|0} = \hat{x}_0$ is required together with the variance-covariance

| | |
|---|---|
| System equation: <br> Measurement equation: | $x_{t+1} = F_t x_t + D_t u_t + w_t$ <br> $y_t = H_t x_t + v_t$ |
| Initial conditions and <br> other assumptions: | $E\{x_0\} = \hat{x}_0,\ E\{\tilde{x}_0 \tilde{x}_0^T\} = P_0$ <br> $E\{w_t\} = 0;\ \text{var}\{w_t\} = Q$ <br> $E\{v_t\} = 0;\ \text{var}\{v_t\} = R$ <br> $E\{w_t v_{t+\tau}^T\} = 0$ for $\tau = 0, \pm1, \pm2, \dots$ . |
| *At an observation*: <br> Innovation error: <br> Kalman gain: <br> State update: <br> State error covariance: | $v_t = y_t - H_t \hat{x}_{t|t-1}$ <br> $K_t = P_{t|t-1} H_t^T (H_t P_{t|t-1} H_t^T + R)^{-1}$ <br> $\hat{x}_{t|t} = \hat{x}_{t|t-1} + K_t v_t$ <br> $P_{t|t} = (I - K_t H_t) P_{t|t-1}$ |
| *At a forecast*: <br> State forecast: <br> State forecast error <br> covariance: | $\hat{x}_{t+1|t} = F_t \hat{x}_{t|t} + D_t u_t$ <br><br> $P_{t+1|t} = F_t P_{t|t} F_t^T + Q$ |

Table 20.3.1: Summary of discrete linear Kalman filter algorithm.

matrix of its error of estimate $P_{0|0}$; these can be supplied using whatever prior information is available. The recursive cycle of filter operations may be summarized as follows: given an observation at time $t$, calculate the innovation vector $\nu_t$ (equation (20.3.8)), the Kalman gain $K_t$ (equation (20.3.13)), the filtered or updated state vector $\hat{x}_{t|t}$ (equation (20.3.6)), and the updated state error covariance matrix $P_{t|t}$ (equation (20.3.14)). The state forecast $\hat{x}_{t+1|t}$ is given by equation (20.3.15) and the state forecast error covariance $P_{t+1|t}$ by equation (20.3.17).

Table 20.3.1 summarizes the discrete linear Kalman filter algorithm developed above but with one minor modification: a vector of exogenous inputs $u_t$ with a weighting matrix $D_t$ is included in the system equation. This may represent a known deterministic input or a control variable used to control the state trajectory; it appears in the filter equations only when making a forecast of the state.

## 20.4.   TESTS OF FILTER PERFORMANCE

### 20.4.1.   Innovation Properties

The innovation sequence, $\{\nu_1, \nu_2, \ldots, \nu_t\}$ defined by equation (20.3.8) can be used to test the performance of the filter. By definition

$$\nu_t = y_t - H_t x_t - H_t(\hat{x}_{t|t-1} - x_t) \tag{20.4.1}$$

$$= v_t - H_t(\hat{x}_{t|t-1} - x_t) \tag{20.4.2}$$

which again illustrates that the innovation depends on the error in the state forecast based on the observation sequence $\{y_1, y_2, \ldots, y_{t-1}\}$ and on the observation at time $t$. The innovation sequence has mean and variance

$$E[\nu_t] = 0 \tag{20.4.3}$$

$$\mathrm{var}\,[\nu_t] = E[\nu_t \nu_t^T] = R + H_t P_{t|t-1} H_t^T ; \tag{20.4.4}$$

furthermore, if the system and measurement noises have Gaussian distributions, then $\nu_t$ will be Gaussian. Jazwinski (1970) shows that if the algorithm is filtering optimally, then $\nu_t$ is uncorrelated with $\nu_s$ and $y_s$ for $s \leq t$. This optimal filter performance implies that the parameters $P_0$, $F_t$, $D_t$, $H_t$, $Q$ and $R$ are correctly specified. The three relationships: $E[\nu_t] = 0$, $E[\nu_t \nu_s^T] = 0$ and $E[\nu_t y_s^T] = 0$ for $s \leq t$, are called the *innovation properties* and provide a basis of testing filter performance.

Statistical tests, based upon the $Q$ and $S$ tests described under the heading of diagnostic checking in Box and Jenkins (1976) can be used to test whether the innovation properties hold. Let us assume that the elements of $\{y_t\}$ and $\{\nu_t\}$ are scalar and represent, respectively, the observed outputs and the calculated innovations over times $0, 1, \ldots, t$. Let $\bar{\nu}$ be the sample mean of $\{\nu_t\}$ and $\bar{y}$ and $s_y^2$ the sample mean and variance of $\{y_t\}$.

The following sample covariance functions are calculated:

$$r_l^2 = \frac{1}{t-l} \sum_{s=l+1}^{t} (v_s - \bar{v})(v_{s-l} - \bar{v}) \qquad l = 0, 1, \ldots, L, \qquad (20.4.5)$$

$$c_l^2 = \frac{1}{t-l} \sum_{s=l+1}^{t} (v_s - \bar{v})(y_{s-l} - \bar{y}) \qquad l = 1, 2, \ldots, L. \qquad (20.4.6)$$

From these, the following statistics are calculated:

$$M = \bar{v}/r_0, \qquad (20.4.7)$$

$$Q = t \sum_{l=1}^{L} (r_l^2 / r_0^2), \qquad (20.4.8)$$

$$S = t \sum_{l=1}^{L} (c_l^2 / \sqrt{(r_0^2 s_y^2)}). \qquad (20.4.9)$$

If the innovation properties hold, then $M$ has Student's distribution with $t-1$ degrees of freedom [see § 2.5.5], and $Q$ and $S$ are distributed as chi-square [see § 2.5.4(a)] with $m$ degrees of freedom. Here $m = L - p$, $p$ being the number of parameters in the overall model. The number of terms $L$ in the above equations is chosen to exceed the memory of the model. A reasonable choice is to take $L$ such that $F^L$ (the product of $L$ consecutive transition matrices) is a matrix all of whose elements are very small.

Although these tests are useful, they do not guarantee that a model is adequate. Certain consistent discrepancies may not be detected: a visual inspection of the innovations may be worthwhile (§ 20.7).

### 20.4.2.   Filter Divergence

The filter algorithm presented in Table 20.3.1 is straightforward and easy to apply as long as the required parameters are specified.

Filter divergence can derive from the situation where the parameters ($F_t$, $D_t$, $H_t$, $P_0$, $R$ and $Q$) are not known exactly; this may be due to simplification of the system dynamics or to poor parameter estimation. The result is that the filter estimates the states incorrectly causing divergence from their true values. The problem is accentuated when the noise terms in the system equation are small; in this case, the error covariance becomes small, thus the Kalman gain is small and subsequent observations have little effect on the estimate.

In actual applications, the onset of divergence manifests itself through the innovations. The result is a filter that is no longer optimal, where $P_{t|t}$ does not correctly measure the estimation error variance and where the error $\tilde{x}_{t|t}$ becomes progressively larger as $t$ increases.

One way of coping with filter divergence is to modify the filter equation so that recent measurements have a larger influence on the filter estimates and as measurements get older their influence decreases. This can be achieved by applying a set of weights

$$1, e^{-1/\tau}, e^{-2/\tau}, \ldots, e^{-s/\tau} \qquad (20.4.10)$$

to the observations $y_t, y_{t-1}, y_{t-2}, \ldots$ where $\tau$ represents the 'memory' of the filter. Table 20.4.1, presents such an exponentially-weighted memory filter. If the parameters in the system and measurements equations are known exactly, then the filter of Table 20.4.1 will be inferior to that of Table 20.3.1, since the estimation error $P_{t|t}$ will be larger than that obtained in the Kalman filter. However, its use can prevent divergence when inaccuracies in the parameters are present.

---

*At an observation:*

$$\nu_t = y_t - H_t \hat{x}_{t|t-1}$$
$$K_t = P_{t|t-1} H_t^T (H_t P_{t|t-1} H_t^T + e^{-1/\tau} R)^{-1}$$
$$\hat{x}_{t|t} = \hat{x}_{t|t-1} + K_t \nu_t$$
$$P_{t|t} = e^{-1/\tau} (I - K_t H_t) P_{t|t-1}$$

*At a forecast (time t):*

$$\hat{x}_{t+1|t} = F_t \hat{x}_{t|t} + D_t u_t$$
$$P_{t+1|t} = F_t P_{t|t} F_t^T + Q$$

---

Table 20.4.1: Weighted memory linear discrete Kalman filter algorithm.

## 20.5. KALMAN FILTERING FOR CONTINUOUS TIME AND NON-LINEAR SYSTEMS

### 20.5.1. Continuous Time Systems

Until now only discrete time systems have been considered. While some systems progress in discrete time steps (e.g. stock exchange prices), many processes actually evolve continuously in time (e.g. streamflow), described by a process $x(t)$ for any real $t \geq 0$. A process $x(t)$ which measures all properties of interest in a continuously evolving system will be called a system state if it is possible to write a system equation:

$$\dot{x}(t) = f(x(t), t) + g(\xi(t), t) \qquad (20.5.1)$$

where $\dot{x}(t)$ is the derivative of $x(t)$ with respect to time $t$. Here $\xi(t)$ is a (continuous) noise process and from (20.5.1) the evolution of $x(.)$ at times following $t$ is completely described by $x(t)$ and by noise which is unpredictable from $x(.)$. In some cases $x(t)$ is accompanied by continuous time (noisy) measurements $y(t)$ (e.g. reception of a radio broadcast and other communication theory applications), although in most cases only discrete time measurements, $y_t$, $t = 1, 2, 3, \ldots$ will be available.

A continuous time system is linear and Gaussian if the system equation is:

$$dx(t) = f(t)x(t)\, dt + g(t)\, dB(t) \qquad t \geq 0 \qquad (20.5.2)$$

where $B(t)$ is a Brownian motion (Gaussian) process [see II, Example 22.1.1] and $dB(t)$ denotes its infinitesimal increments: ($dB(t)$ is the so-called continuous Gaussian white noise). Associated with this linear system is a discrete time measurement process $y_t$, with Gaussian measurement noise, $v_t$:

$$y_t = H_t x(t) + v_t. \qquad (20.5.3)$$

Given the observations $y_1, \ldots, y_t$, estimates of $x(t)$ (filtering), or $x(s)$, $s < t$ (smoothing) or $x(s)$, $s > t$ (prediction) might be required. If, however, only estimates of $x(.)$ at the discrete times $0, 1, \ldots, t$ are required then this problem can be written as the standard discrete linear Kalman filtering problem. To do this, equation (20.5.2) is solved for $x(t)$, to obtain:

$$x(t) = F_{t-1}x(t-1) + w_t \qquad (20.5.4)$$

where

$$F_{t-1} = F(t-1, t) = \exp\left\{ -\int_{t-1}^{t} f(z)\, dz \right\}$$

and

$$w_t = \int_{t-1}^{t} F(\tau, t)g(\tau)\, dB(\tau),$$

in which $\{w_t\}$ is a Gaussian noise sequence. The standard linear Kalman filter can then be applied to derive the usual state estimates together with their error variance-covariance matrices. For further details the reader is referred to Jazwinski (1970) and Gelb (1974).

### 20.5.2. Non-linear Systems: the Extended Kalman Filter

In many practical cases, the systems under study are not linear; usually the state of such systems evolves in continuous time according to a non-linear first order differential equation of the type:

$$dx(t) = f(x(t))\, dt + g(t)\, d\xi(t) \qquad (20.5.5)$$

while the measurement mechanism might also be non-linear and described by:

$$y(t) = h(x(t)) + v(t). \qquad (20.5.6)$$

The filtering problem of estimating $x(t)$ from continuous measurements $y(t)$ or discrete observations $y_t$ can be solved theoretically and an optimum non-linear filter can be specified. However, the calculations involved in trying to evaluate such a filter numerically are highly intractable. Thus such non-linear problems are almost invariably only solved approximately. One such approximation, the extended Kalman filter, which will be described here, allows estimates of $x(t)$, given $y_1, \ldots, y_t$ to be obtained by applying a Kalman filter to a first order Taylor expansion of (20.5.5) and (20.5.6) (see Jazwinski, 1970: § 8.3).

The expansion is taken about a solution $\bar{x}(t)$ of the deterministic part of equation (20.5.5), that is,

$$d\bar{x}(t) = f(\bar{x}(t)) \, dt \qquad (20.5.7)$$

The expanded forms of equations (20.5.5) and (20.5.6) are then

$$dx(t) = f(\bar{x}(t)) \, dt + f'(\bar{x}(t))(x(t) - \bar{x}(t)) \, dt + g(t) \, d\xi(t) \qquad (20.5.8)$$

$$y(t) = h(\bar{x}(t)) + h'(\bar{x}(t))(x(t) - \bar{x}(t)) + v(t). \qquad (20.5.9)$$

Here $f'(\bar{x}(t))$ and $h'(\bar{x}(t))$ are matrices of derivatives of the functions $f$ and $h$, evaluated at $\bar{x}(t)$.

Substituting equation (20.5.7) into (20.5.8) gives

$$d(x(t) - \bar{x}(t)) = f'(\bar{x}(t))(x(t) - \bar{x}(t)) \, dt + g(t) \, d\xi(t). \qquad (20.5.10)$$

Defining the new variables

$$y^*(t) = (y(t) - h(\bar{x}(t))),$$

$$x^*(t) = (x(t) - \bar{x}(t)),$$

$$H_t^* = h'(\bar{x}(t)),$$

$$F_t^* = f'(\bar{x}(t)),$$

the pair of equations (20.5.10) and (20.5.9) can be written as

$$dx^*(t) = F_t^* x^*(t) \, dt + g(t) \, d\xi(t),$$

$$y^*(t) = H_t^* x^*(t) + v(t).$$

This is a linear continuous state space model, recognizable as being of the form given by equations (20.5.2) and (20.5.3). Note that all functions of $\bar{x}(t)$ can be computed to any required degree of accuracy using, for example, the Runge–Kutta algorithm for numerical integration [see III, § 8.2.2].

Care must be exercised when the extended Kalman filter is utilized for nonlinear systems, as very little can be proved theoretically about its performance. Since the Taylor series is truncated after one term, the approximation of the non-linear system is better when the error terms $(x_t - \hat{x}_{t|t})$ and $(x_t - \hat{x}_{t|t-1})$ are small; thus, when the signal-to-noise ratio is high one would expect fewer difficulties with this approach. Once the filter is operating, the error covariance matrices $P_{t|t}$ and $P_{t|t-1}$ can be used as guides to whether $(x_t - \hat{x}_{t|t})^2$ and $(x_t - \hat{x}_{t|t-1})^2$ are small.

Unlike the Kalman filter for linear systems, $P_{t|t}$ and $P_{t|t-1}$ are coupled to the filter equations through $\hat{x}_{t|t}$ and $\hat{x}_{t|t-1}$. In general, these quantities cannot be calculated off-line and do not reach steady-state values as is the case for linear systems with time-invariant parameters. As for the linear Kalman filter, the performance of the extended Kalman filter can be checked using the innovations; the closer these are to white noise, the more nearly optimal the filter. The reader is referred to Ljung (1979) for a discussion of the stability of extended Kalman filters.

Filters have been developed for the case where two expansion terms are kept in the Taylor series; the resulting filters are referred to as second order extended Kalman filters. An application of this type of filter is described by Moore and Weiss (1980b).

There are few guidelines to indicate which algorithm will perform best for a particular problem; each case must be studied and evaluated separately. The reader should be cautioned that filter divergence and poor performance is more often encountered with non-linear systems; thus the design of the filter must be carried out with care.

## 20.6  MODEL IDENTIFICATION AND PARAMETER ESTIMATION

### 20.6.1.  General

In previous sections, the application of the Kalman filter to linear and non-linear systems has been described; in these cases, knowledge of the structural form of the state space model, and of the model parameters, was assumed. There may, however, be many practical problems where these assumptions may not hold and where one of the following situations may obtain: (a) the model structure is known but the parameters are unknown; (b) both the model structure and the parameters are unknown. Here, the field of application sketched out briefly in section 20.1, where a model is required to forecast streamflow from rainfall in real-time, will be used to illustrate the different situations which can arise. In a river basin, the movement of water over the land surface, in the soil zone, in aquifers and in stream channels follows well defined physical laws based on the conservation of mass and momentum. Sets of non-linear partial differential equations describing these processes can be formulated, coupled, and solved on a finite difference grid over the basin, since analytical solutions cannot be obtained (Beven and O'Connell, 1982). The computational demands of such distributed parameter models are high, and, while in principle they can be written in state-space form, the dimensionality of the state vector and the computational resources required to integrate the equations make this approach for all intents impracticable.

By describing the processes whereby rainfall is transformed into streamflow as being spatially lumped, it is possible to formulate lumped parameter non-linear differential equation models with a quasi-physical structure which can represent adequately the transformation of rainfall into streamflow. Such differential equation models typically derive from a continuity equation and an empirical equation describing the relationship between streamflow and the amount of water stored in the basin, the form of which can be established from observed streamflow behaviour. Such models can be formulated within the state space framework of equations (20.5.7) and (20.5.8); while the structure of the model will thus be known a priori, the parameters of the model will not, and will need to be estimated from the available data. A number of different techniques can be used for this purpose, as outlined in section 20.6.3(b); a case study illustrating this approach is described in section 20.7.2.

If nothing is assumed *a priori* about the structure of the model relating rainfall to streamflow, then it is possible to employ model identification procedures based on the observed covariance structure of the system output (streamflow); using these procedures, the dimensionality of the matrices $F_t$, $H_t$, $Q$, and $R$ in the state-space representation

$$x_{t+1} = F_t x_t + w_t \tag{20.6.1}$$

$$y_t = H_t x_t + v_t, \tag{20.6.2}$$

and their structure, can be specified. If this approach is used to model the relationship between rainfall and streamflow, then the following illustrates what such a linear state-space model might look like:

$$
\begin{bmatrix} q_{t+1} \\ p_{t+1} \\ p_t \\ p_{t-1} \end{bmatrix}
=
\begin{bmatrix} \delta_1 & \omega_0 & \omega_1 & \omega_2 \\ 0 & \alpha_1 & \alpha_2 & \alpha_3 \\ 0 & 1 & 0 & 0 \\ 0 & 0 & 1 & 0 \end{bmatrix}
\begin{bmatrix} q_t \\ p_t \\ p_{t-1} \\ p_{t-2} \end{bmatrix}
+
\begin{bmatrix} w_1 \\ w_2 \\ 0 \\ 0 \end{bmatrix}_t
\tag{20.6.3}
$$

$$x_{t+1} \quad = \qquad\qquad F_t \qquad\qquad\qquad x_t \quad + \quad w_t$$

$$
\begin{bmatrix} q_t \\ p_t \end{bmatrix}
=
\begin{bmatrix} 1 & 0 & 0 & 0 \\ 0 & 1 & 0 & 0 \end{bmatrix}
\begin{bmatrix} q_t \\ p_t \\ p_{t-1} \\ p_{t-2} \end{bmatrix}
+
\begin{bmatrix} v_1 \\ v_2 \end{bmatrix}_t
\tag{20.6.4}
$$

$$y_t \quad = \qquad\qquad H_t \qquad\qquad\qquad x_t \quad + \quad v_t$$

In the above formulation, both rainfall and streamflow are represented by linear stochastic difference equation models; the Kalman filter can thus be used to filter the hypothesized measurement errors in both variables, and predictions of streamflow will thus be based on the filtered estimates of rainfall. A limitation of such models is that they represent the non-linear response of streamflow to rainfall as being linear; however, by representing the response as being piece-wise linear, the above form of state space model can be retained, which means that model identification and parameter estimation procedures designed for linear state-space models can be employed. A case study illustrating this approach is described in section 20.7.1, while the techniques to be used for identifying the structure of the model and estimating its parameters are described in sections 20.6.2 and 20.6.3.

In identifying the structure of a state-space model and estimating its parameters, certain conditions must, in theory, be satisfied. For example, theorems concerning the behaviour of the Kalman filter are based on the assumption that the underlying model is observable, controllable and stable. Essentially, these conditions determine what, if anything, can be gained from filtering the data; as might be expected, this is to a large degree determined by the structure of the underlying system model. The concept of system observability defines whether or not the states can be determined from the measurements; this condition is satisfied if the matrix

$$[H_t^T \vdots F_t^T H_t^T \vdots \ldots \vdots (F_t^T)^{n-1} H_t^T]$$

is of rank $n$, where $n$ is the dimension of the state vector. Controllability relates to the conditions under which it is possible to control the state of a deterministic linear dynamic system i.e. when a control variable is included in the system equation (Table 1). The state can be controlled if the matrix

$$[D_t \vdots F_t D_t \vdots \ldots \vdots (F_t)^{n-1} D_t]$$

is of rank $n$.

Stability relates to the behaviour of the state estimates when measurements are suppressed. Suppose that the system behaviour is described by a first order differential equation (20.5.2); then the system is asymptotically stable if $\hat{x}(t) \to 0$ as $t \to \infty$ for any initial condition $\hat{x}(0)$. This is a rather informal definition of stability: the theoretical conditions guaranteeing stability are rather more restrictive, and in many practical applications, these ·conditions are not fulfilled, yet Kalman filters, designed in the normal way, operate satisfactorily. This is attributable to the fact that the solution of the system differential equation frequently tends to zero over a finite time interval of interest, even though it may not be asymptotically stable in the strict sense of the definition. Gelb (1974) suggests that the key issues pertaining to various forms of instability are those associated with modelling errors and implementation considerations, and provides an extensive discussion of these aspects.

Another consideration which may be relevant to some systems is whether or not the system inputs suitably excite or perturb the system so that model identification and parameter estimation is possible. In hydrological modelling the system may only be excited sporadically when rainfall occurs, and so some model parts/parameters may not be estimable; this may be particularly true in the case of large multiple parameter simulation models. Most estimation algorithms assume continuous and persistent excitation of the system; if this is not the case, parameter estimates with large variances may result.

### 20.6.2.   Model Identification

Given the sample stationary covariance of a time series, a state space model can be constructed whose output has approximately the same statistics though the precise form of the model will at first be unknown. In this respect state space models are akin to autoregressive moving average (ARMA) models [see § 18.9]

$$y_t + \Phi_1 y_{t-1} + \ldots + \Phi_p y_{t-p} = \Theta_0 \nu_t + \Theta_1 \nu_{t-1} + \ldots + \Theta_q \nu_{t-q} \qquad (20.6.5)$$

where the $\nu_t$ are assumed to be identically and independently distributed, and is exactly equivalent to the innovation term in the Kalman filter, and $\Phi_1, \ldots, \Phi_p$ and $\Theta_0, \ldots, \Theta_q$ are $(m \times m)$ matrices of parameters. In fact there is an intimate relationship between the two representations. In terms of predictors the Kalman filter can be written as

$$y_t = H\hat{x}_{t|t-1} + \nu_t \qquad (20.6.6)$$

$$\hat{x}_{t+1|t} = F\hat{x}_{t|t-1} + FP_{t+1|t} H^T (HP_{t+1|t} H^T + R)^{-1} \nu_t \qquad (20.6.7)$$

with error variance covariance matrix

$$P_{t+1|t} = FP_{t|t-1} F^T + G_t C_t G_t^T + Q$$

where $G_t = FK_t$ and $C_t = E\{\nu_t \nu_t^T\}$. Here and in the ensuing development it is assumed that the system is time invariant and so the time subscripts on the matrices $H$ and $F$ are dropped. Under these conditions $P_{t+1|t}$ reaches a steady state value $P$, so the gain is constant and may be written $K$. The so-called innovations form of the model is then

$$\hat{x}_{t+1} = F\hat{x}_t + K\nu_t, \qquad (20.6.8)$$

$$y_t = H\hat{x}_t + \nu_t, \qquad (20.6.9)$$

where $\hat{x}_{t+1}$ is interpreted as the one step ahead predictor $\hat{x}_{t+1|t}$. The equivalence of the ARMA and state-space forms can now be demonstrated as follows. Using the forward shift operator, $z$ such that $z(x_t) = x_{t+1}$, the ARMA model (20.6.5) can be written as

$$y_t = \{A(z^{-1})\}^{-1} B(z^{-1}) \nu_t \qquad (20.6.10)$$

where

$$A(z^{-1}) = (I + \Phi_1 z^{-1} + \ldots + \Phi_p z^{-p}):$$

$$B(z^{-1}) = (\Theta_0 + \Theta_1 z^{-1} + \ldots + \Theta_q z^{-q}).$$

By substituting (20.6.8) into (20.6.9), the state space representation is

$$y_t = H(I - Fz^{-1})^{-1} Gz^{-1} \nu_t + \nu_t. \qquad (20.6.11)$$

These representations are equivalent in that given any pair $\{A, B\}$ it is possible to find a set $\{H, F, G\}$ which gives the same covariance structure, and conversely. Notice that stationarity requires that the roots of $A$ lie outside the unit circle, or equivalently that the eigenvalues of $F$ be less than 1 in modulus (this is equivalent to model stability discussed in section 20.6.1). The inverse ARMA model is

$$\nu_t = \{B(z^{-1})\}^{-1} A(z^{-1}) y_t:$$

it may be verified that the inverse of the state space model written as

$$y_t = [I + H(z - F)^{-1} G] \nu_t \qquad (20.6.12)$$

is

$$\nu_t = [I - H[z - (F - GH)]^{-1} G] y_t. \qquad (20.6.13)$$

This inverse exists only if $(F - GH)$ has eigenvalues [see I, chapter 7] less than 1 in modulus, or equivalently if $B$ has roots outside the unit circle (the invertibility condition for ARMA models).

We now state the relationship between ARMA and state-space models for a univariate series. For the ARMA representation

$$y_t + \phi_1 y_{t-1} + \ldots + \phi_p y_{t-p} = \nu_t + \theta_1 \nu_{t-1} + \ldots + \theta_q \nu_{t-q}, \qquad (20.6.14)$$

an equivalent state space model is

$$\hat{x}_{t+1} = \begin{bmatrix} 0 & 1 & 0\ldots & & 0 \\ 0 & 0 & 1\ldots & & 0 \\ & & \cdots & & \\ 0 & 0 & & \cdots & 1 \\ -\phi_n & & & \cdots & -\phi_1 \end{bmatrix} \hat{x}_t + \begin{bmatrix} g_1 \\ . \\ . \\ . \\ g_n \end{bmatrix} \nu_t \qquad (20.6.15)$$

$$y_t = [1 \quad 0 \quad \ldots \quad 0]\hat{x}_t + \nu_t \qquad (20.6.16)$$

where $n = \max[p, q]$ and $\phi_i \equiv 0$ if $i > p$. The $g_i$ are the elements of the impulse response relating the $\nu$'s to the $y$'s. Notice the sparsity of $F$ and $H$; this model is in canonical form, that is, it has a particular, efficiently parameterized structure. There are other canonical forms, for example the above may also be written as

$$y_t = [1 \quad 0 \quad \ldots \quad 0]\hat{x}_t + \nu_t \qquad (20.6.17)$$

$$\hat{x}_{t+1} = \begin{bmatrix} -\phi_1 & 1 & 0 & \ldots & 0 \\ . & 0 & 1 & \ldots & 0 \\ . & & & & \\ . & & \cdots & & \\ -\phi_n & 0 & \ldots & 0 & 1 \end{bmatrix} \hat{x}_t + \begin{bmatrix} \theta_1 \\ . \\ . \\ . \\ \theta_n \end{bmatrix} \nu_t. \qquad (20.6.18)$$

In some ways this is preferable because the parameters which appear are precisely those in the ARMA representation. Canonical forms can be derived by choosing a suitable matrix $M$ and rewriting the model as

$$M\hat{x}_{t+1} = MFM^{-1} M\hat{x}_t + MG\nu_t,$$

$$y_t = HM^{-1} M\hat{x}_t + \nu_t. \qquad (20.6.19)$$

Clearly $M$ can be found in the above case using the relationship between $\{\theta_1 \ldots \theta_n\}$ and $\{g_1 \ldots g_n\}$; however, the states have a different interpretation for the two forms.

Choosing ARMA models of suitable dimension remains something of an art, but clearly once a suitable structure is identified, an equivalent state space representation can be written down and the Kalman filter used to give predictions.

Once the innovations form of the state-space model is found, it may be desirable to return to the original state space model containing system and observation noise; given $\{C, G\}$, this would involve extracting $\{Q, R\}$ using the relationships

$$C = R + HPH^T \qquad (20.6.20)$$

$$G = FPH^T(HPH^T + R)^{-1} \qquad (20.6.21)$$

$$P = FPF^T - GCG^T + Q, \qquad (20.6.22)$$

bearing in mind that $\{P, Q, R\}$ must all be positive semi-definite. In general the equations give an infinity of solutions for $Q$, while in some cases it may be impossible to find $\{P, Q, R\}$ satisfying the required constraints. The former problem may be overcome by restricting the system noise so that

$$x_{t+1} = Fx_t + \Gamma w_t \qquad (20.6.23)$$

and

$$y_t = Hx_t + v_t, \qquad (20.6.24)$$

where $\Gamma$ is a matrix having the same number of columns as there are rows in $y_t$. This model can be viewed as an ARMA model (the system equation) with observation noise $v_t$. The original system noise $w_t$, with variance matrix $Q$, is replaced by $\Gamma w_t'$, having variance matrix $\Gamma Q' \Gamma^T$. In general, this matrix has fewer independent elements than $Q$ and, subject to constraints on some of the elements of $\Gamma$, equation (20.6.22) has a finite number of solutions for $\Gamma$ and $Q'$ when $Q$ is replaced by $\Gamma Q' \Gamma$. Only one of these solutions gives an invertible ARMA model in equation (20.6.23).

Apart from the methods suggested by Box and Jenkins (1976) and others for identifying ARMA structures, Akaike (1974) has developed a method which directly identifies a state space structure in canonical form. The method uses canonical correlation analysis, and has been applied for example, by Cooper and Wood (1982a). The method seems particularly useful for multiple time series, where the canonical forms, which are obscured in the multivariate ARMA model, arise in a natural way. The salient points of the procedure are given here, although readers should refer to Akaike (1974) and Cooper and Wood (1982a) for details.

The model structure to be identified is given by equations (20.6.8) and (20.6.9). The procedure is used to find a suitable dimension for $F$ (and implicitly $H$ and $G$), and provides preliminary estimates of the free parameters of these matrices. The model identified is in canonical form; for a $k$-dimensional observation vector $y_t$, there are $k$ rows of $[H^T : F^T]^T$ which have free parameters, with the remaining rows containing zeros and ones. The matrix $G$ is full, being made up of elements of the impulse response of the model.

The first step in identification is to find an initial estimate of the impulse response, that is, the scalars or matrices $A_1, A_2, A_3, \ldots$ such that

$$y_t = v_t + A_1 v_{t-1} + A_2 v_{t-2} + \ldots . \qquad (20.6.25)$$

The method of doing this follows Box and Jenkins (1976). A high order autoregressive model

$$y_t + B_1 y_{t-1} + \ldots + B_k y_{t-k} = \varepsilon_t. \qquad (20.6.26)$$

is fitted to the series $y_t$ using the Yule–Walker equations. By assuming $\varepsilon_t \equiv v_t$, substituting for $y_t$ in terms of $\{v_t, v_{t-1}, \ldots\}$ and equating coefficients of $\{v_t, v_{t-1}, \ldots\}$ it can be seen that $\{A_1, A_2, A_3, \ldots\}$ are found by solving the equations

$$-A_1 = B_1$$

$$-A_2 = B_1 A_1 + B_2 \qquad (20.6.27)$$

$$-A_3 = B_1 A_2 + B_2 A_1 + B_3$$

$$\vdots$$

The elements of $G$ will be chosen from the matrix

$$[A_1^T : A_2^T : A_3^T \ldots]^T.$$

The precise rows of this matrix which are used depend on the next stage of the identification procedure, in which the structures of $H$ and $F$ are found. This is done using canonical variate analysis (Anderson, 1958) [and see § 16.5] on the vectors of past and future observations, $Y^P = \{y_{t-1}, y_{t-2}, \ldots\}$ and $Y^F = \{y_t, y_{t+1}, y_{t+2}, \ldots\}$. For a univariate example, the sample covariance matrices $S_{PP}, S_{PF}, S_{FF}$ are computed for a finite number, say $k$, of past observations, and for $Y$, consisting, successively, of, $[\{y_t\}, \{y_t, y_{t+1}\}, \ldots]$.

The determinantal equation

$$|S_{PF} S_{PP}^{-1} S_{PF}^T - \lambda S_{FF}| = 0 \qquad (20.6.28)$$

is then solved for $\lambda$; if any of the elements of the solution $\lambda_0$ are not significantly different from zero, this implies that there is an associated linear combination of future values which are independent of the past. This vector is found by solving the equation

$$(S_{PF} S_{PP}^{-1} S_{PF}^T - \lambda_0 S_{FF}) L = 0; \qquad (20.6.29)$$

the vector $L = (l_1, l_2, \ldots, l_j)$ is such that

$$l_1 \hat{y}_{t+j|t-1} + l_2 \hat{y}_{t+j-1|t-1} + \ldots + l_j \hat{y}_{t|t-1} = 0 \qquad (20.6.30)$$

and the $(j-1)$th row of $F$ is $\{-l_2/l_1, \ldots, -l_j/l_1\}$, the predictors of $y$ being identified with the states. The remaining previous $i = 1, \ldots, j-2$ rows of $F$ have a 1 in the $(i-1)$th column and zeros elsewhere, corresponding to the equation

$$\hat{y}_{t+j|i-1} = \hat{y}_{t+j|i-2} + A_{i+1} v_{t-1}. \qquad (20.6.31)$$

In the univariate case $H = [1 \ 0 \ \ldots \ 0]$; clearly in this case the elements of $G$ are $\{A_1, A_2, \ldots, A_j\}$. The multivariate problem gives rise to transition matrices with more than one row of parameters, and the interpretation of such models is more complex. Akaike (1974) and Cooper and Wood (1982a) describe the multivariate case in some detail.

### 20.6.3.   Parameter Estimation

#### (a) *Linear estimation*

Once an appropriate state space model has been chosen, its parameters will generally need to be estimated. As is usually the case, this can be done by minimizing a function of the data and the parameters, so as to produce predictions which are close to the observed values. Commonly, the function chosen is the negative logarithm of the likelihood of the data, giving maximum likelihood estimates of the parameters [see Chapter 6]. The likelihood for a set of data is proportional to their joint probability density evaluated at the data point, regarded as a function of the parameters. This density may be factorized as

$$p(y_t, y_{t-1}, \ldots, y_1) = p(y_t|y^{t-1}) p(y_{t-1}|y^{t-2}) \ldots p(y_1) \qquad (20.6.32)$$

where $y^i \equiv \{y_i, y_{i-1}, \ldots, y_1\}$. Now each $p(y_i|y^{i-1})$ is, in the state space model with Gaussian system and measurement noise, the probability density function given by the Kalman filter. The conditioning on the past is slightly obscured because it appears through the states, so, using Bayes' Theorem [see Chapter 15],

$$p(y_i|y^{i-1}) \propto p(y_i|x_i) p(x_i|y^{i-1}), \qquad (20.6.33)$$

and $p(x_i|y^{i-1})$ is $N(\hat{x}_{i|i}, P_{i|i})$, these being functions of $y^{i-1}$ and the parameters only. To find the likelihood we therefore compute the product of the individual densities of the $y_i$ given by the Kalman filter; this may be done numerically. The likelihood is computed for a sequence of sets of parameter values, and the value of the likelihood noted. Successive new sets of parameter values are chosen which are expected to reduce the value of the likelihood until convergence is achieved using some form of numerical minimization algorithm. Alternatively, the gradient of the log likelihood with respect to the parameters may be derived analytically and a Gauss–Newton algorithm used to give improved parameter estimates. The algorithm gives successive estimates from

$$\theta_s = \theta_{s-1} + \left[ \left\{ \frac{d \log L}{d\theta} \right\} \left\{ \frac{d \log L}{d\theta} \right\}^T \right]^{-1} \frac{d \log L}{d\theta}. \qquad (20.6.34)$$

where $L$ is the complete likelihood,

$$L = p(y_1) \prod_{i=2}^{t} p(y_i | y^{i-1}). \tag{20.6.35}$$

This algorithm generally converges to a zero of $d \log L / d\theta$ which, if the matrix of second derivatives is positive definite, is a local maximum of the likelihood.

The analytical calculation of the derivatives is tedious unless the model is in innovations form; in the univariate case, with

$$\hat{x}_{t+1} = F\hat{x}_t + G\nu_t \equiv F\hat{x}_t + G(y_t - H\hat{x}_{t|t-1}) \tag{20.6.36}$$

and

$$y_t = H\hat{x}_t + \nu_t, \tag{20.6.37}$$

the log likelihood for a single observation $y_i$ is given by

$$l_i = -\tfrac{1}{2}\log 2\pi\sigma^2 - \tfrac{1}{2}\nu_i^2 / \sigma^2. \tag{20.6.38}$$

Differentiation with respect to any parameter $\theta_j$ other than $\sigma^2$ gives

$$\frac{dl_i}{d\theta_j} = -\nu_i \frac{\partial \nu_i}{\sigma^2 \, \partial\theta_j} \tag{20.6.39}$$

where

$$\frac{\partial \nu_i}{\partial \theta_j} = -H \frac{\partial \hat{x}_t}{\partial \theta_j} - \frac{\partial H}{\partial \theta_j}\hat{x}_t \tag{20.6.40}$$

and

$$\frac{\partial \hat{x}_t}{\partial \theta_j} = \frac{\partial F}{\partial \theta_j}\hat{x}_{t-1} + F\frac{\partial \hat{x}_{t-1}}{\partial \theta_j} + \frac{\partial G}{\partial \theta_j}\nu_{t-1}$$
$$- G\left(H\frac{\partial \hat{x}_{t-1}}{\partial \theta_j} + \frac{\partial H}{\partial \theta_j}\hat{x}_{t-1}\right). \tag{20.6.41}$$

This last equation gives derivatives of the states in a sequential manner.

When distributional assumptions cannot be made, some function other than the likelihood may be chosen for minimization, but the principles of estimation are the same. However, the function chosen would have to possess a minimum with respect to the parameters, and such a function might be difficult to find. The properties of any estimates obtained might also be hard to establish.

The technique described above is for 'off-line' use, that is the parameter values are fixed during each run through the data. It is also possible in principle to estimate parameters recursively, so that after each observation a small adjustment is made to the parameter estimates. There are many forms of recursive algorithms, and the convergence of some has not been proved rigorously. They fall into the general category of stochastic approximation algorithms (e.g. Nevelson and Hasminskii, 1973). The general form of these is

$$\hat{\theta}_{t+1} = \hat{\theta}_t + A^{-1}f(y_t, \hat{\theta}_t) \tag{20.6.42}$$

where $f(y_t, \hat{\theta}_t)$ represents some deviation of $y_t$ from its predicted value, and $A$ is a weighting matrix such that $A^{-1}$ tends to zero. For suitable $A$, the algorithm converges to a value of $\theta$ for which the average deviation $f(y_t, \hat{\theta}_t)$ is zero. The most heuristically appealing form is similar to the Gauss–Newton algorithm already given; this modified form is for the case where $l_i$ is a Gaussian density:

$$\hat{\theta}_{t+1} = \hat{\theta}_t + \sum_{i=1}^{t} \left[ \left\{ \frac{d \log l_i}{d\theta} \right\} \left\{ \frac{d \log l_i^T}{d\theta} \right\} \right]^{-1} \frac{d \log l_t}{d\theta}. \qquad (20.6.43)$$

Notice that the matrix $A$ is the sum of the $d \log l_i$ terms, while only the most recent $d \log l_t / d\theta$ is in the term $f(y_t, \hat{\theta}_t)$.

Various other heuristic techniques have been proposed. In the state space model, if the variances $Q$ and $R$ are known then $f(y_t, \hat{\theta}_t)$, can be replaced by $dv^2/d\theta$ (which has a minimum with respect to $F$) where $v_t$ is the innovation, $(y_t - \hat{y}_{t|t-1})$. In this case

$$\hat{\theta}_{t+1} = \hat{\theta}_t + B(y_t - \hat{y}_{t|t-1}), \qquad (20.6.44)$$

where the matrix $B$ now appears as a form of 'gain'. The algorithm may be written as

$$\hat{x}_{t+1} = F\hat{x}_t + Gv_t \qquad (20.6.45)$$

$$\hat{\theta}_{t+1} = \hat{\theta}_t + Bv_t \qquad (20.6.46)$$

$$y_t = H\hat{x}_t + v_t. \qquad (20.6.47)$$

This joint state-parameter estimation scheme can be implemented using the extended Kalman filter as outlined in section 20.6.3(b) below. The matrix $B$ is derived using covariance relationships. This method cannot be used for estimating $Q$ and $R$, for which quadratic terms in $y_t$ are required. Mehra (1969), Sage and Husa (1969), Martin and Stubberud (1976), Brewer (1976), Todini (1978) and others have suggested stochastic approximation-like schemes for estimating $Q$ and $R$. Todini (1978) uses algorithms of the form

$$\hat{R}_t = \hat{R}_{t-1} + B_R[v_t v_t^T - (\hat{R}_{t-1} + HP_{t|t-1}H^T)] \qquad (20.6.48)$$

$$\hat{Q}_t = \hat{Q}_{t-1} + B_Q[Kv_t v_t^T K^T + P_{t|t} - P_{t|t-1}] \qquad (20.6.49)$$

where $B_R$ and $B_Q$ are taken to be $1/t$, and for estimable models has shown that the method works satisfactorily using Monte Carlo simulation experiments. The principles only have been given here: details are provided in Todini (1978) and Todini and O'Connell (1980). For the remaining parameters, an algorithm of the form

$$\hat{\theta}_{t+1} = \hat{\theta}_t + B_\theta(y_t - \hat{y}_{t|t-1}) \qquad (20.6.50)$$

is employed where $B_\theta$ is again given by a Kalman filter employed for parameter estimation. A more appealing algorithm for estimating all the parameters, based on derivatives of the likelihood, has been presented by Ljung (1979); a convergence proof is also supplied.

### (b) Non-linear estimation

Thus far, parameter estimation techniques have been discussed with emphasis on the simplest model, the linear discrete state space model. For non-linear models no new principles are required, but since we will be dealing with an approximation to the true model, it is with respect to this that the parameter estimates will be optimized. Generally, given discrete observations, the following objective function may be used:

$$L_0 = \left[ -\frac{m}{2} \log 2\pi |V| - \tfrac{1}{2}\{y_i - E(y_i)\}^T V^{-1}\{y_i - E(y_i)\} \right] \quad (20.6.51)$$

where $V$ is the variance covariance matrix of the $y$'s. If the $y_i$ are normally distributed then this will give maximum likelihood estimates; otherwise it is still a useful objective function. The general recursive algorithm given by (20.6.42) can also be used with non-linear models, although numerical estimates of $d \log l_i / d\theta$ may be required. Cooper (1982) gives examples of the application of this very general algorithm to some non-linear problems.

A further possible approach to parameter estimation, for both linear and non-linear models, is provided by the extended Kalman filter. By defining an augmented state vector as

$$x_t^* = [x_t^T : \theta_t^T]^T \quad (20.6.52)$$

where $\theta_t$ is a vector of parameters, the system and measurement equations (linear or non-linear) can be redefined in terms of the augmented state vector $x_t^*$; even where the system equation is originally linear (cf. equation (20.2.1)) the system equation for the augmented state vector will be non-linear because of product terms involving the elements of $x$ with the elements of $\theta$. The extended Kalman filter can be applied to solve this non-linear estimation problem as outlined in section 20.5.2.

## 20.7.   APPLICATIONS

### 20.7.1.   An Application of the Discrete Linear Kalman Filter

### (a) Introduction

For this application it is assumed that the structure and parameters of the state space model are unknown a priori, and so the techniques described in sections 20.6.2 and 20.6.3 for the identification and estimation of discrete linear state space models are applied to derive a model which can be used to provide forecasts of flow from rainfall in real-time. As noted in section 20.6.1, the response of streamflow to rainfall is non-linear, largely because a river basin responds differently to rainfall depending on its state of 'wetness'. However, a piece-wise linear approach has been shown to give satisfactory results in a number of cases; this can be applied by splitting the original rainfall series into two or more input series each of which characterizes the response of the basin to rainfall under different states of wetness (Todini and Wallis,

1977, 1978). Thus, while the response of the resulting model to rainfall will be non-linear, the estimation problem remains linear.

For the example chosen, daily rainfall and runoff data were available for the period 1 October, 1962–31 March, 1970 for the Hillsborough River near Zephyrville, Florida. Three separate rainfall inputs were identified in advance, each representing the response of the basin under different conditions as outlined above. The model sought is of the form

$$y_t = H\hat{x}_t + v_t \tag{20.7.1}$$

$$\hat{x}_{t+1} = F\hat{x}_t + Du_t + Gv_t, \tag{20.7.2}$$

where $y_t$ is the output, runoff, and $u_t$ a matrix comprising the three separate rainfall inputs. The dimensionality and structural form of $H$ and $F$ are to be identified, and the free parameters of $H$, $F$, $D$ and $G$ then estimated. This problem must be broken down into two sections, because a model relating $y_t$ and $u_t$ of the form given cannot be found directly unless $u_t$ is a white noise process. Firstly, a state-space model for the rainfall is written in innovations form as

$$\hat{x}_{t+1} = F\hat{x}_t + Ga_t \tag{20.7.3}$$

$$u_t = H\hat{x}_t + a_t, \tag{20.7.4}$$

where $a_t$ is white noise; the $a_t$ series is then used to derive predictions of flow using a further state space model

$$y_t = H'\hat{x}'_t + v_t \tag{20.7.5}$$

$$\hat{x}'_{t+1} = F'\hat{x}'_t + Da_t + G'v_t. \tag{20.7.6}$$

### (b) *Identification and parameter estimation*

THE STATE-SPACE MODEL FOR THE RAINFALL. The first step is to prewhiten the inputs. This is done by fitting a high order autoregressive model

$$u_t + B_1 u_{t-1} + B_2 u_{t-2} + \ldots + B_k u_{t-k} = \varepsilon_t. \tag{20.7.7}$$

Taking covariances of this equation with $\{u_t^T, u_{t-1}^T, \ldots, u_{t-k}^T\}^T$ gives the Yule–Walker equations

$$\begin{bmatrix} C_1 \\ \vdots \\ C_k \end{bmatrix}^T = [B_1, \ldots, B_k] \begin{bmatrix} C_0 & C_1 & \cdots & C_{k-1} \\ \vdots & & & \vdots \\ C_{k-1}^T & & \cdots & C_0 \end{bmatrix} \tag{20.7.8}$$

which can be solved for $\{B_1 \ldots B_k\}$ when each $C_i$ is replaced by its sample estimate. The $B$'s are used to find a first estimate of the impulse response matrices $\{A_1, \ldots, A_k\}$ of the representation

$$u_t = \varepsilon_t + A_1 \varepsilon_{t-1} + \ldots + A_k \varepsilon_{t-k} \tag{20.7.9}$$

using the relationships given by equation (20.6.27).

The structure of a suitable state space model for the three rainfall series is found using canonical variate analysis, with the sequence $\{u_{t-1}^{(1)}, u_{t-1}^{(2)}, u_{t-1}^{(3)}, u_{t-2}^{(1)}, \ldots, u_{t-k}^{(3)}\}$ as the vector of past observations; $k$ is required to be suitably large and was taken as 12 in this case. Following the procedure of Cooper and Wood (1982a) we find that

$$F = \begin{bmatrix} 0\cdot325 & 0\cdot003 & -0\cdot221 & 0 & 0 & 0 & 0 \\ 0 & 0 & 0 & 1 & 0 & 0 & 0 \\ 0 & 0 & 0 & 0 & 1 & 0 & 0 \\ 0 & 0 & 0 & 0 & 0 & 1 & 0 \\ 0 & 0 & 0 & 0 & 0 & 0 & 1 \\ 0\cdot154 & 0\cdot382 & -0\cdot612 & -0\cdot772 & -0\cdot041 & 0\cdot582 & 0\cdot907 \\ -0\cdot236 & -0\cdot857 & 0\cdot801 & 1\cdot645 & 0\cdot701 & 0\cdot055 & 1\cdot007 \end{bmatrix}$$

$$H = [I : 0]$$

$$G = \begin{bmatrix} 0\cdot182 & -0\cdot103 & -0\cdot023 \\ 0\cdot014 & 0\cdot268 & -0\cdot024 \\ 0\cdot004 & 0\cdot228 & 0\cdot231 \\ 0\cdot013 & 0\cdot129 & -0.017 \\ 0\cdot010 & 0\cdot359 & 0\cdot187 \\ 0\cdot006 & 0\cdot043 & -0\cdot015 \\ 0\cdot004 & 0\cdot288 & 0\cdot129 \end{bmatrix}$$

The rows of $G$ are rows $\{1\ 2\ 3\ 5\ 6\ 8\ 9\}$ of the matrix $\{A_1, A_2, A_3\}^T$. Note that there are 17 free parameters in $F$, compared with 27 for a general third order autoregressive model in three variables. The values given here are rather crude parameter estimates; they may be used as starting values for an iterative Gauss–Newton procedure provided the model is stable, that is, the eigenvalues of $F$ are less than 1 in modulus. In this case they are not, and the preliminary estimates must be perturbed to provide useful starting values. At present no rules for choosing how this is done can be given. The model with 17 parameters given by the identification procedure has been reduced further by setting to zero some of the parameters of $F$ and $G$ which appear small. Gauss–Newton estimation then gives

$$F = \begin{bmatrix} 0\cdot384 & 0 & -0\cdot246 & 0 & 0 & 0 & 0 \\ 0 & 0 & 0 & 1 & 0 & 0 & 0 \\ 0 & 0 & 0 & 0 & 1 & 0 & 0 \\ 0 & 0 & 0 & 0 & 0 & 1 & 0 \\ 0 & 0 & 0 & 0 & 0 & 0 & 1 \\ -0\cdot007 & 0\cdot127 & -0\cdot557 & -0\cdot023 & 0\cdot610 & -0\cdot194 & -0\cdot022 \\ 0\cdot005 & -0\cdot015 & 1\cdot294 & -0\cdot007 & 0\cdot160 & -0\cdot890 & -0\cdot778 \end{bmatrix}$$

$$G = \begin{bmatrix} 0\cdot1687 & -0\cdot093 & 0 \\ 0 & 0\cdot254 & 0 \\ 0 & 0\cdot311 & 0\cdot186 \\ 0 & 0\cdot124 & 0 \\ 0 & 0\cdot288 & 0\cdot171 \\ 0 & 0 & 0 \\ 0 & 0\cdot229 & 0\cdot153 \end{bmatrix}$$

The values in $G$ are similar to those given by the identification procedure, while those for $F$ are rather different. Given the estimated $F$ and $G$, and knowing the structure of $H$, the rainfall measurements are transformed to a white noise process $a_t$ using

$$\hat{x}_t = F\hat{x}_{t-1} + Ga_{t-1} \tag{20.7.10}$$

$$a_t = u_t - H\hat{x}_t. \tag{20.7.11}$$

THE STATE-SPACE MODEL RELATING STREAMFLOW TO RAIN-FALL.  A model relating runoff to rainfall can now be identified; this has the structure given by equations (20.7.5) and (20.7.6). First, a high order impulse response model

$$y_t = A'_1 a_t + \ldots + A'_{12} a_{t-11} + \eta_t \tag{20.7.12}$$

is fitted using the $a_t$'s derived from (20.7.11) above, each $A'_i$ being a $(1 \times 3)$ vector; this is done by solving the covariance equations of this expression with $(a_t^T, a_{t-1}^T \ldots a_{t-11}^T)$. Initial estimates of $A'_1$ to $A'_4$ are given in Table 20.7.1; these are required to determine the parameters of the $D$ matrix in equation (20.7.6).

| Impulse response | $a_t^{(1)}$ | $a_t^{(2)}$ | $a_t^{(3)}$ |
|---|---|---|---|
| $A_1$ | 0·000 | 0·003 | 0·009 |
| $A_2$ | 0·029 | 0·049 | 0·076 |
| $A_3$ | 0·030 | 0·072 | 0·092 |
| $A_4$ | 0·021 | 0·080 | 0·096 |

Table 20.7.1: Impulse response parameters for the input series.

An impulse response model for

$$\eta_t = C_0 \varepsilon_t + \ldots + C_{11} \varepsilon_{t-11} \tag{20.7.13}$$

is now required to estimate the $G'$ matrix in equation (20.7.6). The $\eta_t$ process can be generated from (20.7.12), given $A'_1, \ldots, A'_{12}$, since both the $y_t$ and $a_t$ series are known. The model is found in the same way as that for $u_t$, namely by fitting a high order autoregressive model and inverting it. Using this technique initial estimates of $\{C_0, \ldots, C_{11}\}$ are given by

$$[1{\cdot}00 \;\; 1{\cdot}06 \;\; 0{\cdot}77 \;\; 0{\cdot}62 \;\; 0{\cdot}52 \;\; 0{\cdot}42 \;\; 0{\cdot}30 \;\; 0{\cdot}24 \;\; 0{\cdot}23 \;\; 0{\cdot}22 \;\; 0{\cdot}20 \;\; 0{\cdot}22];$$

These are approximate values for the appropriate elements of $G'$.

The final stage of the identification procedure is to carry out canonical variate analysis using as the vector of past observations

$$\{y_{t-1}, a_{t-1}^{(1)}, a_{t-1}^{(2)}, a_{t-1}^{(3)}, \ldots, y_{t-12}, a_{t-12}^{(1)}, a_{t-12}^{(2)}, a_{t-12}^{(3)}\}.$$

For this example, the future vector consisted of, successively, $\{y_t\}$, $\{y_t, y_{t+1}\}$, $\{y_t, y_{t+1}, y_{t+2}\}$, and $\{y_t, y_{t+2}, y_{t+3}\}$; the future vector is expanded until the results of the canonical variate analysis indicate that a good fit has been obtained: details are given in Cooper and Wood (1982a).

The identified model relating streamflow to rainfall is then given by

$$F' = \begin{bmatrix} 0 & 1 & 0 & 0 \\ 0 & 0 & 1 & 0 \\ 0 & 0 & 0 & 1 \\ 0{\cdot}004 & 0{\cdot}048 & -0{\cdot}720 & 1{\cdot}636 \end{bmatrix}$$

$$D = \begin{bmatrix} 0{\cdot}000 & 0{\cdot}003 & 0{\cdot}009 \\ 0{\cdot}029 & 0{\cdot}049 & 0{\cdot}076 \\ 0{\cdot}030 & 0{\cdot}072 & 0{\cdot}092 \\ 0{\cdot}022 & 0{\cdot}080 & 0{\cdot}076 \end{bmatrix}$$

$$G' = \begin{bmatrix} 1{\cdot}06 \\ 0{\cdot}77 \\ 0{\cdot}62 \\ 0{\cdot}52 \end{bmatrix} \qquad H' = [1 \;\; 0 \;\; 0 \;\; 0].$$

Parameter estimation, omitting those parameters which appeared to be near zero, and using the above as initial estimates, gave the final model

$$F' = \begin{bmatrix} 0 & 1 & 0 & 0 \\ 0 & 0 & 1 & 0 \\ 0 & 0 & 0 & 1 \\ 0 & 0{\cdot}067 & -0{\cdot}646 & 1{\cdot}536 \end{bmatrix}$$

$$D = \begin{bmatrix} 0{\cdot}0 & 0{\cdot}0 & 0{\cdot}0 \\ 0{\cdot}027 & 0{\cdot}045 & 0{\cdot}065 \\ 0{\cdot}029 & 0{\cdot}070 & 0{\cdot}085 \\ 0{\cdot}021 & 0{\cdot}071 & 0{\cdot}071 \end{bmatrix}$$

$$G' = \begin{bmatrix} 1{\cdot}093 \\ 0{\cdot}770 \\ 0{\cdot}615 \\ 0{\cdot}482 \end{bmatrix}$$

The models for rainfall (equations (20.7.3) and (20.7.4)) and for streamflow given rainfall (equations (20.7.5) and (20.7.6)) may be combined to give a single joint state space model for rainfall and streamflow. It is easy to verify that this takes the form

$$\begin{bmatrix} x'_{t+1} \\ x_{t+1} \end{bmatrix} = \begin{bmatrix} F' & 0 \\ 0 & F \end{bmatrix} \begin{bmatrix} x'_t \\ x_t \end{bmatrix} + \begin{bmatrix} G' & D \\ 0 & G \end{bmatrix} \begin{bmatrix} \nu_t \\ a_t \end{bmatrix}$$

$$\begin{bmatrix} y_t \\ u_t \end{bmatrix} = \begin{bmatrix} H' & 0 \\ 0 & H \end{bmatrix} \begin{bmatrix} x'_t \\ x_t \end{bmatrix} + \begin{bmatrix} \nu_t \\ a_t \end{bmatrix}.$$

The structure of this joint model is such that future rainfall is independent of past streamflows, while the Kalman filter operates on both rainfall and streamflow.

### (c) *Results*

The model described above was applied to the Hillsborough river data (see § 20.7.1(a)) and its performance assessed by comparing the observed behaviour of the innovations with that expected under optimal conditions. Figure 20.7.1(a) shows daily rainfall over the catchment for a sub-period, 21 July to 29 September, 1964 while Figure 20.7.1(b) shows observed and predicted (one step ahead) streamflow over the same period. The sample autocorrelations of the innovations over the complete period 1 October, 1962 to 31 March,

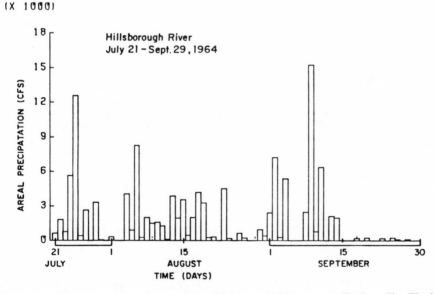

Figure 20.7.1(a): Daily rainfall for the Hillsborough River, near Zephyrville, Florida for the period July 21–September 29, 1964 (after Cooper and Wood, 1982b).

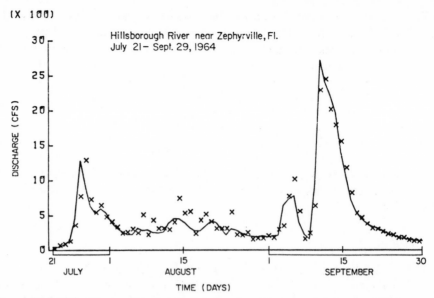

Figure 20.7.1(b): Observed (—) and one day ahead forecasts(×) for the Hillsborough River near Zephyrville, Florida, July 21–September 29, 1964 (after Cooper and Wood, 1982b).

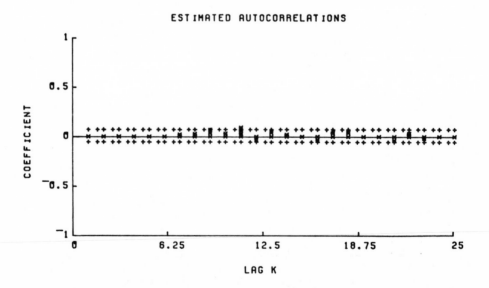

Figure 20.7.2: Autocorrelation function of one step ahead forecast errors over the period October 1, 1962–March 31, 1970 for the Hillsborough River near Zephyrville, Florida (after Cooper and Wood, 1982b).

1970 are shown in Figure 20.7.2, together with limits of twice their standard error. Clearly, judged by the autocorrelations of the innovations the model is adequate. This is to be expected since the model is designed to account for second order moments over the sample period. However, a more detailed examination of the innovations (for example by studying Figure 20.7.1(b)), suggests that the model may not provide a good description over the periods of heavy rainfall which are of most interest. Notably, there seems to be underprediction on the rising limb, that is, as the flow rate increases, and overprediction just after the flow has peaked. These features are masked when the autocorrelations are computed. They arise because the type of model chosen is, as it stands, inherently incapable of modelling some of the characteristics of the relationship between describing this streamflow and rainfall; in particular the assumptions made about the noise characteristics are questionable. This point is discussed further in section 20.7.2(d).

### 20.7.2. An Application of the Extended Kalman Filter

#### (a) *The hydrological model*

The second case study considered here also relates to the real-time forecasting of streamflow from rainfall on the Hirnant sub-catchment (area $33 \cdot 9 \text{ km}^2$) of the River Dee in North Wales, UK (Moore and Weiss, 1980b); here, however, in contrast to the case study described in section 20.7.1, the forecasting model is formulated as a first order non-linear differential equation. Thus, the model structure is prescribed a priori on hydrological grounds, and so the problem is that of using the extended Kalman filter for state estimation and prediction.

The differential equation model is written as

$$\frac{dq}{dt} = a(cp^* - q)q^b, \qquad (20.7.14)$$

where $q$ is flow, and $p^*$ is a rainfall input defined as

$$p_t^* = \sum_{j=0}^{s-1} \omega_j p_{t-j-\tau}, \qquad (20.7.15)$$

with

$$\sum_{j=0}^{s-1} \omega_j = 1,$$

where $\tau$ is a pure time delay.

This lagging and smoothing of the rainfall input is a traditional technique employed in hydrological modelling; the values of the moving average parameters are determined separately from the estimation parameters $a$, $b$ and $c$ of the differential equation model. The parameter $c$ is a coefficient which accounts for 'losses' on converting rainfall into streamflow, due, for example, to the replenishment of soil moisture in the catchment, while the parameters $a$ and $b$ essentially govern the nature of the dynamic response of streamflow to the rainfall input. Further details of the formulation of the model are given in Moore and Weiss (1980a, b).

### (b) *State-space formulation*

It is now required to take account of the fact that the model described above cannot be a perfect representation of the real-world situation but is only an approximation to it. Let $t$ denote continuous time and let $k$ be an index for discrete time such that $t_k$ denotes the value of $t$ which coincides with the $k$th discrete time point. Let the flow at time $t_k$ be known and denoted by $q_k$, and let $h_k \equiv h_k(t)$ be a function satisfying

$$\frac{dh_k}{dt} = a(cp^* - h_k)h_k^b, \qquad t_k \le t \le t_{k+1} \tag{20.7.16}$$

with $h_k(t_k) = q_k$. The true but not-yet-known flow at time $t_{k+1}$, namely $q_{k+1}$, will deviate from the extrapolated estimate $h_k(t_{k+1})$ by a quantity $v_{k+1}$ (Figure 20.7.3), so that

$$q_{k+1} = h_k(t_{k+1}) + v_{k+1}. \tag{20.7.17}$$

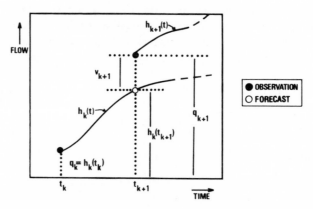

Figure 20.7.3: Change of forecast function at an observation for non-linear hydrological model.

It will be assumed, for $t_k$ equally spaced, that $v_{k+1}$ are independent and identically distributed Gaussian random variables with mean zero and variance $R$. Thus, $v_{k+1}$ is considered to be a random term accounting for imperfections in the model dynamics, and noise entering the model. When the parameters are known this model can be used to give a forecast of the flow at time $t_{k+1}$ as

$$\hat{q}_{k+1} = h_k(t_{k+1}). \tag{20.7.18}$$

The practical problem posed is one of parameter estimation, which once overcome will permit predictions of future flows via solution of the differential equation for $h_k(t_{k+1})$.

The state of the system at any time $t_k$ is described by the three parameters $a$, $b$, and $c$, which will be denoted by the state vector $x_k$ defined as

$$x_k = (a, b, c)^T. \qquad (20.7.19)$$

This state vector is assumed a priori to be normally distributed with expected value $x_0$, and variance $P_0$ at the initial time $t_0$. Since the parameters are assumed to be time-invariant, the system equation is written as

$$x_{k+1} = x_k. \qquad (20.7.20)$$

In order to conform with previous notation the measured flow at time $t_k$ is denoted by $y_k$, so

$$y_k = q_k \qquad (20.7.21)$$

and then (20.7.17) yields the measurement equation

$$y_{k+1} = h_k(t_{k+1}) + v_{k+1}. \qquad (20.7.22)$$

In order to stress the dependence of $h_k$ on the value of the parameters $x$, and on the initial condition $h_k(t_k) = y_k = q_k$, $h_k$ is replaced by $h(t, x, y_k)$. The observation equation is then

$$y_{k+1} = h(t_{k+1}, x_{k+1}, y_k) + v_{k+1} \qquad (20.7.23)$$

with $x_0$ assumed independent of the sequence $\{v_k\}$.

Equations (20.7.20) and (20.7.23) constitute the state-space formulation of the model. This approach of defining the state vector in terms of the parameters to be estimated and the quantity to be predicted has been used by Mayne (1964), Graupe (1972), and Szóllosi–Nagy (1975), but only for linear models.

As part of the state-space formulation it is assumed that river flow is measured without error. This assumption has been fundamental in developing the state-space equations as outlined above. One physical justification for this is that observations of streamflow, being spatially integrated measurements of water volume, as opposed to point sampled measurements of rainfall volume, will be considerably more noise free. Rainfall measurement errors will add to those attributed to the inadequacy of the dynamic model itself, and will be compensated for by the stochastic disturbance term $v_k$. Previous experience with obtaining filtered estimates of river flow in addition to parameter estimates via an augmented state vector (see § 20.6.3(b)) proved unrewarding, and added weight to the belief that measured flow values would be best considered error free. (Moore and Weiss, 1980b).

(c) *Parameter estimation*

The parameters defining the temporally distributed input $\omega_j$ ($j = 1, 2, \ldots, s$) and $\tau$, the pure time delay are obtained by a prewhitening procedure similar to that described by Box and Jenkins (1976). The estimation of the parameters *a*, *b* and *c* constitutes the main problem and is approached by applying the extended Kalman filter to the above state-space formulation of the model.

In the present context where non-linearities occur only in the observation equation, the extended Kalman filter derives simply from the Kalman filter algorithm for linear systems after linearising the measurement equation about the current estimate of *x* as a first order Taylor series as outlined in section 20.5.2. For this application the measurement matrix $H_k$ (here a row vector of dimension three) is that of the linearized measurement equation and contains the first partial derivatives of *h* with respect to the three parameters. The *i*th element of $H_k$ is thus defined as

$$(H_k)_i = \frac{\partial h(t_k, \hat{x}_{k|k-1}, y_{k-1})}{\partial x_i}; \qquad (20.7.24)$$

the procedures employed in calculating the derivatives $\partial h/\partial x_i$ are described in Moore and Weiss (1980a,b). An exponentially age-weighted filter has been employed to avoid filter divergence (§ 20.4.2); the resulting algorithm is summarized in Table 20.7.2. Note that by introducing a fading memory filter, the assumption of time invariance made in the system equation (20.7.20) has been relaxed. While also guarding against filter divergence, this formulation can adapt to system non-stationarity, due for example to changing streamflow response with season, or with catchment 'wetness'. Inspection of the variation of the parameters over time can suggest further alterations to the model structure which could allow the cause of the variation to be modelled and the event to be anticipated in advance.

| | |
|---|---|
| System equation: | $x_{k+1} = x_k$ |
| Measurement equation: | $y_k = h(t_k, x_k, y_{k-1}) + v_k$ |
| *At an observation*: | |
| State prediction | $\hat{x}_{k|k} = \hat{x}_{k|k-1} + K_k(y_k - \hat{y}_{k|k-1})$ |
| Error covariance matrix | $P_{k|k} = e^{-(t_{k+1}-t_k)/T}(I - K_k H_k)P_{k|k-1}$ |
| Kalman gain | $K_k = P_{k|k-1}H_k^T(H_k P_{k|k-1}H_k^T + R)$ |
| *At a forecast*: | |
| State prediction | $\hat{x}_{k+1|k} = \hat{x}_{k|k}$ |
| Output prediction | $\hat{y}_{k+1|k} = h(t_{k+1}, \hat{x}_{k+1|k}, y_k)$ |
| Error covariance matrix | $P_{k+1|k} = P_{k|k}$ |

Table 20.7.2: Extended Kalman filter algorithm for non-linear hydrological model.

(d) *Results*

Half-hourly rainfall and flow data for the period November 1–December 30, 1972 were used for the state estimation and prediction problem described above; plots of the data are shown in Figure 20.7.4(a),(b). The rainfall input

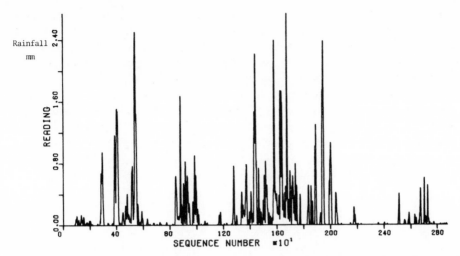

Figure 20.7.4(a). Half hourly rainfall totals in mm for the Hirnant sub-catchment of
the River Dee, North Wales for the period 1 November–30 December, 1972.

$p_t^*$ in (20.7.15) was defined as $p_t^* = p_{t-1}$ i.e. a simple pure time delay of one
half hour was found to be adequate.

Initial values for the parameters forming the state vector, $x$, were chosen
fairly arbitrarily as $a = 0.2$, $b = 0.5$ and $c = 0.66$. The degree of confidence
attached to these initial estimates, expressed by the state error covariance
matrix, $P_{0|0}$, was chosen to have diagonal elements equal to $0.05$, and zero
off-diagonal elements. The variance of the measurement noise, $R$, was fixed

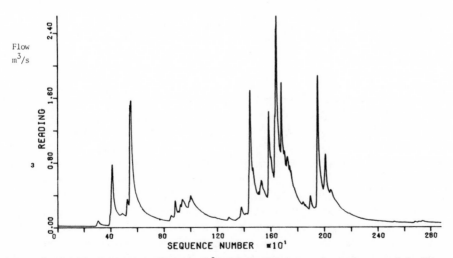

Figure 20.7.4(b). Half hourly flows in m³/s for the Hirnant sub-catchment of the River
Dee in North Wales for the period 1 November–30 December 1972.

equal to 0·01. Note that it is the relative values of $P$ and $R$ which affect how the filter behaves (cf. equation (20.3.18)) rather than their absolute values which are of less practical interest. The time constant of the exponentially age-weighted filter was selected such that the filter degraded the importance of observations by half after three days.

The autocorrelation function of the one-step (half hour) ahead forecast errors (or innovations) is shown in Figure 20.7.5. The innovations should be uncorrelated if the filter is optimal; however, Figure 20.7.5 shows that there are some small but significant correlations at low lags, and these are reflected

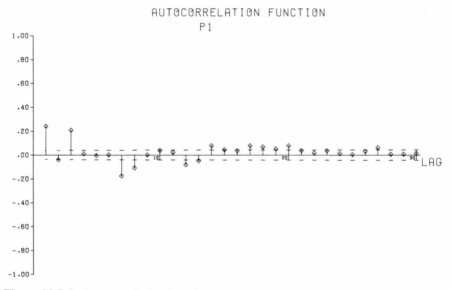

Figure 20.7.5: Autocorrelation function of one step ahead forecast errors over period 1 November–30 December, 1972 for the Hirnant sub-catchment of the River Dee, North Wales.

in a statistically significant value of the $Q$ statistic (equation (20.4.8)) of 288 ($\chi^2_{.05} = 40·11$ for 27 degrees of freedom). The $S$ statistic (equation (20.4.9)) also indicated some residual cross-correlation between the innovations and the rainfall input with a value of 191. These results indicate some sub-optimality of the filter which may result from using a linear approximation to a non-linear system, and/or from noise in the true system which does not comply with the assumptions of the state-space model. Plots of the one hour (two-step) ahead forecasts and forecast errors over a three day period illustrate the nature of the problem (Figures 20.7.6 and 20.7.7); on the rising limbs of flood events, the flows are underestimated, and subsequently overestimated at the peak and initially on the falling limbs. Inspection of Figure 20.7.1(b) shows that the same problem occurs in the case of the linear Kalman filter, even though the

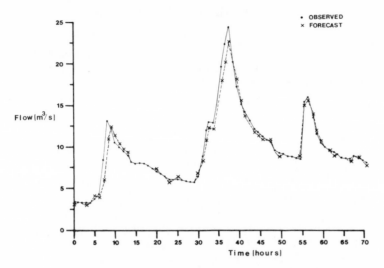

Figure 20.7.6: One hour ahead forecasts for the Hirnant sub-catchment of the River Dee over the period 1 November–30 December 1972.

autocorrelation function (Figure 20.7.2) suggests the innovations to be white. These results may be attributed to the fact that the noise entering hydrological systems is not white; during flood events, the noise tends to be concentrated in 'lumps' adjacent to flood events, and to be more akin to shot noise (e.g. Parzen, 1962) than white noise. Nonetheless, from an operational standpoint, the extended Kalman filter still performs well as an updating and forecasting tool, and so the degree of sub-optimality may not in practice prove a deterrent to operational use. Thus, formal statistical tests should not be regarded as a touch-stone for judging model and filter performance; they may, in fact, be blind to some deficiencies. The user must satisfy himself that the model and filter are providing results which are acceptable or not acceptable from the standpoint of operational use.

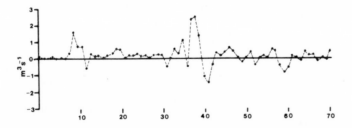

Figure 20.7.7: One hour ahead forecast errors for the Hirnant sub-catchment of the River Dee over the period 1 November–30 December, 1972.

## 20.8. CONCLUDING REMARKS

It is hoped that the foregoing treatise on the Kalman filter, which concentrates on concepts and ideas rather than on rigorous mathematical proofs, will have imparted to the reader and prospective user a basic understanding of this elegant algorithm, and yet at the same time appraised him of the practical difficulties which may be encountered in applying it. We have described the filter in some detail for the straightforward case of a discrete linear dynamic system with known parameters so that the reader will, it is hoped, acquire a good understanding of the basic technique; extensions to handle continuous time and non-linear systems and model identification and parameter estimation have been described more in outline, and the reader may wish to consult the references cited for more details.

If the system under study can be described by a first order linear vector differential or difference equation, and the parameters in such a description, including the variances of the system and measurement noises, are known, then the application of the filter is straightforward, and its performance should be optimal, provided that the model structure and parameters are correctly known. In the case of non-linear systems, the extended Kalman filter can be employed, although it must be emphasized that the algorithm is applied to an approximation to the true system and so optimality is not guaranteed.

Where nothing is known *a priori* about the model structure and its parameters, a method of choosing a possible model is given. This is a linear discrete state-space model in innovations form whose theoretical first and second order moments closely approximate the appropriate sample statistics. The structure of the model is such that predictions of the process behaviour can easily be obtained using the Kalman filter. Where the approximation is good, and these moments adequately describe the statistical features of the process which are of interest, this may be a very useful procedure. If important aspects of the data are not highlighted in the sample covariances, as in some hydrological examples, these models may be poorer descriptors. This aspect is brought out by the applications where a visual inspection of the forecast errors reveals that the original assumptions about the properties of the system and measurement noises are not justified. However, this is not an infrequent result in modelling and predicting the behaviour of real-world systems and the user must judge whether the degree of approximation afforded by the model is adequate for the particular problem at hand.

P.E.O.

## 20.9. FURTHER READING AND REFERENCES

Anderson, T. W., (1958). *An Introduction to Multivariate Statistical Analysis*, John Wiley, New York.

Akaike, H. (1974). Markovian Representation of Stochastic Processes and its Application to the Analysis of Autoregressive Moving Average Processes, *Ann. Inst. Stat. Math.* **26**, 363–387.

Athans, M. (1974). The Importance of Kalman Filtering Methods for Economic Systems, *Annals Econ. Social Measurement* **3**, 49–64.

Beven, K. J. and O'Connell, P. E. (1982). On the Role of Physically-based Distributed Modelling in Hydrology, *Institute of Hydrology Report No. 81*, Wallingford.

Bohlin, T. (1976). Four Cases of Identification of Changing Systems for Discrete Time Series. In: *System Identification, Advances and Case Studies*. (Eds. R. K. Mehra and K. G. Lainiotis), Academic Press, New York.

Box, G. E. P. and Jenkins, G. M. (1976). *Time Series Analysis, Forecasting and Control*, Holden-Day, San Francisco.

Bucy, R. S. and Joseph, P. D. (1968). *Filtering for Stochastic Processes with Applications to Guidance*, Interscience, New York.

Chiu, C. L. (ed.) (1978). Applications of Kalman Filter to Hydrology, Hydraulics and Water Resources, *Proc. AGU Chapman Conference, Pittsburgh*, University of Pittsburgh.

Cooper, D. M. (1982). Adaptive Parameter Estimation for Non-linear Hydrological Models with General Loss Functions, *J. Hydrol.* **58**, 29–45.

Cooper, D. M. and Wood, E. F. (1982a). Identification of Multivariate Time Series and Multivariate Input–Output Models, *Water Resources Research* **18**(4), 937–946.

Cooper, D. M. and Wood, E. F. (1982b). Parameter Estimation of Multiple Input–Output Time Series Models: Application to Rainfall-runoff Processes, *Water Resources Research* **18**(5), 1352–1364.

Gelb, A. (1974). *Applied Optimal Estimation*, MIT Press, Cambridge, Massachusetts.

Graupe, D. (1972). *Identification of Systems*, Van Nostrand–Reinhold, New York.

Harrison, P. J. and Stevens, C. F. (1976). Bayesian Forecasting (with discussion), *JRSS* B **38**(3), 205–247.

Jazwinski, A. H. (1970). *Stochastic Processes and Filtering Theory*, Academic Press, New York.

Kailath, T. (1974). A View of Three Decades of Linear Filtering Theory, *IEEE* IT–20, 146–180.

Kalman, R. E. (1960). A New Approach to Linear Filtering and Prediction Problems, *Trans. ASME, J. Basic Eng.* **82**, 35–45.

Kalman, R. E. and Bucy, R. S. (1961). New Results in Linear Filtering and Prediction Theory, *Trans. ASME, J. Basic Eng.* **83**, 95–107.

Kolmogorov, A. N. (1941). Interpolation and Extrapolation of Stationary Random Sequences, *Bull. Moscow University, Ser. Math.* **5**, Moscow.

Kwakernaak, H. and Sivan, R. (1972). *Linear Optimal Control Systems*, Wiley, New York.

Ljung, L. (1979). Asymptotic Behaviour of the Extended Kalman Filter as a Parameter Estimator for Linear Systems, *IEEE* AC–24(1), 36–50.

Mayne, D. Q. (1965). Optimal Non-stationary Estimation of the Parameters of a Linear System with Gaussian Inputs, *Jour. Electronic Control* **14**, 101–112.

Mehra, R. K. (1973). Identification and Control in Econometric Systems, Similarities and Differences, *Second Workshop on Economic and Control Systems*, Chicago, June (also *Annals of Economic and Social Measurement*, January 1974).

Mehra, R. K. and Wells, C. H. (1971). Dynamic Modelling and Estimation of Carbon in a Basic Oxygen Furnace, *Third International IFAC/IFIP Conference*, Helsinki, 2–5 June.

Moore, R. J. and Weiss, G. (1980a). Real-time Parameter Estimation of a Non-linear Catchment Model Using Extended Kalman Filters. In: Wood, E. F. and Szollosi-Nagy, A. (eds.), op. cit.

Nevelson, M. B. and Hasminskii, R. Z. (1973). Stochastic Approximation and Recursive Estimation, *Am. Math. Soc.*, Providence, R.I.

O'Connell, P. E. (ed.) (1980). *Real-time Hydrological Forecasting and Control*, Institute of Hydrology, Wallingford, England.

Parzen, E. (1962). *Stochastic Processes*, Holden-Day, San Francisco.

Schmidt, S. F. (1966). Application of State-space Methods to Navigation Problems, *Advan. Control Systems* **3**, 293–340.

Snyder, D. L. (1969). *The State Variable Approach to Continuous Estimation with Applications to Analog Communication Theory*, MIT Press, Cambridge, Massachusetts.

Stevens, C. F. (1974). On the Variability of Demand for Families of Items, *Oper. Res. Quart.* **25**, 156–178.

Szollosi-Nagy, A. (1975). An Adaptive Identification and Prediction Algorithm for the Real-time Forecasting of Hydrologic Time Series, *Research Memorandum* RM-75-22, IIASA, Laxenburg.

Takeuchi, K. (1976a). Applications of the Kalman Filter to Cyclone Forecasting 1. Methodology, 2. Typhoon Forecasting, *Research Memorandum* RM-76-9, International Institute for Applied Systems Analysis, Laxenburg.

Takeuchi, K. (1977b). Application of the Kalman Filter to Cyclone Forecasting 3. Hurricane Forecasting, 4. Additional Typhoon Forecasting, *Research Memorandum* RN-76-62, International Institute for Applied Systems Analysis, Laxenburg.

Todini, E. (1978). Mutually Interactive State-parameter (MISP) Estimation. In: Chiu, C. L. (ed.), op. cit.

Todini, E., O'Connell, P. E. and Jones, D. A. (1980). Basic methodology: Kalman filter estimation problems. In: O'Connell, P. E. (ed.), op. cit.

Todini, E. and Wallis, J. R. (1977). Using CLS for Daily or Longer Period Rainfall–Runoff Modelling. In: *Mathematical Models in Surface Water Hydrology* (eds. T. A. Ciriani, U. Maione and J. R. Wallis), London, J. Wiley and Sons, 149–168.

Todini, E. and Wallis, J. R. (1978). A Real-time Rainfall–Runoff Model for an On-line Flood Warning System. In: Chiu, C. L. (ed.), op. cit.

Weiss, G. (1980). Basic Methodology: the Kalman Filter. In: O'Connell, P. E. (ed.), op. cit.

Wiener, N. (1949). *The Extrapolation, Interpolation and Smoothing Stationary Time Series*, Wiley, New York.

Willems, J. C. (1978). Recursive Filtering, *Statistica Neerlandica* **32**(1), 1–39.

Wood, E. F. and Szollosi-Nagy, A. (1980). Real-time Forecasting/Control of Water Resource Systems, *IIASA Proceedings*, Vol. 8, Pergamon, Oxford.

# Bibliography

## BIBLIOGRAPHY

### Catalogues and Guides to the Literature

Anderson, D. W., Das Gupta, S. and Styan, G. D. H. (1972). *A Bibliography of Multivariate Statistical Analysis*, Oliver & Boyd.

Lancaster, H. O. (1968). *Bibliography of Statistical Bibliographies*, Oliver & Boyd.

Savage, I. R. (1962). *A Bibliography of Nonparametric Statistics*, Harvard University Press.

Subrahmaniam, K. and Subrahmaniam, K. (1973). *Multivariate Analysis: Selected and Abstracted Bibliography, 1957–1972,* Dekker.

Buckland, W. R. and Fox, R. A. (1963). *Bibliography of Basic Texts and Monographs on Statistical Methods 1945–1960,* Second edition, Oliver & Boyd.

## BIBLIOGRAPHY B

### Dictionaries, Encyclopedias and Handbooks

Burington, R. S. and May, D. C. (1970). *Handbook of Probability and Statistics with Tables,* Second edition, McGraw-Hill.

Burrington, G. A. (1972). *How to Find Out About Statistics,* Pergamon Press.

Freund, J. and Williams, F. (1966). *Dictionary of Statistical Terms,* McGraw-Hill.

Kendall, M. G. and Buckland, W. R. (1971). *A Dictionary of Statistical Terms,* Third edition, Oliver & Boyd.

Kendall, M. G. and Buckland, W. R. (1960). *A Dictionary of Statistical Terms,* Supplement. (A combined glossary in English, French, German, Italian and Spanish), Oliver & Boyd.

Kotz, S. (1964). *Russian–English Dictionary of Statistical Terms and Expressions, and Russian Reader in Statistics,* University of N. Carolina Press.

Kotz, S. and Johnson, N. L. (editors) (1982). *Encyclopedia of Statistical Sciences,* 8 volumes, Wiley.

Kruskal, W. H. and Tanner, J. M. (editors) (1968). *International Encyclopedia of Statistics,* 2 volumes, The Free Press (New York) and Collier Macmillan (London).

Walsh, J. E. (1962, 1965). *Handbook of Nonparametric Statistics,* 2 vols., Van Nostrand.

# BIBLIOGRAPHY C

## General Texts Covering a Wide Field (not separately listed in Chapter References)

Barnett, V. (1982). *Comparative Statistical Inference* Second edition, Wiley.

Breiman, L. (1973). *Statistics With a View Toward Application* Houghton–Mifflin.

Cox, D. R. and Hinkley, D. V. (1974). *Theoretical Statistics*, Chapman & Hall.

Cramér, H. (1946). *Mathematical Methods of Statistics*, Princeton University Press.

Davies, D. L. (editor) (1957). *Statistical Methods in Research and Production*, Oliver & Boyd.

de Finetti, B. (1974, 1975). *Theory of Probability*, Vols I, II, Wiley.

Fisher, R. A. (1970). *Statistical Methods for Research Workers*, Fourteenth edition, Macmillan.

Fisher, R. A. (1959). *Statistical Methods and Scientific Inferences* Second edition, Oliver & Boyd.

Graybill, F. A. (1976). *Theory and Application of the Linear Model*, Duxbury Press, Massachusetts.

Hald, A. (1957). *Statistical Theory with Engineering Applications*, Wiley.

Hogg, R. V. and Craig, A. T. (1965). *Introduction to Mathematical Statistics*, Collier-Macmillan.

Kalbfleisch, J. G. (1979). *Probability and Statistical Inference*, I and II (2 vols), Springer.

Kendall, M. G. and Stuart, A. (1969, 1973, 1976). *The Advanced Theory of Statistics.* Third edition, three vols. Vol. 1 (1969). Distribution Theory; Vol. 2 (1973). Inference and Relationship; Vol. 3 (1976). Design and Analysis, and Time Series, Griffin.

Krumbein, W. C. and Graybill, F. A. (1965). *An Introduction to Statistical Models in Geology*, McGraw-Hill.

Lindley, D. V. (1965). *Introduction to Probability and Statistics from a Bayesian Viewpoint*, (2 vols), Cambridge University Press.

Mood, A. M., Graybill, F. A. and Boes, D. C. (1974). *Introduction to the Theory of Statistics*, Third edition, McGraw-Hill.

Olkin, I., Gleser, L. J. and Derman, C. (1980). *Probability Models and Applications*, Macmillan.

Rao, C. R. (1965). *Linear Statistical Inference and Its Applications*, Wiley.

Silvey, S. D. (1975). *Statistical Inferences*, Chapman & Hall.

Wetherill, G. B. (1981). *Intermediate Statistical Methods*, Chapman & Hall.

Wilks, S. S. (1961). *Mathematical Statistics*, Wiley.

Zacks, S. (1971). *The Theory of Statistical Inference*, Wiley.

# BIBLIOGRAPHY D

## Historical Material

Box, Joan F. (1978). *R. A. Fisher; the Life of a Scientist*, Wiley.

Fisher, R. A. (1950). *Contributions to Mathematical Statistics*, Wiley.

Pearson, E. S. (editor) (1978). *The History of Statistics in the Seventeenth and Eighteenth Centuries. Lectures by Karl Pearson 1921–1933*, Griffin.

Pearson, E. S. and Kendall, M. G. (editors) (1970). *Studies in the History of Statistics and Probability*, Griffin.

Westergaard, H. (1932). *Contributions to the History of Statistics*, P. S. King, London.

## BIBLIOGRAPHY E

### Guide to Statistical Tables

Greenwood, A. and Hartley, H. O. (1962). *Guide to Tables in Mathematical Statistics*, Princeton University Press.

## BIBLIOGRAPHY F

### Tables of Random Deviates

Barnett, V. D. (1964). *Random Negative Exponential Deviates: Tracts for Computers*, No. XXVII, Cambridge University Press.

Clark, C. E. and Holz, B. (1960). *Experimentally Distributed Random Numbers*, Johns Hopkins Press.

Fieller, E. C., Lewis, T. and Pearson, E. S. (1957). *Correlated Random Normal Deviates: Tracts for Computers*, No. XXVI, Cambridge University Press.

Newman, T. G. and Odell, P. C. (1971). *The Generation of Random Variates*, Griffin.

The RAND Corporation (1953). *A Million Random Digits with 100,000 Normal Deviates*, The Free Press, New York, and Collier-Macmillan, London.

Wold, H. (1954). *Random Normal Deviates: Tracts for Computers*, No. XXV. Cambridge University Press.

## BIBLIOGRAPHY G

### Tables of Statistical Functions

Abramowitz, M. and Stegun, I. R. (editors) (1970). *Handbook of Mathematical Functions with Formulas, Graphs and Mathematical Tables.* (§ 26. Probability Functions), National Bureau of Standards, Washington, D.C.

Beyer, W. H. (editor) (1966). *Handbook of Tables for Probability and Statistics*, Chemical Rubber Co., Cleveland.

Burington, R. S. and May, D. C. (1970). *Handbook of Probability and Statistics with Tables.* Second edition, McGraw-Hill.

Fisher, R. A. and Yates, F. (1974). *Statistical Tables for Biological Agricultural and Medical Research.* Sixth edition, Longman.

General Electric Company (1962), *Tables of Individual and Cumulative Terms of the Poisson Distribution*, Van Nostrand.

Hald, A. *Statistical Tables and Formulas*, Wiley.

Harter, H. C. and Owen D. B. (editors) (1970, 1974, 1975). *Selected Tables in Mathematical Statistics*, 3 Vols. Vol. 1 (second printing with revisions) 1970; Vol. 2, 1974; Vol. 3, 1975.

Harvard Computer Laboratory, Staff of (1953). *Tables of the Cumulative Binomial Probability Distribution*, Harvard University Press.

Isaacs, G. L., Christ, D. E., Norvick M. R. and Jackson, P. H. (1974). *Tables for Bayesian Statisticians*, University of Iowa.

Liebermann, G. J. and Owen, D. B. (1961). *Tables of the Hypergeometric Distribution*, Stanford University Press.

Lindley, D. V. and Miller, J. C. P. (1966). *Cambridge Elementary Statistical Tables*, Cambridge University Press.

Murdoch, J. and Barnes, J. A. (1968). *Statistical Tables for Science, Engineering and Management*, Macmillan.

Neave, H. R. (1978). *Statistical Tables*, George Allen & Unwin.

Owen, D. B. (1962). *Handbook of Statistical Tables*, Pergamon Press, and Addison-Wesley.

Pearson, E. S. and Hartley, H. O. (editors) (1966). *Biometrika Tables for Statisticians*, Vol. 1, Third edition, Cambridge University Press.

Resnikoff, G. J. and Liebermann, G. J. (1957). *Tables of the Non-Central t-Distribution*, Stanford University Press.

Romig, H. G. (1947). *50–100 Binomial Tables*, Wiley.

Thompson, Catherine M. (1941). Tables of the Percentage Points of the Incomplete Beta Function. *Biometrika*, **37**, 168.

Williamson, E. and Bretherton, M. H. (1963). *Tables of the Negative Binomial Probability Distribution*, Wiley.

## BIBLIOGRAPHY H

### Special Topics

Arkin, H. (1963). *Handbook of Sampling for Auditing and Accounting*, 2 vols., McGraw-Hill.

Arthanari, T. S. and Dodge, Y. (1981). *Mathematical Programming in Statistics*, Wiley.

Barnett, V. (1974). *Elements of Sampling Theory*, English University Press.

Barnett, V. and Lewis, T. (1978). *Outliers in Statistical Data*, Wiley.

Cochran, W. G. (1963). *Sampling Techniques*, Second edition, Wiley.

Denning, W. E. (1950). *Some Theory of Sampling*, Wiley.

Gumbel, E. J. (1958). *Statistics of Extremes*, Columbia University Press.

Hald, A. (1981). *Statistical Theory of Sampling Inspection by Attributes*, Academic Press.

Hansen, M. H., Hurwitz, W. A. and Madow, W. G. (1953). *Sample Survey Methods and Theory*, 2 vols., Wiley.

Mardia, K. V. (1972). *Statistics of Directional Data*, Academic Press.

Savage, L. J. (1954). *The Foundations of Statistics*, Wiley.

Stuart, A. (1976). *Basic Ideas of Scientific Sampling*, Second edition, Griffin.

Tryon, R. C. and Bailey, D. E. (1970). *Cluster Analyses*, McGraw-Hill.

Wetherill, G. B. (1969). *Sampling Inspection and Quality Control*, Methuen.

Yates, F. (1968). *Sampling Methods for Censuses and Surveys*, Third edition, Griffin.

# Appendix — Statistical Tables

## APPENDIX T1

**Cumulative Binomial Tables Probabilities** [see II: § 5.2.2]

The tabulated quantity is

$$P\{R(n, \theta) \geq r\} = \sum_{s=r}^{n} \binom{n}{s} \theta^s (1-\theta)^{n-s}, \qquad r = 0, 1, \ldots, n,$$

where $R(n, \theta)$ denotes a random variable which has the Binomial $(n, \theta)$ distribution. Thus $P\{R(n, \theta) \geq r\}$ is the probability of $r$ *or more* successes in $n$ independent trials when the probability of a success in a particular trial is $\theta$. The tables give values of $P$ for $\theta = 0.01(0.01)0.10(0.05)0.50$. For values of $\theta$ exceeding $0.50$, use the relation

$$P\{R(n, \theta) \geq r\} = P\{R(n, 1-\theta) \leq n-r\}$$

$$= 1 - P\{R(n, 1-\theta) \geq n-r+1\}, \qquad r = 1, 2, \ldots, n.$$

For example,

$$P\{R(20, 0.7) \geq 12\} = 1 - P\{R(20, 0.3) \geq 9\}$$

$$= 1 - 0.1133 = 0.8867.$$

Related quantities that can be obtained from the table are

$$P\{R(n, \theta) \leq r\} = 1 - P\{R(n, \theta) \geq r+1\}, \qquad r = 0, 1, \ldots, n-1,$$

and

$$P\{R(n, \theta) = r\} = P\{R(n, \theta) \geq r\} - P\{R(n, \theta) \geq r+1\}, \qquad r = 0, 1, \ldots, n-1.$$

A1

## Cumulative Binomial probabilities

(Reproduced by permission of Macmillan Publishers Ltd. from *Statistical Tables for Science, Engineering and Management* by J. Murdoch and J. A. Barnes.)

$\theta$ = probability of success in a single trial; $n$ = number of trials. The table gives the probability of obtaining *r or more* successes in $n$ independent trials, i.e.

$$\sum_{x=r}^{n} \binom{n}{x} \theta^x (1-\theta)^{n-x}$$

When there is no entry for a particular pair of values of *r* and $\theta$, this indicates that the appropriate probability is less than 0·000 05. Similarly, except for the case $r = 0$, when the entry is exact, a tabulated value of 1·0000 represents a probability greater than 0·999 95.

| $\theta$ | | 0·01 | 0·02 | 0·03 | 0·04 | 0·05 | 0·06 | 0·07 | 0·08 | 0·09 |
|---|---|---|---|---|---|---|---|---|---|---|
| $n=2$ | $r=0$ | 1·0000 | 1·0000 | 1·0000 | 1·0000 | 1·0000 | 1·0000 | 1·0000 | 1·0000 | 1·0000 |
| | 1 | 0·0199 | 0·0396 | 0·0591 | 0·0784 | 0·0975 | 0·1164 | 0·1351 | 0·1536 | 0·1719 |
| | 2 | 0·0001 | 0·0004 | 0·0009 | 0·0016 | 0·0025 | 0·0036 | 0·0049 | 0·0064 | 0·0081 |
| $n=5$ | $r=0$ | 1·0000 | 1·0000 | 1·0000 | 1·0000 | 1·0000 | 1·0000 | 1·0000 | 1·0000 | 1·0000 |
| | 1 | 0·0490 | 0·0961 | 0·1413 | 0·1846 | 0·2262 | 0·2661 | 0·3043 | 0·3409 | 0·3760 |
| | 2 | 0·0010 | 0·0038 | 0·0085 | 0·0148 | 0·0226 | 0·0319 | 0·0425 | 0·0544 | 0·0674 |
| | 3 | | 0·0001 | 0·0003 | 0·0006 | 0·0012 | 0·0020 | 0·0031 | 0·0045 | 0·0063 |
| | 4 | | | | | | 0·0001 | 0·0001 | 0·0002 | 0·0003 |
| $n=10$ | $r=0$ | 1·0000 | 1·0000 | 1·0000 | 1·0000 | 1·0000 | 1·0000 | 1·0000 | 1·0000 | 1·0000 |
| | 1 | 0·0956 | 0·1829 | 0·2626 | 0·3352 | 0·4013 | 0·4614 | 0·5160 | 0·5656 | 0·6106 |
| | 2 | 0·0043 | 0·0162 | 0·0345 | 0·0582 | 0·0861 | 0·1176 | 0·1517 | 0·1879 | 0·2254 |
| | 3 | 0·0001 | 0·0009 | 0·0028 | 0·0062 | 0·0115 | 0·0188 | 0·0283 | 0·0401 | 0·0540 |
| | 4 | | | 0·0001 | 0·0004 | 0·0010 | 0·0020 | 0·0036 | 0·0058 | 0·0088 |
| | 5 | | | | | 0·0001 | 0·0002 | 0·0003 | 0·0006 | 0·0010 |
| | 6 | | | | | | | | | 0·0001 |

**$n = 20$**

| $r$ | | | | | | | | | |
|---|---|---|---|---|---|---|---|---|---|
| 0 | 1·0000 | 1·0000 | 1·0000 | 1·0000 | 1·0000 | 1·0000 | 1·0000 | 1·0000 | 1·0000 |
| 1 | 0·1821 | 0·3324 | 0·4562 | 0·5580 | 0·6415 | 0·7099 | 0·7658 | 0·8113 | 0·8484 |
| 2 | 0·0169 | 0·0599 | 0·1198 | 0·1897 | 0·2642 | 0·3395 | 0·4131 | 0·4831 | 0·5484 |
| 3 | 0·0010 | 0·0071 | 0·0210 | 0·0439 | 0·0755 | 0·1150 | 0·1610 | 0·2121 | 0·2666 |
| 4 | | 0·0006 | 0·0027 | 0·0074 | 0·0159 | 0·0290 | 0·0471 | 0·0706 | 0·0993 |
| 5 | | | 0·0003 | 0·0010 | 0·0026 | 0·0056 | 0·0107 | 0·0183 | 0·0290 |
| 6 | | | | 0·0001 | 0·0003 | 0·0009 | 0·0019 | 0·0038 | 0·0068 |
| 7 | | | | | | 0·0001 | 0·0003 | 0·0006 | 0·0013 |
| 8 | | | | | | | | 0·0001 | 0·0002 |

**$n = 50$**

| $r$ | | | | | | | | | |
|---|---|---|---|---|---|---|---|---|---|
| 0 | 1·0000 | 1·0000 | 1·0000 | 1·0000 | 1·0000 | 1·0000 | 1·0000 | 1·0000 | 1·0000 |
| 1 | 0·3950 | 0·6358 | 0·7819 | 0·8701 | 0·9231 | 0·9547 | 0·9734 | 0·9845 | 0·9910 |
| 2 | 0·0894 | 0·2642 | 0·4447 | 0·5995 | 0·7206 | 0·8100 | 0·8735 | 0·9173 | 0·9468 |
| 3 | 0·0138 | 0·0784 | 0·1892 | 0·3233 | 0·4595 | 0·5838 | 0·6892 | 0·7740 | 0·8395 |
| 4 | 0·0016 | 0·0178 | 0·0628 | 0·1391 | 0·2396 | 0·3527 | 0·4673 | 0·5747 | 0·6697 |
| 5 | 0·0001 | 0·0032 | 0·0168 | 0·0490 | 0·1036 | 0·1794 | 0·2710 | 0·3710 | 0·4723 |
| 6 | | 0·0005 | 0·0037 | 0·0144 | 0·0378 | 0·0776 | 0·1350 | 0·2081 | 0·2928 |
| 7 | | 0·0001 | 0·0007 | 0·0036 | 0·0118 | 0·0289 | 0·0583 | 0·1019 | 0·1596 |
| 8 | | | 0·0001 | 0·0008 | 0·0032 | 0·0094 | 0·0220 | 0·0438 | 0·0768 |
| 9 | | | | 0·0001 | 0·0008 | 0·0027 | 0·0073 | 0·0167 | 0·0328 |
| 10 | | | | | 0·0002 | 0·0007 | 0·0022 | 0·0056 | 0·0125 |
| 11 | | | | | | 0·0002 | 0·0006 | 0·0017 | 0·0043 |
| 12 | | | | | | | 0·0001 | 0·0005 | 0·0013 |
| 13 | | | | | | | | 0·0001 | 0·0004 |
| 14 | | | | | | | | | 0·0001 |

| | θ | 0·10 | 0·15 | 0·20 | 0·25 | 0·30 | 0·35 | 0·40 | 0·45 | 0·50 |
|---|---|---|---|---|---|---|---|---|---|---|
| n = 2 | r = 0 | 1·0000 | 1·0000 | 1·0000 | 1·0000 | 1·0000 | 1·0000 | 1·0000 | 1·0000 | 1·0000 |
| | 1 | 0·1900 | 0·2775 | 0·3600 | 0·4375 | 0·5100 | 0·5775 | 0·6400 | 0·6975 | 0·7500 |
| | 2 | 0·0100 | 0·0225 | 0·0400 | 0·0625 | 0·0900 | 0·1225 | 0·1600 | 0·2025 | 0·2500 |
| n = 5 | r = 0 | 1·0000 | 1·0000 | 1·0000 | 1·0000 | 1·0000 | 1·0000 | 1·0000 | 1·0000 | 1·0000 |
| | 1 | 0·4095 | 0·5563 | 0·6723 | 0·7627 | 0·8319 | 0·8840 | 0·9222 | 0·9497 | 0·9688 |
| | 2 | 0·0815 | 0·1648 | 0·2627 | 0·3672 | 0·4718 | 0·5716 | 0·6630 | 0·7438 | 0·8125 |
| | 3 | 0·0086 | 0·0266 | 0·0579 | 0·1035 | 0·1631 | 0·2352 | 0·3174 | 0·4069 | 0·5000 |
| | 4 | 0·0005 | 0·0022 | 0·0067 | 0·0156 | 0·0308 | 0·0540 | 0·0870 | 0·1312 | 0·1875 |
| | 5 | | 0·0001 | 0̄·0003 | 0·0010 | 0·0024 | 0·0053 | 0·0102 | 0·0185 | 0·0313 |
| n = 10 | r = 0 | 1·0000 | 1·0000 | 1·0000 | 1·0000 | 1·0000 | 1·0000 | 1·0000 | 1·0000 | 1·0000 |
| | 1 | 0·6513 | 0·8031 | 0·8926 | 0·9437 | 0·9718 | 0·9865 | 0·9940 | 0·9975 | 0·9990 |
| | 2 | 0·2639 | 0·4557 | 0·6242 | 0·7560 | 0·8507 | 0·9140 | 0·9536 | 0·9767 | 0·9893 |
| | 3 | 0·0702 | 0·1798 | 0·3222 | 0·4744 | 0·6172 | 0·7384 | 0·8327 | 0·9004 | 0·9453 |
| | 4 | 0·0128 | 0·0500 | 0·1209 | 0·2241 | 0·3504 | 0·4862 | 0·6177 | 0·7430 | 0·8281 |
| | 5 | 0·0016 | 0·0099 | 0·0328 | 0·0781 | 0·1503 | 0·2485 | 0·3669 | 0·4956 | 0·6230 |
| | 6 | 0·0001 | 0·0014 | 0·0064 | 0·0197 | 0·0473 | 0·0949 | 0·1662 | 0·2616 | 0·3770 |
| | 7 | | 0·0001 | 0·0009 | 0·0035 | 0·0106 | 0·0260 | 0·0548 | 0·1020 | 0·1719 |
| | 8 | | | 0·0001 | 0·0004 | 0·0016 | 0·0048 | 0·0123 | 0·0274 | 0·0547 |
| | 9 | | | | | 0·0001 | 0·0005 | 0·0017 | 0·0045 | 0·0107 |
| | 10 | | | | | | | 0·0001 | 0·0003 | 0·0010 |
| n = 20 | r = 0 | 1·0000 | 1·0000 | 1·0000 | 1·0000 | 1·0000 | 1·0000 | 1·0000 | 1·0000 | 1·0000 |
| | 1 | 0·8784 | 0·9612 | 0·9885 | 0·9968 | 0·9992 | 0·9998 | 1·0000 | 1·0000 | 1·0000 |
| | 2 | 0·6083 | 0·8244 | 0·9308 | 0·9757 | 0·9924 | 0·9979 | 0·9995 | 0·9999 | 1·0000 |
| | 3 | 0·3231 | 0·5951 | 0·7939 | 0·9087 | 0·9645 | 0·9879 | 0·9964 | 0·9991 | 0·9998 |
| | 4 | 0·1330 | 0·3523 | 0·5886 | 0·7748 | 0·8929 | 0·9556 | 0·9840 | 0·9951 | 0·9987 |

| r | | | | | | | | | |
|---|---|---|---|---|---|---|---|---|---|
| 5 | 0·9941 | 0·9811 | 0·9490 | 0·8818 | 0·7625 | 0·5852 | 0·3704 | 0·1702 | 0·0432 |
| 6 | 0·9793 | 0·9447 | 0·8744 | 0·7546 | 0·5836 | 0·3828 | 0·1958 | 0·0673 | 0·0113 |
| 7 | 0·9423 | 0·8701 | 0·7500 | 0·5834 | 0·3920 | 0·2142 | 0·0867 | 0·0219 | 0·0024 |
| 8 | 0·8684 | 0·7480 | 0·5841 | 0·3990 | 0·2277 | 0·1018 | 0·0321 | 0·0059 | 0·0004 |
| 9 | 0·7483 | 0·5857 | 0·4044 | ·02376 | 0·1133 | 0·0409 | 0·0100 | 0·0013 | 0·0001 |
| 10 | 0·5881 | 0·4086 | 0·2447 | 0·1218 | 0·0480 | 0·0139 | 0·0026 | 0·0002 | |
| 11 | 0·4119 | 0·2493 | 0·1275 | 0·0532 | 0·0171 | 0·0039 | 0·0006 | | |
| 12 | 0·2517 | 0·1308 | 0·0565 | 0·0196 | 0·0051 | 0·0009 | 0·0001 | | |
| 13 | 0·1316 | 0·0580 | 0·0210 | 0·0060 | 0·0013 | 0·0002 | | | |
| 14 | 0·0577 | 0·0214 | 0·0065 | 0·0015 | 0·0003 | | | | |
| 15 | 0·0207 | 0·0064 | 0·0016 | 0·0003 | | | | | |
| 16 | 0·0059 | 0·0015 | 0·0003 | | | | | | |
| 17 | 0·0013 | 0·0003 | | | | | | | |
| 18 | 0·0002 | | | | | | | | |

*n* = 50

| r | | | | | | | | | |
|---|---|---|---|---|---|---|---|---|---|
| 0 | 1·0000 | 1·0000 | 1·0000 | 1·0000 | 1·0000 | 1·0000 | 1·0000 | 1·0000 | 1·0000 |
| 1 | 1·0000 | 1·0000 | 1·0000 | 1·0000 | 1·0000 | 1·0000 | 1·0000 | 0·9997 | 0·9948 |
| 2 | 1·0000 | 1·0000 | 1·0000 | 1·0000 | 1·0000 | 1·0000 | 0·9998 | 0·9971 | 0·9662 |
| 3 | 1·0000 | 1·0000 | 1·0000 | 1·0000 | 1·0000 | 0·9999 | 0·9987 | 0·9858 | 0·8883 |
| 4 | 1·0000 | 1·0000 | 1·0000 | 1·0000 | 1·0000 | 0·9995 | 0·9943 | 0·9540 | 0·7497 |
| 5 | 1·0000 | 1·0000 | 1·0000 | 1·0000 | 0·9998 | 0·9979 | 0·9815 | 0·8879 | 0·5688 |
| 6 | 1·0000 | 1·0000 | 1·0000 | 0·9999 | 0·9993 | 0·9930 | 0·9520 | 0·7806 | 0·3839 |
| 7 | 1·0000 | 1·0000 | 1·0000 | 0·9998 | 0·9975 | 0·9806 | 0·8966 | 0·6387 | 0·2298 |
| 8 | 1·0000 | 1·0000 | 0·9999 | 0·9992 | 0·9927 | 0·9547 | 0·8096 | 0·4812 | 0·1221 |
| 9 | 1·0000 | 1·0000 | 0·9998 | 0·9975 | 0·9817 | 0·9084 | 0·6927 | 0·3319 | 0·0579 |
| 10 | 1·0000 | 0·9999 | 0·9992 | 0·9933 | 0·9598 | 0·8363 | 0·5563 | 0·2089 | 0·0245 |
| 11 | 1·0000 | 0·9998 | 0·9978 | 0·9840 | 0·9211 | 0·7378 | 0·4164 | 0·1199 | 0·0094 |
| 12 | 1·0000 | 0·9994 | 0·9943 | 0·9658 | 0·8610 | 0·6184 | 0·2893 | 0·0628 | 0·0032 |
| 13 | 0·9998 | 0·9982 | 0·9867 | 0·9339 | 0·7771 | 0·4890 | 0·1861 | 0·0301 | 0·0010 |
| 14 | 0·9995 | 0·9955 | 0·9720 | 0·8837 | 0·6721 | 0·3630 | 0·1106 | 0·0132 | 0·0003 |

| $\theta$ | 0·10 | 0·15 | 0·20 | 0·25 | 0·30 | 0·35 | 0·40 | 0·45 | 0·50 |
|---|---|---|---|---|---|---|---|---|---|
| 15 | 0·0001 | 0·0053 | 0·0607 | 0·2519 | 0·5532 | 0·8122 | 0·9460 | 0·9896 | 0·9987 |
| 16 |  | 0·0019 | 0·0308 | 0·1631 | 0·4308 | 0·7199 | 0·9045 | 0·9780 | 0·9967 |
| 17 |  | 0·0007 | 0·0144 | 0·0983 | 0·3161 | 0·6111 | 0·8439 | 0·9573 | 0·9923 |
| 18 |  | 0·0002 | 0·0063 | 0·0551 | 0·2178 | 0·4940 | 0·7631 | 0·9235 | 0·9836 |
| 19 |  | 0·0001 | 0·0025 | 0·0287 | 0·1406 | 0·3784 | 0·6644 | 0·8727 | 0·9675 |
| 20 |  |  | 0·0009 | 0·0139 | 0·0848 | 0·2736 | 0·5535 | 0·8026 | 0·9405 |
| 21 |  |  | 0·0003 | 0·0063 | 0·0478 | 0·1861 | 0·4390 | 0·7138 | 0·8987 |
| 22 |  |  | 0·0001 | 0·0026 | 0·0251 | 0·1187 | 0·3299 | 0·6100 | 0·8389 |
| 23 |  |  |  | 0·0010 | 0·0123 | 0·0710 | 0·2340 | 0·4981 | 0·7601 |
| 24 |  |  |  | 0·0004 | 0·0056 | 0·0396 | 0·1562 | 0·3866 | 0·6641 |
| 25 |  |  |  | 0·0001 | 0·0024 | 0·0207 | 0·0978 | 0·2840 | 0·5561 |
| 26 |  |  |  |  | 0·0009 | 0·0100 | 0·0573 | 0·1966 | 0·4439 |
| 27 |  |  |  |  | 0·0003 | 0·0045 | 0·0314 | 0·1279 | 0·3359 |
| 28 |  |  |  |  | 0·0001 | 0·0019 | 0·0160 | 0·0780 | 0·2399 |
| 29 |  |  |  |  |  | 0·0007 | 0·0076 | 0·0444 | 0·1611 |
| 30 |  |  |  |  |  | 0·0003 | 0·0034 | 0·0235 | 0·1013 |
| 31 |  |  |  |  |  | 0·0001 | 0·0014 | 0·0116 | 0·0595 |
| 32 |  |  |  |  |  |  | 0·0005 | 0·0053 | 0·0325 |
| 33 |  |  |  |  |  |  | 0·0002 | 0·0022 | 0·0164 |
| 34 |  |  |  |  |  |  | 0·0001 | 0·0009 | 0·0077 |
| 35 |  |  |  |  |  |  |  | 0·0003 | 0·0033 |
| 36 |  |  |  |  |  |  |  | 0·0001 | 0·0013 |
| 37 |  |  |  |  |  |  |  |  | 0·0005 |
| 38 |  |  |  |  |  |  |  |  | 0·0002 |

## APPENDIX T2

**Cumulative Poisson Probabilities** [see II: § 5.4]

The tabulated quantity is

$$P\{S(\lambda) \geq r\} = \sum_{s=r}^{\infty} e^{-\lambda}\lambda^s/s!,$$

where $S(\lambda)$ denotes a random variable which has the Poisson $(\lambda)$ distribution. Thus $P\{S(\lambda) \geq r\}$ is the probability of *r or more* occurrences. Related quantities are

$$P\{S(\lambda) \leq r\} = 1 - P\{S(\lambda) \geq r+1\}, \qquad r = 0, 1, \ldots$$

and

$$P\{S(\lambda) = r\} = P\{S(\lambda) \geq r\} - P\{S(\lambda) \geq r+1\}, \qquad r = 0, 1, \ldots.$$

## Cumulative Poisson probabilities

(Reproduced by permission of Macmillan Publishers Ltd. from *Statistical Tables for Science, Engineering and Management* by J. Murdoch and J. A. Barnes.)

The table gives the probability that *r or more* random events are contained in an interval when the average number of such events per interval is $\lambda$, i.e.

$$\sum_{x=r}^{\infty} e^{-\lambda} \frac{\lambda^x}{x!}$$

Where there is no entry for a particular pair of values of *r* and $\lambda$, this indicates that the appropriate probability is less than 0·000 05. Similarly, except for the case *r* = 0 when the entry is exact, a tabulated value of 1·0000 represents a probability greater than 0·999 95.

| $\lambda$ | 0·1 | 0·2 | 0·3 | 0·4 | 0·5 | 0·6 | 0·7 | 0·8 | 0·9 | 1·0 |
|---|---|---|---|---|---|---|---|---|---|---|
| *r* = 0 | 1·0000 | 1·0000 | 1·0000 | 1·0000 | 1·0000 | 1·0000 | 1·0000 | 1·0000 | 1·0000 | 1·0000 |
| 1 | 0·0952 | 0·1813 | 0·2592 | 0·3297 | 0·3935 | 0·4512 | 0·5034 | 0·5507 | 0·5934 | 0·6321 |
| 2 | 0·0047 | 0·0175 | 0·0369 | 0·0616 | 0·0902 | 0·1219 | 0·1558 | 0·1912 | 0·2275 | 0·2642 |
| 3 | 0·0002 | 0·0011 | 0·0036 | 0·0079 | 0·0144 | 0·0231 | 0·0341 | 0·0474 | 0·0629 | 0·0803 |
| 4 | | 0·0001 | 0·0003 | 0·0008 | 0·0018 | 0·0034 | 0·0058 | 0·0091 | 0·0135 | 0·0190 |
| 5 | | | | 0·0001 | 0·0002 | 0·0004 | 0·0008 | 0·0014 | 0·0023 | 0·0037 |
| 6 | | | | | | | 0·0001 | 0·0002 | 0·0003 | 0·0006 |
| 7 | | | | | | | | | | 0·0001 |

| $\lambda$ | 1·1 | 1·2 | 1·3 | 1·4 | 1·5 | 1·6 | 1·7 | 1·8 | 1·9 | 2·0 |
|---|---|---|---|---|---|---|---|---|---|---|
| *r* = 0 | 1·0000 | 1·0000 | 1·0000 | 1·0000 | 1·0000 | 1·0000 | 1·0000 | 1·0000 | 1·0000 | 1·0000 |
| 1 | 0·6671 | 0·6988 | 0·7275 | 0·7534 | 0·7769 | 0·7981 | 0·8173 | 0·8347 | 0·8504 | 0·8647 |
| 2 | 0·3010 | 0·3374 | 0·3732 | 0·4082 | 0·4422 | 0·4751 | 0·5068 | 0·5372 | 0·5663 | 0·5940 |
| 3 | 0·0996 | 0·1205 | 0·1429 | 0·1665 | 0·1912 | 0·2166 | 0·2428 | 0·2694 | 0·2963 | 0·3233 |
| 4 | 0·0257 | 0·0338 | 0·0431 | 0·0537 | 0·0656 | 0·0788 | 0·0932 | 0·1087 | 0·1253 | 0·1429 |
| 5 | 0·0054 | 0·0077 | 0·0107 | 0·0143 | 0·0186 | 0·0237 | 0·0296 | 0·0364 | 0·0441 | 0·0527 |
| 6 | 0·0010 | 0·0015 | 0·0022 | 0·0032 | 0·0045 | 0·0060 | 0·0080 | 0·0104 | 0·0132 | 0·0166 |
| 7 | 0·0001 | 0·0003 | 0·0004 | 0·0006 | 0·0009 | 0·0013 | 0·0019 | 0·0026 | 0·0034 | 0·0045 |
| 8 | | | 0·0001 | 0·0001 | 0·0002 | 0·0003 | 0·0004 | 0·0006 | 0·0008 | 0·0011 |
| 9 | | | | | | | 0·0001 | 0·0001 | 0·0002 | 0·0002 |

| λ | 2·1 | 2·2 | 2·3 | 2·4 | 2·5 | 2·6 | 2·7 | 2·8 | 2·9 | 3·0 |
|---|---|---|---|---|---|---|---|---|---|---|
| r=0 | 1·0000 | 1·0000 | 1·0000 | 1·0000 | 1·0000 | 1·0000 | 1·0000 | 1·0000 | 1·0000 | 1·0000 |
| 1 | 0·8775 | 0·8892 | 0·8997 | 0·9093 | 0·9179 | 0·9257 | 0·9328 | 0·9392 | 0·9450 | 0·9502 |
| 2 | 0·6204 | 0·6454 | 0·6691 | 0·6916 | 0·7127 | 0·7326 | 0·7513 | 0·7689 | 0·7854 | 0·8009 |
| 3 | 0·3504 | 0·3773 | 0·4040 | 0·4303 | 0·4562 | 0·4816 | 0·5064 | 0·5305 | 0·5540 | 0·5768 |
| 4 | 0·1614 | 0·1806 | 0·2007 | 0·2213 | 0·2424 | 0·2640 | 0·2859 | 0·3081 | 0·3304 | 0·3528 |
| 5 | 0·0621 | 0·0725 | 0·0838 | 0·0959 | 0·1088 | 0·1226 | 0·1371 | 0·1523 | 0·1682 | 0·1847 |
| 6 | 0·0204 | 0·0249 | 0·0300 | 0·0357 | 0·0420 | 0·0490 | 0·0567 | 0·0651 | 0·0742 | 0·0839 |
| 7 | 0·0059 | 0·0075 | 0·0094 | 0·0116 | 0·0142 | 0·0172 | 0·0206 | 0·0244 | 0·0287 | 0·0335 |
| 8 | 0·0015 | 0·0020 | 0·0026 | 0·0033 | 0·0042 | 0·0053 | 0·0066 | 0·0081 | 0·0099 | 0·0119 |
| 9 | 0·0003 | 0·0005 | 0·0006 | 0·0009 | 0·0011 | 0·0015 | 0·0019 | 0·0024 | 0·0031 | 0·0038 |
| 10 | 0·0001 | 0·0001 | 0·0001 | 0·0002 | 0·0003 | 0·0004 | 0·0005 | 0·0007 | 0·0009 | 0·0011 |
| 11 | | | | | 0·0001 | 0·0001 | 0·0001 | 0·0002 | 0·0002 | 0·0003 |
| 12 | | | | | | | | | 0·0001 | 0·0001 |

| λ | 3·1 | 3·2 | 3·3 | 3·4 | 3·5 | 3·6 | 3·7 | 3·8 | 3·9 | 4·0 |
|---|---|---|---|---|---|---|---|---|---|---|
| r=0 | 1·0000 | 1·0000 | 1·0000 | 1·0000 | 1·0000 | 1·0000 | 1·0000 | 1·0000 | 1·0000 | 1·0000 |
| 1 | 0·9550 | 0·9592 | 0·9631 | 0·9666 | 0·9698 | 0·9727 | 0·9753 | 0·9776 | 0·9798 | 0·9817 |
| 2 | 0·8153 | 0·8288 | 0·8414 | 0·8532 | 0·8641 | 0·8743 | 0·8838 | 0·8926 | 0·9008 | 0·9084 |
| 3 | 0·5988 | 0·6201 | 0·6406 | 0·6603 | 0·6792 | 0·6973 | 0·7146 | 0·7311 | 0·7469 | 0·7619 |
| 4 | 0·3752 | 0·3975 | 0·4197 | 0·4416 | 0·4634 | 0·4848 | 0·5058 | 0·5265 | 0·5468 | 0·5665 |
| 5 | 0·2018 | 0·2194 | 0·2374 | 0·2558 | 0·2746 | 0·2936 | 0·3128 | 0·3322 | 0·3516 | 0·3712 |
| 6 | 0·0943 | 0·1054 | 0·1171 | 0·1295 | 0·1424 | 0·1559 | 0·1699 | 0·1844 | 0·1994 | 0·2149 |
| 7 | 0·0388 | 0·0446 | 0·0510 | 0·0579 | 0·0653 | 0·0733 | 0·0818 | 0·0909 | 0·1005 | 0·1107 |
| 8 | 0·0142 | 0·0168 | 0·0198 | 0·0231 | 0·0267 | 0·0308 | 0·0352 | 0·0401 | 0·0454 | 0·0511 |
| 9 | 0·0047 | 0·0057 | 0·0069 | 0·0083 | 0·0099 | 0·0117 | 0·0137 | 0·0160 | 0·0185 | 0·0214 |
| 10 | 0·0014 | 0·0018 | 0·0022 | 0·0027 | 0·0033 | 0·0040 | 0·0048 | 0·0058 | 0·0069 | 0·0081 |
| 11 | 0·0004 | 0·0005 | 0·0006 | 0·0008 | 0·0010 | 0·0013 | 0·0016 | 0·0019 | 0·0023 | 0·0028 |
| 12 | 0·0001 | 0·0001 | 0·0002 | 0·0002 | 0·0003 | 0·0004 | 0·0005 | 0·0006 | 0·0007 | 0·0009 |
| 13 | | | | 0·0001 | 0·0001 | 0·0001 | 0·0001 | 0·0002 | 0·0002 | 0·0003 |
| 14 | | | | | | | | | 0·0001 | 0·0001 |

| $\lambda$ | 4·1 | 4·2 | 4·3 | 4·4 | 4·5 | 4·6 | 4·7 | 4·8 | 4·9 | 5·0 |
|---|---|---|---|---|---|---|---|---|---|---|
| $r=0$ | 1·0000 | 1·0000 | 1·0000 | 1·0000 | 1·0000 | 1·0000 | 1·0000 | 1·0000 | 1·0000 | 1·0000 |
| 1 | 0·9834 | 0·9850 | 0·9864 | 0·9877 | 0·9889 | 0·9899 | 0·9909 | 0·9918 | 0·9926 | 0·9933 |
| 2 | 0·9155 | 0·9220 | 0·9281 | 0·9337 | 0·9389 | 0·9437 | 0·9482 | 0·9523 | 0·9561 | 0·9596 |
| 3 | 0·7762 | 0·7898 | 0·8026 | 0·8149 | 0·8264 | 0·8374 | 0·8477 | 0·8575 | 0·8667 | 0·8753 |
| 4 | 0·5858 | 0·6046 | 0·6228 | 0·6408 | 0·6577 | 0·6743 | 0·6903 | 0·7058 | 0·7207 | 0·7350 |
| 5 | 0·3907 | 0·4102 | 0·4296 | 04488 | 0·4679 | 0·4868 | 0·5054 | 0·5237 | 0·5418 | 0·5595 |
| 6 | 0·2307 | 0·2469 | 0·2633 | 0·2801 | 0·2971 | 0·3142 | 0·3316 | 0·3490 | 0·3665 | 0·3840 |
| 7 | 0·1214 | 0·1325 | 0·1442 | 0·1564 | 0·1689 | 0·1820 | 0·1954 | 0·2092 | 0·2233 | 0·2378 |
| 8 | 0·0573 | 0·0639 | 0·0710 | 0·0786 | 0·0866 | 0·0951 | 0·1040 | 0·1133 | 0·1231 | 0·1334 |
| 9 | 0·0245 | 0·0279 | 0·0317 | 0·0358 | 0·0403 | 0·0451 | 0·0503 | 0·0558 | 0·0618 | 0·0681 |
| 10 | 0·0095 | 0·0111 | 0·0129 | 0·0149 | 0·0171 | 0·0195 | 0·0222 | 0·0251 | 0·0283 | 0·0318 |
| 11 | 0·0034 | 0·0041 | 0·0048 | 0·0057 | 0·0067 | 0·0078 | 0·0090 | 0·0104 | 0·0120 | 0·0137 |
| 12 | 0·0011 | 0·0014 | 0·0017 | 0·0020 | 0·0024 | 0·0029 | 0·0034 | 0·0040 | 0·0047 | 0·0055 |
| 13 | 0·0003 | 0·0004 | 0·0005 | 0·0007 | 0·0008 | 0·0010 | 0·0012 | 0·0014 | 0·0017 | 0·0020 |
| 14 | 0·0001 | 0·0001 | 0·0002 | 0·0002 | 0·0003 | 0·0003 | 0·0004 | 0·0005 | 0·0006 | 0·0007 |
| 15 | | | | 0·0001 | 0·0001 | 0·0001 | 0·0001 | 0·0001 | 0·0002 | 0·0002 |
| 16 | | | | | | | | | 0·0001 | 0·0001 |

| $\lambda$ | 5·2 | 5·4 | 5·6 | 5·8 | 6·0 | 6·2 | 6·4 | 6·6 | 6·8 | 7·0 |
|---|---|---|---|---|---|---|---|---|---|---|
| $r = 0$ | 1·0000 | 1·0000 | 1·0000 | 1·0000 | 1·0000 | 1·0000 | 1·0000 | 1·0000 | 1·0000 | 1·0000 |
| 1 | 0·9945 | 0·9955 | 0·9963 | 0·9970 | 0·9975 | 0·9980 | 0·9983 | 0·9986 | 0·9989 | 0·9991 |
| 2 | 0·9658 | 0·9711 | 0·9756 | 0·9794 | 0·9826 | 0·9854 | 0·9877 | 0·9897 | 0·9913 | 0·9927 |
| 3 | 0·8912 | 0·9052 | 0·9176 | 0·9285 | 0·9380 | 0·9464 | 0·9537 | 0·9600 | 0·9656 | 0·9704 |
| 4 | 0·7619 | 0·7867 | 0·8094 | 0·8300 | 0·8488 | 0·8658 | 0·8811 | 0·8948 | 0·9072 | 0·9182 |
| 5 | 0·5939 | 0·6267 | 0·6579 | 0·6873 | 0·7149 | 0·7408 | 0·7649 | 0·7873 | 0·8080 | 0·8270 |
| 6 | 0·4191 | 0·4539 | 0·4881 | 0·5217 | 0·5543 | 0·5859 | 0·6163 | 0·6453 | 0·6730 | 0·6993 |
| 7 | 0·2676 | 0·2983 | 0·3297 | 0·3616 | 0·3937 | 0·4258 | 0·4577 | 0·4892 | 0·5201 | 0·5503 |
| 8 | 0·1551 | 0·1783 | 0·2030 | 0·2290 | 0·2560 | 0·2840 | 0·3127 | 0·3419 | 0·3715 | 0·4013 |
| 9 | 0·0819 | 0·0974 | 0·1143 | 0·1328 | 0·1528 | 0·1741 | 0·1967 | 0·2204 | 0·2452 | 0·2709 |
| 10 | 0·0397 | 0·0488 | 0·0591 | 0·0708 | 0·0839 | 0·0984 | 0·1142 | 0·1314 | 0·1498 | 0·1695 |
| 11 | 0·0177 | 0·0225 | 0·0282 | 0·0349 | 0·0426 | 0·0514 | 0·0614 | 0·0726 | 0·0849 | 0·0985 |
| 12 | 0·0073 | 0·0096 | 0·0125 | 0·0160 | 0·0201 | 0·0250 | 0·0307 | 0·0373 | 0·0448 | 0·0534 |
| 13 | 0·0028 | 0·0038 | 0·0051 | 0·0068 | 0·0088 | 0·0113 | 0·0143 | 0·0179 | 0·0221 | 0·0270 |
| 14 | 0·0010 | 0·0014 | 0·0020 | 0·0027 | 0·0036 | 0·0048 | 0·0063 | 0·0080 | 0·0102 | 0·0128 |
| 15 | 0·0003 | 0·0002 | 0·0007 | 0·0010 | 0·0014 | 0·0019 | 0·0026 | 0·0034 | 0·0044 | 0·0057 |
| 16 | 0·0001 | 0·0002 | 0·0002 | 0·0004 | 0·0005 | 0·0007 | 0·0010 | 0·0014 | 0·0018 | 0·0024 |
| 17 |  | 0·0001 | 0·0001 | 0·0001 | 0·0002 | 0·0003 | 0·0004 | 0·0005 | 0·0007 | 0·0010 |
| 18 |  |  |  |  | 0·0001 | 0·0001 | 0·0001 | 0·0002 | 0·0003 | 0·0004 |
| 19 |  |  |  |  |  |  |  | 0·0001 | 0·0001 | 0·0001 |

| λ | 7·2 | 7·4 | 7·6 | 7·8 | 8·0 | 8·2 | 8·4 | 8·6 | 8·8 | 9·0 |
|---|---|---|---|---|---|---|---|---|---|---|
| r = 0 | 1·0000 | 1·0000 | 1·0000 | 1·0000 | 1·0000 | 1·0000 | 1·0000 | 1·0000 | 1·0000 | 1·0000 |
| 1 | 0·9993 | 0·9994 | 0·9995 | 0·9996 | 0·9997 | 0·9997 | 0·9998 | 0·9998 | 0·9998 | 0·9999 |
| 2 | 0·9939 | 0·9949 | 0·9957 | 0·9964 | 0·9970 | 0·9975 | 0·9979 | 0·9982 | 0·9985 | 0·9988 |
| 3 | 0·9745 | 0·9781 | 0·9812 | 0·9839 | 0·9862 | 0·9882 | 0·9900 | 0·9914 | 0·9927 | 0·9938 |
| 4 | 0·9281 | 0·9368 | 0·9446 | 0·9515 | 0·9576 | 0·9630 | 0·9677 | 0·9719 | 0·9756 | 0·9788 |
| 5 | 0·8445 | 0·8605 | 0·8751 | 0·8883 | 0·9004 | 0·9113 | 0·9211 | 0·9299 | 0·9379 | 0·9450 |
| 6 | 0·7241 | 0·7474 | 0·7693 | 0·7897 | 0·8088 | 0·8264 | 0·8427 | 0·8578 | 0·8716 | 0·8843 |
| 7 | 0·5796 | 0·6080 | 0·6354 | 0·6616 | 0·6866 | 0·7104 | 0·7330 | 0·7543 | 0·7744 | 0·7932 |
| 8 | 0·4311 | 0·4607 | 0·4900 | 0·5188 | 0·5470 | 0·5746 | 0·6013 | 0·6272 | 0·6522 | 0·6761 |
| 9 | 0·2973 | 0·3243 | 0·3518 | 0·3796 | 0·4075 | 0·4353 | 0·4631 | 0·4906 | 0·5177 | 0·5443 |
| 10 | 0·1904 | 0·2123 | 0·2351 | 0·2589 | 0·2834 | 0·3085 | 0·3341 | 0·3600 | 0·3863 | 0·4126 |
| 11 | 0·1133 | 0·1293 | 0·1465 | 0·1648 | 0·1841 | 0·2045 | 0·2257 | 0·2478 | 0·2706 | 0·2940 |
| 12 | 0·0629 | 0·0735 | 0·0852 | 0·0980 | 0·1119 | 0·1269 | 0·1429 | 0·1600 | 0·1780 | 0·1970 |
| 13 | 0·0327 | 0·0391 | 0·0464 | 0·0546 | 0·0638 | 0·0739 | 0·0850 | 0·0971 | 0·1102 | 0·1242 |
| 14 | 0·0159 | 0·0195 | 0·0238 | 0·0286 | 0·0342 | 0·0405 | 0·0476 | 0·0555 | 0·0642 | 0·0739 |
| 15 | 0·0073 | 0·0092 | 0·0114 | 0·0141 | 0·0173 | 0·0209 | 0·0251 | 0·0299 | 0·0353 | 0·0415 |
| 16 | 0·0031 | 0·0041 | 0·0052 | 0·0066 | 0·0082 | 0·0102 | 0·0125 | 0·0152 | 0·0184 | 0·0220 |
| 17 | 0·0013 | 0·0017 | 0·0022 | 0·0029 | 0·0037 | 0·0047 | 0·0059 | 0·0074 | 0·0091 | 0·0111 |
| 18 | 0·0005 | 0·0007 | 0·0009 | 0·0012 | 0·0016 | 0·0021 | 0·0027 | 0·0034 | 0·0043 | 0·0053 |
| 19 | 0·0002 | 0·0003 | 0·0004 | 0·0005 | 0·0006 | 0·0009 | 0·0011 | 0·0015 | 0·0019 | 0·0024 |
| 20 | 0·0001 | 0·0001 | 0·0001 | 0·0002 | 0·0003 | 0·0003 | 00005 | 0·0006 | 0·0008 | 0·0011 |
| 21 | | | | 0·0001 | 0·0001 | 0·0001 | 0·0002 | 0·0002 | 0·0003 | 0·0004 |
| 22 | | | | | | | 0·0001 | 0·0001 | 0·0001 | 0·0002 |
| 23 | | | | | | | | | | 0·0001 |

| $\lambda$ | 9·2 | 9·4 | 9·6 | 9·8 | 10·0 | 11·0 | 12·0 | 13·0 | 14·0 | 15·0 |
|---|---|---|---|---|---|---|---|---|---|---|
| r = 0 | 1·0000 | 1·0000 | 1·0000 | 1·0000 | 1·0000 | 1·0000 | 1·0000 | 1·0000 | 1·0000 | 1·0000 |
| 1 | 0·9999 | 0·9999 | 0·9999 | 0·9999 | 1·0000 | 1·0000 | 1·0000 | 1·0000 | 1·0000 | 1·0000 |
| 2 | 0·9990 | 0·9991 | 0·9993 | 0·9994 | 0·9995 | 0·9998 | 0·9999 | 1·0000 | 1·0000 | 1·0000 |
| 3 | 0·9947 | 0·9955 | 0·9962 | 0·9967 | 0·9972 | 0·9988 | 0·9995 | 0·9998 | 0·9999 | 1·0000 |
| 4 | 0·9816 | 0·9840 | 0·9862 | 0·9880 | 0·9897 | 0·9951 | 0·9977 | 0·9990 | 0·9995 | 0·9998 |
| 5 | 0·9514 | 0·9571 | 0·9622 | 0·9667 | 0·9707 | 0·9849 | 0·9924 | 0·9963 | 0·9982 | 0·9991 |
| 6 | 0·8959 | 0·9065 | 0·9162 | 0·9250 | 0·9329 | 0·9625 | 0·9797 | 0·9893 | 0·9945 | 0·9972 |
| 7 | 0·8108 | 0·8273 | 0·8426 | 0·8567 | 0·8699 | 0·9214 | 0·9542 | 0·9741 | 0·9858 | 0·9924 |
| 8 | 0·6990 | 0·7208 | 0·7416 | 0·7612 | 0·7798 | 0·8568 | 0·9105 | 0·9460 | 0·9684 | 0·9820 |
| 9 | 0·5704 | 0·5958 | 0·6204 | 0·6442 | 0·6672 | 0·7680 | 0·8450 | 0·9002 | 0·9379 | 0·9626 |
| 10 | 0·4389 | 0·4651 | 0·4911 | 0·5168 | 0·5421 | 0·6595 | 0·7576 | 0·8342 | 0·8906 | 0·9301 |
| 11 | 0·3180 | 0·3424 | 0·3671 | 0·3920 | 0·4170 | 0·5401 | 0·6528 | 0·7483 | 0·8243 | 0·8815 |
| 12 | 0·2168 | 0·2374 | 0·2588 | 0·2807 | 0·3032 | 0·4207 | 0·5384 | 0·6468 | 0·7400 | 0·8152 |
| 13 | 0·1393 | 0·1552 | 0·1721 | 0·1899 | 0·2084 | 0·3113 | 0·4240 | 0·5369 | 0·6415 | 0·7324 |
| 14 | 0·0844 | 0·0958 | 0·1081 | 0·1214 | 0·1355 | 0·2187 | 0·3185 | 0·4270 | 0·5356 | 0·6368 |
| 15 | 0·0483 | 0·0559 | 0·0643 | 0·0735 | 0·0835 | 0·1460 | 0·2280 | 0·3249 | 0·4296 | 0·5343 |
| 16 | 0·0262 | 0·0309 | 0·0362 | 0·0421 | 0·0487 | 0·0926 | 0·1556 | 0·2364 | 0·3306 | 0·4319 |
| 17 | 0·0135 | 0·0162 | 0·0194 | 0·0230 | 0·0270 | 0·0559 | 0·1013 | 0·1645 | 0·2441 | 0·3359 |
| 18 | 0·0066 | 0·0081 | 0·0098 | 0·0119 | 0·0143 | 0·0322 | 0·0630 | 0·1095 | 0·1728 | 0·2511 |
| 19 | 0·0031 | 0·0038 | 0·0048 | 0·0059 | 0·0072 | 0·0177 | 0·0374 | 0·0698 | 0·1174 | 0·1805 |
| 20 | 0·0014 | 0·0017 | 0·0022 | 0·0028 | 0·0035 | 0·0093 | 0·0213 | 0·0427 | 0·0765 | 0·1248 |
| 21 | 0·0006 | 0·0008 | 0·0010 | 0·0012 | 0·0016 | 0·0047 | 0·0116 | 0·0250 | 0·0479 | 0·0830 |
| 22 | 0·0002 | 0·0003 | 0·0004 | 0·0005 | 0·0007 | 0·0023 | 0·0061 | 0·0141 | 0·0288 | 0·0531 |
| 23 | 0·0001 | 0·0001 | 0·0002 | 0·0002 | 0·0003 | 0·0010 | 0·0030 | 0·0076 | 0·0167 | 0·0327 |
| 24 | | | 0·0001 | 0·0001 | 0·0001 | 0·0005 | 0·0015 | 0·0040 | 0·0093 | 0·0195 |
| 25 | | | | | | 0·0002 | 0·0007 | 0·0020 | 0·0050 | 0·0112 |
| 26 | | | | | | 0·0001 | 0·0003 | 0·0010 | 0·0026 | 0·0062 |
| 27 | | | | | | | 0·0001 | 0·0005 | 0·0013 | 0·0033 |
| 28 | | | | | | | 0·0001 | 0·0002 | 0·0006 | 0·0017 |
| 29 | | | | | | | | 0·0001 | 0·0003 | 0·0009 |
| 30 | | | | | | | | | 0·0001 | 0·0004 |
| 31 | | | | | | | | | 0·0001 | 0·0002 |
| 32 | | | | | | | | | | 0·0001 |

## APPENDIX T3

**Cumulative Standard Normal Probabilities** [see § 2.5 and II: § 11.4]

The tabulated quantity is

$$P(U \geq u) = \int_u^\infty (2\pi)^{-1/2} e^{-(1/2)y^2} \, dy$$

$$= 1 - \Phi(u),$$

where $\Phi(u)$ is the c.d.f. of the standard Normal variable $U$.
For example,

$$1 - \Phi(2{\cdot}32) = 0{\cdot}010170.$$

Entries in bold type take the same decimal prefix as entries in the following row. For example,

$$1 - \Phi(2{\cdot}36) = 0{\cdot}0091375.$$

The table gives values of $1 - \Phi(u)$ for $u \geq 0$. For negative values of $u$, use the relation

$$\Phi(u) = 1 - \Phi(-u).$$

For example,

$$\Phi(-2{\cdot}36) = 1 - \Phi(2{\cdot}36) = 0{\cdot}0091375.$$

If $X$ is Normally distributed with expected value $\mu$ and standard deviation $\sigma$ (variance $\sigma^2$),

$$P(X \leq x) = \Phi\left(\frac{x - \mu}{\sigma}\right), \qquad P(X \geq x) = 1 - \Phi\left(\frac{x - \mu}{\sigma}\right).$$

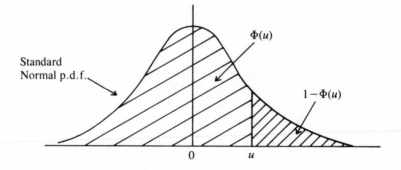

Standard
Normal p.d.f.

$\Phi(u)$

$1 - \Phi(u)$

0      $u$

## The Normal probability integral $1 - \Phi(x)$

(Reproduced by permission of Longman Group Ltd. from *Statistical Tables for Biological, Agricultural and Medical Research* by R. A. Fisher and F. Yates, 1974.)

| $x$ | | 0 | 1 | 2 | 3 | 4 | 5 | 6 | 7 | 8 | 9 |
|-----|---|---|---|---|---|---|---|---|---|---|---|
| 0·0 | 0·0 | 50000 | 49601 | 49202 | 48803 | 48405 | 48006 | 47608 | 47210 | 46812 | 46414 |
| 0·1 | | 46017 | 45620 | 45224 | 44828 | 44433 | 44038 | 43644 | 43251 | 42858 | 42465 |
| 0·2 | | 42074 | 41683 | 41294 | 40905 | 40517 | 40129 | 39743 | 39358 | 38974 | 38591 |
| 0·3 | | 38209 | 37828 | 37448 | 37070 | 36693 | 36317 | 35942 | 35569 | 35197 | 34827 |
| 0·4 | | 34458 | 34090 | 33724 | 33360 | 32997 | 32636 | 32276 | 31918 | 31561 | 31207 |
| 0·5 | | 30854 | 30503 | 30153 | 29806 | 29460 | 29116 | 28774 | 28434 | 28096 | 27760 |
| 0·6 | | 27425 | 27093 | 26763 | 26435 | 26109 | 25785 | 25463 | 25143 | 24825 | 24510 |
| 0·7 | | 24196 | 23885 | 23576 | 23270 | 22965 | 22663 | 22363 | 22065 | 21770 | 21476 |
| 0·8 | | 21186 | 20897 | 20611 | 20327 | 20045 | 19766 | 19489 | 19215 | 18943 | 18673 |
| 0·9 | | 18406 | 18141 | 17879 | 17619 | 17361 | 17106 | 16853 | 16602 | 16354 | 16109 |
| 1·0 | 0·0 | 15866 | 15625 | 15386 | 15151 | 14917 | 14686 | 14457 | 14231 | 14007 | 13786 |
| 1·1 | | 13567 | 13350 | 13136 | 12924 | 12714 | 12507 | 12302 | 12100 | 11900 | 11702 |
| 1·2 | | 11507 | 11314 | 11123 | 10935 | 10749 | 10565 | 10383 | 10204 | 10027 | **98525** |
| 1·3 | | 96800 | 95098 | 93418 | 91759 | 90123 | 88508 | 86915 | 85343 | 83793 | 82264 |
| 1·4 | | 80757 | 79270 | 77804 | 76359 | 74934 | 73529 | 72145 | 70781 | 69437 | 68112 |
| 1·5 | | 66807 | 65522 | 64255 | 63008 | 61780 | 60571 | 59380 | 58208 | 57053 | 55917 |
| 1·6 | | 54799 | 53699 | 52616 | 51551 | 50503 | 49471 | 48457 | 47460 | 46479 | 45514 |
| 1·7 | | 44565 | 43633 | 42716 | 41815 | 40930 | 40059 | 39204 | 38364 | 37538 | 36727 |
| 1·8 | | 35930 | 35148 | 34380 | 33625 | 32884 | 32157 | 31443 | 30742 | 30054 | 29379 |
| 1·9 | | 28717 | 28067 | 27429 | 26803 | 26190 | 25588 | 24998 | 24419 | 23852 | 23295 |
| 2·0 | | 22750 | 22216 | 21692 | 21178 | 20675 | 20182 | 19699 | 19226 | 18763 | 18309 |
| 2·1 | | 17864 | 17429 | 17003 | 16586 | 16177 | 15778 | 15386 | 15003 | 14629 | 14262 |
| 2·2 | | 13903 | 13553 | 13209 | 12874 | 12545 | 12224 | 11911 | 11604 | 11304 | 11011 |
| 2·3 | | 10724 | 10444 | 10170 | **99031** | **96419** | **93867** | **91375** | **88940** | **86563** | **84242** |
| 2·4 | $0·0^2$ | 81975 | 79763 | 77603 | 75494 | 73436 | 71428 | 69469 | 67557 | 65691 | 63872 |

| x | | 0 | 1 | 2 | 3 | 4 | 5 | 6 | 7 | 8 | 9 |
|---|---|---|---|---|---|---|---|---|---|---|---|
| 2·5 | | 62097 | 60366 | 58677 | 57031 | 55426 | 53861 | 52336 | 50849 | 49400 | 47988 |
| 2·6 | | 46612 | 45271 | 43965 | 42692 | 41453 | 40246 | 39070 | 37926 | 36811 | 35726 |
| 2·7 | | 34670 | 33642 | 32641 | 31667 | 30720 | 29798 | 28901 | 28028 | 27179 | 26354 |
| 2·8 | | 25551 | 24771 | 24012 | 23274 | 22557 | 21860 | 21182 | 20524 | 19884 | 19262 |
| 2·9 | | 18658 | 18071 | 17502 | 16948 | 16411 | 15889 | 15382 | 14890 | 14412 | 13949 |
| 3·0 | 0·0³ | 13499 | 13062 | 12639 | 12228 | 11829 | 11442 | 11067 | 10703 | 10350 | 10008 |
| 3·1 | | 96760 | 93544 | 90426 | 87403 | 84474 | 81635 | 78885 | 76219 | 73638 | 71136 |
| 3·2 | | 68714 | 66367 | 64095 | 61895 | 59765 | 57703 | 55706 | 53774 | 51904 | 50094 |
| 3·3 | | 48342 | 46648 | 45009 | 43423 | 41889 | 40406 | 38971 | 37584 | 36243 | 34946 |
| 3·4 | | 33693 | 32481 | 31311 | 30179 | 29086 | 28029 | 27009 | 26023 | 25071 | 24151 |
| 3·5 | 0·0⁴ | 23263 | 22405 | 21577 | 20778 | 20006 | 19262 | 18543 | 17849 | 17180 | 16534 |
| 3·6 | | 15911 | 15310 | 14730 | 14171 | 13632 | 13112 | 12611 | 12128 | 11662 | 11213 |
| 3·7 | | 10780 | 10363 | **99611** | **95740** | **92010** | **88417** | **84957** | **81624** | **78414** | **75324** |
| 3·8 | | 72348 | 69483 | 66726 | 64072 | 61517 | 59059 | 56694 | 54418 | 52228 | 50122 |
| 3·9 | | 48096 | 46148 | 44274 | 42473 | 40741 | 39076 | 37475 | 35936 | 34458 | 33037 |
| 4·0 | 0·0⁵ | 31671 | 30359 | 29099 | 27888 | 26726 | 25609 | 24536 | 23507 | 22518 | 21569 |
| 4·1 | | 20658 | 19783 | 18944 | 18138 | 17365 | 16624 | 15912 | 15230 | 14575 | 13948 |
| 4·2 | | 13346 | 12769 | 12215 | 11685 | 11176 | 10689 | 10221 | **97736** | **93447** | **89337** |
| 4·3 | | 85399 | 81627 | 78015 | 74555 | 71241 | 68069 | 65031 | 62123 | 59340 | 56675 |
| 4·4 | | 54125 | 51685 | 49350 | 47117 | 44979 | 42935 | 40980 | 39110 | 37322 | 35612 |
| 4·5 | 0·0⁶ | 33977 | 32414 | 30920 | 29492 | 28127 | 26823 | 25577 | 24386 | 23249 | 22162 |
| 4·6 | | 21125 | 20133 | 19187 | 18283 | 17420 | 16597 | 15810 | 15060 | 14344 | 13660 |
| 4·7 | | 13008 | 12386 | 11792 | 11226 | 10686 | 10171 | **96796** | **92113** | **87648** | **83391** |
| 4·8 | | 79333 | 75465 | 71779 | 68267 | 64920 | 61731 | 58693 | 55799 | 53043 | 50418 |
| 4·9 | | 47918 | 45538 | 43272 | 41115 | 39061 | 37107 | 35247 | 33476 | 31792 | 30190 |

## APPENDIX T4

### Percentage Points of the Standard Normal Distribution

The table gives the $100\alpha$ percentage points $u_\alpha$ of the standard Normal distribution, that is values $u_\alpha$ such that

$$P(U \geq u_\alpha) = \alpha.$$

In the notation of T3,

$$1 - \Phi(u_\alpha) = \alpha,$$

or

$$u_\alpha = \Phi^{-1}(1 - \alpha).$$

A central interval of content $1 - \alpha$ requires upper and lower tails of size $\frac{1}{2}\alpha$, and the interval is $(-u_{\alpha/2}, u_{\alpha/2})$

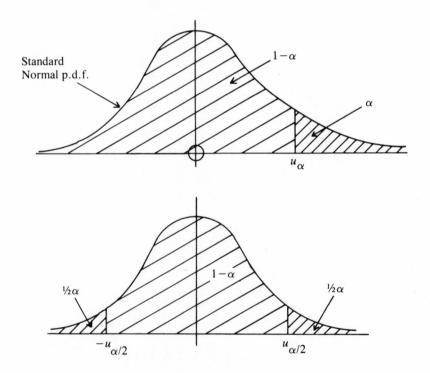

Percentage points of the Normal distribution

(Reproduced by permission of Macmillan Publishers Ltd. from *Statistical Tables for Science, Engineering and Management* by J. Murdoch and J. A. Barnes.)

| $\alpha$ | $u_\alpha$ | $\alpha$ | $u_\alpha$ | $\alpha$ | $u_\alpha$ | $\alpha$ | $u_\alpha$ | $\alpha$ | $u_\alpha$ | $\alpha$ | $u_\alpha$ |
|---|---|---|---|---|---|---|---|---|---|---|---|
| 0·50 | 0·0000 | 0·050 | 1·6449 | 0·030 | 1·8808 | 0·020 | 2·0537 | 0·010 | 2·3263 | 0·050 | 1·6449 |
| 0·45 | 0·1257 | 0·048 | 1·6646 | 0·029 | 1·8957 | 0·019 | 2·0749 | 0·009 | 2·3656 | 0·010 | 2·3263 |
| 0·40 | 0·2533 | 0·046 | 1·6849 | 0·028 | 1·9110 | 0·018 | 2·0969 | 0·008 | 2·4089 | 0·001 | 3·0902 |
| 0·35 | 0·3853 | 0·044 | 1·7060 | 0·027 | 1·9268 | 0·017 | 2·1201 | 0·007 | 2·4573 | 0·0001 | 3·7190 |
| 0·30 | 0·5244 | 0·042 | 1·7279 | 0·026 | 1·9431 | 0·016 | 2·1444 | 0·006 | 2·5121 | 0·00001 | 4·2649 |
| 0·25 | 0·6745 | 0·040 | 1·7507 | 0·025 | 1·9600 | 0·015 | 2·1701 | 0·005 | 2·5758 | 0·025 | 1·9600 |
| 0·20 | 0·8416 | 0·038 | 1·7744 | 0·024 | 1·9774 | 0·014 | 2·1973 | 0·004 | 2·6521 | 0·005 | 2·5758 |
| 0·15 | 1·0364 | 0·036 | 1·7991 | 0·023 | 1·9954 | 0·013 | 2·2262 | 0·003 | 2·7478 | 0·0005 | 3·2905 |
| 0·10 | 1·2816 | 0·034 | 1·8250 | 0·022 | 2·0141 | 0·012 | 2·2571 | 0·002 | 2·8782 | 0·00005 | 3·8906 |
| 0·05 | 1·6449 | 0·032 | 1·8522 | 0·021 | 2·0335 | 0·011 | 2·2904 | 0·001 | 3·0902 | 0·000005 | 4·4172 |

## APPENDIX T5

**Cumulative Probabilities of Student's Distribution (the '*t*-distribution')**
[see § 2.5.5]

The tabulated quantity is

$$p_t(\nu) = P\{T(\nu) \le t\}$$

where the random variable $T(\nu)$ has Student's distribution on $\nu$ degrees of freedom.

The probability in the upper tail beyond $t$ is $1 - p_t(\nu)$.

The table gives values of $p_t(\nu)$ for $\nu = 1(1)24$, 30, 40, 60, 120, $\infty$. For each of the tabulated values of $\nu$, $p_t(\nu)$ is given for non-negative values of $t$. For negative values use

$$p_{-t}(\nu) = 1 - p_t(\nu).$$

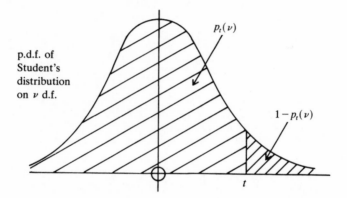

p.d.f. of Student's distribution on $\nu$ d.f.

$p_t(\nu)$

$1 - p_t(\nu)$

$t$

Probability integral, $p_t(\nu)$, of the $t$-distribution

| $t$ \ $\nu$ | 1 | 2 | 3 | 4 | 5 | 6 | 7 | 8 | 9 | 10 |
|---|---|---|---|---|---|---|---|---|---|---|
| 0·0 | 0·50000 | 0·50000 | 0·50000 | 0·50000 | 0·50000 | 0·50000 | 0·50000 | 0·50000 | 0·50000 | 0·50000 |
| 0·1 | 0·53173 | 0·53527 | 0·53667 | 0·53742 | 0·53788 | 0·53820 | 0·53843 | 0·53860 | 0·53873 | 0·53884 |
| 0·2 | 0·56283 | 0·57002 | 0·57286 | 0·57438 | 0·57532 | 0·57596 | 0·57642 | 0·57676 | 0·57704 | 0·57726 |
| 0·3 | 0·59277 | 0·60376 | 0·60812 | 0·61044 | 0·61188 | 0·61285 | 0·61356 | 0·61409 | 0·61450 | 0·61484 |
| 0·4 | 0·62112 | 0·63608 | 0·64203 | 0·64520 | 0·64716 | 0·64850 | 0·64946 | 0·65019 | 0·65076 | 0·65122 |
| 0·5 | 0·64758 | 0·66667 | 0·67428 | 0·67834 | 0·68085 | 0·68256 | 0·68380 | 0·68473 | 0·68546 | 0·68605 |
| 0·6 | 0·67202 | 0·69529 | 0·70460 | 0·70958 | 0·71267 | 0·71477 | 0·71629 | 0·71745 | 0·71835 | 0·71907 |
| 0·7 | 0·69440 | 0·72181 | 0·73284 | 0·73875 | 0·74243 | 0·74493 | 0·74674 | 0·74811 | 0·74919 | 0·75006 |
| 0·8 | 0·71478 | 0·74618 | 0·75890 | 0·76574 | 0·76999 | 0·77289 | 0·77500 | 0·77659 | 0·77784 | 0·77885 |
| 0·9 | 0·73326 | 0·76845 | 0·78277 | 0·79050 | 0·79531 | 0·79860 | 0·80099 | 0·80280 | 0·80422 | 0·80536 |
| 1·0 | 0·75000 | 0·78868 | 0·80450 | 0·81305 | 0·81839 | 0·82204 | 0·82469 | 0·82670 | 0·82828 | 0·82955 |
| 1·1 | 0·76515 | 0·80698 | 0·82416 | 0·83346 | 0·83927 | 0·84325 | 0·84614 | 0·84834 | 0·85006 | 0·85145 |
| 1·2 | 0·77886 | 0·82349 | 0·84187 | 0·85182 | 0·85805 | 0·86232 | 0·86541 | 0·86777 | 0·86961 | 0·87110 |
| 1·3 | 0·79129 | 0·83838 | 0·85777 | 0·86827 | 0·87485 | 0·87935 | 0·88262 | 0·88510 | 0·88705 | 0·88862 |
| 1·4 | 0·80257 | 0·85177 | 0·87200 | 0·88295 | 0·88980 | 0·89448 | 0·89788 | 0·90046 | 0·90249 | 0·90412 |
| 1·5 | 0·81283 | 0·86380 | 0·88471 | 0·89600 | 0·90305 | 0·90786 | 0·91135 | 0·91400 | 0·91608 | 0·91775 |
| 1·6 | 0·82219 | 0·87464 | 0·89605 | 0·90758 | 0·91475 | 0·91964 | 0·92318 | 0·92587 | 0·92797 | 0·92966 |
| 1·7 | 0·83075 | 0·88439 | 0·90615 | 0·91782 | 0·92506 | 0·92998 | 0·93354 | 0·93622 | 0·93833 | 0·94002 |
| 1·8 | 0·83859 | 0·89317 | 0·91516 | 0·92688 | 0·93412 | 0·93902 | 0·94256 | 0·94522 | 0·94731 | 0·94897 |
| 1·9 | 0·84579 | 0·90109 | 0·92318 | 0·93488 | 0·94207 | 0·94691 | 0·95040 | 0·95302 | 0·95506 | 0·95669 |
| 2·0 | 0·85242 | 0·90825 | 0·93034 | 0·94194 | 0·94903 | 0·95379 | 0·95719 | 0·95974 | 0·96172 | 0·96331 |
| 2·1 | 0·85854 | 0·91473 | 0·93672 | 0·94817 | 0·95512 | 0·95976 | 0·96306 | 0·96553 | 0·96744 | 0·96896 |
| 2·2 | 0·86420 | 0·92060 | 0·94241 | 0·95367 | 0·96045 | 0·96495 | 0·96813 | 0·97050 | 0·97233 | 0·97378 |
| 2·3 | 0·86945 | 0·92593 | 0·94751 | 0·95853 | 0·96511 | 0·96945 | 0·97250 | 0·97476 | 0·97650 | 0·97787 |
| 2·4 | 0·87433 | 0·93077 | 0·95206 | 0·96282 | 0·96919 | 0·97335 | 0·97627 | 0·97841 | 0·98005 | 0·98134 |
| 2·5 | 0·87888 | 0·93519 | 0·95615 | 0·96662 | 0·97275 | 0·97674 | 0·97950 | 0·98153 | 0·98307 | 0·98428 |
| 2·6 | 0·88313 | 0·93923 | 0·95981 | 0·96998 | 0·97587 | 0·97967 | 0·98229 | 0·98419 | 0·98563 | 0·98675 |
| 2·7 | 0·88709 | 0·94292 | 0·96311 | 0·97295 | 0·97861 | 0·98221 | 0·98468 | 0·98646 | 0·98780 | 0·98884 |

| | | | | | | | | | | |
|---|---|---|---|---|---|---|---|---|---|---|
| 2·8 | 0·89081 | 0·94630 | 0·96607 | 0·97559 | 0·98100 | 0·98442 | 0·98674 | 0·98840 | 0·98964 | 0·99060 |
| 2·9 | 0·89430 | 0·94941 | 0·96875 | 0·97794 | 0·98310 | 0·98633 | 0·98851 | 0·99005 | 0·99120 | 0·99208 |
| 3·0 | 0·89758 | 0·95227 | 0·97116 | 0·98003 | 0·98495 | 0·98800 | 0·99003 | 0·99146 | 0·99252 | 0·99333 |
| 3·1 | 0·90067 | 0·95490 | 0·97335 | 0·98189 | 0·98657 | 0·98944 | 0·99134 | 0·99267 | 0·99364 | 0·99437 |
| 3·2 | 0·90359 | 0·95733 | 0·97533 | 0·98355 | 0·98800 | 0·99070 | 0·99247 | 0·99369 | 0·99459 | 0·99525 |
| 3·3 | 0·90634 | 0·95958 | 0·97713 | 0·98503 | 0·98926 | 0·99180 | 0·99344 | 0·99457 | 0·99539 | 0·99599 |
| 3·4 | 0·90895 | 0·96166 | 0·97877 | 0·98636 | 0·99037 | 0·99275 | 0·99428 | 0·99532 | 0·99606 | 0·99661 |
| 3·5 | 0·91141 | 0·96358 | 0·98026 | 0·98755 | 0·99136 | 0·99359 | 0·99500 | 0·99596 | 0·99664 | 0·99714 |
| 3·6 | 0·91376 | 0·96538 | 0·98162 | 0·98862 | 0·99223 | 0·99432 | 0·99563 | 0·99651 | 0·99713 | 0·99758 |
| 3·7 | 0·91598 | 0·96705 | 0·98286 | 0·98958 | 0·99300 | 0·99496 | 0·99617 | 0·99698 | 0·99754 | 0·99795 |
| 3·8 | 0·91809 | 0·96860 | 0·98400 | 0·99045 | 0·99369 | 0·99552 | 0·99664 | 0·99738 | 0·99789 | 0·99826 |
| 3·9 | 0·92010 | 0·97005 | 0·98504 | 0·99123 | 0·99430 | 0·99601 | 0·99705 | 0·99773 | 0·99819 | 0·99852 |
| 4·0 | 0·92202 | 0·97141 | 0·98600 | 0·99193 | 0·99484 | 0·99644 | 0·99741 | 0·99803 | 0·99845 | 0·99874 |
| 4·2 | 0·92560 | 0·97386 | 0·98768 | 0·99315 | 0·99575 | 0·99716 | 0·99798 | 0·99850 | 0·99885 | 0·99909 |
| 4·4 | 0·92887 | 0·97602 | 0·98912 | 0·99415 | 0·99649 | 0·99772 | 0·99842 | 0·99886 | 0·99914 | 0·99933 |
| 4·6 | 0·93186 | 0·97792 | 0·99034 | 0·99498 | 0·99708 | 0·99815 | 0·99876 | 0·99912 | 0·99936 | 0·99951 |
| 4·8 | 0·93462 | 0·97962 | 0·99140 | 0·99568 | 0·99756 | 0·99850 | 0·99902 | 0·99932 | 0·99951 | 0·99964 |
| 5·0 | 0·93717 | 0·98113 | 0·99230 | 0·99625 | 0·99795 | 0·99877 | 0·99922 | 0·99947 | 0·99963 | 0·99973 |
| 5·2 | 0·93952 | 0·98248 | 0·99309 | 0·99674 | 0·99827 | 0·99899 | 0·99937 | 0·99959 | 0·99972 | 0·99980 |
| 5·4 | 0·94171 | 0·98369 | 0·99378 | 0·99715 | 0·99853 | 0·99917 | 0·99950 | 0·99968 | 0·99978 | 0·99985 |
| 5·6 | 0·94375 | 0·98478 | 0·99437 | 0·99750 | 0·99875 | 0·99931 | 0·99959 | 0·99975 | 0·99983 | 0·99989 |
| 5·8 | 0·94565 | 0·98577 | 0·99490 | 0·99780 | 0·99893 | 0·99942 | 0·99967 | 0·99980 | 0·99987 | 0·99991 |
| 6·0 | 0·94743 | 0·98666 | 0·99536 | 0·99806 | 0·99908 | 0·99952 | 0·99973 | 0·99984 | 0·99990 | 0·99993 |
| 6·2 | 0·94910 | 0·98748 | 0·99577 | 0·99828 | 0·99920 | 0·99959 | 0·99978 | 0·99987 | 0·99992 | 0·99995 |
| 6·4 | 0·95066 | 0·98822 | 0·99614 | 0·99847 | 0·99931 | 0·99966 | 0·99982 | 0·99990 | 0·99994 | 0·99996 |
| 6·6 | 0·95214 | 0·98890 | 0·99646 | 0·99863 | 0·99940 | 0·99971 | 0·99985 | 0·99992 | 0·99995 | 0·99997 |
| 6·8 | 0·95352 | 0·98953 | 0·99675 | 0·99878 | 0·99948 | 0·99975 | 0·99987 | 0·99993 | 0·99996 | 0·99998 |
| 7·0 | 0·95483 | 0·99010 | 0·99701 | 0·99890 | 0·99954 | 0·99979 | 0·99990 | 0·99994 | 0·99997 | 0·99998 |
| 7·2 | 0·95607 | 0·99063 | 0·99724 | 0·99901 | 0·99960 | 0·99982 | 0·99991 | 0·99995 | 0·99997 | 0·99999 |
| 7·4 | 0·95724 | 0·99111 | 0·99745 | 0·99911 | 0·99964 | 0·99984 | 0·99993 | 0·99996 | 0·99998 | 0·99999 |
| 7·6 | 0·95836 | 0·99156 | 0·99764 | 0·99920 | 0·99969 | 0·99986 | 0·99994 | 0·99997 | 0·99998 | 0·99999 |
| 7·8 | 0·95941 | 0·99198 | 0·99781 | 0·99927 | 0·99972 | 0·99988 | 0·99995 | 0·99997 | 0·99999 | 0·99999 |
| 8·0 | 0·96042 | 0·99237 | 0·99796 | 0·99934 | 0·99975 | 0·99990 | 0·99996 | 0·99998 | 0·99999 | 0·99999 |

Probability integral of the *t*-distribution (continued)

| $t$ \\ $\nu$ | 11 | 12 | 13 | 14 | 15 | 16 | 17 | 18 | 19 | 20 |
|---|---|---|---|---|---|---|---|---|---|---|
| 0·0 | 0·50000 | 0·50000 | 0·50000 | 0·50000 | 0·50000 | 0·50000 | 0·50000 | 0·50000 | 0·50000 | 0·50000 |
| 0·1 | 0·53893 | 0·53900 | 0·53907 | 0·53912 | 0·53917 | 0·53921 | 0·53924 | 0·53928 | 0·53930 | 0·53933 |
| 0·2 | 0·57744 | 0·57759 | 0·57771 | 0·57782 | 0·57792 | 0·57800 | 0·57807 | 0·57814 | 0·57820 | 0·57825 |
| 0·3 | 0·61511 | 0·61534 | 0·61554 | 0·61571 | 0·61585 | 0·61598 | 0·61609 | 0·61619 | 0·61628 | 0·61636 |
| 0·4 | 0·65159 | 0·65191 | 0·65217 | 0·65240 | 0·65260 | 0·65278 | 0·65293 | 0·65307 | 0·65319 | 0·65330 |
| 0·5 | 0·68654 | 0·68694 | 0·68728 | 0·68758 | 0·68783 | 0·68806 | 0·68826 | 0·68843 | 0·68859 | 0·68873 |
| 0·6 | 0·71967 | 0·72017 | 0·72059 | 0·72095 | 0·72127 | 0·72155 | 0·72179 | 0·72201 | 0·72220 | 0·72238 |
| 0·7 | 0·75077 | 0·75136 | 0·75187 | 0·75230 | 0·75268 | 0·75301 | 0·75330 | 0·75356 | 0·75380 | 0·75400 |
| 0·8 | 0·77968 | 0·78037 | 0·78096 | 0·78146 | 0·78190 | 0·78229 | 0·78263 | 0·78293 | 0·78320 | 0·78344 |
| 0·9 | 0·80630 | 0·80709 | 0·80776 | 0·80833 | 0·80883 | 0·80927 | 0·80965 | 0·81000 | 0·81031 | 0·81058 |
| 1·0 | 0·83060 | 0·83148 | 0·83222 | 0·83286 | 0·83341 | 0·83390 | 0·83433 | 0·83472 | 0·83506 | 0·83537 |
| 1·1 | 0·85259 | 0·85355 | 0·85436 | 0·85506 | 0·85566 | 0·85620 | 0·85667 | 0·85709 | 0·85746 | 0·85780 |
| 1·2 | 0·87233 | 0·87335 | 0·87422 | 0·87497 | 0·87562 | 0·87620 | 0·87670 | 0·87715 | 0·87756 | 0·87792 |
| 1·3 | 0·88991 | 0·89099 | 0·89191 | 0·89270 | 0·89339 | 0·89399 | 0·89452 | 0·89500 | 0·89542 | 0·89581 |
| 1·4 | 0·90546 | 0·90658 | 0·90754 | 0·90836 | 0·90907 | 0·90970 | 0·91025 | 0·91074 | 0·91118 | 0·91158 |
| 1·5 | 0·91912 | 0·92027 | 0·92125 | 0·92209 | 0·92282 | 0·92346 | 0·92402 | 0·92452 | 0·92498 | 0·92538 |
| 1·6 | 0·93105 | 0·93221 | 0·93320 | 0·93404 | 0·93478 | 0·93542 | 0·93599 | 0·93650 | 0·93695 | 0·93736 |
| 1·7 | 0·94140 | 0·94256 | 0·94354 | 0·94439 | 0·94512 | 0·94576 | 0·94632 | 0·94683 | 0·94728 | 0·94768 |
| 1·8 | 0·95034 | 0·95148 | 0·95245 | 0·95328 | 0·95400 | 0·95463 | 0·95518 | 0·95568 | 0·95612 | 0·95652 |
| 1·9 | 0·95802 | 0·95914 | 0·96008 | 0·96089 | 0·96158 | 0·96220 | 0·96273 | 0·96321 | 0·96364 | 0·96403 |
| 2·0 | 0·96460 | 0·96567 | 0·96658 | 0·96736 | 0·96803 | 0·96861 | 0·96913 | 0·96959 | 0·97000 | 0·97037 |
| 2·1 | 0·97020 | 0·97123 | 0·97209 | 0·97283 | 0·97347 | 0·97403 | 0·97452 | 0·97495 | 0·97534 | 0·97569 |
| 2·2 | 0·97496 | 0·97593 | 0·97675 | 0·97745 | 0·97805 | 0·97858 | 0·97904 | 0·97945 | 0·97981 | 0·98014 |
| 2·3 | 0·97898 | 0·97990 | 0·98067 | 0·98132 | 0·98189 | 0·98238 | 0·98281 | 0·98319 | 0·98352 | 0·98383 |
| 2·4 | 0·98238 | 0·98324 | 0·98396 | 0·98457 | 0·98509 | 0·98554 | 0·98594 | 0·98629 | 0·98660 | 0·98688 |
| 2·5 | 0·98525 | 0·98604 | 0·98671 | 0·98727 | 0·98775 | 0·98816 | 0·98853 | 0·98885 | 0·98913 | 0·98938 |
| 2·6 | 0·98765 | 0·98839 | 0·98900 | 0·98951 | 0·98995 | 0·99033 | 0·99066 | 0·99095 | 0·99121 | 0·99144 |

| | .00 | .01 | .02 | .03 | .04 | .05 | .06 | .07 | .08 | .09 |
|---|---|---|---|---|---|---|---|---|---|---|
| 2·7 | 0·98967 | 0·99035 | 0·99090 | 0·99137 | 0·99177 | 0·99211 | 0·99241 | 0·99267 | 0·99290 | 0·99311 |
| 2·8 | 0·99136 | 0·99198 | 0·99249 | 0·99291 | 0·99327 | 0·99358 | 0·99385 | 0·99408 | 0·99429 | 0·99447 |
| 2·9 | 0·99278 | 0·99334 | 0·99380 | 0·99418 | 0·99450 | 0·99478 | 0·99502 | 0·99523 | 0·99541 | 0·99557 |
| 3·0 | 0·99396 | 0·99447 | 0·99488 | 0·99522 | 0·99551 | 0·99576 | 0·99597 | 0·99616 | 0·99632 | 0·99646 |
| 3·1 | 0·99495 | 0·99541 | 0·99578 | 0·99608 | 0·99634 | 0·99656 | 0·99675 | 0·99691 | 0·99705 | 0·99718 |
| 3·2 | 0·99577 | 0·99618 | 0·99652 | 0·99679 | 0·99702 | 0·99721 | 0·99738 | 0·99752 | 0·99764 | 0·99775 |
| 3·3 | 0·99646 | 0·99683 | 0·99713 | 0·99737 | 0·99757 | 0·99774 | 0·99789 | 0·99801 | 0·99812 | 0·99821 |
| 3·4 | 0·99703 | 0·99737 | 0·99763 | 0·99784 | 0·99802 | 0·99817 | 0·99830 | 0·99840 | 0·99850 | 0·99858 |
| 3·5 | 0·99751 | 0·99781 | 0·99804 | 0·99823 | 0·99839 | 0·99852 | 0·99863 | 0·99872 | 0·99880 | 0·99887 |
| 3·6 | 0·99791 | 0·99818 | 0·99838 | 0·99855 | 0·99869 | 0·99880 | 0·99890 | 0·99898 | 0·99905 | 0·99911 |
| 3·7 | 0·99825 | 0·99848 | 0·99867 | 0·99881 | 0·99893 | 0·99903 | 0·99911 | 0·99918 | 0·99924 | 0·99929 |
| 3·8 | 0·99853 | 0·99874 | 0·99890 | 0·99902 | 0·99913 | 0·99921 | 0·99928 | 0·99934 | 0·99939 | 0·99944 |
| 3·9 | 0·99876 | 0·99895 | 0·99909 | 0·99920 | 0·99929 | 0·99936 | 0·99942 | 0·99948 | 0·99952 | 0·99956 |
| 4·0 | 0·99896 | 0·99912 | 0·99924 | 0·99934 | 0·99942 | 0·99948 | 0·99954 | 0·99958 | 0·99962 | 0·99965 |
| 4·2 | 0·99926 | 0·99938 | 0·99948 | 0·99955 | 0·99961 | 0·99966 | 0·99970 | 0·99973 | 0·99976 | 0·99978 |
| 4·4 | 0·99947 | 0·99957 | 0·99964 | 0·99970 | 0·99974 | 0·99978 | 0·99980 | 0·99983 | 0·99985 | 0·99986 |
| 4·6 | 0·99962 | 0·99969 | 0·99975 | 0·99979 | 0·99983 | 0·99985 | 0·99987 | 0·99989 | 0·99990 | 0·99991 |
| 4·8 | 0·99972 | 0·99978 | 0·99983 | 0·99986 | 0·99988 | 0·99990 | 0·99992 | 0·99993 | 0·99994 | 0·99995 |
| 5·0 | 0·99980 | 0·99985 | 0·99988 | 0·99990 | 0·99992 | 0·99993 | 0·99995 | 0·99995 | 0·99996 | 0·99997 |
| 5·2 | 0·99985 | 0·99989 | 0·99992 | 0·99993 | 0·99995 | 0·99996 | 0·99996 | 0·99997 | 0·99997 | 0·99998 |
| 5·4 | 0·99989 | 0·99992 | 0·99994 | 0·99995 | 0·99996 | 0·99997 | 0·99998 | 0·99998 | 0·99998 | 0·99999 |
| 5·6 | 0·99992 | 0·99994 | 0·99996 | 0·99997 | 0·99997 | 0·99998 | 0·99998 | 0·99999 | 0·99999 | 0·99999 |
| 5·8 | 0·99994 | 0·99996 | 0·99997 | 0·99998 | 0·99998 | 0·99999 | 0·99999 | 0·99999 | 0·99999 | 0·99999 |
| 6·0 | 0·99995 | 0·99997 | 0·99998 | 0·99998 | 0·99999 | 0·99999 | 0·99999 | 0·99999 | | |
| 6·2 | 0·99997 | 0·99998 | 0·99998 | 0·99999 | 0·99999 | 0·99999 | | | | |
| 6·4 | 0·99997 | 0·99998 | 0·99999 | 0·99999 | | | | | | |
| 6·6 | 0·99998 | 0·99999 | 0·99999 | | | | | | | |
| 6·8 | 0·99998 | 0·99999 | | | | | | | | |
| 7·0 | 0·99999 | | | | | | | | | |

Probability integral of the *t*-distribution (continued)

| $t$ \ $\nu$ | 20 | 21 | 22 | 23 | 24 | 30 | 40 | 60 | 120 | $\infty$ |
|---|---|---|---|---|---|---|---|---|---|---|
| 0·00 | 0·50000 | 0·50000 | 0·50000 | 0·50000 | 0·50000 | 0·50000 | 0·50000 | 0·50000 | 0·50000 | 0·50000 |
| 0·05 | 0·51969 | 0·51970 | 0·51971 | 0·51972 | 0·51973 | 0·51977 | 0·51981 | 0·51986 | 0·51990 | 0·51994 |
| 0·10 | 0·53933 | 0·53935 | 0·53938 | 0·53939 | 0·53941 | 0·53950 | 0·53958 | 0·53966 | 0·53974 | 0·53983 |
| 0·15 | 0·55887 | 0·55890 | 0·55893 | 0·55896 | 0·55899 | 0·55912 | 0·55924 | 0·55937 | 0·55949 | 0·55962 |
| 0·20 | 0·57825 | 0·57830 | 0·57834 | 0·57838 | 0·57842 | 0·57858 | 0·57875 | 0·57892 | 0·57909 | 0·57926 |
| 0·25 | 0·59743 | 0·59749 | 0·59755 | 0·59760 | 0·59764 | 0·59785 | 0·59807 | 0·59828 | 0·59849 | 0·59871 |
| 0·30 | 0·61636 | 0·61644 | 0·61650 | 0·61656 | 0·61662 | 0·61688 | 0·61713 | 0·61739 | 0·61765 | 0·61791 |
| 0·35 | 0·63500 | 0·63509 | 0·63517 | 0·63524 | 0·63530 | 0·63561 | 0·63591 | 0·63622 | 0·63652 | 0·63683 |
| 0·40 | 0·65330 | 0·65340 | 0·65349 | 0·65358 | 0·65365 | 0·65400 | 0·65436 | 0·65471 | 0·65507 | 0·65542 |
| 0·45 | 0·67122 | 0·67134 | 0·67144 | 0·67154 | 0·67163 | 0·67203 | 0·67243 | 0·67283 | 0·67324 | 0·67364 |
| 0·50 | 0·68873 | 0·68886 | 0·68898 | 0·68909 | 0·68919 | 0·68964 | 0·69009 | 0·69055 | 0·69100 | 0·69146 |
| 0·55 | 0·70579 | 0·70594 | 0·70607 | 0·70619 | 0·70630 | 0·70680 | 0·70731 | 0·70782 | 0·70833 | 0·70884 |
| 0·60 | 0·72238 | 0·72254 | 0·72268 | 0·72281 | 0·72294 | 0·72349 | 0·72405 | 0·72462 | 0·72518 | 0·72575 |
| 0·65 | 0·73846 | 0·73863 | 0·73879 | 0·73893 | 0·73907 | 0·73968 | 0·74030 | 0·74091 | 0·74153 | 0·74215 |
| 0·70 | 0·75400 | 0·75419 | 0·75437 | 0·75453 | 0·75467 | 0·75534 | 0·75601 | 0·75668 | 0·75736 | 0·75804 |
| 0·75 | 0·76901 | 0·76921 | 0·76940 | 0·76957 | 0·76973 | 0·77045 | 0·77118 | 0·77191 | 0·77264 | 0·77337 |
| 0·80 | 0·78344 | 0·78367 | 0·78387 | 0·78405 | 0·78422 | 0·78500 | 0·78578 | 0·78657 | 0·78735 | 0·78814 |
| 0·85 | 0·79731 | 0·79754 | 0·79776 | 0·79796 | 0·79814 | 0·79897 | 0·79981 | 0·80065 | 0·80149 | 0·80234 |
| 0·90 | 0·81058 | 0·81084 | 0·81107 | 0·81128 | 0·81147 | 0·81236 | 0·81325 | 0·81414 | 0·81504 | 0·81594 |
| 0·95 | 0·82327 | 0·82354 | 0·82378 | 0·82401 | 0·82421 | 0·82515 | 0·82609 | 0·82704 | 0·82799 | 0·82894 |
| 1·00 | 0·83537 | 0·83565 | 0·83591 | 0·83614 | 0·83636 | 0·83735 | 0·83834 | 0·83934 | 0·84034 | 0·84134 |
| 1·05 | 0·84688 | 0·84717 | 0·84744 | 0·84769 | 0·84791 | 0·84895 | 0·84999 | 0·85104 | 0·85209 | 0·85314 |
| 1·10 | 0·85780 | 0·85811 | 0·85839 | 0·85864 | 0·85888 | 0·85996 | 0·86105 | 0·86214 | 0·86323 | 0·86433 |
| 1·15 | 0·86814 | 0·86846 | 0·86875 | 0·86902 | 0·86926 | 0·87039 | 0·87151 | 0·87265 | 0·87378 | 0·87493 |
| 1·20 | 0·87792 | 0·87825 | 0·87855 | 0·87882 | 0·87907 | 0·88023 | 0·88140 | 0·88257 | 0·88375 | 0·88493 |

| x | | | | | | | | | | |
|---|---|---|---|---|---|---|---|---|---|---|
| 1·25 | 0·89435 | 0·89313 | 0·89192 | 0·89072 | 0·88952 | 0·88832 | 0·88807 | 0·88778 | 0·88747 | 0·88714 |
| 1·30 | 0·90320 | 0·90195 | 0·90071 | 0·89948 | 0·89825 | 0·89703 | 0·89676 | 0·89647 | 0·89616 | 0·89581 |
| 1·35 | 0·91149 | 0·91022 | 0·90896 | 0·90770 | 0·90644 | 0·90519 | 0·90492 | 0·90463 | 0·90431 | 0·90395 |
| 1·40 | 0·91924 | 0·91795 | 0·91667 | 0·91539 | 0·91411 | 0·91285 | 0·91257 | 0·91227 | 0·91194 | 0·91158 |
| 1·45 | 0·92647 | 0·92517 | 0·92387 | 0·92257 | 0·92128 | 0·92000 | 0·91972 | 0·91942 | 0·91908 | 0·91872 |
| 1·50 | 0·93319 | 0·93188 | 0·93057 | 0·92927 | 0·92797 | 0·92667 | 0·92639 | 0·92608 | 0·92575 | 0·92538 |
| 1·55 | 0·93943 | 0·93811 | 0·93680 | 0·93549 | 0·93419 | 0·93289 | 0·93260 | 0·93230 | 0·93196 | 0·93159 |
| 1·60 | 0·94520 | 0·94389 | 0·94257 | 0·94127 | 0·93996 | 0·93866 | 0·93838 | 0·93807 | 0·93773 | 0·93736 |
| 1·65 | 0·95053 | 0·94922 | 0·94792 | 0·94661 | 0·94531 | 0·94401 | 0·94373 | 0·94342 | 0·94309 | 0·94272 |
| 1·70 | 0·95543 | 0·95414 | 0·95284 | 0·95155 | 0·95026 | 0·94897 | 0·94869 | 0·94839 | 0·94805 | 0·94768 |
| 1·75 | 0·95994 | 0·95866 | 0·95738 | 0·95611 | 0·95483 | 0·95355 | 0·95327 | 0·95297 | 0·95264 | 0·95228 |
| 1·80 | 0·96407 | 0·96281 | 0·96156 | 0·96030 | 0·95904 | 0·95778 | 0·95750 | 0·95720 | 0·95688 | 0·95652 |
| 1·85 | 0·96784 | 0·96661 | 0·96538 | 0·96414 | 0·96291 | 0·96167 | 0·96140 | 0·96110 | 0·96078 | 0·96043 |
| 1·90 | 0·97128 | 0·97008 | 0·96888 | 0·96767 | 0·96646 | 0·96524 | 0·96498 | 0·96469 | 0·96437 | 0·96403 |
| 1·95 | 0·97441 | 0·97325 | 0·97207 | 0·97089 | 0·96971 | 0·96852 | 0·96827 | 0·96798 | 0·96767 | 0·96733 |
| 2·0 | 0·97725 | 0·97612 | 0·97498 | 0·97384 | 0·97269 | 0·97153 | 0·97128 | 0·97100 | 0·97070 | 0·97037 |
| 2·1 | 0·98214 | 0·98109 | 0·98003 | 0·97896 | 0·97788 | 0·97679 | 0·97655 | 0·97629 | 0·97601 | 0·97569 |
| 2·2 | 0·98610 | 0·98514 | 0·98416 | 0·98318 | 0·98218 | 0·98116 | 0·98094 | 0·98070 | 0·98043 | 0·98014 |
| 2·3 | 0·98928 | 0·98841 | 0·98753 | 0·98663 | 0·98571 | 0·98478 | 0·98457 | 0·98435 | 0·98410 | 0·98383 |
| 2·4 | 0·99180 | 0·99103 | 0·99024 | 0·98943 | 0·98860 | 0·98774 | 0·98756 | 0·98735 | 0·98712 | 0·98688 |
| 2·5 | 0·99379 | 0·99312 | 0·99241 | 0·99169 | 0·99094 | 0·99017 | 0·99000 | 0·98982 | 0·98961 | 0·98938 |
| 2·6 | 0·99534 | 0·99475 | 0·99414 | 0·99350 | 0·99284 | 0·99215 | 0·99200 | 0·99183 | 0·99164 | 0·99144 |
| 2·7 | 0·99653 | 0·99603 | 0·99550 | 0·99494 | 0·99436 | 0·99375 | 0·99361 | 0·99346 | 0·99329 | 0·99311 |
| 2·8 | 0·99744 | 0·99702 | 0·99657 | 0·99608 | 0·99557 | 0·99504 | 0·99492 | 0·99478 | 0·99463 | 0·99447 |
| 2·9 | 0·99813 | 0·99778 | 0·99740 | 0·99698 | 0·99654 | 0·99607 | 0·99596 | 0·99585 | 0·99572 | 0·99557 |
| 3·0 | 0·99865 | 0·99836 | 0·99804 | 0·99768 | 0·99730 | 0·99690 | 0·99681 | 0·99670 | 0·99659 | 0·99646 |
| 3·1 | 0·99903 | 0·99879 | 0·99853 | 0·99823 | 0·99791 | 0·99756 | 0·99748 | 0·99739 | 0·99729 | 0·99718 |
| 3·2 | 0·99931 | 0·99912 | 0·99890 | 0·99865 | 0·99838 | 0·99808 | 0·99801 | 0·99793 | 0·99785 | 0·99775 |
| 3·3 | 0·99952 | 0·99936 | 0·99918 | 0·99898 | 0·99875 | 0·99849 | 0·99844 | 0·99837 | 0·99829 | 0·99821 |
| 3·4 | 0·99966 | 0·99954 | 0·99940 | 0·99923 | 0·99904 | 0·99882 | 0·99877 | 0·99871 | 0·99865 | 0·99858 |

| ν \ t | 20 | 21 | 22 | 23 | 24 | 30 | 40 | 60 | 120 | ∞ |
|---|---|---|---|---|---|---|---|---|---|---|
| 3·5 | 0·99887 | 0·99893 | 0·99899 | 0·99904 | 0·99908 | 0·99926 | 0·99942 | 0·99956 | 0·99967 | 0·99977 |
| 3·6 | 0·99911 | 0·99916 | 0·99920 | 0·99925 | 0·99928 | 0·99943 | 0·99957 | 0·99968 | 0·99977 | 0·99984 |
| 3·7 | 0·99929 | 0·99933 | 0·99937 | 0·99941 | 0·99944 | 0·99957 | 0·99967 | 0·99976 | 0·99984 | 0·99989 |
| 3·8 | 0·99944 | 0·99948 | 0·99951 | 0·99954 | 0·99956 | 0·99967 | 0·99976 | 0·99983 | 0·99989 | 0·99993 |
| 3·9 | 0·99956 | 0·99959 | 0·99961 | 0·99964 | 0·99966 | 0·99975 | 0·99982 | 0·99988 | 0·99992 | 0·99995 |
| 4·0 | 0·99965 | 0·99967 | 0·99970 | 0·99972 | 0·99974 | 0·99981 | 0·99987 | 0·99991 | 0·99995 | 0·99997 |
| 5·0 | 0·99997 | 0·99997 | 0·99998 | 0·99998 | 0·99998 | 0·99999 | 0·99999 | | | |

Upper percentage points of $t$

| $1 - p_t(\nu)$ | $\nu = 1$ | 2 | 3 | 4 | 5 | 6 | 7 | 8 | 9 | 10 |
|---|---|---|---|---|---|---|---|---|---|---|
| $10^{-3}$ | 318·3 | 22·33 | 10·21 | 7·17 | 5·89 | 5·21 | 4·79 | 4·50 | 4·30 | 4·14 |
| $10^{-4}$ | 3183 | 70·7 | 22·20 | 13·03 | 9·68 | 8·02 | 7·06 | 6·44 | 6·01 | 5·69 |
| $10^{-5}$ | 31831 | 224 | 47·91 | 23·33 | 15·54 | 12·03 | 10·11 | 8·90 | 8·10 | 7·53 |
| $5 \times 10^{-6}$ | 63652 | 316 | 60·40 | 27·82 | 17·89 | 13·55 | 11·22 | 9·79 | 8·83 | 8·15 |

## APPENDIX T6

**Percentage Points of the $\chi^2$-distribution** [see § 2.5.4(a)]

The table gives the $100\alpha$ percentage points $\chi^2(\alpha; \nu)$ of the $\chi^2$-distribution on $\nu$ degrees of freedom, that is, values $\chi^2(\alpha; \nu)$ such that

$$P\{X_\nu^2 \geq \chi^2(\alpha; \nu)\} = \alpha,$$

where the random variable $X_\nu^2$ has the chi-squared distribution on $\nu$ degrees of freedom.

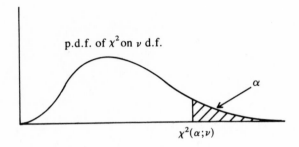

Percentage points of the $\chi^2$ distribution

(Reproduced by permission of Longman Group Ltd. from *Statistical Tables for Biological, Agricultural and Medical Research* by R. A. Fisher and F. Yates, 1974.)

| $\nu$ | Probability | | | | | | | | | | | | | |
|---|---|---|---|---|---|---|---|---|---|---|---|---|---|---|
| | 0·99 | 0·98 | 0·95 | 0·90 | 0·80 | 0·70 | 0·50 | 0·30 | 0·20 | 0·10 | 0·05 | 0·02 | 0·01 | 0·001 |
| 1 | 0·0³157 | 0·0³628 | 0·00393 | 0·0158 | 0·0642 | 0·148 | 0·455 | 1·074 | 1·642 | 2·706 | 3·841 | 5·412 | 6·635 | 10·827 |
| 2 | 0·0201 | 0·404 | 0·103 | 0·211 | 0·446 | 0·713 | 1·386 | 2·408 | 3·219 | 4·605 | 5·991 | 7·824 | 9·210 | 13·815 |
| 3 | 0·115 | 0·185 | 0·352 | 0·584 | 1·005 | 1·424 | 2·366 | 3·665 | 4·642 | 6·251 | 7·815 | 9·837 | 11·345 | 16·266 |
| 4 | 0·297 | 0·429 | 0·711 | 1·064 | 1·649 | 2·195 | 3·357 | 4·878 | 5·989 | 7·779 | 9·488 | 11·688 | 13·277 | 18·467 |
| 5 | 0·554 | 0·752 | 1·145 | 1·610 | 2·343 | 3·000 | 4·351 | 6·064 | 7·289 | 9·236 | 11·070 | 13·388 | 15·086 | 20·515 |
| 6 | 0·872 | 1·134 | 1·635 | 2·204 | 3·070 | 3·828 | 5·348 | 7·231 | 8·558 | 10·645 | 12·592 | 15·033 | 16·812 | 22·457 |
| 7 | 1·239 | 1·564 | 2·167 | 2·833 | 3·822 | 4·671 | 6·346 | 8·383 | 9·803 | 12·017 | 14·067 | 16·622 | 18·475 | 24·322 |
| 8 | 1·646 | 2·032 | 2·733 | 3·490 | 4·594 | 5·527 | 7·344 | 9·524 | 11·030 | 13·362 | 15·507 | 18·168 | 20·090 | 26·125 |
| 9 | 2·088 | 2·532 | 3·325 | 4·168 | 5·380 | 6·393 | 8·343 | 10·656 | 12·242 | 14·684 | 16·919 | 19·679 | 21·666 | 27·877 |
| 10 | 2·558 | 3·059 | 3·940 | 4·865 | 6·179 | 7·267 | 9·342 | 11·781 | 13·442 | 15·987 | 18·307 | 21·161 | 23·209 | 29·588 |
| 11 | 3·053 | 3·609 | 4·575 | 5·578 | 6·989 | 8·148 | 10·341 | 12·899 | 14·631 | 17·275 | 19·675 | 22·618 | 24·725 | 31·264 |
| 12 | 3·571 | 4·178 | 5·226 | 6·304 | 7·807 | 9·034 | 11·340 | 14·011 | 15·812 | 18·549 | 21·026 | 24·054 | 26·217 | 32·909 |
| 13 | 4·107 | 4·765 | 5·892 | 7·042 | 8·634 | 9·926 | 12·340 | 15·119 | 16·985 | 19·812 | 22·362 | 25·472 | 27·688 | 34·528 |
| 14 | 4·660 | 5·368 | 6·571 | 7·790 | 9·467 | 10·821 | 13·339 | 16·222 | 18·151 | 21·064 | 23·685 | 26·873 | 29·141 | 36·123 |
| 15 | 5·229 | 5·985 | 7·261 | 8·547 | 10·307 | 11·721 | 14·339 | 17·322 | 19·311 | 22·307 | 24·996 | 28·259 | 30·578 | 37·697 |
| 16 | 5·812 | 6·614 | 7·962 | 9·312 | 11·152 | 12·624 | 15·338 | 18·418 | 20·465 | 23·542 | 26·296 | 29·633 | 32·000 | 39·252 |
| 17 | 6·408 | 7·255 | 8·672 | 10·085 | 12·002 | 13·531 | 16·338 | 19·511 | 21·615 | 24·769 | 27·587 | 30·995 | 33·409 | 40·790 |
| 18 | 7·015 | 7·906 | 9·390 | 10·865 | 12·857 | 14·440 | 17·338 | 20·601 | 22·760 | 25·989 | 28·869 | 32·346 | 34·805 | 42·312 |
| 19 | 7·633 | 8·567 | 10·117 | 11·651 | 13·716 | 15·352 | 18·338 | 21·689 | 23·900 | 27·204 | 30·144 | 33·687 | 36·191 | 43·820 |
| 20 | 8·260 | 9·237 | 10·851 | 12·443 | 14·578 | 16·266 | 19·337 | 22·775 | 25·038 | 28·412 | 31·410 | 35·020 | 37·566 | 45·315 |
| 21 | 8·897 | 9·915 | 11·591 | 13·240 | 15·445 | 17·182 | 20·337 | 23·858 | 26·171 | 29·615 | 32·671 | 36·343 | 38·932 | 46·797 |
| 22 | 9·542 | 10·600 | 12·338 | 14·041 | 16·314 | 18·101 | 21·337 | 24·939 | 27·301 | 30·813 | 33·924 | 37·659 | 40·289 | 48·268 |
| 23 | 10·196 | 11·293 | 13·091 | 14·848 | 17·187 | 19·021 | 22·337 | 26·018 | 28·429 | 32·007 | 35·172 | 38·968 | 41·638 | 49·728 |
| 24 | 10·856 | 11·992 | 13·848 | 15·659 | 18·062 | 19·943 | 23·337 | 27·096 | 29·553 | 33·196 | 36·415 | 40·270 | 42·980 | 51·179 |
| 25 | 11·524 | 12·697 | 14·611 | 16·473 | 18·940 | 20·867 | 24·337 | 28·172 | 30·675 | 34·382 | 37·652 | 41·566 | 44·314 | 52·620 |

| ν | | | | | | | | | | | | | | |
|---|---|---|---|---|---|---|---|---|---|---|---|---|---|---|
| 26 | 12·198 | 13·409 | 15·379 | 17·292 | 19·820 | 21·792 | 25·336 | 29·246 | 31·795 | 35·563 | 38·885 | 42·856 | 45·642 | 54·052 |
| 27 | 12·879 | 14·125 | 16·151 | 18·114 | 20·703 | 22·719 | 26·336 | 30·319 | 32·912 | 36·741 | 40·113 | 44·140 | 46·963 | 55·476 |
| 28 | 13·565 | 14·847 | 16·928 | 18·939 | 21·588 | 23·647 | 27·336 | 31·391 | 34·027 | 37·916 | 41·337 | 45·419 | 48·278 | 56·893 |
| 29 | 14·256 | 15·574 | 17·708 | 19·768 | 22·475 | 24·577 | 28·336 | 32·461 | 35·139 | 39·087 | 42·557 | 46·693 | 49·588 | 58·302 |
| 30 | 14·953 | 16·306 | 18·493 | 20·599 | 23·364 | 25·508 | 29·336 | 33·530 | 36·250 | 40·256 | 43·773 | 47·962 | 50·892 | 59·703 |
| 32 | 16·362 | 17·783 | 20·072 | 22·271 | 25·148 | 27·373 | 31·336 | 35·665 | 38·466 | 42·585 | 46·194 | 50·487 | 53·486 | 62·487 |
| 34 | 17·789 | 19·275 | 21·664 | 23·952 | 26·938 | 29·242 | 33·336 | 37·795 | 40·676 | 44·903 | 48·602 | 52·995 | 56·061 | 65·247 |
| 36 | 19·233 | 20·783 | 23·269 | 25·643 | 28·735 | 31·115 | 35·336 | 39·922 | 42·879 | 47·212 | 50·999 | 55·489 | 58·619 | 67·985 |
| 38 | 20·691 | 22·304 | 24·884 | 27·343 | 30·537 | 32·992 | 37·335 | 42·045 | 45·076 | 49·513 | 53·384 | 57·969 | 61·162 | 70·703 |
| 40 | 22·164 | 23·838 | 26·509 | 29·051 | 32·345 | 34·872 | 39·335 | 44·165 | 47·269 | 51·805 | 55·759 | 60·436 | 63·691 | 73·402 |
| 42 | 23·650 | 25·383 | 28·144 | 30·765 | 34·157 | 36·755 | 41·335 | 46·282 | 49·456 | 54·090 | 58·124 | 62·892 | 66·206 | 76·084 |
| 44 | 25·148 | 26·939 | 29·787 | 32·487 | 35·974 | 38·641 | 43·335 | 48·396 | 51·639 | 56·369 | 60·481 | 65·337 | 68·710 | 78·750 |
| 46 | 26·657 | 28·504 | 31·439 | 34·215 | 37·795 | 40·529 | 45·335 | 50·507 | 53·818 | 58·641 | 62·830 | 67·771 | 71·201 | 81·400 |
| 48 | 28·177 | 30·080 | 33·098 | 35·949 | 39·621 | 42·420 | 47·335 | 52·616 | 55·993 | 60·907 | 65·171 | 70·197 | 73·683 | 84·037 |
| 50 | 29·707 | 31·664 | 34·764 | 37·689 | 41·449 | 44·313 | 49·335 | 54·723 | 58·164 | 63·167 | 67·505 | 72·613 | 76·154 | 86·661 |
| 52 | 31·246 | 33·256 | 36·437 | 39·433 | 43·281 | 46·209 | 51·335 | 56·827 | 60·332 | 65·422 | 69·832 | 75·021 | 78·616 | 89·272 |
| 54 | 32·793 | 34·856 | 38·116 | 41·183 | 45·117 | 48·106 | 53·335 | 58·930 | 62·496 | 67·673 | 72·153 | 77·422 | 81·069 | 91·872 |
| 56 | 34·350 | 36·464 | 39·801 | 42·937 | 46·955 | 50·005 | 55·335 | 61·031 | 64·658 | 69·919 | 74·468 | 79·815 | 83·513 | 94·461 |
| 58 | 35·913 | 38·078 | 41·492 | 44·696 | 48·797 | 51·906 | 57·335 | 63·129 | 66·816 | 72·160 | 76·778 | 82·201 | 85·950 | 97·039 |
| 60 | 37·485 | 39·699 | 43·188 | 46·459 | 50·641 | 53·809 | 59·335 | 65·227 | 68·972 | 74·397 | 79·082 | 84·580 | 88·379 | 99·607 |
| 62 | 39·063 | 41·327 | 44·889 | 48·226 | 52·487 | 55·714 | 61·335 | 67·322 | 71·125 | 76·630 | 81·381 | 86·953 | 90·802 | 102·166 |
| 64 | 40·649 | 42·960 | 46·595 | 49·996 | 54·336 | 57·620 | 63·335 | 69·416 | 73·276 | 78·860 | 83·675 | 89·320 | 93·217 | 104·716 |
| 66 | 42·240 | 44·599 | 48·305 | 51·770 | 56·188 | 59·527 | 65·335 | 71·508 | 75·424 | 81·085 | 85·965 | 91·681 | 95·626 | 107·258 |
| 68 | 43·838 | 46·244 | 50·020 | 53·548 | 58·042 | 61·436 | 67·335 | 73·600 | 77·571 | 83·308 | 88·250 | 94·037 | 98·028 | 109·791 |
| 70 | 45·442 | 47·893 | 51·739 | 55·329 | 59·898 | 63·346 | 69·334 | 75·689 | 79·715 | 85·527 | 90·531 | 96·388 | 100·425 | 112·317 |

For odd values of ν between 30 and 70 the mean of the tabular values for $\nu-1$ and $\nu+1$ may be taken. For larger values of ν, the expression $\sqrt{2\chi^2} - \sqrt{2\nu - 1}$ may be used as a normal deviate with unit variance.

**Percentage points of the F distribution** [see § 2.5.6]

The tables give the $100\alpha$ percentage point $x_\alpha(m, n)$ of the $F_{m,n}$ distribution with $\alpha = 0.05$, $0.01$ and $0.001$, that is the upper 5%, 1% and 0.1% points of the distribution. The tables give the values of $x_\alpha(m, n)$ exceeding unity. For values less than unity use the result

$$x_\alpha(m, n) = 1/x_{1-\alpha}(n, m).$$

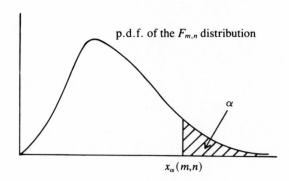

## F distribution: Upper 5% points

(Reproduced by permission of Biometrika Trustees from *Biometrika Tables for Statisticians*, Vol. 1, 3rd edition, 1966.)

| $\nu_2$ \ $\nu_1$ | 1 | 2 | 3 | 4 | 5 | 6 | 7 | 8 | 9 | 10 | 12 | 15 | 20 | 24 | 30 | 40 | 60 | 120 | ∞ |
|---|---|---|---|---|---|---|---|---|---|---|---|---|---|---|---|---|---|---|---|
| 1 | 161.4 | 199.5 | 215.7 | 224.6 | 230.2 | 234.0 | 236.8 | 238.9 | 240.5 | 241.9 | 243.9 | 245.9 | 248.0 | 249.1 | 250.1 | 251.1 | 252.2 | 253.3 | 254.3 |
| 2 | 18.51 | 19.00 | 19.16 | 19.25 | 19.30 | 19.33 | 19.35 | 19.37 | 19.38 | 19.40 | 19.41 | 19.43 | 19.45 | 19.45 | 19.46 | 19.47 | 19.48 | 19.49 | 19.50 |
| 3 | 10.13 | 9.55 | 9.28 | 9.12 | 9.01 | 8.94 | 8.89 | 8.85 | 8.81 | 8.79 | 8.74 | 8.70 | 8.66 | 8.64 | 8.62 | 8.59 | 8.57 | 8.55 | 8.53 |
| 4 | 7.71 | 6.94 | 6.59 | 6.39 | 6.26 | 6.16 | 6.09 | 6.04 | 6.00 | 5.96 | 5.91 | 5.86 | 5.80 | 5.77 | 5.75 | 5.72 | 5.69 | 5.66 | 5.63 |
| 5 | 6.61 | 5.79 | 5.41 | 5.19 | 5.05 | 4.95 | 4.88 | 4.82 | 4.77 | 4.74 | 4.68 | 4.62 | 4.56 | 4.53 | 4.50 | 4.46 | 4.43 | 4.40 | 4.36 |
| 6 | 5.99 | 5.14 | 4.76 | 4.53 | 4.39 | 4.28 | 4.21 | 4.15 | 4.10 | 4.06 | 4.00 | 3.94 | 3.87 | 3.84 | 3.81 | 3.77 | 3.74 | 3.70 | 3.67 |
| 7 | 5.59 | 4.74 | 4.35 | 4.12 | 3.97 | 3.87 | 3.79 | 3.73 | 3.68 | 3.64 | 3.57 | 3.51 | 3.44 | 3.41 | 3.38 | 3.34 | 3.30 | 3.27 | 3.23 |
| 8 | 5.32 | 4.46 | 4.07 | 3.84 | 3.69 | 3.58 | 3.50 | 3.44 | 3.39 | 3.35 | 3.28 | 3.22 | 3.15 | 3.12 | 3.08 | 3.04 | 3.01 | 2.97 | 2.93 |
| 9 | 5.12 | 4.26 | 3.86 | 3.63 | 3.48 | 3.37 | 3.29 | 3.23 | 3.18 | 3.14 | 3.07 | 3.01 | 2.94 | 2.90 | 2.86 | 2.83 | 2.79 | 2.75 | 2.71 |
| 10 | 4.96 | 4.10 | 3.71 | 3.48 | 3.33 | 3.22 | 3.14 | 3.07 | 3.02 | 2.98 | 2.91 | 2.85 | 2.77 | 2.74 | 2.70 | 2.66 | 2.62 | 2.58 | 2.54 |
| 11 | 4.84 | 3.98 | 3.59 | 3.36 | 3.20 | 3.09 | 3.01 | 2.95 | 2.90 | 2.85 | 2.79 | 2.72 | 2.65 | 2.61 | 2.57 | 2.53 | 2.49 | 2.45 | 2.40 |
| 12 | 4.75 | 3.89 | 3.49 | 3.26 | 3.11 | 3.00 | 2.91 | 2.85 | 2.80 | 2.75 | 2.69 | 2.62 | 2.54 | 2.51 | 2.47 | 2.43 | 2.38 | 2.34 | 2.30 |
| 13 | 4.67 | 3.81 | 3.41 | 3.18 | 3.03 | 2.92 | 2.83 | 2.77 | 2.71 | 2.67 | 2.60 | 2.53 | 2.46 | 2.42 | 2.38 | 2.34 | 2.30 | 2.25 | 2.21 |
| 14 | 4.60 | 3.74 | 3.34 | 3.11 | 2.96 | 2.85 | 2.76 | 2.70 | 2.65 | 2.60 | 2.53 | 2.46 | 2.39 | 2.35 | 2.31 | 2.27 | 2.22 | 2.18 | 2.13 |
| 15 | 4.54 | 3.68 | 3.29 | 3.06 | 2.90 | 2.79 | 2.71 | 2.64 | 2.59 | 2.54 | 2.48 | 2.40 | 2.33 | 2.29 | 2.25 | 2.20 | 2.16 | 2.11 | 2.07 |
| 16 | 4.49 | 3.63 | 3.24 | 3.01 | 2.85 | 2.74 | 2.66 | 2.59 | 2.54 | 2.49 | 2.42 | 2.35 | 2.28 | 2.24 | 2.19 | 2.15 | 2.11 | 2.06 | 2.01 |
| 17 | 4.45 | 3.59 | 3.20 | 2.96 | 2.81 | 2.70 | 2.61 | 2.55 | 2.49 | 2.45 | 2.38 | 2.31 | 2.23 | 2.19 | 2.15 | 2.10 | 2.06 | 2.01 | 1.96 |
| 18 | 4.41 | 3.55 | 3.16 | 2.93 | 2.77 | 2.66 | 2.58 | 2.51 | 2.46 | 2.41 | 2.34 | 2.27 | 2.19 | 2.15 | 2.11 | 2.06 | 2.02 | 1.97 | 1.92 |
| 19 | 4.38 | 3.52 | 3.13 | 2.90 | 2.74 | 2.63 | 2.54 | 2.48 | 2.42 | 2.38 | 2.31 | 2.23 | 2.16 | 2.11 | 2.07 | 2.03 | 1.98 | 1.93 | 1.88 |
| 20 | 4.35 | 3.49 | 3.10 | 2.87 | 2.71 | 2.60 | 2.51 | 2.45 | 2.39 | 2.35 | 2.28 | 2.20 | 2.12 | 2.08 | 2.04 | 1.99 | 1.95 | 1.90 | 1.84 |
| 21 | 4.32 | 3.47 | 3.07 | 2.84 | 2.68 | 2.57 | 2.49 | 2.42 | 2.37 | 2.32 | 2.25 | 2.18 | 2.10 | 2.05 | 2.01 | 1.96 | 1.92 | 1.87 | 1.81 |
| 22 | 4.30 | 3.44 | 3.05 | 2.82 | 2.66 | 2.55 | 2.46 | 2.40 | 2.34 | 2.30 | 2.23 | 2.15 | 2.07 | 2.03 | 1.98 | 1.94 | 1.89 | 1.84 | 1.78 |
| 23 | 4.28 | 3.42 | 3.03 | 2.80 | 2.64 | 2.53 | 2.44 | 2.37 | 2.32 | 2.27 | 2.20 | 2.13 | 2.05 | 2.01 | 1.96 | 1.91 | 1.86 | 1.81 | 1.76 |
| 24 | 4.26 | 3.40 | 3.01 | 2.78 | 2.62 | 2.51 | 2.42 | 2.36 | 2.30 | 2.25 | 2.18 | 2.11 | 2.03 | 1.98 | 1.94 | 1.89 | 1.84 | 1.79 | 1.73 |
| 25 | 4.24 | 3.39 | 2.99 | 2.76 | 2.60 | 2.49 | 2.40 | 2.34 | 2.28 | 2.24 | 2.16 | 2.09 | 2.01 | 1.96 | 1.92 | 1.87 | 1.82 | 1.77 | 1.71 |
| 26 | 4.23 | 3.37 | 2.98 | 2.74 | 2.59 | 2.47 | 2.39 | 2.32 | 2.27 | 2.22 | 2.15 | 2.07 | 1.99 | 1.95 | 1.90 | 1.85 | 1.80 | 1.75 | 1.69 |
| 27 | 4.21 | 3.35 | 2.96 | 2.73 | 2.57 | 2.46 | 2.37 | 2.31 | 2.25 | 2.20 | 2.13 | 2.06 | 1.97 | 1.93 | 1.88 | 1.84 | 1.79 | 1.73 | 1.67 |
| 28 | 4.20 | 3.34 | 2.95 | 2.71 | 2.56 | 2.45 | 2.36 | 2.29 | 2.24 | 2.19 | 2.12 | 2.04 | 1.96 | 1.91 | 1.87 | 1.82 | 1.77 | 1.71 | 1.65 |
| 29 | 4.18 | 3.33 | 2.93 | 2.70 | 2.55 | 2.43 | 2.35 | 2.28 | 2.22 | 2.18 | 2.10 | 2.03 | 1.94 | 1.90 | 1.85 | 1.81 | 1.75 | 1.70 | 1.64 |
| 30 | 4.17 | 3.32 | 2.92 | 2.69 | 2.53 | 2.42 | 2.33 | 2.27 | 2.21 | 2.16 | 2.09 | 2.01 | 1.93 | 1.89 | 1.84 | 1.79 | 1.74 | 1.68 | 1.62 |
| 40 | 4.08 | 3.23 | 2.84 | 2.61 | 2.45 | 2.34 | 2.25 | 2.18 | 2.12 | 2.08 | 2.00 | 1.92 | 1.84 | 1.79 | 1.74 | 1.69 | 1.64 | 1.58 | 1.51 |
| 60 | 4.00 | 3.15 | 2.76 | 2.53 | 2.37 | 2.25 | 2.17 | 2.10 | 2.04 | 1.99 | 1.92 | 1.84 | 1.75 | 1.70 | 1.65 | 1.59 | 1.53 | 1.47 | 1.39 |
| 120 | 3.92 | 3.07 | 2.68 | 2.45 | 2.29 | 2.17 | 2.09 | 2.02 | 1.96 | 1.91 | 1.83 | 1.75 | 1.66 | 1.61 | 1.55 | 1.50 | 1.43 | 1.35 | 1.25 |
| ∞ | 3.84 | 3.00 | 2.60 | 2.37 | 2.21 | 2.10 | 2.01 | 1.94 | 1.88 | 1.83 | 1.75 | 1.67 | 1.57 | 1.52 | 1.46 | 1.39 | 1.32 | 1.22 | 1.00 |

*F distribution: Upper 1% points*

| $v_2$ \ $v_1$ | 1 | 2 | 3 | 4 | 5 | 6 | 7 | 8 | 9 | 10 | 12 | 15 | 20 | 24 | 30 | 40 | 60 | 120 | ∞ |
|---|---|---|---|---|---|---|---|---|---|---|---|---|---|---|---|---|---|---|---|
| 1 | 4052 | 4999·5 | 5403 | 5625 | 5764 | 5859 | 5928 | 5981 | 6022 | 6056 | 6106 | 6157 | 6209 | 6235 | 6261 | 6287 | 6313 | 6339 | 6366 |
| 2 | 98·50 | 99·00 | 99·17 | 99·25 | 99·30 | 99·33 | 99·36 | 99·37 | 99·39 | 99·40 | 99·42 | 99·43 | 99·45 | 99·46 | 99·47 | 99·47 | 99·48 | 99·49 | 99·50 |
| 3 | 34·12 | 30·82 | 29·46 | 28·71 | 28·24 | 27·91 | 27·67 | 27·49 | 27·35 | 27·23 | 27·05 | 26·87 | 26·69 | 26·60 | 26·50 | 26·41 | 26·32 | 26·22 | 26·13 |
| 4 | 21·20 | 18·00 | 16·69 | 15·98 | 15·52 | 15·21 | 14·98 | 14·80 | 14·66 | 14·55 | 14·37 | 14·20 | 14·02 | 13·93 | 13·84 | 13·75 | 13·65 | 13·56 | 13·46 |
| 5 | 16·26 | 13·27 | 12·06 | 11·39 | 10·97 | 10·67 | 10·46 | 10·29 | 10·16 | 10·05 | 9·89 | 9·72 | 9·55 | 9·47 | 9·38 | 9·29 | 9·20 | 9·11 | 9·02 |
| 6 | 13·75 | 10·92 | 9·78 | 9·15 | 8·75 | 8·47 | 8·26 | 8·10 | 7·98 | 7·87 | 7·72 | 7·56 | 7·40 | 7·31 | 7·23 | 7·14 | 7·06 | 6·97 | 6·88 |
| 7 | 12·25 | 9·55 | 8·45 | 7·85 | 7·46 | 7·19 | 6·99 | 6·84 | 6·72 | 6·62 | 6·47 | 6·31 | 6·16 | 6·07 | 5·99 | 5·91 | 5·82 | 5·74 | 5·65 |
| 8 | 11·26 | 8·65 | 7·59 | 7·01 | 6·63 | 6·37 | 6·18 | 6·03 | 5·91 | 5·81 | 5·67 | 5·52 | 5·36 | 5·28 | 5·20 | 5·12 | 5·03 | 4·95 | 4·86 |
| 9 | 10·56 | 8·02 | 6·99 | 6·42 | 6·06 | 5·80 | 5·61 | 5·47 | 5·35 | 5·26 | 5·11 | 4·96 | 4·81 | 4·73 | 4·65 | 4·57 | 4·48 | 4·40 | 4·31 |
| 10 | 10·04 | 7·56 | 6·55 | 5·99 | 5·64 | 5·39 | 5·20 | 5·06 | 4·94 | 4·85 | 4·71 | 4·56 | 4·41 | 4·33 | 4·25 | 4·17 | 4·08 | 4·00 | 3·91 |
| 11 | 9·65 | 7·21 | 6·22 | 5·67 | 5·32 | 5·07 | 4·89 | 4·74 | 4·63 | 4·54 | 4·40 | 4·25 | 4·10 | 4·02 | 3·94 | 3·86 | 3·78 | 3·69 | 3·60 |
| 12 | 9·33 | 6·93 | 5·95 | 5·41 | 5·06 | 4·82 | 4·64 | 4·50 | 4·39 | 4·30 | 4·16 | 4·01 | 3·86 | 3·78 | 3·70 | 3·62 | 3·54 | 3·45 | 3·36 |
| 13 | 9·07 | 6·70 | 5·74 | 5·21 | 4·86 | 4·62 | 4·44 | 4·30 | 4·19 | 4·10 | 3·96 | 3·82 | 3·66 | 3·59 | 3·51 | 3·43 | 3·34 | 3·25 | 3·17 |
| 14 | 8·86 | 6·51 | 5·56 | 5·04 | 4·69 | 4·46 | 4·28 | 4·14 | 4·03 | 3·94 | 3·80 | 3·66 | 3·51 | 3·43 | 3·35 | 3·27 | 3·18 | 3·09 | 3·00 |
| 15 | 8·68 | 6·36 | 5·42 | 4·89 | 4·56 | 4·32 | 4·14 | 4·00 | 3·89 | 3·80 | 3·67 | 3·52 | 3·37 | 3·29 | 3·21 | 3·13 | 3·05 | 2·96 | 2·87 |
| 16 | 8·53 | 6·23 | 5·29 | 4·77 | 4·44 | 4·20 | 4·03 | 3·89 | 3·78 | 3·69 | 3·55 | 3·41 | 3·26 | 3·18 | 3·10 | 3·02 | 2·93 | 2·84 | 2·75 |
| 17 | 8·40 | 6·11 | 5·18 | 4·67 | 4·34 | 4·10 | 3·93 | 3·79 | 3·68 | 3·59 | 3·46 | 3·31 | 3·16 | 3·08 | 3·00 | 2·92 | 2·83 | 2·75 | 2·65 |
| 18 | 8·29 | 6·01 | 5·09 | 4·58 | 4·25 | 4·01 | 3·84 | 3·71 | 3·60 | 3·51 | 3·37 | 3·23 | 3·08 | 3·00 | 2·92 | 2·84 | 2·75 | 2·66 | 2·57 |
| 19 | 8·18 | 5·93 | 5·01 | 4·50 | 4·17 | 3·94 | 3·77 | 3·63 | 3·52 | 3·43 | 3·30 | 3·15 | 3·00 | 2·92 | 2·84 | 2·76 | 2·67 | 2·58 | 2·49 |
| 20 | 8·10 | 5·85 | 4·94 | 4·43 | 4·10 | 3·87 | 3·70 | 3·56 | 3·46 | 3·37 | 3·23 | 3·09 | 2·94 | 2·86 | 2·78 | 2·69 | 2·61 | 2·52 | 2·42 |
| 21 | 8·02 | 5·78 | 4·87 | 4·37 | 4·04 | 3·81 | 3·64 | 3·51 | 3·40 | 3·31 | 3·17 | 3·03 | 2·88 | 2·80 | 2·72 | 2·64 | 2·55 | 2·46 | 2·36 |
| 22 | 7·95 | 5·72 | 4·82 | 4·31 | 3·99 | 3·76 | 3·59 | 3·45 | 3·35 | 3·26 | 3·12 | 2·98 | 2·83 | 2·75 | 2·67 | 2·58 | 2·50 | 2·40 | 2·31 |
| 23 | 7·88 | 5·66 | 4·76 | 4·26 | 3·94 | 3·71 | 3·54 | 3·41 | 3·30 | 3·21 | 3·07 | 2·93 | 2·78 | 2·70 | 2·62 | 2·54 | 2·45 | 2·35 | 2·26 |
| 24 | 7·82 | 5·61 | 4·72 | 4·22 | 3·90 | 3·67 | 3·50 | 3·36 | 3·26 | 3·17 | 3·03 | 2·89 | 2·74 | 2·66 | 2·58 | 2·49 | 2·40 | 2·31 | 2·21 |
| 25 | 7·77 | 5·57 | 4·68 | 4·18 | 3·85 | 3·63 | 3·46 | 3·32 | 3·22 | 3·13 | 2·99 | 2·85 | 2·70 | 2·62 | 2·54 | 2·45 | 2·36 | 2·27 | 2·17 |
| 26 | 7·72 | 5·53 | 4·64 | 4·14 | 3·82 | 3·59 | 3·42 | 3·29 | 3·18 | 3·09 | 2·96 | 2·81 | 2·66 | 2·58 | 2·50 | 2·42 | 2·33 | 2·23 | 2·13 |
| 27 | 7·68 | 5·49 | 4·60 | 4·11 | 3·78 | 3·56 | 3·39 | 3·26 | 3·15 | 3·06 | 2·93 | 2·78 | 2·63 | 2·55 | 2·47 | 2·38 | 2·29 | 2·20 | 2·10 |
| 28 | 7·64 | 5·45 | 4·57 | 4·07 | 3·75 | 3·53 | 3·36 | 3·23 | 3·12 | 3·03 | 2·90 | 2·75 | 2·60 | 2·52 | 2·44 | 2·35 | 2·26 | 2·17 | 2·06 |
| 29 | 7·60 | 5·42 | 4·54 | 4·04 | 3·73 | 3·50 | 3·33 | 3·20 | 3·09 | 3·00 | 2·87 | 2·73 | 2·57 | 2·49 | 2·41 | 2·33 | 2·23 | 2·14 | 2·03 |
| 30 | 7·56 | 5·39 | 4·51 | 4·02 | 3·70 | 3·47 | 3·30 | 3·17 | 3·07 | 2·98 | 2·84 | 2·70 | 2·55 | 2·47 | 2·39 | 2·30 | 2·21 | 2·11 | 2·01 |
| 40 | 7·31 | 5·18 | 4·31 | 3·83 | 3·51 | 3·29 | 3·12 | 2·99 | 2·89 | 2·80 | 2·66 | 2·52 | 2·37 | 2·29 | 2·20 | 2·11 | 2·02 | 1·92 | 1·80 |
| 60 | 7·08 | 4·98 | 4·13 | 3·65 | 3·34 | 3·12 | 2·95 | 2·82 | 2·72 | 2·63 | 2·50 | 2·35 | 2·20 | 2·12 | 2·03 | 1·94 | 1·84 | 1·73 | 1·60 |
| 120 | 6·85 | 4·79 | 3·95 | 3·48 | 3·17 | 2·96 | 2·79 | 2·66 | 2·56 | 2·47 | 2·34 | 2·19 | 2·03 | 1·95 | 1·86 | 1·76 | 1·66 | 1·53 | 1·38 |
| ∞ | 6·63 | 4·61 | 3·78 | 3·32 | 3·02 | 2·80 | 2·64 | 2·51 | 2·41 | 2·32 | 2·18 | 2·04 | 1·88 | 1·79 | 1·70 | 1·59 | 1·47 | 1·32 | 1·00 |

F distribution. Upper 0.1% points

| $v_2$ \ $v_1$ | 1 | 2 | 3 | 4 | 5 | 6 | 7 | 8 | 9 | 10 | 12 | 15 | 20 | 24 | 30 | 40 | 60 | 120 | ∞ |
|---|---|---|---|---|---|---|---|---|---|---|---|---|---|---|---|---|---|---|---|
| 1 | 4053* | 5000* | 5404* | 5625* | 5764* | 5859* | 5929* | 5981* | 6023* | 6056* | 6107* | 6158* | 6209* | 6235* | 6261* | 6287* | 6313* | 6340* | 6366* |
| 2 | 998·5 | 999·0 | 999·2 | 999·2 | 999·3 | 999·3 | 999·4 | 999·4 | 999·4 | 999·4 | 999·4 | 999·4 | 999·4 | 999·5 | 999·5 | 999·5 | 999·5 | 999·5 | 999·5 |
| 3 | 167·0 | 148·5 | 141·1 | 137·1 | 134·6 | 132·8 | 131·6 | 130·6 | 129·9 | 129·2 | 128·3 | 127·4 | 126·4 | 125·9 | 125·4 | 125·0 | 124·5 | 124·0 | 123·5 |
| 4 | 74·14 | 61·25 | 56·18 | 53·44 | 51·71 | 50·53 | 49·66 | 49·00 | 48·47 | 48·05 | 47·41 | 46·76 | 46·10 | 45·77 | 45·43 | 45·09 | 44·75 | 44·40 | 44·05 |
| 5 | 47·18 | 37·12 | 33·20 | 31·09 | 29·75 | 28·84 | 28·16 | 27·64 | 27·24 | 26·92 | 26·42 | 25·91 | 25·39 | 25·14 | 24·87 | 24·60 | 24·33 | 24·06 | 23·79 |
| 6 | 35·51 | 27·00 | 23·70 | 21·92 | 20·81 | 20·03 | 19·46 | 19·03 | 18·69 | 18·41 | 17·99 | 17·56 | 17·12 | 16·89 | 16·67 | 16·44 | 16·21 | 15·99 | 15·75 |
| 7 | 29·25 | 21·69 | 18·77 | 17·19 | 16·21 | 15·52 | 15·02 | 14·63 | 14·33 | 14·08 | 13·71 | 13·32 | 12·93 | 12·73 | 12·53 | 12·33 | 12·12 | 11·91 | 11·70 |
| 8 | 25·42 | 18·49 | 15·83 | 14·39 | 13·49 | 12·86 | 12·40 | 12·04 | 11·77 | 11·54 | 11·19 | 10·84 | 10·48 | 10·30 | 10·11 | 9·92 | 9·73 | 9·53 | 9·33 |
| 9 | 22·86 | 16·39 | 13·90 | 12·56 | 11·71 | 11·13 | 10·70 | 10·37 | 10·11 | 9·89 | 9·57 | 9·24 | 8·90 | 8·72 | 8·55 | 8·37 | 8·19 | 8·00 | 7·81 |
| 10 | 21·04 | 14·91 | 12·55 | 11·28 | 10·48 | 9·92 | 9·52 | 9·20 | 8·96 | 8·75 | 8·45 | 8·13 | 7·80 | 7·64 | 7·47 | 7·30 | 7·12 | 6·94 | 6·76 |
| 11 | 19·69 | 13·81 | 11·56 | 10·35 | 9·58 | 9·05 | 8·66 | 8·35 | 8·12 | 7·92 | 7·63 | 7·32 | 7·01 | 6·85 | 6·68 | 6·52 | 6·35 | 6·17 | 6·00 |
| 12 | 18·64 | 12·97 | 10·80 | 9·63 | 8·89 | 8·38 | 8·00 | 7·71 | 7·48 | 7·29 | 7·00 | 6·71 | 6·40 | 6·25 | 6·09 | 5·93 | 5·76 | 5·59 | 5·42 |
| 13 | 17·81 | 12·31 | 10·21 | 9·07 | 8·35 | 7·86 | 7·49 | 7·21 | 6·98 | 6·80 | 6·52 | 6·23 | 5·93 | 5·78 | 5·63 | 5·47 | 5·30 | 5·14 | 4·97 |
| 14 | 17·14 | 11·78 | 9·73 | 8·62 | 7·92 | 7·43 | 7·08 | 6·80 | 6·58 | 6·40 | 6·13 | 5·85 | 5·56 | 5·41 | 5·25 | 5·10 | 4·94 | 4·77 | 4·60 |
| 15 | 16·59 | 11·34 | 9·34 | 8·25 | 7·57 | 7·09 | 6·74 | 6·47 | 6·26 | 6·08 | 5·81 | 5·54 | 5·25 | 5·10 | 4·95 | 4·80 | 4·64 | 4·47 | 4·31 |
| 16 | 16·12 | 10·97 | 9·00 | 7·94 | 7·27 | 6·81 | 6·46 | 6·19 | 5·98 | 5·81 | 5·55 | 5·27 | 4·99 | 4·85 | 4·70 | 4·54 | 4·39 | 4·23 | 4·06 |
| 17 | 15·72 | 10·66 | 8·73 | 7·68 | 7·02 | 6·56 | 6·22 | 5·96 | 5·75 | 5·58 | 5·32 | 5·05 | 4·78 | 4·63 | 4·48 | 4·33 | 4·18 | 4·02 | 3·85 |
| 18 | 15·38 | 10·39 | 8·49 | 7·46 | 6·81 | 6·35 | 6·02 | 5·76 | 5·56 | 5·39 | 5·13 | 4·87 | 4·59 | 4·45 | 4·30 | 4·15 | 4·00 | 3·84 | 3·67 |
| 19 | 15·08 | 10·16 | 8·28 | 7·26 | 6·62 | 6·18 | 5·85 | 5·59 | 5·39 | 5·22 | 4·97 | 4·70 | 4·43 | 4·29 | 4·14 | 3·99 | 3·84 | 3·68 | 3·51 |
| 20 | 14·82 | 9·95 | 8·10 | 7·10 | 6·46 | 6·02 | 5·69 | 5·44 | 5·24 | 5·08 | 4·82 | 4·56 | 4·29 | 4·15 | 4·00 | 3·86 | 3·70 | 3·54 | 3·38 |
| 21 | 14·59 | 9·77 | 7·94 | 6·95 | 6·32 | 5·88 | 5·56 | 5·31 | 5·11 | 4·95 | 4·70 | 4·44 | 4·17 | 4·03 | 3·88 | 3·74 | 3·58 | 3·42 | 3·26 |
| 22 | 14·38 | 9·61 | 7·80 | 6·81 | 6·19 | 5·76 | 5·44 | 5·19 | 4·99 | 4·83 | 4·58 | 4·33 | 4·06 | 3·92 | 3·78 | 3·63 | 3·48 | 3·32 | 3·15 |
| 23 | 14·19 | 9·47 | 7·67 | 6·69 | 6·08 | 5·65 | 5·33 | 5·09 | 4·89 | 4·73 | 4·48 | 4·23 | 3·96 | 3·82 | 3·68 | 3·53 | 3·38 | 3·22 | 3·05 |
| 24 | 14·03 | 9·34 | 7·55 | 6·59 | 5·98 | 5·55 | 5·23 | 4·99 | 4·80 | 4·64 | 4·39 | 4·14 | 3·87 | 3·74 | 3·59 | 3·45 | 3·29 | 3·14 | 2·97 |
| 25 | 13·88 | 9·22 | 7·45 | 6·49 | 5·88 | 5·46 | 5·15 | 4·91 | 4·71 | 4·56 | 4·31 | 4·06 | 3·79 | 3·66 | 3·52 | 3·37 | 3·22 | 3·06 | 2·89 |
| 26 | 13·74 | 9·12 | 7·36 | 6·41 | 5·80 | 5·38 | 5·07 | 4·83 | 4·64 | 4·48 | 4·24 | 3·99 | 3·72 | 3·59 | 3·44 | 3·30 | 3·15 | 2·99 | 2·82 |
| 27 | 13·61 | 9·02 | 7·27 | 6·33 | 5·73 | 5·31 | 5·00 | 4·76 | 4·57 | 4·41 | 4·17 | 3·92 | 3·66 | 3·52 | 3·38 | 3·23 | 3·08 | 2·92 | 2·75 |
| 28 | 13·50 | 8·93 | 7·19 | 6·25 | 5·66 | 5·24 | 4·93 | 4·69 | 4·50 | 4·35 | 4·11 | 3·86 | 3·60 | 3·46 | 3·32 | 3·18 | 3·02 | 2·86 | 2·69 |
| 29 | 13·39 | 8·85 | 7·12 | 6·19 | 5·59 | 5·18 | 4·87 | 4·64 | 4·45 | 4·29 | 4·05 | 3·80 | 3·54 | 3·41 | 3·27 | 3·12 | 2·97 | 2·81 | 2·64 |
| 30 | 13·29 | 8·77 | 7·05 | 6·12 | 5·53 | 5·12 | 4·82 | 4·58 | 4·39 | 4·24 | 4·00 | 3·75 | 3·49 | 3·36 | 3·22 | 3·07 | 2·92 | 2·76 | 2·59 |
| 40 | 12·61 | 8·25 | 6·60 | 5·70 | 5·13 | 4·73 | 4·44 | 4·21 | 4·02 | 3·87 | 3·64 | 3·40 | 3·15 | 3·01 | 2·87 | 2·73 | 2·57 | 2·41 | 2·23 |
| 60 | 11·97 | 7·76 | 6·17 | 5·31 | 4·76 | 4·37 | 4·09 | 3·87 | 3·69 | 3·54 | 3·31 | 3·08 | 2·83 | 2·69 | 2·55 | 2·41 | 2·25 | 2·08 | 1·89 |
| 120 | 11·38 | 7·32 | 5·79 | 4·95 | 4·42 | 4·04 | 3·77 | 3·55 | 3·38 | 3·24 | 3·02 | 2·78 | 2·53 | 2·40 | 2·26 | 2·11 | 1·95 | 1·76 | 1·54 |
| ∞ | 10·83 | 6·91 | 5·42 | 4·62 | 4·10 | 3·74 | 3·47 | 3·27 | 3·10 | 2·96 | 2·74 | 2·51 | 2·27 | 2·13 | 1·99 | 1·84 | 1·66 | 1·45 | 1·00 |

* Multiply these entries by 100.

This 0·1% table is based on the following sources: Colcord & Deming (1935); Fisher & Yates (1953, Table V) used with the permission of the authors and of Messrs Oliver and Boyd; Norton (1952).

## APPENDIX T8

**Random Numbers** [see II: §§ 5.1, 11.1]

The table gives 5000 'Random Digits' which are independent realizations of a random variable $N$ which has the discrete uniform distribution on $(0, 1, \ldots, 9)$; thus

$$P(N = n) = \tfrac{1}{10}, \qquad n = 0, 1, \ldots, 9.$$

The table may be read in any direction, starting at any point. A $d$-decimal random number may be obtained by placing a decimal point before any $d$ consecutive digits. For example, starting at the ninth digit in the seventh row and taking $d = 5$ gives the random number $0 \cdot 31572$. This may be taken as a rounded realization of a continuous uniform $(0, 1)$ random variable $Z$, for which the p.d.f. at $z$ is

$$f(z) = \begin{cases} 1, & 0 \le z \le 1 \\ 0, & \text{otherwise.} \end{cases}$$

Random numbers

(Reproduced by permission of Longman Group Ltd. from *Statistical Tables for Biological, Agricultural and Medical Research* by R. A. Fisher and F. Yates, 1974.)

| | | | | |
|---|---|---|---|---|
| 03 47 43 73 86 | 36 96 47 36 61 | 46 98 63 71 62 | 33 26 16 80 45 | 60 11 14 10 95 |
| 97 74 24 67 62 | 42 81 14 57 20 | 42 53 32 37 32 | 27 07 36 07 51 | 24 51 79 89 73 |
| 16 76 62 27 66 | 56 50 26 71 07 | 32 90 79 78 53 | 13 55 38 58 59 | 88 97 54 14 10 |
| 12 56 85 99 26 | 96 96 68 27 31 | 05 03 72 93 15 | 57 12 10 14 21 | 88 26 49 81 76 |
| 55 59 56 35 64 | 38 54 82 46 22 | 31 62 43 09 90 | 06 18 44 32 53 | 23 83 01 30 30 |
| 16 22 77 94 39 | 49 54 43 54 82 | 17 37 93 23 78 | 87 35 20 96 43 | 84 26 34 91 64 |
| 84 42 17 53 31 | 57 24 55 06 88 | 77 04 74 47 67 | 21 76 33 50 25 | 83 92 12 06 76 |
| 63 01 63 78 59 | 16 95 55 67 19 | 98 10 50 71 75 | 12 86 73 58 07 | 44 39 52 38 79 |
| 33 21 12 34 29 | 78 64 56 07 82 | 52 42 07 44 38 | 15 51 00 13 42 | 99 66 02 79 54 |
| 57 60 86 32 44 | 09 47 27 96 54 | 49 17 46 09 62 | 90 52 84 77 27 | 08 02 73 43 28 |
| 18 18 07 92 46 | 44 17 16 58 09 | 79 83 86 19 62 | 06 76 50 03 10 | 55 23 64 05 05 |
| 26 62 38 97 75 | 84 16 07 44 99 | 83 11 46 32 24 | 20 14 85 88 45 | 10 93 72 88 71 |
| 23 42 40 64 74 | 82 97 77 77 81 | 07 45 32 14 08 | 32 98 94 07 72 | 93 85 79 10 75 |
| 32 36 28 19 95 | 50 92 26 11 97 | 00 56 76 31 38 | 80 22 02 53 53 | 86 60 42 04 53 |
| 37 85 94 35 12 | 83 39 50 08 30 | 42 34 07 96 88 | 54 42 06 87 98 | 35 85 29 48 39 |
| 70 29 17 12 13 | 40 33 20 38 26 | 13 89 51 03 74 | 17 76 37 13 04 | 07 74 21 19 30 |
| 56 62 18 37 35 | 96 83 50 87 75 | 97 12 25 93 47 | 70 33 24 03 54 | 97 77 46 44 80 |
| 99 49 57 22 77 | 88 42 95 45 72 | 16 64 36 16 00 | 04 43 18 66 79 | 94 77 24 21 90 |
| 16 08 15 04 72 | 33 27 14 34 09 | 45 59 34 68 49 | 12 72 07 34 45 | 99 27 72 95 14 |
| 31 16 93 32 43 | 50 27 89 87 19 | 20 15 37 00 49 | 52 85 66 60 44 | 38 68 88 11 80 |
| 68 34 30 13 70 | 55 74 30 77 40 | 44 22 78 84 26 | 04 33 46 09 52 | 68 07 97 06 57 |
| 74 57 25 65 76 | 59 29 97 68 60 | 71 91 38 67 54 | 13 58 18 24 76 | 15 54 55 95 52 |
| 27 42 37 86 53 | 48 55 90 65 72 | 96 57 69 36 10 | 96 46 92 42 45 | 97 60 49 04 91 |
| 00 39 68 29 61 | 66 37 32 20 30 | 77 84 57 03 29 | 10 45 65 04 26 | 11 04 96 67 24 |
| 29 94 98 94 24 | 68 49 69 10 82 | 53 75 91 93 30 | 34 25 20 57 27 | 40 48 73 51 92 |

## Random numbers (cont.)

| | | | | |
|---|---|---|---|---|
| 16 90 82 66 59 | 83 62 64 11 12 | 67 19 00 71 74 | 60 47 21 29 68 | 02 02 37 03 31 |
| 11 27 94 75 06 | 06 09 19 74 66 | 02 94 37 34 02 | 76 70 90 30 86 | 38 45 94 30 38 |
| 35 24 10 16 20 | 33 32 51 26 38 | 79 78 45 04 91 | 16 92 53 56 16 | 02 75 50 95 98 |
| 38 23 16 86 38 | 42 38 97 01 50 | 87 75 66 81 41 | 40 01 74 91 62 | 48 51 84 08 32 |
| 31 96 25 91 47 | 96 44 33 49 13 | 34 86 82 53 91 | 00 52 43 48 85 | 27 55 26 89 62 |
| | | | | |
| 66 67 40 67 14 | 64 05 71 95 86 | 11 05 65 09 68 | 76 83 20 37 90 | 57 16 00 11 66 |
| 14 90 84 45 11 | 75 73 88 05 90 | 52 27 41 14 86 | 22 98 12 22 08 | 07 52 74 95 80 |
| 68 05 51 18 00 | 33 96 02 75 19 | 07 60 62 93 55 | 59 33 82 43 90 | 49 37 38 44 59 |
| 20 46 78 73 90 | 97 51 40 14 02 | 04 02 33 31 08 | 39 54 16 49 36 | 47 95 93 13 30 |
| 64 19 58 97 79 | 15 06 15 93 20 | 01 90 10 75 06 | 40 78 78 89 62 | 02 67 74 17 33 |
| | | | | |
| 05 26 93 70 60 | 22 35 85 15 13 | 92 03 51 59 77 | 59 56 78 06 83 | 52 91 05 70 74 |
| 07 97 10 88 23 | 09 98 42 99 64 | 61 71 62 99 15 | 06 51 29 16 93 | 58 05 77 09 51 |
| 68 71 86 85 85 | 54 87 66 47 54 | 73 32 08 11 12 | 44 95 92 63 16 | 29 56 24 29 48 |
| 26 99 61 65 53 | 58 37 78 80 70 | 42 10 50 67 42 | 32 17 55 85 74 | 94 44 67 16 94 |
| 14 65 52 68 75 | 87 59 36 22 41 | 26 78 63 06 55 | 13 08 27 01 50 | 15 29 39 39 43 |
| | | | | |
| 17 53 77 58 71 | 71 41 61 50 72 | 12 41 94 96 26 | 44 95 27 36 99 | 02 96 74 30 83 |
| 90 26 59 21 19 | 23 52 23 33 12 | 96 93 02 18 39 | 07 02 18 36 07 | 25 99 32 70 23 |
| 41 23 52 55 99 | 31 04 49 69 96 | 10 47 48 45 88 | 13 41 43 89 20 | 97 17 14 49 17 |
| 60 20 50 81 69 | 31 99 73 68 68 | 35 81 33 03 76 | 24 30 12 48 60 | 18 99 10 72 34 |
| 91 25 38 05 90 | 94 58 28 41 36 | 45 37 59 03 09 | 90 35 57 29 12 | 82 62 54 65 60 |
| | | | | |
| 34 50 57 74 37 | 98 80 33 00 91 | 09 77 93 19 82 | 74 94 80 04 04 | 45 07 31 66 49 |
| 85 22 04 39 43 | 73 81 53 94 79 | 33 62 46 86 28 | 08 31 54 46 31 | 53 94 13 38 47 |
| 09 79 13 77 48 | 73 82 97 22 21 | 05 03 27 24 83 | 72 89 44 05 60 | 35 80 39 94 88 |
| 88 75 80 18 14 | 22 95 75 42 49 | 39 32 82 22 49 | 02 48 07 70 37 | 16 04 61 67 87 |
| 90 96 23 70 00 | 39 00 03 06 90 | 55 85 78 38 36 | 94 37 30 69 32 | 90 89 00 76 33 |

## APPENDIX T9

### Random Standard Normal Deviates

The table gives 500 independent realizations, rounded to three decimals, of the Standard Normal random variable $U$ for which the p.d.f. at $u$ is

$$\varphi(u) = \frac{1}{\sqrt{(2\pi)}} e^{-(1/2)u^2}, \qquad -\infty < u < \infty.$$

A realization $u$ of the Standard Normal $U$ may be transformed to a realization $x$ of a Normal $(\mu, \sigma)$ variable by setting $x = \mu + \sigma u$.

Random standard Normal deviates

(Reproduced by permission of Macmillan Publishers Ltd. from *Statistical Tables for Science, Engineering and Management* by J. Murdoch and J. A. Barnes.)

|    | 0 | 1 | 2 | 3 | 4 | 5 | 6 | 7 | 8 | 9 |
|----|------|------|------|------|------|------|------|------|------|------|
| 00 | -0·179 | -0·399 | -0·235 | -0·098 | -0·465 | +1·563 | -1·085 | +0·860 | +0·388 | +0·710 |
| 01 | +0·421 | +1·454 | +0·904 | +0·437 | -2·120 | +1·085 | -0·277 | -2·170 | +0·018 | -0·722 |
| 02 | +0·210 | -0·556 | +0·465 | -1·812 | -2·748 | -0·345 | -0·251 | +0·622 | -1·015 | +0·762 |
| 03 | -1·598 | +0·919 | -0·266 | -0·999 | +0·308 | -0·592 | +0·817 | -0·454 | +1·598 | +0·240 |
| 04 | +1·717 | +1·514 | -0·012 | -0·852 | +0·118 | +0·399 | -0·123 | +0·432 | -0·470 | +0·776 |
| 05 | -0·308 | +0·867 | -0·372 | +0·697 | -1·787 | +0·568 | -0·002 | -0·133 | +0·545 | -0·824 |
| 06 | -0·421 | +0·516 | -0·038 | +1·200 | +0·063 | -0·377 | -1·007 | -0·334 | +1·299 | +0·038 |
| 07 | -0·776 | +0·874 | -1·265 | -0·580 | +0·377 | -0·697 | -2·226 | -1·299 | -0·796 | -0·628 |
| 08 | +0·640 | -0·522 | +0·023 | -0·393 | -1·142 | -2·457 | -1·580 | +1·160 | +0·008 | +0·487 |
| 09 | -0·319 | +0·889 | +1·180 | -0·404 | +1·322 | +0·410 | +1·468 | +0·235 | -0·810 | -1·131 |
| 10 | +0·610 | -0·383 | +1·812 | +0·729 | +0·204 | -0·225 | +0·169 | -0·729 | -0·432 | +0·634 |
| 11 | -0·174 | -0·154 | +0·098 | +0·393 | -3·090 | +1·762 | +1·530 | +0·028 | +0·950 | -0·935 |
| 12 | +2·576 | -0·684 | -1·200 | +0·002 | +0·261 | -0·415 | +0·598 | -0·769 | -0·169 | -1·498 |
| 13 | -1·103 | +1·398 | -0·653 | +1·739 | +0·476 | +0·510 | +0·782 | -0·634 | +0·562 | -0·053 |
| 14 | +1·635 | +0·448 | -1·530 | -0·043 | +2·290 | -0·063 | -1·695 | +0·199 | +1·211 | -1·360 |
| 15 | -0·068 | -0·860 | -0·194 | -1·616 | +0·334 | +0·189 | +0·927 | -1·454 | +0·958 | +0·404 |
| 16 | -1·960 | +1·076 | -0·671 | -0·103 | +1·041 | +2·226 | +1·838 | -0·510 | -1·322 | +2·366 |
| 17 | +0·443 | -0·912 | +0·251 | -0·574 | +1·131 | -0·204 | -0·324 | -0·487 | -1·287 | +0·522 |
| 18 | +1·360 | +0·533 | +1·094 | +0·671 | +0·852 | -2·576 | -0·539 | -0·568 | +0·225 | -0·545 |
| 19 | +0·810 | +0·319 | -1·514 | +0·556 | +1·112 | -0·210 | +0·292 | +0·749 | +0·882 | +0·033 |
| 20 | +0·616 | +1·347 | -1·866 | -0·755 | +0·329 | +0·148 | -0·058 | -0·199 | +0·048 | +1·546 |
| 21 | -0·598 | -2·366 | -0·831 | +0·454 | -0·118 | -1·762 | +0·493 | +1·103 | +0·361 | +0·113 |
| 22 | +0·426 | +1·580 | -1·112 | +0·550 | -1·254 | -0·033 | +0·143 | -1·141 | +0·366 | -0·073 |
| 23 | +0·831 | -0·516 | -1·717 | -0·340 | +1·655 | +0·194 | -0·388 | -0·942 | -1·243 | -0·292 |
| 24 | -0·640 | -0·128 | +1·276 | -1·838 | -0·410 | +0·646 | +2·075 | -0·159 | +1·695 | +0·527 |

Random standard Normal deviates (continued)

| | 0 | 1 | 2 | 3 | 4 | 5 | 6 | 7 | 8 | 9 |
|---|---|---|---|---|---|---|---|---|---|---|
| 25 | −0·927 | +0·838 | −1·546 | +0·246 | −0·742 | −0·143 | +2·457 | +0·043 | −1·058 | −0·867 |
| 26 | +1·232 | +2·170 | +0·088 | −0·803 | +0·574 | +0·058 | +0·282 | +0·356 | +0·350 | −1·927 |
| 27 | +0·935 | +0·665 | +2·034 | −1·995 | +0·703 | −0·083 | −1·468 | +0·078 | −0·966 | −0·303 |
| 28 | −1·739 | −0·622 | −1·563 | +0·313 | +0·220 | −0·586 | −0·272 | +0·789 | −1·335 | +1·440 |
| 29 | +0·990 | −1·483 | +0·154 | −1·372 | −1·896 | +1·385 | −1·041 | +0·974 | +0·482 | −1·211 |
| 30 | −0·189 | −0·240 | +0·133 | −2·290 | −0·616 | −0·437 | +0·459 | −0·499 | +0·845 | +0·383 |
| 31 | +1·866 | −1·398 | +0·068 | +0·053 | −2·034 | +1·426 | +1·254 | +1·067 | +0·592 | +0·174 |
| 32 | −0·018 | +0·628 | +0·230 | +0·659 | −0·298 | +1·927 | −0·282 | +0·769 | −0·690 | +1·675 |
| 33 | −0·646 | −0·350 | +0·324 | −1·675 | +1·190 | −1·076 | +1·287 | −1·426 | +0·345 | −0·215 |
| 34 | −1·150 | −0·220 | −0·533 | +0·912 | −0·710 | −0·904 | −0·817 | −1·160 | −0·919 | −0·659 |
| 35 | +0·103 | −0·361 | +1·024 | −0·6−0·482 | −0·562 | +0·277 | −1·440 | −0·366 | −0·256 | +2·120 |
| 37 | −0·093 | −1·190 | +0·580 | −1·276 | +0·653 | −0·048 | +0·742 | −1·170 | +1·960 | +1·787 |
| 38 | −0·261 | −0·194 | +0·303 | +0·340 | +1·498 | −1·232 | −0·078 | −0·443 | +1·141 | +1·995 |
| 39 | −0·230 | −0·550 | +0·266 | −1·655 | +0·999 | −1·067 | +1·058 | +0·796 | +0·415 | |
| 40 | −0·148 | +0·504 | −0·028 | +0·083 | +0·824 | −1·024 | +1·412 | −0·164 | +1·150 | −0·272 |
| 41 | +1·122 | +0·896 | −0·789 | +0·215 | −0·426 | −1·049 | −0·974 | +0·586 | +1·311 | −0·736 |
| 42 | +0·499 | −1·032 | +0·159 | +0·123 | +2·748 | −0·749 | −0·665 | −1·221 | −1·180 | +1·049 |
| 43 | +0·678 | −0·782 | +0·470 | +0·256 | +0·298 | −0·990 | +0·287 | +0·942 | +0·128 | +1·372 |
| 44 | −1·347 | +3·090 | −0·896 | +0·138 | −0·838 | +0·690 | +1·007 | +0·184 | +0·164 | +0·179 |
| 45 | −1·094 | −0·610 | −0·287 | +0·755 | −0·459 | −1·635 | −0·108 | −0·246 | +1·032 | −0·527 |
| 46 | −0·088 | −0·889 | +0·803 | −1·311 | −0·703 | +1·170 | −0·113 | +0·108 | −0·874 | +0·372 |
| 47 | +0·093 | −0·476 | +1·265 | −0·448 | +1·015 | −0·313 | −0·958 | +0·716 | +1·483 | −0·722 |
| 48 | −0·950 | −0·008 | +0·012 | +0·073 | −0·762 | −0·493 | +1·896 | +0·982 | +1·616 | +1·221 |
| 49 | −0·329 | −0·138 | −0·504 | −0·678 | +1·335 | −2·075 | −1·385 | −0·023 | −0·356 | −0·982 |

## APPENDIX T10

### Confidence Limits for Binomial Parameter [see § 4.7]

Charts providing confidence limits for the parameter $\theta$ in a Binomial $(n, \theta)$ distribution, given a sample fraction $r/n$.

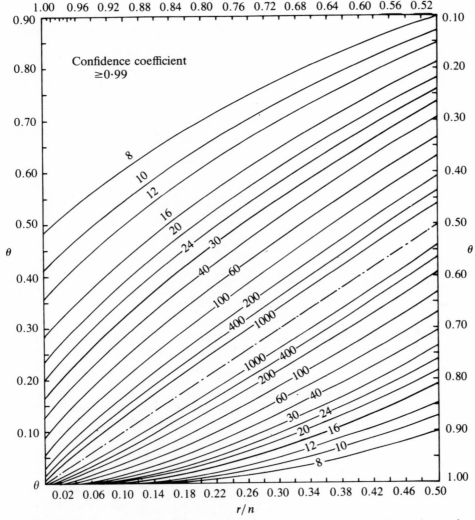

The numbers printed along the curves indicate the sample size, $n$. If for a given value of $r/n$, $\theta'$ and $\theta''$ are the ordinates read from the appropriate lower and upper curves, then $(\theta', \theta'')$ is a confidence interval for $\theta$, with confidence interval $\geq 99\%$.

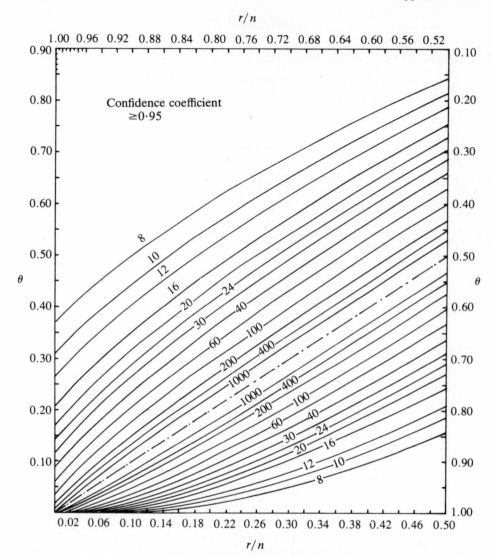

The numbers printed along the curves indicate the sample size, *n*. If for a given value of *r/n*, $\theta'$ and $\theta''$ are the ordinates read from the appropriate lower and upper curves, then $(\theta', \theta'')$ is a confidence interval for $\theta$, with confidence level $\geq 95\%$.

## APPENDIX T11

**Confidence Limits for the Expectation of a Poisson Variable** [see § 4.7]

Confidence limits for the parameter $\lambda$ in a Poisson $(\lambda)$ distribution given $c$ occurrences of the Poisson event. The table gives lower and upper confidence limits for $\lambda$, at confidence level $1 - 2\alpha$.

Confidence limits for the expectation of a Poisson variable

| $1-2\alpha$ | | 0·90 | | 0·95 | | 0·98 | | 0·99 | | 0·998 | |
| --- | --- | --- | --- | --- | --- | --- | --- | --- | --- | --- | --- |
| $\alpha$ | | 0·05 | | 0·025 | | 0·01 | | 0·005 | | 0·001 | |
| $c$ | | Lower | Upper | Lower | Upper | Lower | Upper | Lower | Upper | Lower | Upper |
| 0 | | 0·0000 | 3·00 | 0·0000 | 3·69 | 0·0000 | 4·61 | 0·00000 | 5·30 | 0·00000 | 6·91 |
| 1 | | 0·0513 | 4·74 | 0·0253 | 5·57 | 0·0101 | 6·64 | 0·00501 | 7·43 | 0·00100 | 9·23 |
| 2 | | 0·355 | 6·30 | 0·242 | 7·22 | 0·149 | 8·41 | 0·103 | 9·27 | 0·0454 | 11·23 |
| 3 | | 0·818 | 7·75 | 0·619 | 8·77 | 0·436 | 10·05 | 0·338 | 10·98 | 0·191 | 13·06 |
| 4 | | 1·37 | 9·15 | 1·09 | 10·24 | 0·823 | 11·60 | 0·672 | 12·59 | 0·429 | 14·79 |
| 5 | | 1·97 | 10·51 | 1·62 | 11·67 | 1·28 | 13·11 | 1·08 | 14·15 | 0·739 | 16·45 |
| 6 | | 2·61 | 11·84 | 2·20 | 13·06 | 1·79 | 14·57 | 1·54 | 15·66 | 1·11 | 18·06 |
| 7 | | 3·29 | 13·15 | 2·81 | 14·42 | 2·33 | 16·00 | 2·04 | 17·13 | 1·52 | 19·63 |
| 8 | | 3·98 | 14·43 | 3·45 | 15·76 | 2·91 | 17·40 | 2·57 | 18·58 | 1·97 | 21·16 |
| 9 | | 4·70 | 15·71 | 4·12 | 17·08 | 3·51 | 18·78 | 3·13 | 20·00 | 2·45 | 22·66 |
| 10 | | 5·43 | 16·96 | 4·80 | 18·39 | 4·13 | 20·14 | 3·72 | 21·40 | 2·96 | 24·13 |
| 11 | | 6·17 | 18·21 | 5·49 | 19·68 | 4·77 | 21·49 | 4·32 | 22·78 | 3·49 | 25·59 |
| 12 | | 6·92 | 19·44 | 6·20 | 20·96 | 5·43 | 22·82 | 4·94 | 24·14 | 4·04 | 27·03 |
| 13 | | 7·69 | 20·67 | 6·92 | 22·23 | 6·10 | 24·14 | 5·58 | 25·50 | 4·61 | 28·45 |
| 14 | | 8·46 | 21·89 | 7·65 | 23·49 | 6·78 | 25·45 | 6·23 | 26·84 | 5·20 | 29·85 |
| 15 | | 9·25 | 23·10 | 8·40 | 24·74 | 7·48 | 26·74 | 6·89 | 28·16 | 5·79 | 31·24 |
| 16 | | 10·04 | 24·30 | 9·15 | 25·98 | 8·18 | 28·03 | 7·57 | 29·48 | 6·41 | 32·62 |
| 17 | | 10·83 | 25·50 | 9·90 | 27·22 | 8·89 | 29·31 | 8·25 | 30·79 | 7·03 | 33·99 |
| 18 | | 11·63 | 26·69 | 10·67 | 28·45 | 9·62 | 30·58 | 8·94 | 32·09 | 7·66 | 35·35 |
| 19 | | 12·44 | 27·88 | 11·44 | 29·67 | 10·35 | 31·85 | 9·64 | 33·38 | 8·31 | 36·70 |

| | | | | | | | | | | | |
|---|---|---|---|---|---|---|---|---|---|---|---|
| 20 | 8·96 | 38·04 | 10·35 | 34·67 | 11·08 | 33·10 | 12·22 | 30·89 | 13·25 | 29·06 | 20 |
| 21 | 9·62 | 39·38 | 11·07 | 35·95 | 11·82 | 34·36 | 13·00 | 32·10 | 14·07 | 30·24 | 21 |
| 22 | 10·29 | 40·70 | 11·79 | 37·22 | 12·57 | 35·60 | 13·79 | 33·31 | 14·89 | 31·42 | 22 |
| 23 | 10·96 | 42·02 | 12·52 | 38·48 | 13·33 | 36·84 | 14·58 | 34·51 | 15·72 | 32·59 | 23 |
| 24 | 11·65 | 43·33 | 13·25 | 39·74 | 14·09 | 38·08 | 15·38 | 35·71 | 16·55 | 33·75 | 24 |
| 25 | 12·34 | 44·64 | 14·00 | 41·00 | 14·85 | 39·31 | 16·18 | 36·90 | 17·38 | 34·92 | 25 |
| 26 | 13·03 | 45·94 | 14·74 | 42·25 | 15·62 | 40·53 | 16·98 | 38·10 | 18·22 | 36·08 | 26 |
| 27 | 13·73 | 47·23 | 15·49 | 43·50 | 16·40 | 41·76 | 17·79 | 39·28 | 19·06 | 37·23 | 27 |
| 28 | 14·44 | 48·52 | 16·24 | 44·74 | 17·17 | 42·98 | 18·61 | 40·47 | 19·90 | 38·39 | 28 |
| 29 | 15·15 | 49·80 | 17·00 | 45·98 | 17·96 | 44·19 | 19·42 | 41·65 | 20·75 | 39·54 | 29 |
| 30 | 15·87 | 51·08 | 17·77 | 47·21 | 18·74 | 45·40 | 20·24 | 42·83 | 21·59 | 40·69 | 30 |
| 35 | 19·52 | 57·42 | 21·64 | 53·32 | 22·72 | 51·41 | 24·38 | 48·68 | 25·87 | 46·40 | 35 |
| 40 | 23·26 | 63·66 | 25·59 | 59·36 | 26·77 | 57·35 | 28·58 | 54·47 | 30·20 | 52·07 | 40 |
| 45 | 27·08 | 69·83 | 29·60 | 65·34 | 30·88 | 63·23 | 32·82 | 60·21 | 34·56 | 57·69 | 45 |
| 50 | 30·96 | 75·94 | 33·66 | 71·27 | 35·03 | 69·07 | 37·11 | 65·92 | 38·96 | 63·29 | 50 |

# *Index*